MARKUS VOELTER

ONCE YOU START ASKING

Insights, stories and experiences from ten years of reporting on science and engineering

1st Edition, Version 1.01, April 2020

http://onceyoustartasking.com

ISBN 978-3-00-064949-3

ONCE YOU START ASKING,

almost everything becomes interesting. Any topic contains a wealth of unexpected details that are fascinating to learn about.

ONCE YOU START ASKING,

people who know something about anything will open up and will happily tell you about “their thing”.

ONCE YOU START ASKING,

you uncover new knowledge. The scientific process relies heavily on asking “why” and “how” questions over and over.

ONCE YOU START ASKING,

you notice that most things aren’t quite as simple in practice and you have to engineer a solution that carefully trades off properties.

BUT ONLY ONCE YOU START ASKING.

This book would not exist without the contributions of the following people.

Aaron Parsons
Ace Beall
Adam Spink
Al Bowers
Alexander Kenan
Alexander Müller
Alfred Krabbe
Andreas Gallus
Angelo Holshout
Antoine Mérand
Benjamin Greiner
Benjamin Knispel
Bettina Wurche
Brian Calder
Carl LaRue
Carsten Aulbert
Christian Fischer
Christian Osendorfer
Christoph Affeldt
Cory Bush
David Baker
Davide Sivolella
Despina Hatzifotiadou
Dick Butler
Dörte Mehlert
Duane Carey
Elisabeth Mittelbach
Eric Delesalle
Florian Freistetter
Frank Schlegel
Frank Zimmermann
Friedrich Hoffmeyer
Geoff Crew
Georgia Mansell
Gerard van Belle
Graham Blenkinsop
Gudrun Thäter
Guy Corbett
Heiko Sommer
Heino Falcke
Hervé Normand
Ian Sutton
Igor Nikolic
Jafet Ojeda
Jan Stahl
Jason Markzon
Jason Parisi
Jean-François Clervoy
Jeff Fellmeth
Jeff Nielsen
Jim Payne
Jochen Eislöffel
Jochen Liske
Joerg Hacker
John Hill
Jon Holzman
Jonas Holtkamp
Justin Ball
Justin Bayer
Katrin Stang
Ken Bower
Ken Hays
Kerstin Borras
Klaus Birken
Klaus Ohlmann
Kyle O'Regan
Lars Bensch
Lee Williams
Lothar Bodingbauer
Major Tom
Maria Kapala
Matthias Hirsch
Meike Klischies
Michael Benedikt
Michael Kramer
Michael Tiller
Ned Johnston
Nicolas Barbotin
Nicolas Wöhrl
Nora Ludewig
Norbert Junkes
Oliver Jessop
Oliver Zeile
Pascal Gorke
Peter Bauer
Peter Chandler
Peter Jenni
Phil Harper
Philipp Sinapius
Philipp Windischhofer
Pierre Bauer
Rainer Kresken
Ralf Jedamzik
Ralf Kleiber
Reinhard Prix
Reinhard Reichel
René Heise
Richard Buckland
Richard Graham
Richard Green
Richard Pogge
Rick Ruiz
Robért Glein
Robert Karban
Robert Laing
Roberto Fillbrandt
Rodrigo Amestica
Rolf Landua
Sebastian Ritterbusch
Seth Cannon
Severin Haas
Steffen Braasch
Steve Rickaby
Tamás Szabó
Thomas Bierwirth
Thomas Geiger
Thomas Keilig
Thomas Krichbaum
Thomas Pfrommer
Thomas Westerhoff
Tom Herbst
Tony Flynn
Tony McNally
Ulrich Husemann
Vito Baggiolini
Werner Engel
Werner Scholz
Wim de Boer

Foreword

September 15, 2010, Hamburg. Markus and I are stranded in the container terminal of Hamburg harbor. We have just disembarked from the CMA CGA Quartz, a 250-meter 40,000 ton container vessel, and we are still kind of high from the last couple of hours. In the morning we had to change our plans: it was not possible to visit the floating pilot station off the mouth of the Elbe river because of bad weather. Instead, we got the chance to accompany Ben Lodemann, one of the Elbe pilots, while he was driving the Quartz up the Elbe into Hamburg.

There is a rhythm to producing a feature for omega tau: we decide on a particular topic, search for a guest, agree on the outline of the conversation, negotiate what we can visit on site and find a suitable date for the interview. Then come the logistics: we organize the trip, book a hotel, get a train ticket. We spend quite some time with a topic, even if it is not necessarily about the content all the time. During this preparation the interview, the guest, the location, even the resulting omega tau episode have already "become" something in our minds. There is a vision of how it is going to be—the anticipation of this fascinating, still mysterious, complex thing and the people who know about it in detail. Then the recording begins and it is like a roller-coaster: no possibility of stopping, no time to think, just go with the flow and enjoy. (I am sure Markus will be laughing here: he knows that I am scared of roller-coasters.)

You have to make sure that the recorder works correctly, that the questions build on each other to allow our listeners to follow, that the guests enjoy the interview as well, and that the conversation keeps flowing. You are on duty all the time, even if you spend the whole day or more with your guests.

Eventually the interview and the visit come to an end and you are suddenly spat out of the tunnel in which you were so focused on producing good content. You're back in the real world and need some time to readjust and to digest all the impressions you collected and the information you gathered.

This was very much the case on that day on the Elbe. After the change of plan in the morning we had signed a waiver for liability in case of accidents, hopped onto a pilot vessel, approached the Quartz, followed Ben up the rope ladder dangling down the huge black wall of the vessel, were guided through several staircases, corridors and elevators. We suddenly found ourselves on the bridge of this giant ship, overlooking thousands of containers. We got to know the captain and some of the crew, asked our questions, observed what was happening on the bridge on our way towards the harbor, exchanged batteries in our recorder and checked the question list we had brought along. Finally, the ship was docked to the harbor wall and we were guided down again towards a narrow gangway, and there we were, somewhere in Hamburg's container terminal, with shining eyes and a sense of disbelief in what had just happened. We had no idea how to get out of the terminal. We weren't even sure if we would need to pass some kind of checkpoint. After all, we'd just walked off a ship that had come all the way from Asia.

Once we started asking, once we started omega tau in 2008, we were en-route for experiences like this. Markus especially, who's been the driving force of omega tau from the beginning, has built a whole universe of episodes on scientific, technological and societal topics. Every episode captures a world of its own. At the same time there are clusters of topics and unexpected interconnections between many of them. Throughout the episodes you can trace the struggle to understand and explain the world and to improve and integrate the models we have of it.

This book takes it further. It digs deeper into the questions of science and broadens the theoretical background. It emphasizes the links between topics more than is possible in podcast episodes, which always have to work as standalone products. The book is not about rehashing omega tau episodes in written form, but rather about exploring the questions beneath those we asked in the interviews. Exploiting the visual channel, some questions can be answered much more easily by drawing a sketch or jotting down a formula. It is about the places Markus visited, the work of scientists and engineers there, about how their equipment functions, about the phenomena they observe and the theoretical foundations of those observations.

While reading, I got more and more absorbed by a second and less obvious storyline. This book is also the story of a curious and passionate man whose openness and enthusiasm led him to experiences he'd only been dreaming of for a long time. We can watch him asking what lies behind things and finding the answers, not just by sticking his nose into books and talking to the experts, but also by experiencing real places, real people and their work. We witness not only this very pure and very honest curiosity, but also the joy of finding out and the awe at the wonders of science and engineering.

Once you start reading, you will see that Markus sometimes moves fast, sometimes slowly, sometimes he has a distant view of a topic, sometimes he is very close. Just like a roller-coaster again! And that's a really enjoyable ride—even for me.

One sentence struck me: "The cockpit is a strange place." How can someone whose lifelong passion has been flying, airplanes and spacecraft, and who has spent thousands of hours in a cockpit himself, still consider it a strange place? Maybe it is the same paradox that is also experienced by many scientists and engineers: the more they dig into a topic, the humbler they become, seeing all the questions they cannot answer and the overwhelming number of principles in physics, chemistry and nature that are the building blocks of our universe and our planet. There are strange places everywhere!

In fact, that's good news—that there are so many more questions to ask. But just in case there are ever no questions left, the cold war novel is waiting. I found its beginning very promising.

Nora Ludewig, December 2019

Table of Contents

I

Introduction

1 | *About this book*

WHO SHOULD READ THIS BOOK? This is not an important book: there's nothing new in it, and it does not address the science-related topics that are on everybody's mind today. This book is just entertainment for geeks. Along the way, though, I think it explains some fascinating science and engineering, hopefully in an approachable and entertaining way, and maybe that's important in today's world of science denialism.

This book is for anyone interested in science and engineering in a general sense. People who might wonder how to mount a telescope onto a Jumbo Jet, and why this makes sense in the first place. People who wonder how the LHC accelerates particles to an energy at which new particles form when beams collide. People who ponder about how a collection of telescopes can join together to image a black hole. People who wonder how a glider can fly 1,000 km with only the power of the sun. And why and to what degree we can trust models in science and engineering. Such people should find a few hours of interesting tales and tidbits in this book. It's not too detailed—you won't need a PhD in physics or aerospace engineering—but I hope detailed enough for you to be surprised by astounding or wondrous facts. And without the dumbing down that is all too prevalent these days: I promise not to use childish analogies, you do need a little bit of Geek cred :-)

THE OMEGA TAU PODCAST In 2008 Nora Ludewig and I started a podcast called omega tau. At the time we were still a couple and were looking for something we could do together, something that would be "ours". At the time I had been involved for a few years with Software Engineering Radio, a podcast that published interviews with experts in software engineering and computer science. I thought: why not use this formula and apply it to science in general? I also had a second thought, about flying in an F-16, which you can read about later in the book. As Nora is a science geek just like me, I suggested we start an interview podcast about science, and she agreed to give it a try: we published the first episode on August 15 2008. Today, after 11 years, 340 episodes, over 500 hours and lots of inspiring conversations and personal experiences later, omega tau is still going strong even though we are no longer a couple.

WHY A BOOK Why a book? Why not spend the time recording more episodes? There are several reasons. First, I hope to reach a different audience. People who don't listen much, and who certainly don't listen to long interview podcasts. Perhaps some of them will become listeners after reading the book. Second, I have always loved writing (although not in school); I've written lots of papers and several books in my professional life. Third, I can write on my own schedule—in contrast to podcasts, where I have to adjust to the guests' schedule, rely on them answering emails and so on. I don't like this lack of control and independence, it's something that often irritates me in the context of omega tau. Writing is different—I am much more in control.

HOW THE BOOK IS DIFFERENT FROM THE PODCAST Obvious things first: it is much shorter than the respective episodes. So there is generally less detail, although I go deeper in some areas. Also, except for occasional quotes, I tell things in my words rather than in the words of guests.

While writing the book I noticed that in surprisingly many cases I didn't fully understand what guests were talking about during the interview: at the time it all made sense somehow, and I could relate to it sufficiently to ask reasonable follow-up questions, but when I asked myself "why" or "how" a couple of times, I couldn't answer. Writing the book forced me to understand the topics at a deeper level so that I could explain them consistently. This the reason for digging deeper in some places than the material in the episodes, for example in the context of the multi-beam echo-sounder in Chapter 3, the A320 fly-by-wire system in Chapter 4, or optical interferometry in Chapter 5. I just had to really understand—so then why not write it down? Overall, digging deeper was very satisfying! I generally drew the line on understanding and writing just before serious math would have become necessary.

Which brings me to another difference. A podcast in the style of omega tau is basically a "say it once" exercise: my guest explains something once, and it might be a good or not so good explanation. Depending on how well I can follow up during the interview I might or might not ask for clarification, but it is still basically a "one shot" medium. Writing is different: you draft an initial version, edit for clarity and coherence, and get feedback from reviewers, which prompts more changes. You can work on explaining the causal connections between things much more, and point out similarities between different and seemingly unrelated topics. Oh, and you can add illustrations! I tried to do all of this in this book. I'm sure it didn't always work out perfectly—in which case you will have to start asking questions of the right people yourself :-)

WHY ENGLISH? I could have written this in German, and indeed have been asked why I didn't. Here is the answer. First, I have done all my professional writing in English: I'm really not very good at writing in German. I might also not be very good at writing in English, but you shouldn't notice, because my copyeditor Steve Rickaby fixes things :-) Second, I feel that with all the listener meet-ups and occasional live shows, we'd already gone quite a long way beyond a podcast for the local German-speaking audience. So I decided I wanted to do something for the rest of you. And: if you speak German and have a grasp of the necessary technical vocabulary, I would be happy for you to translate it.

A NOTE ON STYLE I wrote this book in a colloquial style. I think this is much more suitable for explaining technical details than the terse, sciencey prose required for the academic papers I write in my professional life. But while I do inject some of my own impressions from my visits to interesting places such as SOFIA, HMS Enterprise, the Large Binocular Telescope, and various flying-related anecdotes, the chapters are not stories—I don't try to weave an arc of suspense through a whole chapter. There are several reasons for this. First, it's hard to do well, and I'm not a fiction author. But maybe more importantly, I get annoyed by the "storification" of popular science journalism. An article should certainly have a flow and a logical structure, but tying it to the protagonists' life story is unnecessary and often feels contrived. So I'm not doing it.

THE ILLUSTRATIONS I didn't originally plan to have illustrations in this book, but when I gave the first chapter to people for review they suggested it would benefit from a few. Thinking about it, I realized that it's a bit dumb to artificially limit the book to the same lack of visuals that the audio podcast suffers from!

I started looking around for someone who could draw the illustrations, but didn't find anyone whose style I liked and who was affordable. I didn't really consider doing them myself: I've never been much of an artist—I was really bad at art in school. But I've always liked design-type activities; for example, I enjoy working on the websites for the podcast, or this book, or taking interesting-looking photos. So I decided to give it a go myself. I started with simple drawings. I guess you will be able to tell which came earlier on and which came later. Over time I started to enjoy drawing, and I often got into the famous "flow state" when working on them: clearly a good sign.

Ultimately the book has ended up with over 150 illustrations, some quite simple, some relatively sophisticated (and often derived from things I found online). While I'm sure they are not technically great or wonderful works of art, I think they add to the book and help with understanding.

PICTURES I wrote the book with the intent of including photographs. I took some during my visits, and the web is full of pictures of pictures I can legally use in the book, assuming I add the respective copyright notice. I collected lots of suitable pictures during the writing process and wrote short captions.

However, it turns out that it's not quite as easy as I thought. Putting reasonably high-quality pictures into the print edition of this book requires the book to be printed in color (which makes it expensive) or for the pictures to be transformed to grey scales (which removes an important aspect of their appeal). In the e-book, they increase the file size significantly, which, as I learned, leads to problems with some e-book readers.

Ultimately I decided to just publish the pictures online. I can use high quality, it has no impact on cost, and does not lead to technical problems. And everybody has a way of accessing the web all the time anyway. So, check out the book's website at onceyoustartasking.com, and take a look at the pictures I collected there. Or just google some yourself, the web is full of pictures on the topics I cover in the book.

TECHNICAL I wrote this book using Scrivener: its support for structuring and outlining book-length documents has proved very useful. It also has support for exporting content in various formats, including eBooks (.mobi and .epub) and PDF for printed versions. I struggled quite a bit with some of the formatting and export-configuration details. But it never crashed on me or lost data. The one bug that affected me directly has been fixed in a minor update, and until then the developers gave me access to a bug-fix release. Overall I'd use Scrivener again and would recommend it: it's a solid piece of software. In an episode of the Mindscape podcast about how he writes his books, Sean Carroll said something to the effect of: "I don't want to tell you how to write a book as long as you use Scrivener." I agree.

To create the illustrations I got bought an iPad with an Apple pen and ProCreate—any excuse for a new gadget! I exported the drawings as PNGs and then added text on a Mac using Preview—annotating by hand in ProCreate just looked too bad. Another export to PNG and I integrated the illustrations into the Scrivener document by reference to the external files.

I wrote the formulas in Powerpoint using Microsoft's Equation Editor, exported them as PNGs and integrated them into the document as images. There might have been a more elegant way to do this, or even some way to do it natively in Scrivener, but I didn't want to experiment, I just went with what I knew.

The production of the final PDF was slightly roundabout because Scrivener is not very good with layouting, page breaks, and sizing illustrations. This is why I compiled/exported the text to Word, chapter by chapter. There I manually fixed the size of illustrations and formulas and adjusted page breaks to avoid single lines spilling over to a new page. From Word, I printed each chapter into a PDF after manually setting the start page number. To get to the full book PDF, I merged the chapter PDF files with Mac's Preview, and then added front and back matter, which I wrote in Word directly. I created the table of contents manually as well.

HOW TO READ THE BOOK This book has seven chapters (excluding this introduction), with each covering a particular field of science or engineering. Chapters consist of several sections: a section is either based on a particular podcast episode or about a particular topic in a longer episode. The chapter binds the sections together.

You can read each chapter separately and in any order. There are cross-chapter connections, but they are limited and won't really spoil your reading experience if you pick and choose. If you want to know more about any particular topic you can listen to the podcast episodes; the website for the book, onceyoustartasking.com, has an episode list for each chapter.

The various chapters turned out quite different in length. That might not be surprising, because a general topic like Aviation or Telescopes is likely to be longer than one about a specific ship or aircraft. In retrospect, though, I guess it would have been better to strive for more balance in terms of chapter size. The first three chapters also contain much more personal experience than the remaining ones, because these were "adventure holidays" for me personally or cover "a life" in aviation.

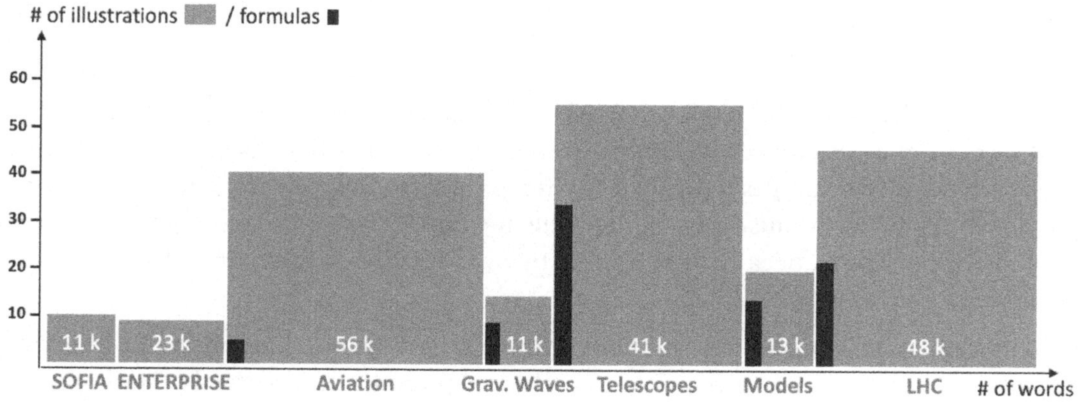

HOW I SELECTED THE TOPICS Chapters 2 and 3, about SOFIA and HMS Enterprise, basically selected themselves. I spent a week with each, and they are so rich in content that they warranted a chapter to themselves. They are also multi-faceted, so the topic can be narrated from different perspectives.

Chapter 4, on aviation, was also an easy choice. Lots of omega tau episodes revolve around aviation, and I also have a lot of general knowledge in aerospace. My personal flying allowed me to put a bit of my own experience into the text. Basically, this was an easy chapter for me to write, so I selected it.

Chapter 7, on models, was also a choice driven mostly by my personal experience and the relative ease with which I would be able write it. In addition, models really do play an important role in science; this was highlighted by many of my guests, so the topic is relevant.

So what about the three science chapters: gravitational waves, telescopes and the LHC? Many factors were in play when choosing those. First, we've covered these topics in many, many episodes, so I had enough detail available. Second, they are about physics, where I have some background. Writing about chemistry, biology or social sciences would have incurred a prohibitively greater amount of work. Third, each of these topics has a relatively recent discovery that I could talk about: the first direct measurement of a gravitational wave, the imaging of the event horizon of a black hole and the discovery of the Higgs boson. These chapters were hard but very satisfying work, and I went into much more detail than I had planned. In fact I had originally intended for this to be a single chapter, tentatively called "Big Science".

Three topic areas were on my shortlist but didn't make it into the book. Space was one of them: omega tau has lots of episodes about spaceflight. However I interviewed a lot of authors, David Woods and Davide Sivolella among them, and writing about space would risk essentially copying their work. I didn't want to do that: you can just get hold of their wonderful books. Another was climate science. We have lots of episodes on particular aspects of this, so it would have made a nice chapter, but I felt that treating this topic well would require a book of its own. And there are already lots of those. The third topic on my shortlist was nuclear fusion, energy and the grid. The only reason this chapter is not in the book is because I simply ran out of stamina.

FACT-CHECKING AND PROOFREADING Have you ever written something that was then edited by someone who wasn't as experienced in the particular field as you are? They might make subtle wording changes to make it "easier to understand", "clearer" or "less cluttered", but then frustrate you because they changed the meaning. You wrote something in a particular way for good reason, because you wanted to emphasize something or make some kind of subtle point.

That editor is basically me writing this book. I (think I) understand things from the interviews, then put them into my own words, assuming that I represent them correctly. But I might not have done. When you try to create a consistent and logical explanation of something in your own words, you may well discover that it requires a much deeper understanding of the subject than you might get from a podcast, even if you're the one who conducted the interview. So I did a lot of additional research, read papers, and in several cases reached out to experts in the field and asked them to explain something to me. With very few exceptions all the people I approached were willing to help, with pointers to the literature, email dialogs or phone calls. The science community really rocks!

I was also surprised by how hard it often is to find out the actual facts. We all know this to be the case for "facts" in political discussions, because there "facts" are often used to substantiate some position or theory, and so they are selected and tweaked to suit whoever uses them. But this problem also appears for the technical characteristics of systems. A great example is the number of filaments within the wire that makes up the superconducting cable in the LHC. I variously found 6,300, 6,480, 7,000, 8,800 and "around 10,000", all in scientific papers or articles in reputable magazines or on reputable websites. Another example is the number of drift tubes in ATLAS tracker: 56,000, 300,000 or 370,000. Sometimes such differences can be explained because the writings represent different stages of knowing—the original design, what was in fact installed, or maybe what is planned for a future update. Sometimes the numbers are correct but their reference unit is wrong. For example, I found that 600 MJ of energy is stored in the magnetic field of the LHC—but I found this figure quoted as the total amount, the amount per sector and the amount per half-sector. For this reason some of the specific numbers in this book might be "wrong".

Ultimately I had every line I wrote reviewed by experts. I think it is irresponsible to write a book as broad as this without such review. Some parts were read by multiple people, who then disagreed over some things, especially about judgements or assessments, so at some point I had to stop the review process and just go with what I had. Most explanations are simplifications to some degree—in several places I summarized a 10-page paper in two paragraphs. This might have lead to imprecisions or miscontextualization, but simplifications and elision of detail is mandatory in such a book.

Ultimately there will still be errors, misleading formulations or things you disagree with in the book; I will point out (and correct) these over time at onceyoustartasking.com, behind the Bugs button.

So let me take this opportunity to thank everybody who was involved in the creation of this book, through input or review.

Special thanks goes out to DSI's Dörte Mehlert, who proofread the chapter for factual imprecisions and errors—and found quite a few—and to Oliver Zeile for helping me understand the various coordinate systems involved in the telescope controller and also reading the whole chapter. I also want to thank Davide Sivolella, Ken Hays, Adam Spink and Angelo Holshout for helping me improve this chapter by providing feedback on earlier versions. DSI asked me to include this text here:

SOFIA, the Stratospheric Observatory for Infrared Astronomy, is a joint project of the Deutsches Zentrum für Luft- und Raumfahrt e.V. (DLR; German Aerospace Center, grants 50OK0901, 50OK1301 and 50OK1701) and the National Aeronautics and Space Administration

(NASA). It is funded on behalf of DLR by the Federal Ministry for Economic Affairs and Energy based on legislation by the German Parliament, the State of Baden–Württemberg and the University of Stuttgart. Scientific operation for Germany is coordinated by the German SOFIA-Institute (DSI) of the University of Stuttgart, in the USA by the Universities Space Research Association (USRA).

For the chapter on HMS Enterprise, most importantly my thanks goes to Phil Harper for fact-checking the complete chapter, and to Richard Buckland of RN Media Communications for working with me to get everything approved. Finally, I want to thank Meike Klischies (Geomar) and Brian Calder (Chase Ocean Engineering Lab, University of New Hampshire) for helping me better understand multi-beam sonars through multiple telcos and feedback on drafts of the respective section.

On to the aviation chapter. Guy Corbett checked the parts on gliding for accurate use of English-language gliding jargon. Several people on the Facebook Segelfliegen group helped me with water ballast; Roberto Fillbrandt was especially helpful. A number of people checked the various sections for correctness, including Pascal Gorke, Philipp Sinapius and Alexander Kenan. Roberto Fillbrandt proofread the whole chapter. Eric Delesalle, AirZeroG chief pilot, proofread my description of parabolic flights. Steffen Braasch and Nick Anderson, both commercial Airbus pilots, helped with the Airbus fly-by-wire system, and Reinhard Reichel of Stuttgart University and my guest in the respective episode gave it a final fact check. Jan Stahl, USAF F-16 pilot, proofread the part on the F-16 and contributed a number of useful clarifications. Davide Sivolella explained the finer points of the Shuttle's reentry and landing. René Heise provided additional details around the Mountain Wave Project, and Jon Holzman, a former US Army helicopter pilot, took a good look at what I wrote about the Apache. During my work on that chapter the grounding of the 737MAX occurred. I wrote a bit about that in the context of flight control systems, and the respective section was kindly reviewed by Michael Tiller, Thomas Geiger, Rick Ruiz, Steffen Braasch and Jeff Fellmeth, all of them either pilots or system engineers. Al Bowers, former Chief Engineer at NASA Armstrong read most of the chapter, found a couple of bugs and, very importantly for me, said that he really liked the chapter. Finally, Werner "micro" Scholz, aerospace engineer, glider designer, fellow glider pilot and omega tau listener also read the whole chapter in its final form and still found several inaccuracies and a few mistakes. Thank you :-)

In the science chapters I once again got a lot of help from people who actually understand the science and the engineering. These chapters were really "on the limit" for me: writing in a way that non-experts could understand while myself not understanding everything in detail and missing some of the fundamentals—for example in optics—would have had no chance of succeeding without expert help. Which luckily I got. Benjamin Knispel of the Max Planck Institut for Gravitational Physics read my text on gravitational wave detectors. He found and reported lots of inaccuracies and outright mistakes, which I then of course fixed. Georgia Mansell of

LIGO proofread the corrected version and still found issues. Reinhard Prix, my guest on relativity, proofread the introduction to this chapter, which explains the basics of relativity. Thanks to all of you!

In the chapter on telescopes. John Hill, the technical director of the LBT, proofread the part about the LBT. Richard W Pogge of Ohio State University helped me understand "his" MODS instrument and proofread the respective section. Tom Herbst, the principal investigator of LINC–NIRVANA, read a draft of the respective section, gave me lots of feedback and spent hours on Skype helping me understand interferometry. ESO's Antoine Mérand and Gerard van Belle of the Lowell Observatory explained the field-of-view limitations of the VLTI, while Markus Wittkowski, also from ESO, gave my reworded text a final check. Heiko Sommer of ESO helped me with finding relevant papers on the ELT and ALMA. Aaron Parsons of UC Berkeley proofread the math on radio interferometry. Rodrigo Amestica of NRAO helped me with the ALMA correlator. Robert Karban, a former ELT control systems engineer, proofread the section on the ELT's control system, while Schott's Ralf Jedamzik checked the part on mirror production. The section on Effelsberg was proofread by Norbert Junkes, my guest on the respective episode. The section on the EHT was proofread by Thomas Krichbaum and Michael Kramer of the Max Planck Institute for Radio Astronomy; I was quite confident about it, but they still found lots of errors and inaccuracies! Geoff Crew of the MIT Haystack Observatory read the EHT part as well and, in addition to finding a few errors, pointed out lots of nuances and aspects I didn't cover. Norbert Junkes also read the short summary of the SKA. The telescopes chapter was reviewed in full by three people: Heiko Sommer and Thomas Bierwirth from ESO, as well as Richard Green, the former director of the Large Binocular Telescope and the Division of Astronomical Sciences at the NSF. This chapter became much bigger and more detailed than I had originally anticipated, so I really appreciate their support in reading and commenting on the whole thing!

The modeling chapter is closer to my own experience. Nonetheless, or maybe because of this, I asked a few of my "modeling buddies" for feedback. Klaus Birken and Tamás Szabó were very helpful. Werner Engel also read parts of the chapter, and I got a printed version with lots of red scribbles from Nora. Thank you :-)

Finally to the LHC chapter. The part on the magnets and beam control was reviewed by Pierre Bauer, my guest on superconductivity and high-performance magnet expert. Vito Baggiolini, my guest on the episode on controlling the LHC beam did a thorough review of the LHC part of the chapter and provided a lot of corrections. Severin Haas who works in the machine protection department at the LHC provided feedback and input on this aspect of the chapter. Robért Glein of the University of Colorado, who works on the CMS Level 1 Trigger FPGAs, contributed lots of details on those. Despina Hatzifotiadou, my guest on ALICE, provided feedback on most of the chapter, including on ALICE, "her" experiment. Peter Jenni of CERN (and one of the fathers of ATLAS) provided feedback on my explanation of the standard model as well as some other parts of the LHC chapter. Ulrich Husemann of KIT gave very thorough feedback on the Higgs chapter and, through several iterations, helped me really understand what happens in a collision. His colleague Wim de Boer clarified the relationship between energy and momentum for relativistic particles. Michael Benedikt, who was a guest on an earlier episode as well, and who leads the Future Circular Collider study group at CERN, proofread the part about future of colliders, and provided lots of useful feedback. Justin Ball and Jason Parisi, my guests on the 2019 fusion episode, jointly read most of the chapter and provided lots of useful feedback, both in terms of technical errors and how to better explain things—they have written books themselves. I also got feedback regarding the overall structure of the chapter from three of my listeners: Jonas Holtkamp, Andreas Gallus, and Frank Schlegel. Thank you! Special thanks goes to Philipp Windischhofer of Oxford University: through several reviews and many email exchanges regarding different parts of the chapter, Philipp filled in holes in my understanding or corrected things where I was just wrong. In a few places I directly used formulations in the text that he suggested in the emails. At one point during the editing process I felt like the chapter was a bit too much, and his reviews and explanation helped me get through it. Thank you very much Philipp!

MORE THANKS Most importantly I want to thank Nora. When I suggested in 2008 that we could start a podcast, I didn't really expect she would be up for it. But she was, and we're still doing this some 11 years later. Although I have always done most of the episodes and ancillary work, omega tau would not be the same without her, and maybe wouldn't even exist. Thank you Nora, for 11 years of omega tau (so far), for the wonderful foreword, for giving the whole book a final read, and for seven years of your life!

I also want to thank my copyeditor, Steve Rickaby. Just as for all my books on computer science topics, he did a great job of improving style and spelling. Steve is an engineer himself, and it seems like he also enjoyed this book from a content perspective [SR: Yes I did :-)]. He pointed out a few errors and non-sequiturs in my text and provided anecdotes from his own experience before he became a professional word wizard. Thank you Steve.

My thanks also goes out to the people who wrote nice words about the book; as of this time these include Nicolas Wöhrl, Jeff Nielsen, Davide Sivollela, Al Bowers, Richard Green, Florian Freistetter and Lothar Bodingbauer.

Next, our listeners. Without an audience there would obviously be no podcast: if nobody cared, we would not produce. More importantly, unless we were able to say "X number of people listen to this stuff" many of our guests wouldn't bother to be on the show. But our listeners also send us praise (which keeps us going), criticism (which helps us improve), suggestions for topics (which broadens our scope), concrete suggestions and sometimes introductions to guests (which makes it much easier for us to find good guests) and specific questions to ask (which adds perspectives we might not even think about when preparing the interviews). Importantly, a subset of our audience supports us financially, which is absolutely crucial to keep omega tau alive. Best of all, some listeners (and a few guests and fellow podcasters) have become genuine friends, enriching our lives even outside the podcast. So thank you, dear listener :-)

Finally, I also want to thank all the guests on omega tau, on whose explanations this book is based: it would not exist without the podcast episodes. Even though I read and watched additional material to help me understand some details, I wouldn't have discovered my lack of knowledge were it not for the podcast interviews.

I want to thank the SOFIA team for allowing me to spend a week in Palmdale, for taking me on two of their science missions and for tolerating my interminable questions. In particular, I want to thank the protagonists of the chapter: Alfred Krabbe, Thomas Keilig, Christian Fischer, Ace Beall, Ken Bower, Maria Kapala and Jochen Eislöffel. Plus of course Dörte Mehlert who made it possible for me to participate as well as Antje Lischke-Weis who shepherded us on our trip to Palmdale. A final thanks goes to Benjamin Greiner who initially got me in touch with Dörte.

Regarding my trip on the Enterprise, I first want to thank Phil Harper, then commanding officer of HMS Enterprise, for allowing me to travel with his ship and crew for a week. My thanks also goes to Kyle O'Regan, for coping with the practical consequences of Phil's decision. Thank you very much for a unique and beautiful week! I also want to thank all the other officers of Enterprise for letting me intrude into their home at sea. A big thank-you also to Oliver Jessop, Graham Blenkinsop, Ian Sutton and Lee Williams for talking to me while the mic was on. Finally, I also want to thank Rainer Kresken and Bettina Wurche for putting me in touch with Phil and thus setting in motion the process that got me onto Enterprise.

Aviation. Lots of guests to thank: Jean-François Clervoy, Eric Delesalle, Hervé Normand, Nicolas Barbotin, Major Tom, Friedrich Hoffmeyer, Seth Cannon, Davide Sivolella, Duane Carey, Tony Flynn, Reinhard Reichel, Peter Chandler, Richard Graham, David Baker, Carl LaRue, Al Bowers, Jan Stahl, Lars Bensch, Jim Payne, Rene Heise, Klaus Ohlmann, Joerg Hacker, Alexander Müller, Dick Butler, Elisabeth Mittelbach, and Katrin Stang. And of course I must thank Jason Markzon, Jafet Ojeda, Ned Johnston, Cory Bush and the whole Thunderbirds team for the best day of my life!

The list of guests for the gravitational waves episode is shorter: Reinhard Prix, Christoph Affeldt, Carsten Aulbert. Thank you for being on the show!

The episodes on telescopes included Robert Laing, Jochen Liske, Heino Falcke, Norbert Junkes, Thomas Westerhoff, Thomas Pfrommer and Richard Green. Richard really deserves a special thank you for a wonderful night at the LBT!

On models, it's Reinhard Reichel again, Igor Nikolic, Ralf Kleiber, Matthias Hirsch, Gudrun Thäter, Sebastian Ritterbusch, Christian Osendorfer, Justin Bayer, Peter Bauer, and Tony McNally.

And finally the episodes on the LHC and the Higgs: Despina Hatzifotiadou, Pierre Bauer, Michael Benedikt, Frank Zimmermann, Kerstin Borras, Vito Baggiolini and Rolf Landua.

WHY ARE THERE NO REFERENCES? You may notice that this book is not full of citations that substantiate what I have written. There are several reasons for this. The most obvious is that a lot of what I have written is taken directly from the podcast. The relevant episodes are mentioned on the book website; adding a reference to them for every non-trivial point I make isn't helpful. I looked at a lot of additional sources and learned things that are not in the episodes, of course, but many of those I've learned over time from a diffuse set of papers, books, emails, web pages and conversations with reviewers or other experts. Often pinning things down to one particular source is not really feasible. In a few places, where I adapted a particular example or when I copied something verbatim, I do cite the source. The quality assurance aspect of citing sources is partially compensated by expert review of everything I wrote, as I explain above. Finally, it's a lot of work to keep track of all the mappings between claims or points and their respective sources. I have to do this when I write for computer science conferences or journals as part of my job, but I didn't want to burden myself with this as part of a "hobby" project.

However, the book website at onceyoustartasking.com lists all of the sources I used—modulo those I forgot—more than 700 in total! Of course I didn't fully read all of them; some I just skimmed, trying to fill a particular gap in my understanding. And there are no additional sources for SOFIA because I had the idea of collecting the sources only after that chapter was finished; there's also nothing for Models, because it's so close to my everyday work that I wrote most of the stuff based on the podcast episodes and things I know anyway.

Finally, please remember that this book is not a scientific treatise, it is intended as entertainment for geeks. There's nothing new in the book; it's all based on information available in papers, websites, presentations and of course podcast episodes. My contribution is hopefully to summarize and explain it in an approachable and entertaining way, while adding some of my own personal experiences.

2 | *How I got here*

The immediate context for writing this book is of course the omega tau podcast, which got me into reporting on science. But there's a bit more to the story, which I want to tell here briefly, mainly to (implicitly) thank the people who were instrumental in getting me here.

CHILDHOOD I have never been much of a sports guy, neither actively nor as a spectator. I do remember the yearly athletics competition in the fourth grade, the Bundesjugendspiele—I remember it because I didn't participate. My father negotiated with the school that I could instead join him on a visit to the Gundremmingen nuclear power plant. My father was a school teacher and on that day he was due to visit the plant with his class, and I really wanted to tag along.

I remember this because the visit was typical of what my brother and I did with my father. As an engineering nerd himself, my father often used his children as an excuse for visiting interesting places. We went to Stuttgart station to watch locomotives—the 103 Series was hot at the time. We visited the huge excavators in Germany's brown-coal strip mines near Düren, where my mother was born, and on the same trip we visited the 100-meter radio telescope in Effelsberg. We went to the airports in Stuttgart and Frankfurt and spent hours on the observation deck watching airplanes take off and land. We went to the Ellwangen barracks to see the Leopard 2 tank, and during the summer holidays a visit to the Deutsche Museum in Munich was mandatory for years. There we inspected old ships, satellites and rockets, huge paper mills or old, mechanical computers—and aircraft, of course. I am pretty sure that all of this laid the basis for my interest in engineering and science.

AVIATION While I was and still am interested in lots of aspects of engineering and science, aviation has always been first among equals, with space a close second. I think it started with a book I dug out of my grandfather's bookshelf in the early 1980s about the moon landing. It was already quite tattered when I found it, and it didn't get any better once I took possession. I also remember being given a Revell 1:144 scale Space Shuttle at some point. It was the first in a long series of plastic models I built between age 8 and 16. I later had a larger Space Shuttle, a 1:32 scale F-15C Eagle and a very nice 1:48 scale F-16C Wild Weasel.

Military airplanes were my "gateway drug" into aviation. I read everything I could about them. I started with German language "general audience" books, but as soon as I learned English at school, primarily from a teacher named Heinz Ohler, I started importing "real" aviation books from the US through specialized book dealers (no Amazon in those days). I also regularly wrote letters to aircraft manufacturers and air forces: "I am a 12-year old German boy who loves airplanes and I would like to get a few pictures of XYZ." Mr. Ohler helped me to write the letters and I often got great pictures back. I remember actual photos, not prints, of beautiful F-16s, adorned with pilots' signatures. I was very proud. The F-16 has always been my favorite military jet, and flying one has been my dream ever since I was that 12-year old boy. When I moved out of my parents' house after university I owned literally meters of books on aviation, including all the issues of the wonderfully detailed (and sadly now defunct) World Air Power Journal. I was seriously infected with the flying virus.

GLIDING I grew up in Steinheim. Like many villages on the Swabian Alb it has its own small glider airfield. My father, my brother and I often visited it to watch takeoffs and landings on Sunday mornings, while my mother was preparing lunch. I think I was only mildly interested initially. I was into military jets and the gliders weren't quite so cool. However my father knew a guy there, a flight instructor called Werner Sick. One day, when I must have been about 13, my father and I inquired if I could get a short flight in a glider. This was no problem, of course, as they did passenger flights all the time. Although this was my first flight in a glider I don't remember much about it—whether it was in an ASK-13 or an ASK-21, and whether Werner Sick was the pilot—but a few weeks later, after I turned 14, I enrolled as a student pilot in the Fliegergruppe Heidenheim/Steinheim, the local flying club.

During that summer of 1988 I spent every non-school minute at the "Heide", as it was called. I basically cancelled all visits to my relatives' birthday parties and I didn't go on vacation with my parents, because of course I had to join the flying club's summer camp. I am really grateful that my parents allowed me this freedom. For my mother especially it wasn't so easy, for years she was anxious when her son—and later, sons—flew, on their own, in these flimsy airplanes. The situation wasn't improved when, one day, I returned home from a visit to the hospital where they checked my back after a particularly hard landing that damaged the club's DG-101 glider.

An airfield is a paradise for a teenager. You get to drive cars long before you can qualify for your driving license, there's a tractor, you can operate a 240-hp winch, there are lots of nice people (and some idiots, as always) and, last but not least, you get to fly really cool airplanes. I soloed in July 1988, after around four months and 90 flights. Many of these training flights were with Werner, but also with other instructors.

Literally the only weekend day I didn't spend at the airfield that year was August 28. My father, brother and I drove to Ramstein to visit what was to become the last military airshow there: you might remember the terrible accident involving the Frecce Tricolori aerobatic team, in which around 70 people died. We felt the heat from the explosion and ran away in panic, but we weren't hurt. Again, my mother suffered, because she heard about the catastrophe in the news, but it took us a long time to call her; no mobile phones at the time, and long queues in front of the few phone booths there.

I had been a member of the Heidenheim/Steinheim flying club for 13 years, and clocked up around 1,000 flying hours on the ASK-21, ASK-23, DG-101, LS-4s, Duo Discus and ASH-26, when I got frustrated with club politics and switched to Fliegergruppe Donzdorf. At the same time, as a consequence of those frustrations, I bought my first glider, a used ASW-27. I'd been working as a software consultant for a few years by that time, and spent almost all my savings on the glider. I flew the "42" for three years, then switched to an ASH-26E self-launcher, the "DS". I flew her for ten years, then in 2014 switched once more to a ASG-29E, the "2M". Today, with around 4,000 hours, I am still a fairly serious cross-country glider pilot. Gliding in particular, and aviation in general, is a huge part of my life. This is reflected in the topics of omega tau and this book.

SCHOOL When I was ten and about to enter fifth grade, I switched from the local elementary school in Steinheim to Werkgymnasium High School in Heidenheim. The Werkgymnasium was rather unusual. It was an all-day school; we had lessons in the afternoon every day except

Wednesdays. In return for this we didn't have to do homework. At the time every other Saturday was also a school day—but not at the Werkgymnasium. This was great, because it gave me more time for flying; I was often the only student pilot on Saturday mornings.

More importantly, the Werkgymnasium had been an "experimental" school of the Baden-Württemberg state. The experiment was about something called Kurs, five additional hours of instruction every week. However, these were not normal lessons, but practical hands-on work. Students had to choose between Kurs A (wood and metalwork, construction), Kurs B (physics, chemistry, biology) and Kurs C (arts and design). I chose Kurs B. Over the years we analyzed substances in a chemistry lab, dissected fish, mapped lichen, operated microscopes and microtomes, learned about meteorology, worked with radioactive materials and learned the basics of electronics. These were my first steps in becoming a scientist/engineer.

A particular benefit of the all-day school was the 90-minute midday break. Lunch was provided by the school and was often awful, so it didn't take up much of the break time: we basically had an hour's worth of free time at school every day. In sixth grade I joined the computer club. This was run by a guy called Uli Frei, a math and physics teacher who was really into computers and programming. We learned to program in Pascal on Stone Age LNW computers. This computer club, with the same teacher and the same core group of students, continued until 13th grade, the end of my school career. For years we tried to build a plotter from scratch, including its mechanics, electronics and software. We never quite finished it, but it was an excuse to "keep the club going". During the last few years the group had its own little room in the school, equipped with computers, soldering irons, Twix bars and a coffee machine. The computer club and Uli Frei launched my career in programming.

The fact that the Werkgymnasium was an experimental school offering the Kurs system also meant that the school had more money than other schools, which translated into really cool equipment. All of it was available to students during the midday break, and also outside school hours: computer rooms, the weather station, the physics lab with its electronics gear and oscilloscopes, a rather good amateur telescope that we could borrow for watching the stars at night, and so on. If you were a student who was interested in science and technology, this was the school for you—also because the teachers let you "do stuff". People like Werner Schölzel, Wolf Hirlinger, Simon Wörz, Reinhard Schmierer, Gerd Rehm and Wolfgang Jäckel really supported your inner geek. My sports teacher Heinz Holzner helped as well, by realizing that I was hopeless at sports and letting me stay in the computer room or physics lab instead of the gym. My art teacher, Reinhard Ehrtmann, also spotted my nonexistent talent for the visual arts and let me spend many hours of art class with the school's desktop publishing equipment.

In 13th grade we were tasked with our own project in Kurs B. I decided on aerodynamics, unsurprisingly: I built a wind tunnel with the help of Uli Frei and a buddy of his who owned a company that installed air-conditioning systems. To run it the room where we planned to put it had to be supplied with three-phase power: the fan was actually quite big. No problem—the school headmaster and a friend of mine who worked at the city administration had it installed. Of course I did some of the aerodynamics experiments with the real glider.

This is how school should be! I owe a lot to my years at the Werkgymnasium and the teachers there. Much more than to my time at university, Fachhochschule Ravensburg-Weingarten, where I would later study technical physics.

COMPUTERS AND WORK Flying is one topic that hugely influenced my life, and computers and programming is the second. I am not sure whether my first exposure to programming was Pascal on the LNW computers at school or Basic on my father's Schneider CPC 464. At home there was a rule that I would get half an hour of computer time per day, mostly because my father needed the computer for work. When I used it for "something useful" though (read: programming instead of playing games) the half hour was often extended. I wrote my first data management system on the Schneider; it was called "Profi-Datei".

A few years later my father got his first PC, which we still shared. I don't quite remember in which grade it was, but the group from the computer club at school all bought the same PC, I think during a Cebit trade show. It was the first computer I had owned personally. We also bought a student version of Borland Pascal 7.0. It came with over a meter of matching manuals and tutorials, which looked very impressive in the bookshelf: I never read any of them, of course. I programmed a very early version of an electronic log book for the flying club and developed an auction management system for a local auction house. I learned the challenges of concurrent access to shared data—painfully—during a real-world auction.

When I finished school I refused to go into the military, and instead did community service. I was lucky to find a job in the finance department of the local hospital. Officially I was the "statistics Zivi", the guy who was responsible for creating various legally required stats. For this I had my own office and my own PC. I got quite good with Excel, but more importantly, I had enough time to learn C++ (and play Flight Simulator). I also programmed a few utilities that sped up data entry and stats creation. Martina Assfalg, my boss, didn't mind if I the time freed this way creatively as long as I did the stats. I also got in touch with the hospital's IT department, and in particular with a guy called Wolfgang Karasch.

When I moved to Weingarten to study technical physics I was still planning to become an applied physics engineer. And as every student I had to make some money while I was studying, so I needed a job. I found one at the Labor Dr Gärtner, a large medical laboratory facility. A guy called Uwe Kutter and his one-man company was developing a new IT system for the lab and was looking for students to help out. My buddy Andreas and I got the jobs. I learned Clipper, and we happily spent all our non-university time in the lab, where they also had a very good staff canteen that we regularly frequented instead of the not-so-great one at the university. Our third semester was a mandatory practical, so of course we did ours at Gärtner, continuing to build the lab's IT system. We both learned a lot (including the importance of database backups) and made quite a bit of money, considering that it was a student job. At this point it started to dawn on me that I might not become a physics engineer.

The development of the lab IT system didn't go quite as well as everybody hoped and the job ended. I got back in touch with Wolfgang Karasch from the Heidenheim hospital, where it turned out that they needed a system to collect and interpret data on the treatments performed by the various hospital departments. It was intended to be only for some of the departments, so why not have it developed by a student? I did that using Borland Delphi and MS SQL Server. It became the backbone of all the data collection in the whole hospital for years to come and financed the rest of my university career. I also did my half of my second practical semester there, developing the hospital's first website; the other half I spent programming Ada on naval radar systems for DASA, the pre-precursor of Airbus Defense and Space.

By this point I had given up the idea of becoming an engineer. I was becoming a software guy; I lost interest in the topics at university and tried to skew it towards software as much as I could. I convinced my computer science professor Wolfgang Georgi to teach C++ instead of Fortran. It was also Georgi who supervised my diploma thesis, which was some kind of C++ and Java-based distribution middleware. I developed that using UML and code generation—a sign of things to come.

I am not going to recount my whole professional career here; you can find that in my CV at voelter.de if you wish. But a few things are worth pointing out, because I had a few projects that led more or less directly to episodes on omega tau. For example, I was involved in the software for the ALMA telescope array as part of my work for ESO, and visited the VLA in New Mexico with them. I also did some consulting work for the ESRF. Before starting omega tau I ran the Software Engineering Radio podcast with a bunch of friends from the software community. I did around 120 interviews with computer scientists before I handed SE Radio over to IEEE. Finally, throughout my career I wrote a lot. I started with software patterns and articles in industry magazines, wrote a couple of books on middleware, modeling and language engineering and later, partly because of my PhD in computer science at TU Delft, wrote quite a few academic papers. All my writing was and continues to be about explaining complex technical subjects. I guess this book would not exist without all the "practice" I got from writing about computer science topics.

WRAP-UP What really shaped my interests were the early years spent visiting cool things with my father, his effectively teaching me to program, the environment and the teachers at school, as well as my experiences in my early flying career. I don't expect teenagers to read this book, but parents might. If, as a parent, you aren't just personally entertained, but also use the book's stories to inspire your children to become engineers and scientists, then it has done its job. Oh, and tell them to listen to omega tau :-)

2

SOFIA: The Airborne Infrared Observatory

I am tired. It is 4 am California time and I have been awake since early the previous morning, having just arrived in the US from Europe. I am jet-lagged. I am sitting in an old but comfy business-class seat close to the nose of a Boeing 747SP. The SP version of the venerable Jumbo Jet has never been widely used and is even rarer today; thirteen are still in active use at the time of this writing. But the one I am flying on is unique: it is the SOFIA airborne observatory. We are on a ten-hour flight across the US to observe the night sky, and I am on board to record interviews for the podcast. I have been recording on and off for several hours, and now I'm taking a short break, drinking Coke Zero to try to stay awake for the next few hours. Our landing in Palmdale is scheduled for around 7 am. SOFIA missions are usually planned to return before daylight, because the rising sun might damage the delicate optics if the telescope's door fails to close because of a technical malfunction. There's a plan B—flying a track that points the telescope, located on the left side of the fuselage, away from the sun. But why take a risk—better to land before sunrise. If the mission requirements dictate landing after sunrise, the mission planners have a software program that tells them which ground heading is safe for the telescope based on the date and time. And which (alternate) runways are available to land with a suitable heading.

1 | *A Flying Telescope*

WHY NOT ON THE GROUND Why put a telescope on an airplane? It's all about how well the telescope can see faint objects in the night sky. Generally, the atmosphere is a problem, because it is unstable due to thermal fluctuations, which leads to a "wobbly image". The less atmosphere there is between the telescope and the target in space, the better. This is why telescopes are typically built on the top of mountains. In addition to being closer to space, mountains usually also have the advantage of being exposed to less light pollution, human-made light sources that spill into the telescope.

SOFIA is short for Stratospheric Observatory For Infrared Astronomy, so its purpose is to observes infrared light. Because of its long wavelengths, between 700 nanometers and 1 millimeter, the "wobbly" atmosphere is not so much of a problem. But there is a even bigger one: infrared radiation is absorbed by water vapor. The atmosphere has lots of water vapor, but luckily most of it is in the troposphere, the lowest 10 kilometers of the atmosphere. It is trapped there because of a thermal inversion at the top of the troposphere—the tropopause. In an inversion, temperature increases with altitude, limiting thermal convection and thus limiting the amount of water vapor transported to higher altitude. The reason why SOFIA is useful as an infrared telescope is that it has the capability to fly above the tropopause, and thus above most of the atmospheric water. The specifics depend on the current altitude of the tropopause, which fluctuates with weather and season, and SOFIA's weight, which in turn depends primarily on the fuel burn during the mission.

OR ON A SATELLITE Putting an infrared telescope on the ground therefore isn't a really great option—for far-infrared light, above 15 micrometers, it is not an option at all. But why not mount it onto a satellite, where it would have even less atmosphere to contend with? This is certainly an alternative: the Herschel satellite was just that. However, there is a major argument against a satellite: once it is in orbit it is effectively impossible to upgrade a satellite (even if the community was willing to repeat the huge effort necessary to fix and upgrade Hubble, there's currently no Space Shuttle available to perform the mission). It turns out that, in addition to the telescope optics, the particular instrument that analyses the incoming light is a major factor for successful science. SOFIA currently uses six different instruments optimized for different scientific objectives, some of them have already been upgraded, and most are scheduled to evolve further over SOFIA's lifetime. You just couldn't do that with a satellite. An additional benefit of an airplane is that you can fly it to different places.

You might think that an airplane is also cheaper than a satellite. However, in fact, the cost is comparable, as a calculation by the SOFIA team showed, around 90,000 USD per observation hour. While a satellite is very expensive to build, validate and launch, the operational costs of flying SOFIA are relatively high as a consequence of the relatively large crew complement, high fuel burn and ongoing maintenance of the relatively old airframe. And the engineering necessary to put the telescope onto the airframe was also no small feat, as we will see in the next section. An additional benefit is that an airplane can observe different parts of the sky by flying from different places on earth. SOFIA normally operates out of California, but also flies over the southern hemisphere during its yearly deployments to Christchurch, New Zealand.

OPERATIONS SOFIA is operated jointly by the US and Germany. The construction phase was managed by NASA and its German counterpart, DLR. Now, during operations, the setup is slightly more complicated. Flight operations are still handled by NASA out of Palmdale, California. This makes sense because NASA operates several other research aircraft from there, among them the DC-8 and the ER-2. However, the telescope, the instruments, and more generally, the science mission, is performed by the SOFIA Science Mission Operations Center out of NASA Ames, which, in turn, is jointly run by the Universities Space Research Association (USRA) and the Deutsche SOFIA Institut (DSI) at the University of Stuttgart.

2 | *The Making of SOFIA*

THE 747 SP Colloquially the "SP" in Boeing 747SP stands for "Short Plane", because it is 15 meters shorter than the regular 747–100/200 on which it is based. In fact, SP actually stands for "Special Performance". It was built for ultra-long range flights, initially for Pan Am; the shorter fuselage means that it is lighter, so it can go further on the same amount of fuel. It's also a bit faster, due to the reduced drag of the shorter fuselage. SOFIA exploits this extra performance mainly for altitude, as higher altitude means better observation conditions for the telescope. SOFIA can reach flight level (FL) 350 (35,000 ft, or about 10.5 km, in a standard atmosphere) directly after take-off, when almost all of the fuel is still in the tanks (as opposed to having been used up). Over a typical ten-hour mission it can climb to FL 410. This kind of performance is rare in aircraft of this size: business jets such as the Gulfstream 550 can cruise at FL 450, while Concorde had an initial climb performance to FL 520. However, only the 747SP combines this performance with a large enough fuselage to accommodate a 2.7 meter mirror.

Forty-five SPs were built in the late 1970s and early 1980s; the thirteen that are still operational are used by governments or private operators. For example, recently Pratt & Whitney retired their engine testbed 747SP. The SOFIA airframe was built in 1977 and NASA purchased it from United Airlines, who had put it into storage. In fact three SPs were used during the project; one was used together with a wooden mockup of the telescope to verify the basic approach of installing the telescope in the airframe's aft section. Aircraft number three has been disassembled for spares; surprisingly many parts of the SP are different from the regular 100/200 variant: the flap system, the vertical and horizontal stabilizers and parts of the landing gear are all different. Considering SOFIA's projected lifetime of 20 years, keeping a stock of spares certainly seems like a good idea for a niche aircraft like the SP.

CONVERSION TO SOFIA The development of SOFIA can be broken into two parts: the modification of the 747SP and installation of the telescope, and the development of the telescope and its control system. As mentioned before, NASA took care of the airframe modification and DLR developed the telescope. I suspect that Conway's Law, which states that a (software) system's structure will resemble the structure of the organization that created it, applies equally outside of software. The actual modification of the airplane was done by L-3 out of Waco, TX. In terms of overall investment, the aircraft/telescope work share was around 80:20.

Thomas Keilig, now the managing director of the DSI, the organization that manages the German contributions to SOFIA, was involved in the conversion of the 747SP, the flight tests and the certification process. He says of this period in SOFIA's life that the main challenge in the conversion of the SP to SOFIA was, unsurprisingly, the 4.1 × 5.5 meter cutout for the telescope door—roughly a quarter of the circumference of the fuselage. Additional beams had to be added to the top and sides of the fuselage to carry the structural load. A thicker skin was also installed to take some of the bending and twisting loads. In terms of aerodynamics, the hole in the rear fuselage could cause turbulence to strike the tail and lead to aeroacoustic vibrations as a consequence of the resulting "organ pipe" geometry of the fuselage. Those in turn would negatively affect the telescope's precision and eat up the lifetime of the SOFIA airframe. An aerodynamic fairing has therefore been installed around the door cutout to smooth the airflow and avoid such problems. It's not feasible to put a glass or plastic pane over the cutout, because there is no material that is completely transparent for the whole range of infrared light observed by SOFIA.

Flight testing revealed that the theory worked in practice: vibrations and turbulence are minimal; the pilots can hardly tell whether the door is open from the way the airplane handles. There is also very little additional drag, which would reveal itself in a reduced maximum speed for a given power setting. During normal airframe development, structural stability is verified by testing components up to and over their stress limit—in other words, one tries to break them. This is not a good idea if you have only one airplane. Instead, over 800 strain gauges were installed all over the fuselage, and in particular around the heavily modified aft area. Data was collected in flight and later analyzed. This showed that SOFIA's fuselage is stiffer than the original.

The heavy telescope at the aft of the fuselage is also problematic for aerodynamic stability. For aircraft to be stable in flight their center of gravity (CoG) has to be at a specific position along their long axis in relation to the center of pressure, the (theoretical) point at which all aerodynamic forces act on an airframe. The 17 ton telescope in the back of SOFIA moves the CoG way aft. To compensate, some of the telescope's systems, such as the hydraulics, have been installed at the front of the airplane. The floor plates in the front of the cabin, usually engineered to be as light as possible, have also been replaced by thick, heavy wolfram plates. This came as quite a surprise to the mechanics of Lufthansa Technik when in 2014 they disassembled SOFIA as part of a D-Check, a thorough inspection of an aircraft's structural components that requires removals of all panels and covers.

Thomas Keilig remembers the flight testing of SOFIA as a fascinating part of his career. It was performed at NASA's Dryden Flight Research Center at Edwards Air Force base, hallowed ground for an "avgeek" like him. The initial flights were performed with a minimum crew, and he and his colleagues had to monitor the telescope systems from the ground. But "sitting in the control room that was formerly used for Shuttle landings was a real treat", remembers Keilig. He was on board for some of the later tests: "Flying a 747 to 2.3 g and 0 g is not something you experience every day." Test flights also investigated the effects on the fuselage of a maximum rate rotation of the telescope: essentially none could be detected. SOFIA is now fully certified based on FAR 25, the Federal Aviation Administration's regulations for transport category aircraft. It can be flown with the same type rating as other SPs.

3 | *Telescope Architecture and Control*

A NASMYTH TELESCOPE SOFIA uses a Nasmyth telescope, which is fundamentally arranged in a dumbbell shape. Its relatively compact vertical size makes it ideal for fitting into an airplane. It consists of three major parts: a set of mirrors for collecting and guiding the infrared radiation, a (replaceable) instrument that analyses that radiation, plus a massive carbon fibre construction that connects these two and serves as the structural backbone for the whole telescope, the so-called "Nasmyth tube". The complete telescope weighs 17 tons; the movable parts held by the Nasmyth tube weigh 10 tons.

An airplane's cabin is pressurized to roughly 10,000 ft when it is at its operational altitude of 30,000 to 40,000 ft. However, the tail section of a commercial airliner is not. The pressure vessel that constitutes the cabin ends with the aft bulkhead, a strong circular structure that can withstand the pressure difference between the cabin and the unpressurised rear part of the fuselage. The same is true for SOFIA, but this bulkhead is moved forward significantly. This is because the telescope has to be in the unpressurised section, the so-called telescope cavity, since the open door exposes this part of the observatory to the external low atmospheric pressure. On the other hand, it was decided to put the instrumentation into the pressurized part of the cabin to simplify in-flight monitoring and diagnostics. So the light that is collected in the unpressurised part somehow has to make its way into the pressurized section for analysis in the instrument. The Nasmyth tube plays an important role here: it is sturdy enough to act as the backbone onto which the telescope and the instrument are mounted, but it also is hollow, so the light can pass through from the mirror assembly to the instrument mounted at the other end. This illustration, adapted from a similar one by DSI, provides an overview.

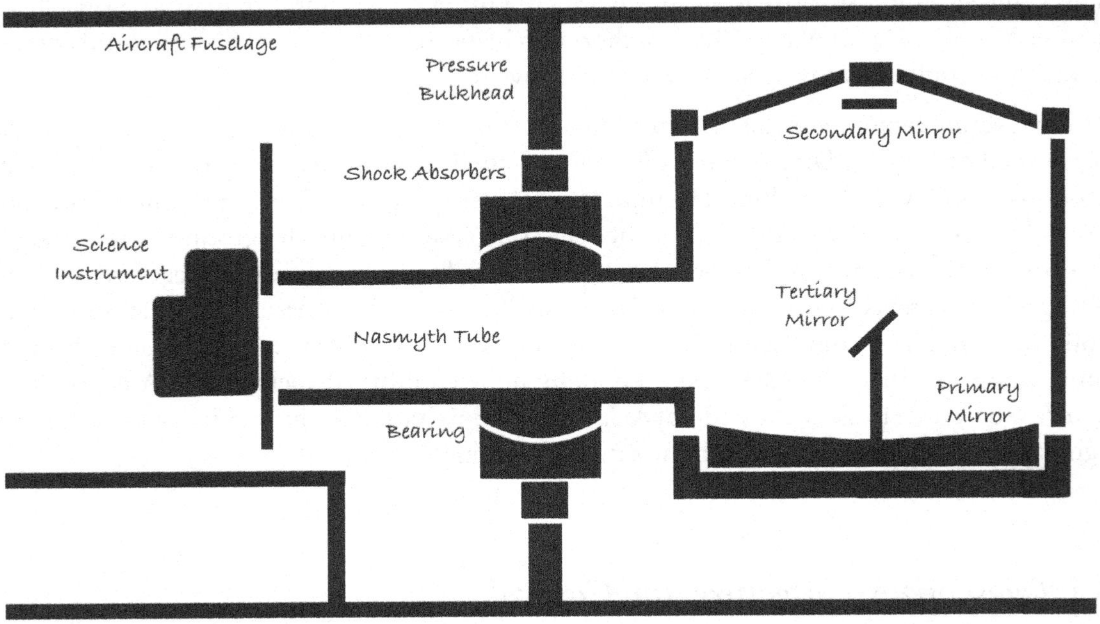

The Nasmyth tube itself is carried by a bearing that installed in the bulkhead that allows the telescope to rotate in elevation between 17 and 65 degrees from the horizon. This corresponds to a rotation around the long axis of the Nasmyth tube and is achieved by a motor-driven cog/wheel assembly. To point the telescope at a particular target, in addition to elevation you also have to control azimuth, the "compass direction" at which the telescope is pointing. The telescope itself has (almost) no means of changing the azimuth; instead the airplane has to change its heading to control the direction of view. The desired observation targets for a particular mission therefore govern SOFIA's flight path for that mission—an interesting challenge for flight planning, as we will see later.

What I wrote above is not quite true: in fact, the telescope can rotate by ±2.5 degrees in azimuth. This is achieved by means of a spherical bearing inside the cog/wheel, which is in turn mounted inside the aft bulkhead. The bearing also provides ±2.5 degrees of control in elevation, as well in a third axis in order to track the sky's rotation and keep the image's orientation stable. These ±2.5 degrees in all three directions are the domain of the fine drive, managed by the telescope control system, which I will explain in the next section.

The telescope itself has three mirrors. The main mirror, at the bottom/rear of the telescope, is 2.7 meters in diameter. This reflects light to the secondary mirror located the top/front of the mirror assembly. The fact that this mirror is in front of the main mirror means that SOFIA, like all reflector telescopes, has a small blind spot in the middle of its field of view. The secondary mirror reflects the light onto two tertiary mirrors positioned about two-thirds of the way along the telescope's axis. These are both arranged at a 45-degree angle and reflect the light through 90 degrees into the Nasmyth tube. The upper of the two tertiary mirrors reflects infrared light into the instrument but it is transparent to visible light; this strikes the lower tertiary mirror and is reflected into the focal plane imager, an optical camera that is used mainly for tracking. Occasionally the focal plane imager is also used for science, making SOFIA also an optical telescope that can be moved around the planet to observe interesting phenomena such as occultations, where the airplane has to be at a very specific point in time and place to make meaningful observations. It's not the world's greatest optical telescope, but it's unique in its ability to move to places of interest.

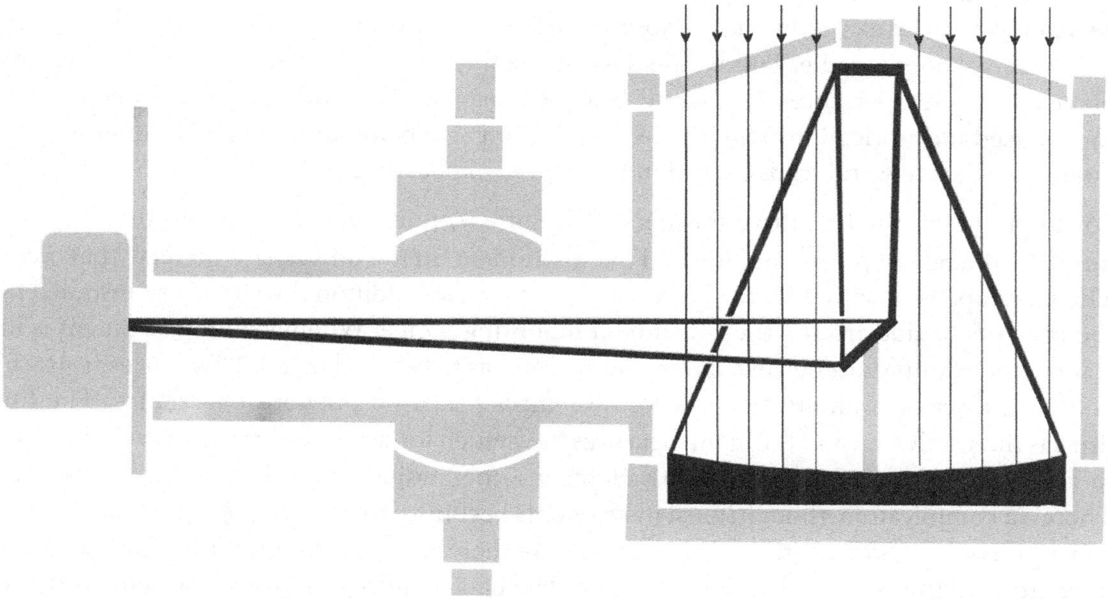

Two more things are worth mentioning here. First, in addition to the cog/wheel drive and the spherical bearing, the mounting of the telescope in the bulkhead also features passive mechanical dampers to filter out high-frequency vibrations—the vibration isolation system. These are composed of air springs, essentially like rubber tires: there are no metal-to-metal contacts between the telescope and the airframe when the telescope is operating. During takeoff and landing, though, the telescope is secured by pressing it against metal hard stops.

The second interesting tidbit is that, as the outside atmospheric pressure extends through the Nasmyth tube to the instrument, the instrument's entry window must be able to bear that pressure differential. It does, although since this window has a very small surface area, the resulting force is relatively small.

ACTUATORS As with any control system, it's all about sensors and actuators. Let's start with the actuators. I mentioned the 1.2 meter spherical bearing and the fine drive before. Let's look at it in detail. As indicated before, it can move the telescope in three directions to follow the sky's rotation and keep an image's orientation stable: elevation, cross-elevation and line-of-sight. This movement is achieved with electromagnetic torquers: the inner, movable sphere of the bearing contains the magnets, and the outer, static cradle contains the coils. This kind of electromagnetic actuation is very precise, but cannot create huge forces. This means that the friction of the bearing must be reduced as much as possible. To achieve this, oil is pressed into the 50-micron gap between the sphere and the cradle. The whole ten-ton movable part of the telescope therefore rests on an oil film! This arrangement is so delicate that the temperature of the sphere and cradle have to be controlled precisely. For example, if the inner sphere gets too cold, it shrinks, reducing the 50-micron gap and risking the metal of the sphere touching the metal of the cradle. This would destroy the fine drive's ability to control the telescope precisely. The temperature of the whole bearing is primarily controlled by the temperature of the oil used for lubrication. The SOFIA's cold cabin also helps, although it is sometimes annoying for the crew and leads to the kind of interesting headwear shown below.

For the forces of the fine-drive torquers to be sufficient, it is also crucial that the two main masses—the telescope on one side and the instrument on the other—are perfectly balanced. The telescope is usually heavier than the instrument, so additional weights are installed on the instrument side around the instrument mounting flange. Whenever an instrument is installed (for example after an exchange) the system has to be rebalanced. This process starts by attaching a set of counterweights whose weight is based on previous experience. The fine drive is then activated and the team measures how much force it has to create in each direction to keep the telescope balanced. The weights are then adjusted until these forces are zero. There's a complication, though: most instruments become lighter during a flight because they consume coolant. Some of the counterweights are therefore installed on a drive that can move outward with time, increasing their leverage. The balancer drives move the weights at 0.0012 to 0.0036 millimeters per second, using an open control loop based on empirically derived values for the instrument's rate of consumption of coolant. All these steps ensure that the ten tons of the telescope can be moved by magnetic torquers. In fact, "the whole telescope can be moved with your pinky once it's balanced", explains SOFIA's telescope engineer, Oliver Zeile.

SENSORS Let's move on to the sensors. The telescope is "attached" to a guide star by optical cameras of which the telescope has several. The one used primarily for tracking is the focal-plane imager, which looks through the Nasmyth tube and the telescope. It has a relatively narrow field of view, around 8.5 arc minutes (although this is still wider than the field of view of the instruments, which range from 8 arc minutes to 6 arc seconds). It is possible that for a particular observation, no optically visible guide star is "in the picture". In this case SOFIA can also track using the fine-field imager, a camera with a slightly wider field of view mounted on the telescope ring; it does not "look" through the telescope. The third optical camera, the wide-field imager, also mounted on the telescope ring, is mainly used for initial targeting, where an even wider field of view is useful to position the telescope relative to the wider neighborhood of the target.

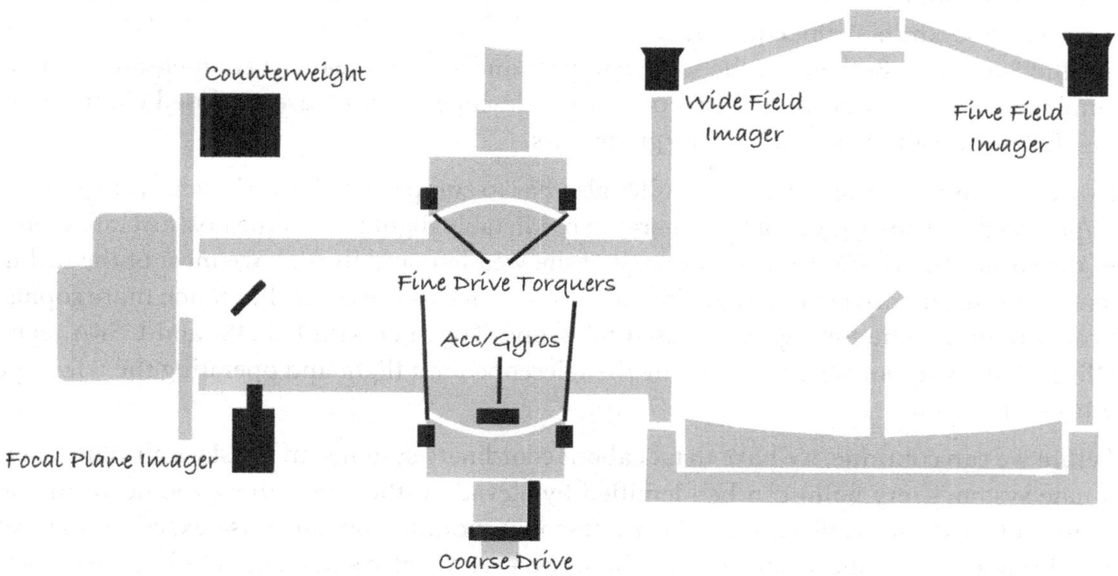

To keep the telescope locked onto a particular target, a controller tries to keep the guide star image seen by the tracking camera at the same location. There is a problem, though. Depending on the brightness of the guide stars, the cameras have an integration time of half a second or even a second. This would would make the control loop very slow and thus imprecise. This is why the telescope also has several fiber-optic gyros that measure rotational speed in all relevant dimensions. Their signals can be used to interpolate between the subsequent guide camera updates.

You might ask why the gyros aren't used exclusively. Gyros drift over time, so a gyro might report a very small angular velocity when in fact the situation is stable. The guide cameras provide an absolute signal and can thus be used to reset any gyro drift computationally; the cameras and the gyros work together to keep the telescope's attitude stable. In addition, from a control theory perspective it is useful to acquire not only the set value and the actual value as inputs to a control loop, but also the derivative of the actual value; some control algorithms take the derivative into account explicitly to improve precision. The gyros provide that derivative much more precisely—by differentiating the velocity with respect to time—than the pixel movement of an image derived from the guide cameras.

CONTROL REQUIREMENTS To achieve good image quality the telescope must be precisely locked onto its observation target during exposure. The longer the exposure, the more important a stable tracking becomes to avoid "blurry" images. A telescope that is mounted on a moving platform must itself move actively to compensate for movement of that platform. Every earthbound telescope does this to compensate for the rotation of the Earth, which is roughly 1.600 km/h at the equator. SOFIA's telescope is built on an aircraft that moves at around 800 km/h. Both are in the same ballpark, and similar means of compensation can be employed. So, like professional ground-based telescopes, SOFIA uses a closed control loop that fundamentally relies on tracking guide stars.

However, the SOFIA telescope controller also has to compensate for turbulence, smaller but unpredictable disturbances of the otherwise predictable flight path of the host aircraft. Flying in the stratosphere, SOFIA is above most of the weather, and thus above most of the turbulence, but turbulence nevertheless does occur. This makes the controller much more sophisticated than the one for a ground based-telescope. Together with his DSI and USRA team, Oliver Zeile is responsible for readying the telescope for a flight and operating the telescope during the flight.

Before we can continue, we have to talk about coordinate systems. In the sky's absolute coordinate system every point can be identified by elevation (the angle above the horizon) and azimuth (the compass direction). Many telescopes mounted on earth use exactly these two axes for pointing. Using a control loop, these two axes are also sufficient to follow the (apparent) rotation of the night's sky.

As we have seen in the previous section, however, SOFIA's telescope is mounted on a bearing in the bulkhead which allows rotation around three axes for pointing and tracking. These three axes form their own, telescope-local coordinate system. One axis, elevation, is similar in both coordinate systems: the telescope can rotate up and down. The other two axes are different, however. They are called cross-elevation and line-of-sight. To understand how they are aligned relative to azimuth, let's run a thought experiment. I suggest you follow along with your hands to visualize the situation. Let's assume we could point the telescope completely horizontally and straight up, changing the elevation between 0 and 90 degrees. Now let's say the airplane changes its heading and thus changes azimuth (in the sky's absolute coordinate system). If we have the telescope pointed at zero elevation, then a change in azimuth maps completely to a change in cross-elevation. If instead the telescope were pointed straight up, a change in azimuth would map to a change only in line-of-sight. For all practical elevations between 17 and 65 degrees, a change in azimuth always maps to a change in both cross-elevation and line-of-sight, using a coordinate transformation that involves the sine and cosine of the elevation angle. The point is: the telescope's control loop, which we will discuss next, works in the elevation/cross-elevation/line-of-sight coordinate system, while the scientists care about elevation and azimuth, and how this changes as the sky rotates.

CONTROL STRATEGY Let's recap: the coarse drive is used to put the telescope into an elevation that is suitable for the observation target. The heading of the airplane provides the correct azimuth. As the airplane and the sky move, the fine drive compensates for both by moving the telescope assembly in the telescope's elevation/cross-elevation/line-of-sight coordinate system relying on the coordinate transformation mentioned earlier. But remember, it only has 2.5 degrees of travel in each direction. So what happens when the fine drive reaches its limits? What actually happens differs for each axis. For elevation it is simple. Say that the fine drive runs into its upper limit. The coarse drive can then simply move up while the fine drive compensates for this movement by moving down by the same amount, effectively resetting its own range of travel. This is done automatically by the telescope controller, continuously, in small steps.

For the other two axes it is more complicated. As we have seen before, they "share" the aircraft's movement in azimuth depending on the elevation. Said differently: if we run into the 2.5 degree limit of the fine drive, the cross-elevation component can be compensated by a change in azimuth: the mission director tells the flight crew to turn the airplane left or right by a degree or two. Integration of the telescope controller with the 747SP's autopilot has been discussed for a long time. This would allow heading changes to be automated, avoiding the need for the mission director to tell the pilots when to change heading. However at the time of my flights this was not available, and the pilots said that they were happy to carry out manual corrections, as during cruise they didn't have very much else to do anyway.

The other component, the rotation around line-of-sight, cannot be compensated transparently by a movement of a drive or the aircraft. Instead, the telescope has to be explicitly "rewound". This means that the orientation of the "picture" seen by the telescope actually changes (which is relevant if the target is not point-symmetric). So, while the reset in elevation and cross-elevation are transparent to the telescope and do not impede the instrument's data collection, this is not true for the line-of-sight rewind, whose timing is therefore coordinated with the instrument. A rewind takes roughly a second, and, depending on the observation's target and the position of the airplane, a line-of-sight rewind is necessary every three to 20 minutes.

What is intriguing about the attitude control of the SOFIA telescope is that it requires no explicit coordination with the aircraft's movements in the sense that it needs to be integrated with the aircraft's systems—or with the coarse drive, for that matter. If the aircraft turns, or if the coarse drive rewinds in elevation, the fine drive continues with its usual job of keeping the target locked, just as for any other disturbance such as turbulence. The telescope is stabilized inertially. In fact, controlling SOFIA's telescope closely resembles the control of a satellite, which is why the team includes satellite people. Oliver's PhD, for example, was in satellite attitude control. There are many videos on YouTube showing the telescope moving inside SOFIA's fuselage. These videos are sped up to make it easier to see the effect, but I was able to see the movement by carefully observing the instrument in the fuselage. Of course it is the fuselage that moves around the telescope: the telescope is stable in space, locked on to its target. Quite impressive, if you think about it.

There is one more complication. The whole reason for mounting an infrared telescope on an airplane is that this allows it to be high enough to avoid problems caused by water vapor in the atmosphere. However, even at SOFIA's altitude, 99% of the radiation collected by the telescope (and thus, the instrument) comes from the atmosphere (and the telescope itself). This atmospheric noise fluctuates on a scale of seconds to minutes. To get the signal-to-noise ratio to a reasonable level and produce a meaningful observation, SOFIA also collects data from the background "dark atmosphere" and subtracts this from the data that includes the signal from the target object. The instrument is aware of the difference in these data sets and processes them correctly. To collect the atmosphere-only data, the attitude of the telescope is periodically shifted to a location in the sky close to the target to capture a signal from the same part of the sky but without the target object. This happens at a frequency of 2 Hz and is achieved by moving the secondary mirror in a process referred to as "chopping". Moving the heavy primary mirror at this high rate would be infeasible. The requirements for chopping limits SOFIA—or any other infrared telescope, as chopping is standard practice—to observing targets that have "empty" space close to them.

Chopping leads to another problem, however. The primary mirror is relatively massive, so one cannot assume that it is at exactly the same temperature over the whole of its surface. Because the secondary mirror, during chopping, "looks" at different parts of the primary mirror, it might see different temperatures at various parts of the primary mirror. Those temperatures affect the signal. "It's almost as if a different telescope were used for the actual target and the chopping target", says Christian Fischer, an instrument engineer. To compensate for this effect, the primary mirror moves as well, in a process known as "nodding". This essentially flips the locations of the actual target and the chop area on the main mirror, canceling out the "two different telescopes" effect. The nodding process happens on a longer timescale and is achieved by the fine drive moving the whole telescope.

This chop/nod scheme explains why two different numbers for the size of SOFIA's primary mirror exist. Physically it is 2.7 meters in diameter, but effectively only 2.5 meters can be used; the rest of the mirror's surface is not "seen" at any setting of the secondary mirror. The "extra" 20 centimeters are necessary for nodding.

During an observation the telescope is controlled by the instrument attached to it: the instrument is in charge of "running the show". The reason for this is that the instrument's data acquisition must be coordinated with whatever the telescope does (chopping, rewinding) for it to be meaningful. Around 50 parameters have to be set for every specific observation, including the position in the sky and the configuration of the chop/nod scheme. Hundreds of such parameter sets have to be generated for each flight. This is done with scripts written in the programming language IDL—similar to Python—which is also used for data reduction.

Here is the overall high-level control algorithm, courtesy of Oliver Zeile. Of course there is lots of math going on in many of these boxes.

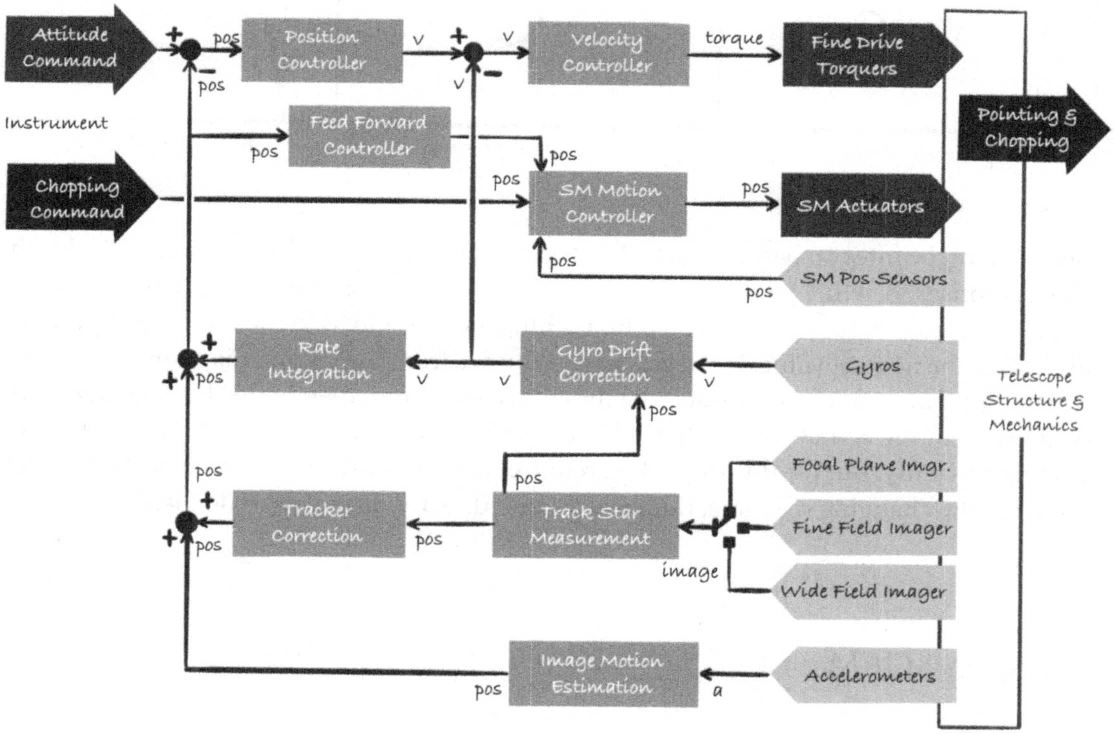

Note how the gyros directly feed data into the velocity controller (and do not just go through the position controller). This allows the algorithm to take into account the first derivative of the quantity the system is controlling, the position, allowing for faster reaction to disturbances. Fundamentally, this is a PID-controller: a control algorithm that takes into account three components. The first one is the proportional component. It creates a response that is proportional to the error between the set value and the measured value. Because it needs an error to generate a response, a pure P controller will never quite catch the actual value up with a new set value. The diagrams below show a signal (dashed line) that changes at some point in time. The actual value oscillates, but when it eventually stabilises, an error will remain.

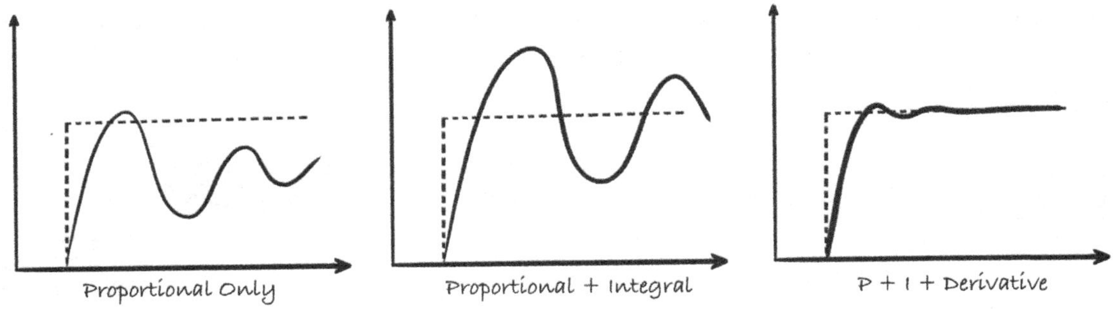

This is where the integral part comes in. It integrates (sums up) the historic error and adds this to the response. So, while an error still exists, it increases its response, which is why it eventually pushes the error to zero. At this point, nothing more is added to the error and the system stabilises at the new set value. However, as you can see from the second diagram above, it still oscillates a lot around the set value. The derivative part "anticipates" this by considering the first derivate of the actual signal. Once all factors are tuned correctly, this dampens the oscillation significantly. But for this to work, it requires access to the derivative. It could be calculated from the changing set value, but measuring it directly, as the gyros do, helps.

4 | *Flying SOFIA*

TAKEOFF It's 19:00 local time, about nine hours earlier than when I took a short rest on that business class seat. I am already tired from jet lag, but I'm also very excited. Not only is my first flight on SOFIA about to begin, but I will have the opportunity to sit in the cockpit for the take-off and initial climb. The flight crew today consists of Ace Beall in the left seat, Jim Less on the right and Chris Farinha, a flight engineer. Yes, you read that right—SOFIA has a flight engineer. Remember that the 747SP is based on the –100 model of the 747, which dates from the days before highly automated two-person cockpits were introduced. In 2012 SOFIA acquired a glass cockpit incorporating advanced avionics to give it the required navigational performance to fly in international airspace. But only the avionics—displays, autopilot, flight management system—were modernized, not the SP's basic systems: its fuel, hydraulics, electrical and environmental systems still date from the 1960s. Even a cursory glance at the cockpit reveals SOFIA to be a unique hybrid of modern avionics and steam gauges.

ACE: Main deck, are you guys ready for engine start?

MAIN DECK: Stand by [confirming with other crew in the cabin]

MAIN DECK: Ok, we're ready.

The cockpit confirms with the mission director whether the cabin and the telescope systems are ready for the next phase of flight. Cockpit and cabin collaborate closely to guarantee a successful mission.

JIM: Ace, ready for before-start engine checklist?

ACE: I am.

JIM: Fuel panel, fuel pumps are on, cross-feeds are open.

JIM: Parking brake.

ACE: Set.

JIM: Chocks.

ACE: Chocks are being removed.

... a few more checklist items ...

JIM: Before-start checklist complete.

JIM (Radio): Palmdale ground, NASA747 heavy, in front of hangar west, request engine start-up.

GROUND (R): NASA747, engine start-up approved.

ACE: Ok, we're gonna start up engine 2.

CHRIS: Valves open.

JIM: We've got RPM, engine's coming up ... Fuel ... 30, 35, ... 45 ... light it up.

Engine start-up is a much more manual process than in more modern airplanes, where everything except pressing the "start" button is automated via FADEC, the digital engine control system. In SOFIA all three cockpit crew work together to light up the engines; they start with number 2, the left inboard engine, then engine 3, the right inboard engine. The outboard engines are not started until almost on the runway, because with Palmdale's narrow taxiways these engines are over dirt; running them would risk ingesting this into the engine, not to mention raising lots of dust behind the airplane. Neither are really helpful.

Except for a smaller plane on approach, SOFIA is the only aircraft in Palmdale this evening. We have the whole airport to ourselves, and air traffic control gives us quite a bit of leeway. Ace suggests taxiing a few minutes early and ATC is ok with that. We taxi to the runway, starting the remaining two engines as we get close. During taxi, Ace, Jim and Chris find an amber light showing green and joke about it. As expected with professional aircrew, they are very formal and structured in their communication when it counts, but funny and relaxed otherwise. More checklists: because of the low degree of automation, checklists are a bit more extensive. The 747SP also requires more extensive monitoring of systems.

After finally requesting and getting permission to enter the runway, SOFIA lines up on runway 25. Planned take-off is still a few minutes away, so we wait patiently. Taking off precisely at the scheduled time is crucial because, once out of Palmdale, SOFIA will climb into the very busy Los Angeles airspace, and sticking to the filed flight plan is important for deconfliction, especially since SOFIA wants to climb straight up, traffic permitting, rather than in steps as most airliners do. Ace and Jim set and check radios, brief the weather based on ATIS (Airport Terminal Information System, automated broadcast by airports about weather, active runways or, for example, taxiway closures), set the altimeter pressure and configure the autopilot to altitude select mode for the initial assigned flight level of 7,000 ft. They select the correct frequencies for VOR navigation aids, brief the departure route and recap V-speeds, the weight- and weather-depend speeds at which, for example, SOFIA can still safely abort a take-off run, rotate the nose or retract the flaps. They talk through scenarios for aborted take-offs, something not to be taken lightly with this very heavy airplane, as well as the routes they'd fly if an engine fails just after take-off: with Palmdale surrounded by mountains, this must be planned carefully.

ACE: Mission, how're you guys doing down there, you ready for take-off?

MAIN DECK: Few more minutes...

ACE: Right, we have ten more minutes till scheduled.

... Take-off checklist ...

ACE: Ok, go ahead, call for take-off.

JIM (R): Tower, NASA747 ready for take-off, runway 25.

TOWER (R): NASA747, cleared for take-off, runway 25.

JIM (R): NASA747, cleared for take-off.

The throttles advance, four beautifully loud JT9D engines run up their power and SOFIA rumbles down the runway, first slowly—80 knots—then faster and faster. Even as a passenger in the cockpit, without hands on the controls, one can feel the weight of the fully laden airplane. We don't have to abort the take-off and pass V1; a few moments later, Jim calls out "Rotate". SOFIA lifts into the air, nose gear first, main gear following. "V2, positive rate" and we're climbing. We're immediately handed off by Palmdale Tower to LA Center. We make a right turn over Palmdale city, still quite low. Again, one can tell that we are within a few tons of the maximum take-off weight for a 747SP: we're not exactly climbing away confidently. I guess if you live in Palmdale you must be used to jet noise. Those engines, at this altitude, must be really loud down there in the city.

CRUISE After we reached cruising altitude and the workload for the cockpit crew has reduced, I chatted with Ace about the peculiarities of flying SOFIA. Ace is a former NASA research pilot who flew the Shuttle Carrier 747 for 24 years. Now retired, he works as a part-time

SOFIA pilot. He flies three to four missions a month, spending the rest of his time flying legacy Learjets for a private operator.

On this flight Ace sits in the left seat, filling the role of the pilot in command, or Captain. Jim is the co-pilot on this flight. However, in contrast to airline operations, "Captain" and "co-pilot" are not qualification levels, they are just roles played on a particular flight. Jim could sit on the left just as well. Ace says that on next week's flight he will sit in the right-hand seat.

I mentioned earlier that the telescope's azimuth is determined by the airplane's true heading: the telescope always points 90 degrees left of the aircraft's heading. Airliners that want to get from point A to point B fly a track, a course over the ground. If there is no crosswind, track and heading—the direction an airplane's nose is pointing—are the same. In the case of crosswind, however, regular airliners fly at a "crab" angle: they point the nose into the wind to keep to the required ground track. SOFIA cannot do that, because to keep the telescope pointing in the direction the scientists need the heading is what counts, not the track. This means that a crosswind will blow the airplane off its track. The mission planners take this into account, of course: when they file a flight plan with ATC they plan the effective course over ground according to the heading required by the observation and the expected wind. However, the wind is not always exactly as predicted. In this case SOFIA has to divert from its planned track—it cannot just adapt a crab angle. The mission management system neatly displays this divergence.

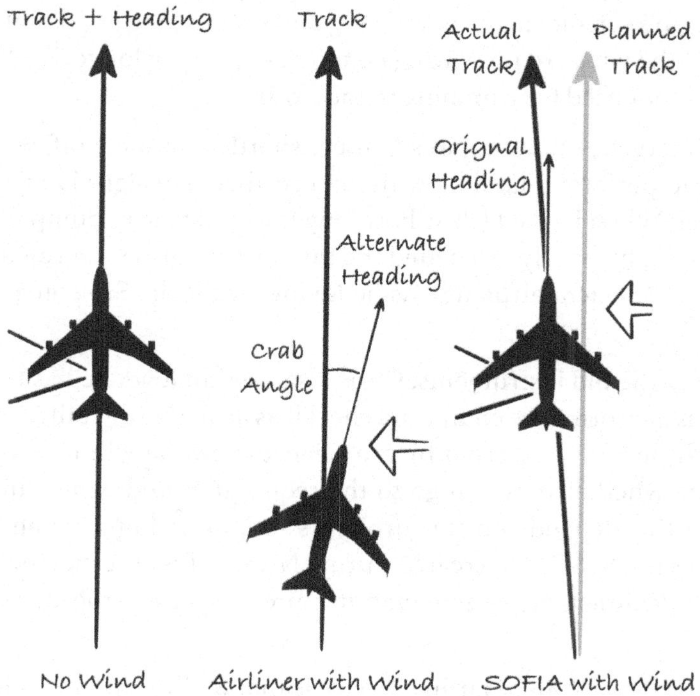

However this must also be coordinated with ATC, who generally don't like airplanes diverting from their filed flight plans. Fortunately the controllers have gotten used to SOFIA's special needs and are usually quite accommodating. Ace recalls that already 90 minutes into the flight they are usually at 39,000 ft. Not too many other airplanes are up there, especially at night, so deviating from the plan requires hardly any deconfliction on the part of ATC.

FLIGHT ENGINEER I also chatted with Chris, the flight engineer. Like Ace he has lots of experience of "classic" airplanes, previously working for 30 years on Lockheed Electras, C-130s and DC-8s. His previous 747 experience comes from Evergreeen's tankers and GE's engine testbed.

The job of the flight engineer is to manage the 747SP's systems. This involves a lot of monitoring and troubleshooting in case of problems. To make this possible he has more detailed information about important systems than the pilots. For example, for the engines, the pilots have engine pressure ratio (EPR), N1 (outer fan speed), exhaust gas temperature (EGT) and fuel flow on their display. The flight engineer's console also shows oil quantities, temperatures and pressures. Interestingly, those indicators are the only ones that have been "glassified" on the flight engineer's panel. But Chris is also involved in routine flying duties. For example, together with the two pilots he starts up the engines, and also manages the fuel. Aircraft like the 747 have several fuel tanks, and the flight engineer can configure which engines take fuel from which tank. This is routinely used to reduce aerodynamic loads on the airframe. In particular, Chris' job is to keep fuel in the outer wing tanks for as long as possible, as the resulting increased weight of the wings reduces wing bending, improving longevity. Unlike other civilian airplanes, fuel is not used for trimming in the 747SP.

In the case of an emergency that requires landing shortly after take-off, when the airplane's weight is too high to perform a landing without a costly overweight landing inspection, aircraft will dump fuel. When I asked Chris how long it would take to dump fuel to get to a permitted landing weight, he briefly calculated the number on his pocket calculator. I guess as a flight engineer on an "ancient airplane" (as he lovingly calls the SP) you always have one of those at the ready.

In quite a contrast to the old instruments, Chris also uses an iPad: he has the complete set of the 747SP's manuals, performance charts and checklists available digitally. For example, when an increase in flight level is suggested by the science crew, he will check the performance charts to determine whether they can go to the required altitude and which engine power setting is required; this depends on the airplane's weight and outside air temperature. All these charts are specific to SOFIA, created during NASA's flight testing of the airplane with input from Pratt & Whitney, the engine manufacturers. Special procedures just for a single airplane!

One difference to using a 747SP in airline service is that SOFIA optimizes for altitude rather than for fuel efficiency, running the engines at maximum continuous thrust for much longer than would an airline. To compensate to some extent, SOFIA's procedures require initial climbs to 26,000 ft to be performed with reduced power. Above that, they go to "plus power": 100% N1 or 900 degrees EGT, whichever is lower. "This would never be done on a civilian airliner", Chris says. SOFIA typically flies close to the maximum altitude the airplane can reach for any given weight. This also leads to a smaller "coffin corner", the speed band between the minimum speed (before the airplane stalls) and maximum speed (before high-speed stall sets in). Those two speeds depend on air density (that is, altitude and temperature), weight and g-loading. At altitude, the minimum speed increases and the maximum speed reduces, narrowing the band that can be used safely. As Chris explains, in airliners this band is typically 50 knots wide. For SOFIA, as a consequence of its striving for high altitude to make the scientists happy, minimum speed can be 251 knots and maximum speed 260 knots—only 9 knots difference. As SOFIA grows lighter over a flight the band gets wider, but initially it does require some concentrated flying.

The crew also uses the iPad as an additional means of navigation; like many general aviation pilots they use ForeFlight, a popular app for flight planning and navigation. Ironically, this provides better navigation and mapping support than the "modern" glass cockpit displays.

5 | *Instruments*

CATEGORIZATION Astronomical instruments can be broadly classified into two groups: imagers and spectrometers. Imagers essentially work like everyday cameras. They have a grid of pixels, and each pixel measures incoming total energy. The brightness of a particular pixel therefore corresponds to the amount of energy (the number of photons) collected over a particular period of time, referred to as "exposure" or "integration time". A pixel collects energy over a particular range of wavelengths; color images can be formed by co-locating detector pixels that collect red, green and blue light and overlaying their data. An imaging sensor that works in the range of visible light is what we know as a camera. But the same approach can be used for other wavelengths, too. For example, an infrared imager records different "colors" of infrared light.

A spectrometer on the other hand can be seen as a single pixel; spatial resolution is achieved by pointing it at different locations to analyze the spectrum of light received by the pixel. This spectrum measures the amount of energy at a particular wavelength, and is usually represented as a diagram that shows energy on the vertical axis and wavelength on the horizontal axis.

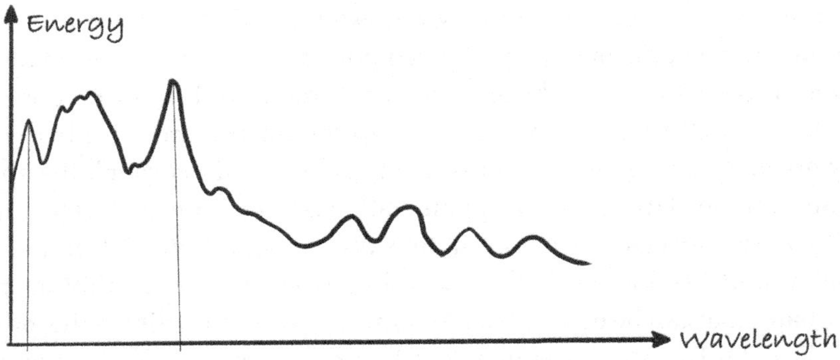

A spectrometer splits the incoming light into its constituent wavelengths and analyzes the energy for each wavelength separately. Both imager and spectrometer have a number of separate detector units: the imager uses them to measure the total energy per sub-area (pixel), whereas the spectrometer uses them to measure the energy per wavelength interval. Splitting the light into its constituent frequency bands for analysis is done using a prism or via a grating. A grating is a surface etched with many parallel lines that diffract light in different directions according to its wavelength: it works in the same way as a prism, but is reflective instead of transparent. Because of how gratings and prisms work, they often produce a line of spectra; they support, if you will, imaging along one single linear dimension. This can be used to “image” a cross-section of an observation target.

Imagers and spectrometers can also be combined in an instrument called an imaging spectrometer. It measures separate spectra in two spatial dimensions and thus creates an “image” of the target. It is a compromise of course: because it creates a spectrum for each pixel, such instruments usually have fewer pixels than an imager, and the spectral wavelength resolution for each pixel may be lower than for a spectrometer.

FIFI-LS FIFI-LS, one of the instruments for SOFIA, is such a hybrid. The abbreviation stands for Far-Infrared Field-Imagining Line Spectrometer. It was originally conceived and designed by the Max Planck Institute for Extraterrestrial Physics in Garching, Germany, and its development has been completed by the University of Stuttgart. It is now one of the instruments operated by the SOFIA Mission Operations center and flies on SOFIA regularly. I chatted about FIFI-LS with instrument scientist Christian Fischer. Like everything on SOFIA, it is a marvel of engineering.

FIFI-LS operates at wavelengths of between 50 and 200 microns, the “cold” universe. It has 50 separate detectors. They cover two channels: a higher energy channel of 51–120 microns and a lower energy channel of 115–200 microns. Each channel has 25 detectors arranged in 5 × 5 square, creating a 5 × 5 pixel “image”. Splitting the light into its constituent frequency bands for analysis is achieved via a grating. A grating is a surface etched with many parallel lines that diffract light in different directions according to its wavelength: it works in the same way as a prism, but is reflective instead of transparent. The light of different wavelengths separated in this way strikes 16 photoconductors to measure their incoming energy. The 25 pixels for each channel (with 16 detectors each) are arranged linearly, which makes the grating simpler. This means that the 5 × 5 grid used to cover the telescope's observed region of the sky has to be linearized by a collection of mirrors, known as the "slice tower".

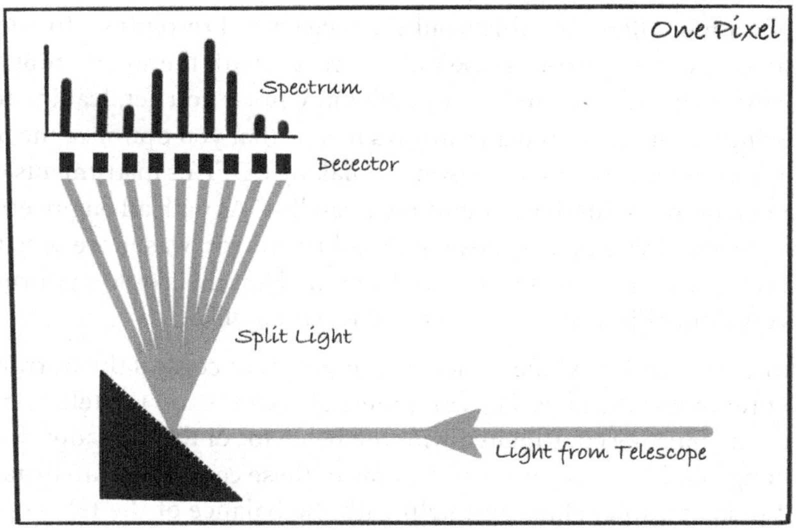

The detectors themselves are also a challenge. Photoconductors detect incoming photons when electrons are knocked out of their atomic structure. These electrons then create an electrical charge that can be measured. The photoconductor materials are selected such that the energy necessary to knock out an electron corresponds with the range of energy that particular detector is intended to detect. Gallium-doped germanium is used in FIFI-LS.

That is the principle, but in practice a whole bunch of tricks have to be used to make this work. For example, the detectors are compressed by a mechanical spring to about 80% of their breaking point to change the crystal structure in such a way as to make it easier for electrons to be knocked out. The system also requires some serious cooling. The photons the detector is designed to collect are of such low energy that the heat energy in the surrounding material and housing is in the same range. The detectors themselves must therefore be cooled to 1.7 K, the optics to 6 K and the remainder of the instrument's mechanics to 70 K. This is achieved using superfluid helium, gaseous helium and boiling liquid nitrogen respectively. Very good optical isolation is required to avoid what Christian refers to as "light bulbs in the instrument".

These coolants have to be refilled for every mission, something that would clearly not be possible for a satellite. Even though less cooling would be necessary in space due to the lower ambient temperature, such an arrangement limits the lifetime of a space-borne infrared telescope. For example, the Herschel telescope had an (expected and actual) lifetime of three years, which leaves SOFIA as the only operational infrared observatory today.

Optimizing detectors for high sensitivity by improving their signal-to-noise ratio is worthwhile. If this is not done, longer integration times are necessary to detect a source of a given brightness. Since observation time corresponds to SOFIA flight time, this is of course costly and reduces the capacity of the observatory. Alternatively one could make the pixels bigger, so that they collect more light, but this would reduce spatial resolution. In addition, as integration time increases, the signal-to-noise ratio increases with the square root of the integration time: the longer the integration interval, the more noise you get, leading to diminishing returns. Improving sensitivity in practice always means that you optimize the instrument for a particular purpose. This is the main reason for making SOFIA's instruments exchangeable, and makes it much more useful than a build-once satellite. As with all engineering, this is also a trade-off: the Herschel satellite, because of its colder mirror, was more sensitive. Together with its very stable position at the Earth–Sun L2 point, Herschel could perform very long integrations on very faint objects that are not feasible with SOFIA.

The instrument comes with a whole bunch of computers to control the mirrors, grating and cooling and to preprocess the data. The computers also command the telescope, as data capture and processing must be coordinated with the behavior of the telescope, for example, for chopping/nodding and line-of-sight rewind. Some of these computers are installed on the instrument itself to keep cables short and help with the balance of the telescope's dumb-bell shape. Others are installed in a 19" rack closer to the instrument engineers' workstations.

Whenever the instrument is (re-)installed on the telescope it has to be recalibrated. Ideally this calibration is performed on the ground in the lab, using the telescope simulator. This can create light of precisely defined wavelengths to align the exact angles of the gratings. However, every installation of the instrument onto the telescope's instrument flange leads to a slightly different position of the pixels relative to the telescope's orientation. Instead of trying to replicate the mounting in exactly the same way every time, the alignment error between the instrument's pixels and the telescope's area of view are measured and software is used to correct for the mismatch during an observation. This notion of measuring and correcting—rather than constructively precise—mechanics is an approach I have seen many times in such high-precision scientific instruments, such as CERN's Large Hadron Collider.

Finally, each of the pixels has to be calibrated precisely, because each of them has a slightly different sensitivity. The technique used here is "flat-fielding": all 25 pixels look at the same "flat" source, such as dark sky, and the differences in flux for the various pixels is recorded for subsequent correction of the measurements. In megapixel consumer digital cameras this is not necessary, because interpolation (averaging) is used to obtain a homogeneous picture. For the 5 × 5 pixels in an instrument like FIFI-LS averaging is not feasible.

In addition to the detectors, springs, gratings, mirrors and the cooling infrastructure, the instrument must also be able to cope with the constant vibrations and bumps of an airplane in flight. It must also be certificated to be crash-resistant: you can't have the whole thing explode into pieces in the event of a crash! FIFI-LS is certificated to withstand 9 g. This makes the instrument a rather massive beast: 600 kg and roughly a cubic meter in size.

OTHER INSTRUMENTS SOFIA currently operates a suite of six instruments: EXES (Echelon-Cross-Echelle Spectrograph), FORCAST (Faint Object InfraRed CAmera for the SOFIA Telescope), GREAT (German REceiver for Astronomy at Terahertz Frequencies), HAWC+ (High-resolution Airborne Wideband Camera) and of course FIFI-LS. The optical focal plane imager also counts as an instrument in itself. Most instruments, such as FIFI-LS, are owned and operated by SOFIA, although some are owned by external principal investigators; GREAT is an example (which also wins extra bonus points for the best acronym). Three first-generation instruments, CASIMIR, SAFIRE, and AIRES were initially selected for SOFIA, but were then halted at different stages of development. A number of future instruments are in various stages of proposal and investigation.

6 | *Science*

TARGETS What is SOFIA primarily investigating? Alfred Krabbe, Director of the DSI, summarizes it as "the cold universe"—the interstellar medium, the gas between the stars, the dust from which stars are born. The interstellar medium is invisible to the eye, and has low energy, radiating primarily in infrared wavelengths. Research topics studied by SOFIA include star formation, black holes and accretion activities, the process of forming planets from interstellar dust.

Spectrometers such as FIFI-LS are crucial here. From experiments on Earth, as well as from theory, we know that particular physical processes create specific elements or put those elements into an excited energy state. When the excited electrons "fall back" into their original energy bands, their excess energy is released. This occurs in the form of radiation at correspond- ing wavelengths that can be seen as an energy peak—a line—in their spectrum. Conversely, radiation can be absorbed by particular elements to cause excitation states. This results in an absorption spectrum, one in which there is significantly reduced energy at a specific wavelength.

The scientific theorization is often indirect. For example, we know from theory that during star formation very high-energy ultraviolet radiation is produced. This ultraviolet radiation is extremely hard to observe directly, because it interacts with lots of other elements and thus "reacts away" before it can reach a telescope mirror. One such reaction is the ionization of carbon. The carbon excited in the temperature and pressure regime that is typical for star forming region has a charactistic line in its spectrum at 158 microns, the "C+ line". This spectral line therefore provides an (indirect) indicator of star formation. Maria Kapala, one of the scientists who flew on SOFIA, provides more detail below. By measuring the intensity of the radiation at a particular wavelength it becomes possible to do quantitative measurements. Jochen Eislöffel, the other scientists aboard SOFIA, gives an example of this later in this section.

A specific process might produce several elements, indicated by several spectral lines. Because their relative separation in wavelength is known, we can recognize this fingerprint even if it occurs at different absolute wave- lengths. Why would this appear at different absolute wavelengths? This is because whole spectra are shifted to higher wavelengths (lower frequencies) if their source is moving away from us, and to lower wavelengths if they are moving towards the observer. These effects are known as red shift and blue shift respectively, and result from the Doppler effect. This therefore allows us to use spectra to detect the motion of observed sources, such as the rotation of galaxies. In particular, with a spectrometer that provides spatial resolution such as FIFI-LS, you can "see" both in a single image—one side of a galaxy rotating towards us, the other away from us.

MARIA Maria Kapala, who was at that time a postdoc at the University of Cape Town, had her observation proposal for SOFIA accepted and was flying on the same missions as I was. I chatted with her about her science. Her work centers around a better understanding the occurrence of excited carbon, C+. As this is used as an indicator of star formation, it is very important to understand exactly when and why carbon is ionized to C+. There are oddities. For example, observations with the LWS instrument on the ISO satellite revealed that there are regions in large spiral galaxies where there's enough C+ to indicate star formation, but there are no new stars. So something else must excite carbon to its C+ state. To better understand this phenomenon, Maria uses the spatial resolution provided by SOFIA to observe Andromeda M31, the spiral galaxy closest to us. In some senses this is fundamental work, because it improves the reliability of a process that is used as a tool in astronomy. "Maybe this is the reason why my proposal was accepted for flying on SOFIA", Maria wondered. Maria's observation was split over two flights, as her work requires a very long integration time, and for reasons of flight planning this could not be fitted into one flight. Remember that the heading SOFIA flies is directly correlated with the targets it can observe. The total integration time would have required SOFIA to fly a particular heading for a very long time. So long, in fact, that there would not have been enough fuel to fly back home.

I heard rumors about extending a single leg in one direction by flying from California to Hawaii and then back the next day, but I suspect this idea is more a consequence of the scientists' vacation plans :-)

After her first flight Maria was frustrated, as the signal-to-noise ratio was much worse than predicted and her data was not useful at all. Since this was the first project for which she was the principal investigator, and therefore responsible for its success, it really darkened her mood. After the second flight, however, she was happy: "The overall signal-to-noise ratio was much better. Even though I have not yet processed the data, and the scientific results are not yet available, I am sure the data is useful."

JOCHEN Jochen Eislöffel of the Thüringische Landessternwarte was the second scientist who flew on science missions 248 and 249. He also investigates star formation, but to understand his work we have to recap a little bit of theory. Stars form from interstellar molecular clouds of gas and dust through a process called "accretion". Starting from random disturbances in dust density, the slightly higher gravitational force of an area starts to attract additional matter, which accretes around a nucleus. The more matter accretes, the stronger the local gravitational pull becomes. This process accelerates, density increases, and at some point the pressure and temperature created from the resulting clump's own gravitational pull becomes great enough for fusion to ignite, and a new star is born.

There is a catch, however. When particles strike the forming star they impart their angular momentum to it, making it rotate. The dust particles might already have angular momentum, for example because the whole dust cloud is rotating, typically because the part of the galaxy in which the cloud resides is also rotating. Or a dust particle might strike the forming star off-center, tangentially, as when someone jumps onto a merry-go-round and thereby accelerates its rotation. Theory predicts that, over time, the new star's rotation should become so fast that the centrifugal forces become stronger than the gravitational force, pulling the growing star apart. As stars clearly exist, though, this does not happen. Why? Where is the angular momentum going?

Young stars also exhibit a bipolar jet, a beam of gas that streams away from both poles of their rotational axis. This happens for all accreting objects that are disrupted by centrifugal forces, such as brown dwarfs, quasars and super-massive black holes. At this point it is not clear how much matter is lost and how it is accelerated, although it is likely that magnetic fields and rotation play a role. However, there must also be a connection with the dispersal of angular momentum. Jochen wants to better understand this connection, and in particular, the amount of matter that is ejected by the jet.

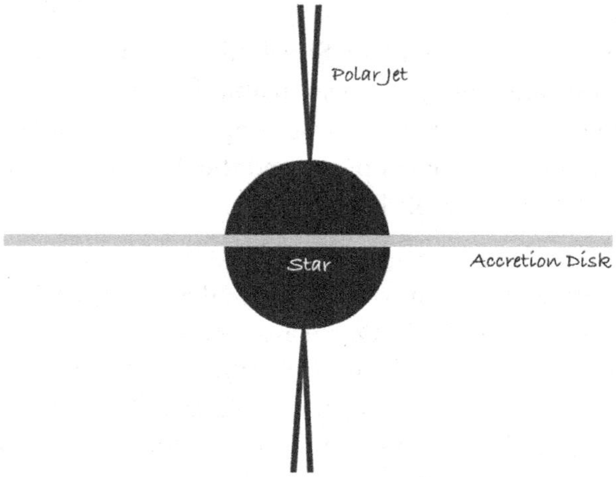

So how does he do this? The high-energy jets "spew out" into the surrounding gas cloud, shock-exciting the material there as in a neon tube. Various atoms and molecules in the gas cloud and the jet itself become ionized, which in turn creates radiation at specific wavelengths that we can observe in the star's spectrum. From the intensity of the radiation in a particular spectral band we can therefore estimate the type and amount of the various elements present. Red–blue shifts can also tell us something about the velocity of the jet. Taken together, this data reveals—at least approximately—the mass flow of the jet, which is in turn related to angular momentum. This provides the "big picture". Jochen's observations are focused on detecting a specific line of oxygen at 63 microns that cannot be observed from the ground: only SOFIA has a chance. Like Maria, Jochen's observation was split across two flights. After the second he was happy: "We detected what we wanted to see."

You might ask why the scientists are on board their observation flight. They don't operate the telescope, or even the instrument: specialists do that. However, they can make on-the-spot decisions if something unforeseen happens. Let's say that an object of interest is fainter than expected, and that the plan was to observe the object with three different filters. Instead of getting no useful data at all because the signal-to-noise for all three filters is too low, they could make a decision to only use two filters and integrate for longer, improving the signal-to-noise ratio to a useful level. Or if a technical problem occurs with the telescope and observation time is cut short, they can decide how best to use the remaining time. Jochen, by the way, was lucky: they scheduled ten minutes to set up the telescope and center it on his object. It only took two minutes, however, so he got an additional eight minutes of integration time.

Scientists can also plug their laptops into SOFIA's onboard network and access the "quick look" data, data that has gone through a preliminary simplified reduction pipeline, and make real-time decisions based on that data. What they cannot do, of course, is change their target, because this would require a change to SOFIA's flight plan, which is usually not possible. Flying on SOFIA is also a bit of an adventure for the scientists, a little bit of a perk of the job. Maria said that flying on SOFIA was like being a kid in a candy shop: "It's the second-best thing to going into space."

The scientist's real work starts after the flight: they have to process the data. From SOFIA/FIFI-LS they get a 3D data cube: the 2D position + wavelength + intensity. They get raw measurement values, but also the cleaned-up data that takes into account the instrument's calibration and the chop/nod corrections. It is now up to the scientists to make sense of the data and integrate it into their theories of "how stuff works" in the Universe. They have exclusive access to the data from their observations for roughly a year, after which it is made public through the SOFIA science archive. SOFIA is funded by the tax-payer, so the public has a right to access its work eventually.

I followed up with Jochen about his research while I was writing this essay. Unfortunately progress had not been in quite the way Jochen hoped for. It turned out that the calibration pipeline for the instrument wasn't quite ready. "It felt more like being a beta tester; I looked at the data cubes and noticed remaining issues. Over several iterations." Since 2017 the pipeline has worked well, however, and a new doctoral student began working with Jochen in October to carry out the science originally planned.

SCHEDULE In 2015, SOFIA flew three missions a week; the other four days were used for maintenance and ground work. Due to requirements for crew rest time, increasing to four flight days a week would cause a disproportionate increase in cost because of the need for additional crew members. SOFIA has always been strapped for cash, so this was not possible at the time. SOFIA also has two extended periods every year for maintenance of the airplane and installation of upgrades to the telescope and systems.

In 2014, SOFIA visited Lufthansa Technik for a D-Check. Lufthansa Technik is one of the few maintenance providers that still have experience with and access platforms for the 747SP. It came back to Hamburg for a C-Check in late 2017. When I talked to the SOFIA guys during that C-Check they mentioned that they now sometimes perform four flight days per week.

7 | *A SOFIA Mission*

DURING THE DAY Any mission essentially starts immediately after the previous one. Minor problems with the aircraft and the instrument and telescope are fixed by the maintenance crew shortly after landing. A typical sequence is shown below.

T – 10 hours. Preparation of the telescope starts. Technicians check the telescope compartment for loose parts or other problems. The telescope door remains closed at this time; the cavity is accessed through a small door from within the fuselage. The telescope cavity is a clean room environment, so the technicians wear hairnets. After this check the telescope cavity is off limits for the remainder of the mission. Next comes fueling of the airplane, which takes roughly 90 minutes. If SOFIA hadn't been left outside from the previous mission, this is also when it is towed out of its hangar, as fueling inside is not permitted.

T – 5 hours. The instrument crew refill the cooling fluids and the telescope crew start the bearing's oil pump to cool the bearing down to operational temperature.

T – 4 hours. The daytime telescope operators prepare the telescope. They activate the telescope's control system and SOFIA's MCCS (mission control and communication system), check the guide cameras and the chopper/nodder. They also ensure that the software is ready, for example by checking that all relevant guide stars are loaded into the system's star catalog.

T – 3 hours. Flight briefing. This is similar to flight briefing for airliners: the flight crew discusses the aircraft status, weather, flight plan and any other particularities of the flying aspect of the mission.

T – 2 hours. Everyone who will fly on the mission—between 11 and 25 people—gather in SOFIA's briefing room for the mission briefing. This typically takes around 45 minutes and covers the science objectives, the flight track and expected altitudes, the weather situation (in particular the predicted water vapor in the upper atmosphere), problems with the aircraft, telescope or instrument, a general safety briefing, as well as—importantly—rules for dealing with the fridge and the coffee maker ("Please please please remove your stuff from the fridge after landing." Same problems everywhere :-)). The briefing is run by the mission director but various people may be present: if the scientists are flying with their mission they will present the scientific objectives, while weather is briefed by meteorologists.

ROUTE PLANNING Before continuing with the mission countdown, let's zoom out a bit and look at the planning phase. I talked to flight planner Ken Bower about this aspect of SOFIA. At the beginning of a new observation season a call for proposals is published. In response scientists from all over the world submit proposals that are then reviewed by a SOFIA time allocation committee and ranked according to scientific merit. The accepted proposals are then anonymized, leaving only the sky location of interest, the required integration time and the observation's importance. Additional constraints are included, such as the time at which an observation has to occur if the phenomenon of interest is time-sensitive. Examples include measuring the size of exoplanets passing in front of their parent star, or maybe the moons of Jupiter being as far away from Jupiter as possible. Other projects might have relative constraints between successive observations: one example is the observation of a nova in successive states of decay.

The list of anonymized proposals is Ken's input for flight planning. Coming up with a set of flight tracks is then a huge puzzle. Let's look at the constraints that have to be taken into account beyond those given by the scientists.

How high the target is over the horizon at a given date and time; SOFIA can only observe between 17 and 65 degrees elevation.

Specific observation conditions, which might have to be scheduled towards the end of a mission when SOFIA flies higher (because of reduced aircraft weight) and where the telescope works slightly better because, after hours in the stratosphere, it is in perfect thermal equilibrium.

Observations that rely on faint guide stars: tracking those might be easier in the middle of the night when the sky is darker.

Avoiding non-observing legs: like any modern observatory, SOFIA is expensive to operate, so it is mandatory that it be used as fully as possible.

Ending SOFIA missions at the departure airport to ensure that the track is closed. Together with the fact that the telescope always looks to the left side of the fuselage, this is a significant constraint.

Areas into which SOFIA cannot fly. For example, it has no permission to fly over Mexico, which is relatively close to southern California where SOFIA is based. Area 51 and other restricted airspace are also off limits. Since SOFIA is operated by NASA's Armstrong Flight Research Center at Edwards Air Force Base, they can use the Edwards airspace after coordination with military ATC.

SOFIA usually flies around ten hours, the maximum time that can be flown with a single crew. On average, these ten hours of flight equate to around eight hours of observations (the rest is climb, preparation, descent and the occasional technical problem). During the summer, when nights get shorter, it is a challenge to fly in darkness for eight hours. SOFIA uses several tricks to make this work: taking off before sunset and flying the climb with door closed while the sun is still up; initially flying east, which makes sunset earlier (but puts a constraint on possible observation targets), then flying west at the end of the mission, pushing sunrise later. If everything goes perfectly this can allow up to nine and a half hours of observation. The trick of last resort is to go to New Zealand for the summer and schedule the ten hour mission into the 14-hour night. Since this is the southern hemisphere, this also puts constraints on possible targets.

The minimization of instrument exchanges. Each takes time for the actual exchange, but also requires additional effort for recalibration.

How do Ken and his colleagues attack this multi-objective optimization problem? He uses Excel to keep track of the flight plans and sky map software to understand which targets are visible and when. He also has a program that tells him the heading the plane has to fly to be able to look at a particular position in the sky. The actual creation of flight tracks however is a manual, iterative process: "I create a few plans, and then, when I get stuck because I cannot make a plan work, I start trading targets between plans." It's based largely on instinct and experience—and hard work, around 30 hours of planning per flight on average.

Once Ken has come up with a set of plans, the scientists get them for review. They point out mistakes, "...or more likely, requirements they forgot to mentioned initially", says Ken. Another iteration of flight planning follows. "Then the mission people look at them. They may say 'This is infeasible, we don't trust this airspace remaining open till then'". Ken makes more changes. The review phase takes roughly a month.

You might wonder why this process can't be automated. In fact the team has a full-time software developer working on just this problem, but it is impractical to automate completely. One reason for this is that the search space explodes geometrically. Another is that it is not clear exactly what the goal of automation would be: more overall science minutes, or fewer but higher-quality science minutes? An efficient plan for low-priority targets, or a less efficient plan for higher-priority targets? If you cannot formalize the goal, it is very hard to get a machine to do the optimization. It's also easy to forget a constraint, giving you a formally correct but ultimately ridiculous result. Like a traffic-light system optimized for "throughput" but without a fairness constraint, an optimizer might decide always to only let one direction drive. In non-obvious cases Ken prepares alternative sets of flight plans, and the SOFIA science director makes the final decision. Ken and the software developer are instead working on a human–machine joint optimization process, which they hope will reduce the work to around five hours per flight rather than the current 30.

There's yet more preparation. For example, the FIFI-LS instrument team writes a 30-page plan for every flight that details the execution of the observation. The high operating cost of observatories such as SOFIA makes it cost-effective to invest a lot of time in an elaborate plan that more or less guarantees optimal use of the observatory.

BEFORE THE FLIGHT T – 1 hour. At this point the mission and flight crews board the airplane. The telescope and instrument operators take over from their daytime counterparts. The flight crew prepares for start-up of the airplane. Guests like myself put their stuff in the fridge and everyone in the cabin gets a safety briefing not unlike that on an airliner. A few additional details are covered, such as the handling of the portable emergency oxygen system, which is required because there are no masks to drop from the ceiling. Finally, we go to our seats, strap in and wait for take-off.

This is probably a good time to discuss the cabin layout between the telescope cavity at the aft end of the fuselage and the set of business-class seats for passengers at the front. Starting from the back and facing rearward towards the telescope, on the left are the two telescope operators. They have a couple of screens for monitoring telescope parameters and logs. Slightly further forward and more to the right are two workstations for the instrument operators; next to them is the 19-inch rack with the instrument control system. Further forward, directly behind the instrument guys, is the console for the mission director and flight planner. This has a couple of screens, one showing the overall system status of the telescope and instrument, the other typically showing the planned and actual tracks. Left of this console is a group of four seats around a table to provide a working areas for the scientists. Making our way towards the front of the cabin there's an additional console with three seats for guests, such as the educators who regularly fly on SOFIA, as well as another group of four seats around a table.

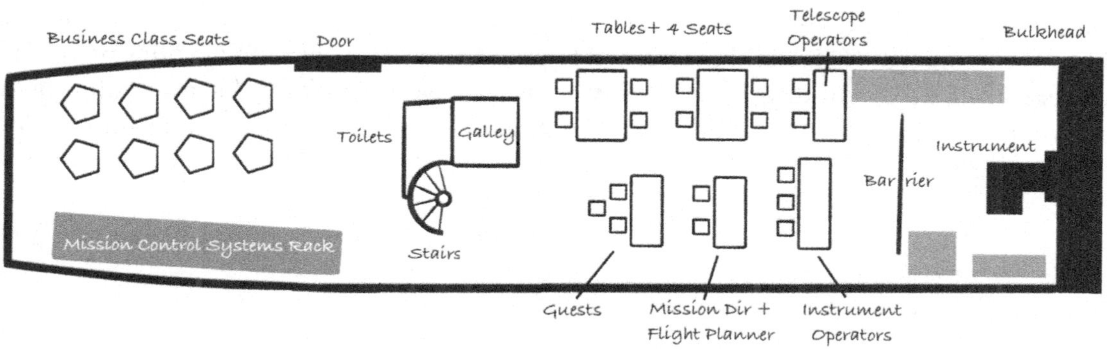

All seats except the passenger seats in the front have access to the communication system, a multi-channel intercom system into which you plug your microphone and headphones. You can select several channels on which to listen and one channel on which to talk, based on your role. For example, there's a channel for the telescope and instrument guys who "run" the observation, and another where the mission director talks with the flight crew. As you wander around the airplane you can plug your headset in wherever you are. The area between the telescope operators and the instrument and bulkhead is off-limits to anyone who does not have a good reason to be there. Most seats have a full four-point harness rather than a standard airline-style seatbelt. This feels good :-)

T – 30 minutes. It's time to start the engines and begin the flight. We've witnessed that part in a previous section, so let's fast forward to reaching the first observation altitude at around thirty-something-thousand feet. The mission director is now running the show. On our flight the mission director is Charlie Kaminski. He's been a mission director for five years, from the very beginning of SOFIA's flight operations. Like all the others on the airplane he's quite experienced, managing telescopes on Hawaii's Mauna Kea for 20 years, among them the IRTF infrared telescope. He then ran the SPIREX infrared telescope at the South Pole, spending ten years at McMurdo station and other places and wintering there twice.

Before the flight, his job is to run the briefing and ensure that all plans are available and agreed. During the flight he's responsible for the safety of everyone in the cabin, as well as for the success of the observation and the efficient use of the observatory. Charlie collaborates with the two safety technicians for the former, and with Ken, the flight planner, for the latter. Together they discuss and decide on climbs, turns and any changes to the flight track that might become necessary due to weather, malfunctions or marginal optimizations of the observations.

Charlie: "I try to keep track of the big picture. I talk to everybody, make sure they have everything they need to do their job." He also coordinates with the flight crew upstairs—it's a 747, after all—who are responsible for the airplane. If you want to go up to the cockpit you have to ask Charlie's permission—he calls up and asks if people can come up.

In reality it is probably coffee that runs the show: it helps to keep everyone warm and awake. I heard rumors that NASA initially wanted to prohibit the consumption of coffee on SOFIA, under some kind of anti-drug policy. Many of the crew would probably fall asleep during the second half of the night were it not for coffee. I'm not sure if this is true, but it's funny nonetheless.

8 | *The End*

LANDING It's around 5 am. Now that the excitement of the flight is wearing off I am getting really, really, tired. I'm not flying in the cockpit for the landing, but I have one of the console seats, so I can plug my headset in and listen to the pilots talk to ATC. After staying high as long as possible, caging the telescope and closing the door, we're now on our way down through the airspaces of Oakland Center, LA Center and Joshua approach—associated with the military at Edwards Air Force Base—towards Palmdale. At some point we fly through some turbulence from the Sierras, leading to a little bit of bumpines. Charlie asks the cockpit crew to warm up the cabin now that the telescope is caged. Everyone is quite cold and craves a bit of warmth.

Palmdale ATIS tells us that the tower is closed. We're supposed to talk to Joshua approach if we want to talk to a human. This sounds strange to European ears, but in the US it is common even for airfields used by passenger aircraft to be untowered at times. I assume it's rare, though, for a 747 to land at an untowered airport. It is the responsibility of the aircrews of the aircraft in the area to coordinate directly with each other: you have to transmit every intention and change to heading or altitude on the airport's frequency. There seems to be no-one around to hear our transmissions. The sun is about to rise, and there's a beautiful view from the window as we descend into the late-fall desert of Palmdale. After asking Charlie whether the cabin is ready for landing, and another cascade of checklists, Ace and Jim perform a direct visual approach into Palmdale, touching down with a smooth landing. After the landing, however, it became interesting—at least for an avgeek like me.

JIM: Hm, it indicates that [engine] three is still in reverse [thrust mode].

ACE: Mission, can you have somebody look out the window and check if number three's still in reverse?

...

CHARLIE: Yep, the reverser of three is still extended.

Ace intuitively wants to shut down the number three engine, but, being professional pilots, they still refer to the "books" (the manuals that define every operational aspect of an airplane) to check what they are supposed to do. The manuals confirm shut down.

ACE: We're gonna shut down, the dust storm [created by the reverser] should be over.

This of course leads to all kinds of other questions. Since this is an inboard engine, and they had to shut down the outboard engines to avoid ingesting and blowing dust, would the single remaining engine produce enough power to push them up the slightly upslope taxiway? Could they make a left turn with only the number two engine on the left side producing thrust? Ace remarks that "We'll shut down as we cross runway 22, so we got a bit of steam going up the hill". Worst case, he says, "We'll stop after crossing 22 and they'll just have to tow us in." That isn't necessary though: the single engine produces enough power. They also have enough hydraulics for the brakes (which they tested before they actually needed them on the apron), and low enough brake temperatures for a safe stop in front of the hangar. But more questions arise: what about the electrical load now that they have only one generator?

CHARLIE: Do we have to shut down the CECS?

CHRIS: No, we should be able to make it work.

CHRIS: The book says we can pull up to 81 kW for less than 5 minutes, and 54 continuous. I see 55.

JIM: Isn't it 54?

CHRIS: Yeah, you're right, it's 54. We're good.

Inboard landing lights off. 53 kW. "If you guys have any lights down there you can turn off, please do". Down to 52. Radios off. 50. "We've shut down a couple of computers and the galley." 49. Reducing the electrical load evolves to a kind of sport, which the team won.

CLOSING Let me close SOFIA with some impressions of NASA's Palmdale facility. SOFIA is run by NASA's Armstrong Flight Research Center. They operate a whole suite of aircraft, including a DC-8, ER-2s, a couple of business jets and of course SOFIA, all of them research aircraft. However, the aircraft don't research flying itself, but are instead used as airborne platforms for research into other domains. For SOFIA it's astronomy, but the others are used mainly for atmospheric research. Palmdale is an outstation of the Armstrong Flight Research Center (AFRC), located at Edwards Air Force Base a couple of dozen kilometers northeast.

There NASA operates a fleet of F-15s, F-18s, several UAVs and sub-scale models, jointly with a whole range of Air Force aircraft. I had the pleasure of visiting that part of AFRC in 2017 as well.

Both at Edwards and Palmdale I noticed two things I found really interesting. Firstly, you can see, feel and almost smell the passion and professionalism of everybody working for NASA. All aircraft are clean and polished to a degree you wouldn't find at any airline—not quite to over-the-top Thunderbirds or Blue Angels level, but close. Some of the aircraft are quite old, but they are in an immaculate state.

The second thing that impressed me was the flight operations. We've all spent time at commercial airports, so we know how large passenger airplanes are handled, with tow trucks, passenger bridges and gates. I've also spent lots of time on general aviation airfields, where people operate gliders, small piston-engined aircraft and ultralights. Things work different there. More hands-on, less "proceduralized", but not unsafe. NASA's Palmdale operation was a curious mix. When we landed with SOFIA we taxied the huge 747 onto an apron between two hangars. We turned off the engines and a guy in a T-shirt came walking towards us. He signaled to the pilots, Jim opened the cockpit window and just started chatting about the flight and whether everything was ok with the aircraft, just like that. When we left, two other guys towed an ER-2 out of the hangar in preparation for a mission, just as we pull out our Duo Discus glider on the morning of a nice soaring day. Somehow this put a smile on my face.

GOOD BYE After the telescope door was closed, and while SOFIA was descending back to its base, Charlie wrote his flash report. This provides a brief summary of the mission for SOFIA management and the observatory technicians. Were the science objectives met? Was anything wrong with the equipment? The pilots did the same for the aircraft. After landing they handed both reports over to ground personnel, officially ending science mission 248. Tired but satisfied, we got 100% of our planned data that night. And as real avgeeks, we witnessed the third flight of the week from the end of the runway and enjoyed the magnificent sound of the four JT9Ds.

3 *Charting the Seas with HMS Enterprise*

I don't like being hungry, and I suspect Kyle here feels the same way. We've been standing in front of a deserted terminal building in the harbor of Limassol, Cyprus, for close to thirty minutes. We are trying to get into—or rather, through—it so that we can board HMS Enterprise, which is berthed in the harbor. For Kyle getting through won't be a problem, since he is an officer in Enterprise and has the paperwork to prove it. I however am a civilian, and the lady we are talking to cannot quite square a civilian with a military ship. Here's the problem: if we have to wait any longer, we'll miss dinner!

It is mid-December 2017 and I have just flown into Larnaca from Munich. Kyle O'Regan, Navigation Officer Two in Enterprise, met me at the airport and we took a taxi to Limassol. I'm scheduled to spend a week in Enterprise on her trip from Cyprus to Malta to record interviews for omega tau. Even though the Enterprise's crew had filed all the paperwork for their civilian visitor with the Cypriot authorities, things didn't go quite as smoothly as expected.

Kyle, confirming my suspicion that he was also hungry, called the ship on the phone to make sure they would put two portions of turkey and fries aside, ensuring we'd get something whenever we arrived. They promised they would, and we were both relieved. About forty-five minutes and several phone calls by the lady later we were allowed to pass through the terminal building. We crossed the jetty towards the ship and walked up the gangway. A friendly, smiling guard with a rifle welcomed us to the ship. We made our way to the mess straight away, where we were promptly handed our dinner.

Afterwards Kyle showed me where I would live for the week: a two-person cabin I would share with him. He assigned the top bunk to me. His clearly showed that he was an experienced sailor, because it sported his personal feather bed instead of the standard issue bedclothes. It looked very comfy!

After making myself at home Kyle took me to the captain's quarters, where Commander Philip Harper, Captain of HMS Enterprise, welcomed me to the ship. Phil has been in the Royal Navy for 27 years and has served on everything from submarines to both large and small surface vessels. He's the first non-hydrographer to command a Royal Navy hydrographic ship in 220 years. He was kind enough to invite me to the Enterprise to record a podcast episode after an amusing Twitter exchange initiated by a mutual friend, Rainer Kresken. It was great to finally meet him in person.

I had arrived in Enterprise, and I was massively looking forward to the week!

1 | *Introduction*

OVERVIEW "Every ship can be a survey ship once" is a one-liner frequently heard in the survey community. It humorously refers to a ship running aground somewhere, thus locating shallow water or an undersea mountain the hard way. Enterprise does this systematically—her primary mission is hydrography: how deep is the water, the nature of the seabed, how much sediment between seabed and bedrock, the composition of the water column in terms of salinity, temperature and pressure. The Enterprise's equipment, which we will discuss later, is optimised for working on the continental shelf at depths of up to around 200 meters. Deep-water hydrography is performed by HMS Scott, a larger, freighter-like vessel that can host the more powerful sonar systems required for deeper water. The data obtained by the Enterprise, her sister ship HMS Echo, plus the rest of the UK hydrographic fleet is collected and processed by the UK's Hydrographic Office. Its data is used primarily for military purposes, but everything that relates to the safety of navigation also ends up on civilian charts.

HMS Enterprise was built in 2004. She displaces 3,740 tons and is 90 meters long. Her beam, the widest point on the ship, is 16 meters, which makes her "short and fat", as Phil Harper put it, I am sure lovingly. In terms of size, construction and equipment she is deployable all over the world, including the Arctic, Antarctic and the deep oceans. For example, her length of 90 meters is governed by stability considerations in the North Atlantic's typical wave lengths.

To make progress in charting the vast oceans of our planet, Enterprise is optimized for endurance, both in terms of design and in how her crew rotation is organised. She can stay at sea for weeks on end, "An unfortunately long time", as I heard someone say, and is routinely deployed on the world's oceans for 90% of the year, in contrast to the 40% that is typical for warships.

One reason for the endurance is her relatively large size for a crew of 50—she can store lots of supplies and fuel. The other reason for essentially continuous operation is that Enterprise has a total crew complement of 75, of which only 50 are on the ship at any given time, the other 25 being at home on leave. Conversely, two thirds of her crew are always available at any given time.

In contrast to many modern warships, Enterprise does not use gas turbines for propulsion. Instead, she uses a very efficient diesel-electric system that also contributes to her endurance. Three diesel engines, which can be selectively started, produce electrical power that powers electrical engines in two Azipods mounted under her stern. Azipods are gondolas that house an electrical motor and a propeller. They can be rotated through 360 degrees and combine the functions of propulsion (propeller) and steering (rudder). While this kind of "podded" propulsion is available from several manufacturers, the term "Azipod" is trademarked by ABB, which is why we capitalise it in this text. The two Azipods, together with a bow thruster, give her very good maneuverability, which, in addition to making it easy to get into and out of ports, also allows her to control her position precisely, important for surveying. "Enterprise is more like a shopping trolley than a car." The downside to this is that the Enterprise is not fast. After a repaint of her hull during a dry dock refit in 2018, the (very enjoyable) Twitter account @HMSEnterprise proudly showed a picture of the speed indicator: 16.9 knots. On the other hand, thanks to the Azipods, she can go backwards at 12 knots.

Enterprise has a secondary role as a mother ship for mine-countermeasure vessels. This is not because of some relationship between hydrography and mine hunting, but simply because, as a consequence of her relatively large size and comparatively small crew, she can carry lots of stores and staff to "run" a mine-hunting fleet. When I visited the Enterprise in late 2017 she was in the middle of a one-year mission as the flagship of the Standing NATO Mine Countermeasures Group 2.

As a warship, Enterprise also has some light armament: two 20mm guns, forward and aft of the bridge, as well as three miniguns and four machine guns spread around the ship. While this is suitable for self-defense during survey operations, in wartime the ship relies on protection by a task group, as do many other warships that fulfill supporting roles.

LEAVING CYPRUS The forklift lifts the gangway and lays it down on the jetty. Lines still connect the ship to its berth.

> **PHIL HARPER:** Ok, let go both forward spring and back spring. [Bow thruster] Thrust to port step 4. Port [Azipod power] lever ahead 1.
>
> **QUARTERMASTER:** Thrust to port step 4, port ahead 1; [and a bit later] Thrusting to port 4, port Azipod ahead 1.

The lines are released by staff from the civilian service company that handles logistics for Enterprise while she is in port. The ship begins, very slowly, to move backwards. Her stern starts moving away from the pier, while the bow thruster and the Azipods keep the bow still. Effectively, the Enterprise backs out of her berth; she had parked for a few days in a corner of the port basin, its walls directly on the port side and a few meters astern.

PHIL: Thrust to starboard step 8, train both Azipods green 160 [degrees to the starboard side], set both levers ahead 8.

Enterprise's stern is slowly rotating away from the pier as we increase the distance from the basin walls. Phil commands "Three short blasts" and a crew member sounds the ship's siren three times to show we are operating astern propulsion—the maritime equivalent of a reversing alarm in trucks or construction vehicle. We then start a 180 degree rotation so that we can leave the harbor basin bow-first. Once we've turned around, we accelerate slowly and leave Limassol.

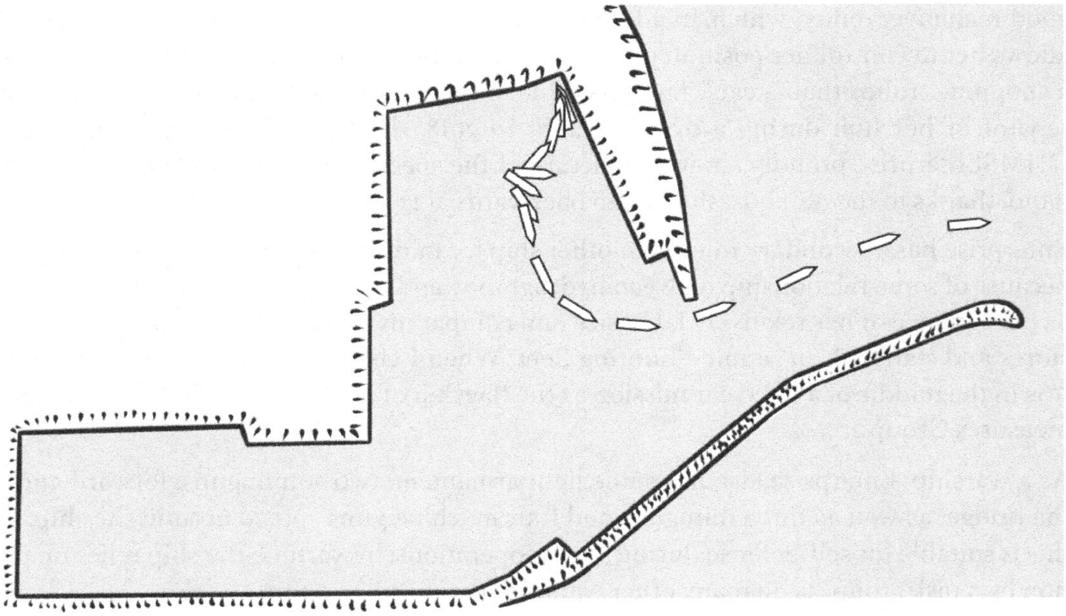

SONAR OFFICER: Depth 12 meters and steady.

NAVIGATION OFFICER: WECDIS concurs.

The seaman manning the sonar reports the depth to be 12 meters, while the officer on the electronic chart system confirms that this corresponds to the depth indicated by the chart for the current location—one of many cross-checks. We slowly accelerate towards the exit of the harbor basin.

OFFICER: Kilo bears 300. Hotel bearing 230.

A junior officer uses a compass to determine the bearing to predetermined reference points around the harbor basin, including ones identified as Kilo and Hotel. He calls them out aloud so that the officer manning the WECDIS charting system can enter the bearings into the system and thus allow it to "fix" Enterprise's position. The Enterprise has GPS, of course, indeed a very precise one for survey work, but being able to determine its position without relying on external infrastructure is useful for a military vessel—another cross-check. Based on this (now confirmed) position, the navigation officer reports that the "Clearing line is 90 yards to port and 50 yards to starboard. Next wheel-over in six minutes." The clearing line demarcates the area near the shore where the water is dangerously shallow; a "wheel-over" is a change of course. A few minutes later, as we pass the breakwater, the pilot boat approaches and we disembark the Pilot. We are on our own now, in open water, starting our voyage to the port of Valletta on the island of Malta.

PREPARATION FOR LEAVING Enterprise had been berthed in Limassol for several days. She took on supplies and rotated her crew. When I arrived the evening before the day of departure the "old" crew was still on board. They had left earlier today for the airport by bus to fly back home to the UK. A few hours before we departed I had to visit an office in the terminal building to complete the paperwork that documented my departure from Cyprus.

Phil explains that a ship has two lives: in port and at sea. While berthed an officer of the day is responsible for the activities on the ship: who comes and leaves, and more generally, the safety of the ship and her people. He carries out his duty, together with the guard I mentioned before, from a small office at the top of the gangway. In port many of Enterprise's systems are shut down or operating at reduced capacity. At sea, in contrast, Enterprise's life is structured by the 24-hour, seven-days-a-week, four-hour watch rotation and the ship is controlled from the bridge. The challenge of a departure day is to transform the first organisational structure into the second: getting all the machinery going, having all the people in place, and generally not forgetting anything. "Luckily, these days we all have cell phones so you cannot forget to disconnect the phone lines from ship to port", Phil says, as if he remembers a particular episode from earlier times.

Of course the military has a process for this transition. The visit of a warship to a foreign country requires diplomatic clearance: a plan is created and agreed with the host country. The plan includes the expected times of arrival and departure, and thus sets the basic parameters of the visit. As the time of departure comes closer the Captain signs a set of sailing orders. These define when the ship should be at specific states of readiness, the departure route, and who will fill which roles during the operation of the plan: it basically defines a countdown with predefined checkpoints.

About two hours before undocking the crew calls the harbor's vessel tracking service, known as VTS, a sort of "air traffic control" for ships in a harbor area. They recap the previously filed departure plan and confirm that it is still valid from the harbor's perspective. Sometimes a large commercial ship might come in unexpectedly and a departure must be postponed briefly. The analogy with air traffic control only goes so far, though: VTS does not provide radar vectors, but in larger or complex ports they give advice on how to deconflict ships based on the bigger picture obtained via radar or AIS, the maritime equivalent of an airplane's transponder. In some ports ships require VTS permission to enter the port area.

The call to VTS is typically also used to confirm the pilotage arrangements. Pilots are mandatory in nearly every port. They are master mariners themselves, and have a lot of experience with the local conditions; in particular, they speak the local language. On commercial ships they typically take command of the ship, and might even actively take charge of the ship's controls, the rudder and the engines. Royal Navy vessels do not allow this, so pilots act as advisor to the Captain. While a port might require the use of pilots—and captains value their advice and support in any case—the Royal Navy trains to be able to get into and out of every port, day or night, without the use of pilots, buoys, GPS or other external support.

About an hour before the planned departure the engineers perform a "flash-up", starting the engines, checking controls, valves and sensor readings. Right after that the bridge crew performs a check of the steering. In the case of the Enterprise they specifically check the rotation of the Azipods. These checks are crucial: if there is a problem that cannot be fixed quickly, the schedule is in jeopardy.

Around the same time the navigation officer briefs the bridge crew, as well as members of the support crew who, for example, handle the lines. The goal is to make everybody aware of how the departure will be executed: when the Pilot will join, whether tugs will be used, in which order lines will be released, what the departure route will be.

Around thirty minutes before the scheduled departure the Pilot comes aboard. A fast-paced discussion between the Captain and the Pilot ensues, about the technical capabilities of the ship and in particular its ability to maneuver. It's also a bit of a judgement of each party's confidence and skills. Phil politely makes it clear that he, Phil, will command the ship: "Ship handling during departure and arrival is the most fun part of my job. I've been doing this for most of my 27 year Navy career."

> **Phil Harper (to Pilot):** It's going to be an easy departure today, we've got Azipods, two of them, both working, and a bow thruster. We'll bring the stern out with the Azipods, keeping the bow off the berth with the thruster. We'll move backwards and out, then turn around. Please tell the tug to stay 100 meters off, we'll attach her only if we have a breakdown in the propulsion.

Ports often require the use of tugs even if, as in the case of Enterprise, they aren't really necessary, because Enterprise is adequately maneuverable on her own. The Pilot forwards Phil's request to the tug, and we will later see her "dance" around Enterprise at around 100 meters distance, as requested.

It is now about 20 minutes before departure. An announcement comes over the ship's loudspeaker system: "Special sea duty-men close up [seamen/officers with specific roles for entering/leaving harbor to go to their stations]. Assume damage control state 3 [this refers to tightening the ship against nuclear, chemical and biological contaminants; state 3 is peacetime cruising] condition Yankee [close the doors marked X and Y]. Assume force protection posture Bravo [refers to the current threat level, for example, from terrorism]. Hands to harbor stations, hands to harbor stations [seamen to ropes]. Close all upper deck screen doors and hatches. Rig on the upper deck and bridge will be PCS [personal clothing shirt] and baseball caps."

Just before we leave the current position is fixed to cross-check the GPS and gyros. "The northernmost point of the floating dock is at 180. The corner of the basin is 211.5, the left hand edge of port building, southernmost point I guess, is bearing 252." The fixed position agrees with the GPS and gyro compasses, which have accuracy errors due to their mechanical nature.

PHIL HARPER: Ok, let go both forward spring and back spring [refers to particular ropes that connect the ship to the jetty]. Thrust to port step 4. Set port lever ahead 1.

We are leaving a few minutes ahead of schedule. Since, as Phil says, "The military has a bit of an obsession with being on time", I guess this is a good thing.

WRAP-UP This marks the beginning of my first extended sea voyage. I've been on ferries across Lake Constance and the English channel, and spent a few hours on a container ship and a harbor tug, both to record podcast episodes. But I've never spent days on end on a ship. Phil suggests I take seasickness pills, so I do. I still feel fine when we disembark the Pilot outside the breakwater after thirty minutes, but only now does the ship start its slow roll and pitch as it enters the open ocean.

2 | *On the Bridge*

THE BRIDGE The bridge is the most interesting place on the ship. It is where "the Con" is located, the person who commands the operation of a ship. The bridge also serves as the command center for the mission, such as surveying. It is however not where all the engineering systems are managed: this is done from the machinery control room. The bridge is also the highest regularly accessible place on the ship, and its big windows provide a great view. No surprise then that I tried to spend as much time as I could on the bridge, especially when interesting things were happening.

I spent some time there with Kyle during one of his watches. "Captain's feet are on the bridge", I hear him announce as Phil enters the bridge directly from his quarters. It is common to announce the Captain's presence because this changes the command structure on the bridge. I am surprised to hear the word "feet" in the announcement though. Kyle explain that once, during a dark night, something came up and Phil entered the bridge in his bathrobe, without shoes. Someone then stepped on his feet—someone who did wear shoes. "It really hurt", says Phil. Since then there has been a kind of inside joke that the Captain's feet are separately announced if Phil enter the bridge without shoes.

Enterprise was steadily humming along on the current leg, so Kyle had some time to explain the equipment on the bridge.

NAVIGATION EQUIPMENT Much of the equipment is used to precisely and reliably understand Enterprise's position, its track and route, as well as the sea surrounding it. This equipment is located mostly on the starboard side of the bridge. Three systems are primarily relevant here: WECDIS, the radars and AIS.

WECDIS is the military ("Warship") version of ECDIS, the Electronic Chart Display and Information System. It displays electronic charts that show water depth, areas of shallow water, locations of wrecks, the shore line and navigational fixes on the shore, as well as specially designated areas such as those pre-approved for exercises. In addition, the military version, WECDIS, can also overlay various military grids, coordinate systems and map projections. WECDIS is fed with and displays data obtained from various sensors, such as the ship's heading (obtained from gyros and compass), its track (obtained from two GPS receivers), the speed over ground (computed from subsequent positions), as well as wind speed and direction. A trackball allows the operator to point at any feature on the chart, such as a wreck, and query the system for detailed information.

WECDIS is configured with the planned route, including waypoints, heading and speed for each leg, expected time of arrival for each waypoint, as well as positions and durations for special exercises (for example, man overboard) or activities, such as training in the use of the side-scan sonar. WECDIS then displays the actual trip in relation to the plan in terms of geographical deviation and timing. It also maintains a countdown until the next wheel-over, the time when a change in steering will be required.

Because ECDIS is approved as the primary means of navigation by the International Maritime Organisation (IMO), it has to have a specific screen size and be adjustable in brightness for use both day and night. The Enterprise has two independent WECDIS units, plus a laptop with the same software, so consequently is allowed to be a "paperless ship", as Kyle puts it. "We do not carry paper charts anymore."

Radar is the other major navigational sensor. While it can be used to detect the shore and thus help with fixing the ship's position independent of GPS, radar is mainly used as a tool for deconflicting with other traffic. Radar primarily detects the position of targets in polar coordinates (direction and range) relative to the ship. From the change of their position over time it can calculate targets' course and speed. The Enterprise's radar systems come with automatic plotting functionality which, for each detected target, assigns an ID and plots a map with this data for each target. It also calculates an expected closest point of approach, as well as the time when this will happen based on the velocity vectors of each target and the Enterprise. The system highlights those targets whose closest point of approach is predicted to be below a particular threshold and which thus require special attention. Enterprise has two radars that work on different frequencies, optimized for shorter and longer ranges respectively. They can detect other ships up to around 25 nautical miles, and can "see" land out to around 90 nautical miles. Of course this is only true if the land is high enough to rise over the horizon. For a distance of 90 nautical miles—166 km—and assuming the radar is installed 20 meters above the water, the land has to be 1,750 meters high!

Enterprise has AIS, similar to civilian ships, which is the third major navigational system. AIS is short for Automatic Identification System and actively broadcasts information about a vessel. While conceptually similar to a transponder and ADS-B in aircraft, it broadcasts a much richer set of information, mainly because shipping is generally slower and can thus manage with a lower update frequency, which in turn creates bandwidth for more data. AIS transmits a ship's name, its registration number at the IMO, as well as the type of ship (freighter, passenger ship, military vessel) and its dimensions. In terms of navigational data, it transmits the status of the ship (at anchor, underway with engines, not under command), its position relative to the WGS 84 coordinate system used in GPS, as well as the positional accuracy, the course and speed over ground and the rate of turn. It also transmits data about a ship's voyage, including destination and expected time of arrival, as well as the current draught. This data is packaged into different message types that are transmitted at different intervals. For example, the navigational data is transmitted every two seconds to three minutes, depending on the ship's speed. The static data is transmitted at least every six minutes. AIS is received by other ships, as well as vessel tracking services. It is also rebroadcast via satellite and can be seen world-wide, for example via services such as vesseltracker.com or marinetraffic.com. On the bridge of Enterprise received AIS data is displayed on a dedicated system, but is also fed into WECDIS. Because AIS relies on a ship's willingness to actively broadcast, it is not quite as reliable as radar for deconfliction—radar requires no cooperation from "opposing" ships. On the other hand, it provides much richer information.

Phil told me that military vessels have traditionally been hesitant to transmit AIS. "The ocean is a big place, and when turning off AIS, we can effectively disappear." However, on the flip side, a ship that is otherwise detected (visually, radar) and does not transmit AIS effectively reveals itself as a military vessel, so there is an increasing trend for military ships to turn AIS on. When approaching ports it is often a requirement anyway, including for military ships.

DRIVING EQUIPMENT The center-port side of the bridge hosts the position for the quartermaster, the person who actually "drives" the ship based on the orders of the conning officer.

Enterprise has a propulsion system that consists of two Azipods. Each has a propeller that drives the ship forward, but they can also be turned through 360 degrees to act as rudders. One way of controlling the ship is thus to vary the power and direction of each of the two Azipods separately. To do this, the bridge has two Azipod controllers. They look like thrust levers in airplanes (they can be moved forward and backward to adjust power), but each lever can also be rotated through 360 degrees to adjust the direction of thrust. Controlling the ship with these levers is referred to as "maneuvering mode", and is used whenever precise control of the ship is required, as the name suggests.

For example, sailing into a harbor, approaching the berth and the actual berthing is usually done using maneuvering mode. Similar Azipod-controller-based steering positions are available on the two bridge wings for berthing, as well as in the Azipod compartment in the belly of the ship; the latter are used as a means of steering the Enterprise in the case of an emergency, when connections from the bridge might be severed. The Enterprise's bow thruster is also regularly used in maneuvering mode; the lever to control its power output and direction is located adjacent to the two Azipod controllers.

In situations in which less precise control of the ship is required, for example en-route, "wheel and RPM mode" is typically used. Here the ship's heading is controlled with a traditional wheel that controls the "rudder position" (which, of course, is the direction of the two Azipods). However, since the wheel does not control a rudder mechanically, it is much smaller than what you might picture from classic ships, around 25 cm in diameter. In wheel/RPM mode the power output of the propellers is controlled via thrust levers.

The third mode is called "DP joystick". This refers to a set of steering modes that are controlled by Enterprise's dynamic positioning system. In this case it is actually "controlling" as opposed to "steering", because the DP system gets sensor feedback about the ship's position and heading and actively works to achieve whatever input was given by commanding the Azipods and the thruster. In the first sub-mode, "DP-joystick-joystick", the ship maintains its current position while compensating for wind and currents. The quartermaster can then use a joystick controller to precisely (and very slowly) move the ship in all four directions. They can also enter a numeric offset (move 20 meters left, or along a particular heading) and the ship will move there automatically, again maintaining its position once it arrives. We will see this mode in action later when sailing into Valletta. "DP-joystick-autopilot" mode can be used to maintain a heading and speed, similar to an airplane autopilot's attitude/altitude/heading hold modes. It is often engaged en-route on long straight legs. The third sub-mode available from the DP system is "DP-joystick-track-line". This controls the ship to move precisely along preplanned survey lines created on and fed from the Enterprise's PL10 survey planning system. In contrast to the autopilot, this mode actively works to stay on the line and does not "just" hold a course and speed.

Based on what I saw during my stay in Enterprise, it is obviously much easier to control the ship precisely using the dynamic positioning system: you specify an offset or a rotation and the ship just does it. In contrast, in maneuvering mode, when the Azipod controllers are used, it was not uncommon to hear several—even dozens—of commands a minute to control the ship. When I asked Phil about this he basically agreed, but also said that, of course, the dynamic positioning system could fail at any time and then you have to be able to at least stabilise the ship using the Azipod controllers, especially in a tight environment such as in Valletta harbor, which we'll discuss later. He added, humorously, that "DP is boring".

Talking about safety: the quartermaster's steering position has a rip cord. Kyle explains: "Similar to small boats or jet-skis, quartermasters attach themselves to a ripcord in rough seas, so if they fall off their seat all propulsion is turned off and the ship immediately stops." Seeing my slightly surprised expression, after a while he adds, smiling: "At least this is what we tell newbies. It's a great prank, and some actually believe it." To help with believability, a dummy cord actually does exist at the quartermaster's position.

There are of course also various indicators, such as the power each Azipod draws from the ship's engines. 1,700 kW each is the maximum: at 1,000 kW each Enterprise will make roughly 12 knots. The spinning reserve indicates how much more power the generators currently running can provide. As the power required by the Azipods increases (because the ship is commanded to go faster), this value will decrease; when a reserve of 180 kW is reached an additional generator is started automatically by the ship's management system (assuming that not all are yet running). The steering position also has a master caution panel that indicates failures of major components such as the Azipod drive and steering. Finally, the steering position sports basic indicators for speed and heading.

CENTER PEDESTAL Let's recap the layout of the bridge: equipment that provides detailed situational awareness—WECDIS, radar, AIS—is on the starboard side of the bridge, while the ship is driven from the port side. This requires at least two people on the bridge during "interesting" maneuvers. None of them has the full picture: operating a military ship is a team effort. These two positions—and others—can be seen as doing the legwork for the officer who "has the Con" at any given time, usually the Captain or the executive officer, but often also another senior officer or a junior officer in training. The commanding officer, when not walking around on the bridge, rests in a comfortable and slightly elevated seat behind the center WECDIS display (which famously has a "Drive it like you stole it" sticker attached). Their responsibility is to understand the big picture and supervise the major decisions of the officer of the watch. The people on WECDIS and the radars supply data to the commanding officer, while the quartermaster "implements" the steering commands. These can be rather general: "Steer 120 degrees, set thrust ahead 95%", or really detailed: "Train port pod red 120, set port lever ahead 7", and potentially given every few seconds, for example, during berthing.

This approach to operating a ship requires a lot of people. When I spent a few hours on the bridge of a container vessel in the port of Hamburg there were only two people on the bridge, the Captain and the Pilot. I asked Phil why the Royal Navy uses this more people-intensive approach. He said that, generally, it is more resilient, because it relies less on automation and electronic systems. He elaborated particularly on giving steering commands verbally: "The main reason is that I can have people check what I'm doing. Navs usually stands next to me on the bridge. Before any non-trivial maneuver I will first explain to him what I am generally trying to achieve. Then I give the steering commands with which I plan to achieve it. If he thinks the order doesn't make sense he can intervene at the time of the order, not just when he sees its effect." I suspect it also has the benefit of allowing officers in training to better understand how the ship is controlled. And finally, the economic pressure of making do with as few people as possible does not apply to the military in the same way that it applies to commercial shipping. Says Phil: "The ship is manned with 50 people not because we need them to get in or out of port, but because I need that many at action stations. So, since I've got them, I use them."

At the center of the bridge is the center pedestal, which holds a large compass. In fact the compass is only an indicator (albeit mechanical), and works off the gyros installed in the ship's keel or the magnetic compass mounted on top of the bridge. When showing the gyro signal, the compass indicates the true (geographic) heading; when based on the magnetic compass it shows the magnetic heading. In the latter case compass errors have to be corrected manually. These include the magnetic variation—the difference between magnetic north and geographic north—which changes with position and time because the magnetic poles move, as well as the deviation—the effect of the magnetic signature of the ship on the compass—which changes with heading.

This compass plays a crucial role in navigation, because it is used to fix land-based reference points. Fixing means that a crew member aims the compass at the reference point and determines its bearing. This bearing is then entered into WECDIS, which, by triangulating from many such headings, computes Enterprise's position. This way of determining the ship's position is completely independent of external systems such as GPS, and, as mentioned before, this more resilient approach makes sense for a military vessel because GPS can be jammed. Of course it is not completely infallible: the weather must be good enough to be able to see suitable reference points and the points must be clearly identifiable. Good candidates are lighthouses and tall buildings, but can also just be an unambiguous landmark, such as "Southeast edge of the harbor basin". The crew generates the reference points themselves at the planning stage by using chart features to create a reference set file. Taking fixes requires an extra person, but as we have seen, people are available.

The bridge also features a set of indicators mounted in front of the center pedestal above the front windows. These show heading, speed, barometric pressure, wind direction and speed, as well as the RPM and orientation of the two Azipods. They are optimally placed to serve the commanding officer, but they are large enough to allow most people on the bridge to glance at them for essential information about the status of the ship.

OTHER EQUIPMENT A few more systems are worth mentioning. There is a panel that indicates the status of the gyros and the compass, and a periscope with which the compass above the bridge can be read directly instead of through the indicator on the center pedestal. There are VHF radios for communicating with port authorities (156–174 MHz) as well as with civilian aviation (118–136.975 MHz), and UHF radios for military communication, as well as a satellite phone. There are telephone-like intra-ship communication systems, and there are broadcast systems for various predefined comms loops (whole ship, surveying, machine spaces). There are also sound-powered phones for especially critical communication links. Relying on dynamic microphones, sound-powered phones can maintain predefined circuits without any external power. Finally, there is the "main broadcast", an announcement system that is transmitted through the whole ship. These announcements are known in the Royal Navy as "pipes", a term dating from the days when orders were passed using pipes or whistles. Among other things this is used to wake everyone up at 0700 hours.

There is a slave screen for the IPMS, the ship's integrated platform management system, which provides a view of what the engineers see on their management consoles in the machinery control room. Example information includes fuel reserves, energy produced by the engines and the amount of ballast.

There is a panel for the navigation lights and one for the siren. The siren has to be sounded, for example, if a man is overboard or if the ship is unable to maneuver. In reduced visibility the siren has to be sounded every two minutes, which can also be scheduled automatically. When anchored, though, it is the ship's bell that has to be sounded at regular intervals. The bell is not automated. "When I was in training, I once had to do the bell for hours", Kyle remembers, not so fondly.

The fire management system shows the status of hundreds of heat and smoke detectors. Next to it is a panel that allows the bridge to crash-stop the ventilation system to prevent smoke from being spread through the ship and to close fire doors. Fire is among the most critical risks on any ship.

Two rooms are located at the aft end of the bridge. One is the survey room. As working the equipment during a survey session has to be carefully coordinated with operating the ship, it makes sense to put this room close to the bridge. Survey data analysis happens in a separate office below decks.

The other room is the chart house. Traditionally this has been the place where the navigation officer would spread his charts to plan and track the voyage. This is now all done in WECDIS: Enterprise does not carry paper charts. However, the chart house still serves as the "office" of the navigation officer where they plan routes—on a laptop. It also hosts the emergency

WECDIS-on-a-laptop system. There are also additional computers that contain all kinds of reference handbooks, including a tidal atlas, sailing directions for many regions of the world, as well as the Book of Reference 45, the Royal Navy's manual on all matters to do with navigation and seamanship. Finally, the chart house has sextant, a device for measuring the angle between stars and the horizon when navigating by the stars. While not routinely used, celestial navigation serves as an ultimate backup if all else fails. The sextant is a handheld manual device, not a fully-automated star-tracker like those on, for example, the SR-71 airplane. Kyle hypothesized that a fully-automated system would probably be too expensive and that, considering that the ship moves so slowly, and because—in contrast to an airplane—more people are available, a manual solution is good enough.

WRAP-UP Standing on the bridge watching the nautical miles tick by slowly, just looking out the window, was a fun pastime. But I also liked to spend time outside, usually on the upper deck towards the aft end of the ship, looking down onto the quarterdeck, watching the sea, the sky, the sun or stars, the wake produced by the engines and the ship, and feeling her slow roll. During the first few days of our trip the weather was reasonably good and not too cold, so standing outside with just a coat on was pleasant. On the other hand, not much was going on, so I could see how watching the sea could become boring after a while. ("Objection! I'm not bored of it yet, and I've been doing it my whole adult life."—Phil)

3 | *Surveying*

MISSION Enterprise's primary mission is hydrographic surveying: collecting the data that goes into navigational charts and charts of the sea floor. The most important information conveyed by such charts is the depth profile. Knowing the location of shallow areas, underwater mountains rising from the sea floor, or wrecks is critically important for the safety of navigation, both for surface ships as well as for submarines. The nature of the sea floor—rocks or sediment—as well as how it changes over time can also be relevant information, but we will focus on depth mapping first. This section is a little intensive in terms of detail and the use of formulae: I hope you don't mind.

LEAD LINES Historically the sea's depth at a particular location was measured using a lead line, a long line with a lead weight attached. This was lowered into the water, and when the "pull" of the weight ceased because it reached the sea floor the current length of the line indicated the depth. This, of course, is a very slow process and not very suitable for mapping extended areas, or for very deep water, where the lead line might never reach bottom because it drifts away. Phil adds: "There are deep-sea lines. HMS Challenger mapped the Challenger

Deep in the Marianas Trench with one back in the late 19th century. A single measurement could take most of a day to complete!" The deepest location she discovered was 8,148 meters.

SONAR BASICS A more efficient approach relies on sonar. Sonar is an acronym for SOund NAvigation and Ranging", and is comparable to radar or lidar, but uses acoustic waves instead of electromagnetic waves. Acoustic waves travel around five times faster in water than in air, the specifics depending on a number of parameters we will cover later. Radio waves, on the other hand, hardly propagate at all in water, especially the high frequencies used in radars or radios. This is why submarines either have to surface to use their radios or use extremely low frequencies for communication—below 100 Hz—with a corresponding low bandwidth. Aircraft like the E-6B Mercury can trail kilometer-long antennas to communicate at these frequencies, flying to the location of submarines to exchange messages. These are dubbed "TACAMO" aircraft: TAke Charge And Move Out.

Fundamentally a sonar sends out short acoustic "pings", and their echo, as produced by a solid surface, is recorded via hydrophones and then analyzed in various ways. The most-used form of analysis is to measure the time between sending the ping and receiving the echo. Knowing the speed of sound in water, the distance d between the transmitter and the reflecting surface can be calculated as $d = c \cdot t/2$, with c being the speed of sound and t being the round-trip time of the ping and echo. It is therefore possible to determine depth by reflecting a ping from the seafloor. A device that uses sonar this way is called an echo-sounder.

Basic echo-sounders send out a relatively non-directional ping, one that extends as a cone under the ship, insonifying a whole area of seafloor. The ping might then be reflected by any number of solid objects, not just the point directly under the ship. The echo-sounder really measures the length of the hypotenuse of a triangle given by the vertical distance of the point of reflection (that is, depth under the ship) and its lateral offset. If an echo is received late, this can either be because of greater depth or because of a greater lateral offset.

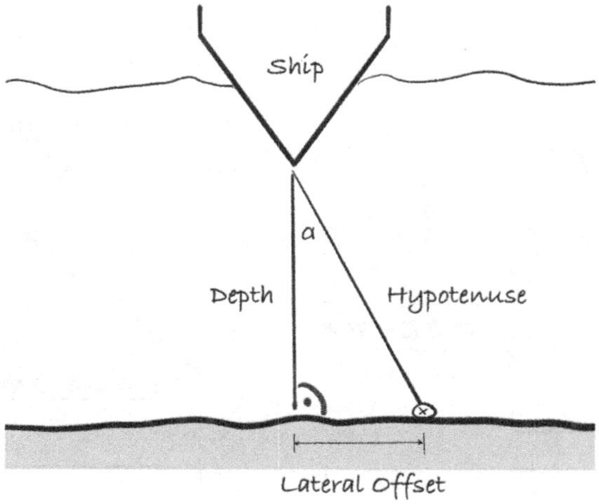

BEAMFORMING To avoid this problem, echo-sounders use various techniques to limit the beam's width and focus it in a particular direction. By directing the beam vertically below the ship, reflections from laterally-offset surfaces are avoided, and the measured time is proportional to depth, as intended. Practically, however, the beam is not perfect, and reflections from side lobes at other surfaces are still possible. These are much weaker than the primary reflection, which usually allows the echo-sounder to discriminate between them and the primary reflection. Such single-beam echo-sounders are used by most ships these days to measure how much water they have under their keel. However, because they only measure one point at a time, they are still not suitable for mapping whole areas of seabed.

MULTIBEAM SONARS For large-scale mapping, Enterprise, like other survey ships, uses a multi-beam echo-sounder. The fundamental idea is to measure the depth of many points at the same time.

A ping's reflection results in an echo that is not sharply defined; it is a continuous sound that stretches over a period of time. This is because the ping is reflected "everywhere", and each of the infinitely many points creates it own little echo, with a reflection time that is proportional to its hypothenuse. The idea behind a multi-beam sonar is to break down the continuous signal into "chunks" that correspond to particular, discrete receiving angles.

Once the direction of a particular echo is known, it becomes possible to calculate the depth (vertical distance) based on the measured distance (length of the hypotenuse) and the cosine of the receiver angle. The lateral offset can be computed as well using the sine of the angle.

In practice the angular discrimination is only needed in one dimension. The pings are shaped to be narrow along the ship's track and wide across track, so a single ping results in an across-track line of depth measurements called a "swath". The Enterprise's multi-beam echo-sounder, a Kongsberg EM 710, can "look" to up to 70 degrees (the view angle can be configured) to either side of vertical. As the ship moves forward, a sequence of swaths is recorded, resulting in a grid of depth measurements.

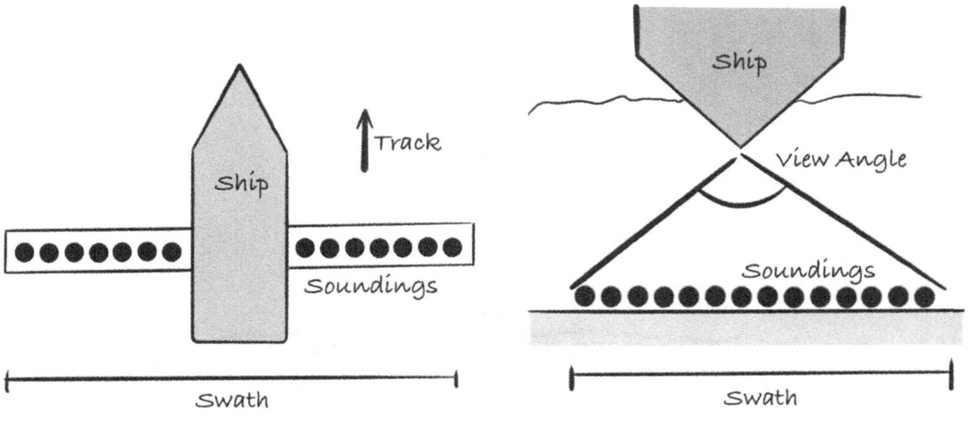

The echo's angle is determined via beamforming, by which the receiver's sensitivity is increased in a particular direction. We all know this from microphones, where it is called the "directional characteristic", common characteristics being omnidirectional, cardioid or hypercardioid.

However, we really want to beam-form in many discrete directions at the same time, so that we can record distinct echoes from many different directions—the "chunks" mentioned above. According to its specifications, the EM 710 can record up to 400 discrete echoes, giving an angular resolution of up to 0.3 degrees.

ANGLE VS. DELAY So how does beamforming actually work? A multi-beam echo-sounder is fundamentally arranged in a T-shape. The vertical part of the T is aligned with the long axis of the ship, the horizontal part perpendicular to it along the ship's transverse axis. The vertical/long part contains the transmitters, the horizontal/transverse part is an array of receiver hydrophones spaced at equal distances. Let's say we want to detect an echo that comes from a particular angle α and that the wavefront of the ping is planar. The incoming wavefront arrives at each of the hydrophones at slightly different times, where the reception delay between any two hydrophones is constant, because the wavefront is planar. The duration of the delay depends on the speed of sound (which we assume to be constant for now) and sin(α), plus of course on the spacing of the hydrophones, a known parameter. The dependency is easy to understand: if the wavefront comes from directly below (α = 0), the delay is zero, because all hydrophones receive the reflected signal at the same time. Conversely, if α = 90 degrees, the delay is maximal, because the distance s between two hydrophones has to be traveled fully by the wavefront. In general, the delay is given as $(s/c) \cdot \sin(\alpha)$.

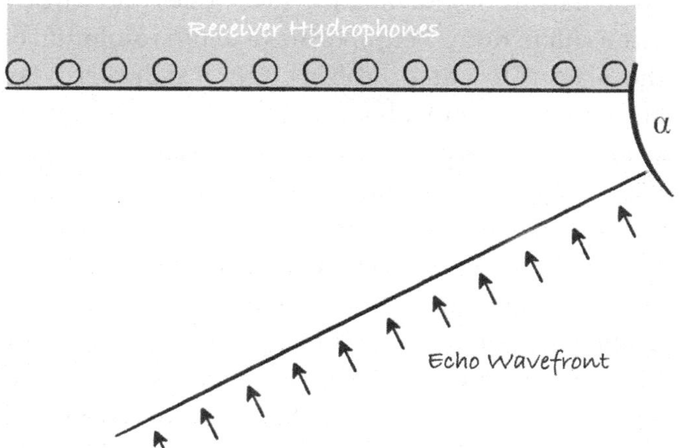

FROM DELAY TO FREQUENCY The sequence of hydrophone detections can itself be seen as a signal D with a particular frequency f = 1 / delay. So to compute the angle of the incoming wavefront α, we have to measure f. Here is a problem: at any given time we might receive echoes from multiple angles. So in terms of the signal D we measure, we get a complex waveform that is created by the interaction of many constituent frequencies. We have to find out what these are. This is where the Fourier transform comes in. A Fourier transform is a mathematical transformation that takes a signal in time, D in our case, and decomposes it into its constituent frequencies. So basically it tells us "how much" of each frequency is in a particular signal. It can be understood as producing a diagram of energy over frequency. If a particular frequency f is dominant, we see this as a peak in this diagram. Since $f \approx \alpha$, this tells us α.

FOURIER OVER TIME In practice, to simplify and speed up signal processing, the angle is not actually measured—instead, the angles at which the beams are formed, the discretisation of the overall angular range, is predefined. The travel time t_α for these predefined angles α_i is then measured. So how do we get to t_α? To achieve this, we look at how the signal D evolves over time. We discretise time into slices (the length of these influences the precision of the depth measurement) and perform a Fourier transformation for each slice. A peak in a slice from time t at frequency f indicates a reception of an echo from the corresponding angle α at time t. The intensity of the reflection and the "sharpness" of the peak depends on where and how the ping is reflected and how the sound is scattered along the way. So we look at sequences of slices and observe how the size of the various peaks evolves. If, after growing, it starts to shrink again, we have found a maximum. The occurrence of a maximum at a particular frequency f in the slice that corresponds to time t constitutes the detection of an echo from the direction α that corresponds to f.

THEORY VS. PRACTICE Those are the basics: beamforming is done through lots of signal processing of received signals from individual hydrophones and involves Fourier transformations and time-discretisation. Kongsberg specifies a depth resolution of up to two centimeters as a result of all this magic. Of course the exact technical details of how this is achieved is usually not something the vendors talk about much. But there's quite a bit more to it than what I have explained above. For example, if the sea floor is close enough, the assumption that the echo's wavefront is planar becomes invalid—it's actually spherical, the so-called "near field". This makes processing much harder. A technique called beam-focussing can be applied here; you can imagine it working in a similar way to focusing in a telescope. The assumption that the speed of sound is constant and identical for all of the hydrophones might also not be true. For example, turbulence created from a ship's propellers or when turning can create air bubbles under the ship in the path of the sonar; air, of course, has a very different speed of sound. This is why only those soundings are used that are recorded when the ship's movement is stable; measurements made during turns are usually discarded during post-processing. In addition, the speed of sound directly below the echo-sounder is monitored continuously. But the problem of gas bubbles in the water column disturbing measurements can also be turned into a benefit, for example when trying to find underwater gas vents.

In practice the theoretical maximum angular resolution of 0.3 degrees cannot really be achieved. While the beam axes are spaced 0.3 degrees apart, the beams overlap because they diffuse with depth. At the nadir (looking straight down) they are generally at least 0.5 degrees wide, and for the outer beams, sometimes quite a bit more.

ATTENUATION The selection of the frequency for the pings is interesting as well, because the attenuation of sound in water is dependent on frequency. Generally, higher frequencies are attenuated more, and the difference is significant: at 10 °C the attenuation over the range from 1 kHz to 100 kHz goes roughly from 0.1 to 100 decibels per km—and remember, decibels are logarithmic units, so the attenuation in terms of energy is even greater. In fresh water the relationship is linear, but in salt water the effect has additional contributions from boric acid and magnesium sulphate and is thus not completely linear, but the general trend holds.

Assuming a given sensitivity in the receiving hydrophones, the higher the attenuation, the more transmission power is required, which is undesirable for all kinds of engineering reasons. So you want to use lower frequencies with lower attenuation. On the other hand, lower frequencies of course have a longer wavelength, as given by $\lambda = c / f$. For $f = 1$ kHz, with $c = 1{,}500$ m/s, you get a wavelength of 60 cm. So, picture this: at 1 kHz it takes a full millisecond for a wave to "pass" and it extends over 60 cm. This is clearly not helpful for the use case of a multi-beam sonar, where it is all about detecting a ping quickly and precisely. So this suggests higher frequencies. In addition, because of this long wavelength, the resolution of the depth measurement is lower.

In practice, as with any engineered solution, a compromise between the forces that shape the solution is adopted. For example, the EM 710 produces pings in the range of 40 to 100 kHz, and this is adjustable. So for shallow water (where the distance of travel for the ping is less and attenuation does not matter so much), a higher frequency is used. This results in a higher granularity of measurement, better depth precision and more pings per second (because the pings are more sharply defined). For deeper water, where attenuation is of more concern, lower frequencies are used. The resulting ping lengths are in the range of 0.1–1 millisecond. This results in reduced precision, but in deeper water this is generally not a big concern. However it makes precisely resolving fine structures, such as wrecks, in deep water, more of a challenge.

INTERNAL MODEL To help with signal processing, echo-sounders maintain an internal model of the seafloor. Based on previous measurements, they "predict" when they expect the echo. If multiple echoes are detected from a particular direction, the model helps to decide which should be used (usually the strongest one). Since these algorithms are based on heuristics and other assumptions implicit in the model, errors can occur. It is the job of the hydrographer to spot such errors and either correct or discard the particular measurement. The need for the internal model also means that the system requires some time to stabilise at the beginning of a campaign, to allow it to build an initial model.

INSTALLATION AND STABILIZATION The Enterprise's multi-beam sonar is installed in a pod under the keel. The pod contains a multitude of sensors, for example for measuring the speed of sound in water, and is around four meters long. The receiver array for the EM 710 are around 50 cm in length. Since it is part of the ship's hull, the sonar heaves, rolls and yaws along with the ship. This means that, for example in roll, the measured points move from left to right. To map the seabed accurately this has to be taken into account, either by physically tilting the receiver or as part of the beamforming signal processing. To be able to do this the echo-sounder is fed with data from an inertial measurement unit that records the movement of the ship around all three axes.

GRID DENSITY How dense is the measurement grid? The answer depends on several parameters, among them the echo-sounder's capabilities, the water depth and the speed of the ship. The echo-sounder's capabilities determine the angular resolution, as well as the field of view. The depth then determines how wide the swath will be for a given field of view angle, and how far the soundings will be spaced laterally for a given angular resolution. Generally, the deeper the water the wider the swath, but at some point the swath width starts to decrease as the absorption of sound in the water starts to reduce the signal-to-noise ratio, to the point at which the receivers can no longer detect a reflected signal. The change-over point and shape depend on the makeup of the sea floor: rock and sand reflect different amounts of energy and therefore affect the signal-to-noise ratio.

The forward speed of the ship is the main determinant of the along-track spacing of soundings. However it also depends on depth, because for greater depths the ping travels for a longer time (and, at a given speed, longitudinal distance), with the result that you get fewer measurements for a particular time increment. Let's say we are mapping 1,000 meter-deep water. The speed of sound is roughly 1,500 m/s, so it would take 1.3 seconds for the round trip in the vertical and around three seconds for the beam at 65 degrees (hypotenuse). The beams also have an along-track beam-width, so the ship can make headway and still cover all the sea floor with sound as long as it doesn't move more per ping cycle than the along-track footprint of the beams.

Some numbers from the user manual for the EM 710 are of interest. At a depth of 50 meters you can achieve a ping rate of about five per second. At a speed of eight knots you get an along-track spacing of 0.8 meters. In deep water, say 800 meters, the spacing extends to ten meters. You've got a swath width of 173 meters and 2,300 meters respectively. So, with around 400 distinguishable beams, this results in a lateral spacing of 0.4 meters and 5.7 meters respectively. A slower headway increases the along-track grid density, while using a narrower field of view reduces the lateral resolution; the EM 710 can be configured for 90, 120 and 140 degrees. There is however a limit imposed by the 0.3 degrees maximum angular resolution.

The S-44 standard of the International Hydrographic Organisation defines different levels of grid density and depth accuracy. For governments this standard is a requirement; for private parties, such as port authorities, it is not, but still serves as a guideline. Depth precision of 0.3 meters is required up to a depth of 30 meters; beyond that 1% relative precision is required.

In terms of density the standard defines four different categories of area. "Special" areas are those with minimum keel clearance. Category One are harbors, harbor approaches and other recommended tracks. Category Two is everything outside Special with depths of less than 200 meters, and Category Three is everything else. For Special areas the standard requires nine grid points per cubic meter—per cubic meter instead of square meter makes sense, because it implies more measurements if the ground banks steeply, which is especially relevant for shallow water. It is clear that from what we have seen above that this is really tough to fulfill and might require multiple passes. One would probably also not use an EM710 for this; dedicated shallow-water sonars that ping at frequencies around 400 kHz are more suitable. Remember that attenuation is not so much a problem for shallow water. For the other areas, with the right equipment, the density requirements are such that typically seven to ten knots of mapping speed is realistic. Nonetheless, humankind so far has only charted around 10% of the seabed: mostly (and predictably) coastal areas, critical shipping routes, and, probably, areas of strategic interest for armed forces.

POSITION For the grid to be useful at all the ship's position must be known accurately. The Enterprise has a geodetic (high precision) GPS receiver. A GPS receiver triangulates its position relative to a set of GPS satellites. Each satellite broadcasts a signal that contains data about its orbit and the precise time at which the signal was sent. A receiver can calculate a satellite's position from the orbital data. The receiver can then calculate the travel time of the signal from the satellite based on the timestamp in the signal and the time of reception (as measured relative to signals received from other GPS satellites). These two pieces of data allow a receiver to calculate its distance to the known position of the satellite. Doing this with several satellites (the more the more accurate) allows receivers to determine their own position. (As an aside, this design of GPS is pretty ingenious, because it allows the system to scale indefinitely: the load on the ground-based and space-based infrastructure does not increase if more receivers use the system.)

For GPS to be accurate several things have to be true: the satellite's orbital data must be accurate (it isn't, because the orbit drifts), the atomic clocks of all satellites must be synced (they aren't, because they also drift slightly), and the signal must travel to the receiver in a straight line (it doesn't, mainly due to distortions in the ionosphere, and it might also be reflected by buildings, but this is not so much a problem in the open ocean). To compensate for these errors, various countries and organisations operate "GPS augmentation" systems. These systems use ground stations to measure these errors precisely and transmit them to receivers, which take the corrections into account to improve the accuracy of their computed position.

One such system is the US Department of Defense's Wide-Area Augmentation System (WAAS), which transmits the required correction data via (additional) satellites. The Enterprise receives the WAAS data to determine its position to an accuracy of less than one meter. The precision of its own position determination translates directly to the accuracy of the depth map.

SPEED OF SOUND As we have seen, the measurement of depth relies on knowing the speed of sound in water. We have already discussed the problem of air bubbles directly under the echo-sounder, but the speed of sound also varies in undisturbed water. This is due to water pressure, which changes with depth, temperature and salinity. For practically occurring values of temperature and salinity, the speed of sound varies between 1,450 and 1,550 meters per second, or about seven percent—so therefore also would any depth measurement. Close to the surface the effects of temperature and salinity dominate, while at greater depth these are relatively constant and the change in pressure becomes more relevant.

These values must be known in order to obtain precise depth data. Before conducting a survey, therefore, Enterprise uses a probe lowered via a winch to measure the depth profile of these parameters. The probe is used again regularly, and whenever the pod-mounted sensors or patterns in the data suggest a change in the water's characteristics. Speed measurements are fed into the echo-sounder's computer system, which then calibrates the raw measurements into a precise position-referenced depth map of the sea floor.

DATA PROCESSING The data obtained by the Enterprise and the rest of the UK hydrographic fleet is ultimately delivered to the UK Hydrographic Office, which is in charge of creating the actual charts. Because every measurement campaign results in dozens of gigabytes of data, delivery is made via hard disk when Enterprise is in port. Before the data is delivered it is cleaned up and quality-assured on the ship. The crew assesses whether the required grid density and depth precision has been achieved, and removes or corrects the occasional measurement error that result from, for example, attitude errors of the ship, or from errors in the heuristical models in the echo-sounder. This is a partly manual and partly automated process. The manual work looks at single swath lines, as well as adjacent lines, to check whole areas for plausibility, and is performed mostly using visualisations of the data. Outliers are reasonably easy to spot. The automated work relies on mostly statistical corrections, for example using a filter known as CUBE (which is short for Combined Uncertainty and Bathymetric Estimator). It basically averages adjacent soundings based on the errors associated with each particular measurement.

The hydrography room in the Enterprise features a set of computer workstations running commercial software packages for visualising and analysing the data. The data formats are standardised to allow exchange between the tools, which can show a top-down map of the seabed in three dimensions with depth encoded as color, as well as cross-sections. The latter is especially impressive when used on wreck data, as it allows you to move around and "cut through" a wreck interactively with the mouse.

OTHER SENSORS The multi-beam sonar is the Enterprise's bread-and-butter tool and is used most, but they have other tools as well. Together with Graham I visited Enterprise's towed body workshop, also known affectionately as the "hangar deck" because of all the strangely airplane- and rocket-shaped towed bodies stored there. Located just forward of the quarterdeck, it doubles as the gym, so you'll find all kinds of sports equipment there.

The properties of the water, such as temperature, salinity and chemical composition, are relevant beyond their use for determining the speed of sound during depth mapping. For example, wide-area differences in salinity are a major factor for large-scale ocean currents. Water also tends to layer according to temperature, and these layers are relevant, for example, for submarines (remember Tom Clancy's submarines hiding below the thermocline?). Using the winched probe to perform depth profiles one at a time is not efficient for larger areas. To do this more efficiently, Enterprise carries the "undulating oceanographic recorder". This looks like a little airplane: it has wings and even a unpowered propeller at its tail for stabilisation purposes. It is towed behind Enterprise, moving up and down in depth based on a preprogrammed profile. Because of the ship's forward speed the recorder effectively "undulates" along a sine wave, continuously measuring a range of hydrographic parameters and transmitting them back to Enterprise via a datalink built into to the towing cable.

More sonars! The side-scan sonar creates an image of the sea floor, also along a moving swath, just like the multi-beam sonar. However, it is primarily used to create "visual" maps of the seafloor. Because it is towed underwater it is not impacted by wave action and creates a stable picture. Spikes on the seabed will typically result in "shadows" because of the slant angle of the sonar. For depth mapping this can be used to ensure that there are no surprises between soundings—you basically want to see a contiguous surface. If you don't, you go back and investigate the anomaly.

Because of the varied sound reflection properties of different materials on the sea floor—sediment, rock, metal—the images allow spatial distinction of these different materials. Remember our discussion of ping peak width for the multi-beam sonar? It is affected by different materials. To calibrate the visual representation with hard data, Enterprise also has a grabber that can be lowered via a winch to take samples from the seabed.

While an exact position is not so important for the undulating "airplane", it is important for the side-scan sonar. A first approximation of the towed body's position is the ship's position plus the length of the towing cable. However, since the side-scan "fish" also sinks depending on the speed of the ship, the length of the cable corresponds more to the hypotenuse of a triangle given by the horizontal distance and the depth. However, it is not helpful to simply measure the angle at which the cable "leaves" the ship because, of course, the cable curves. A precise position is therefore established using a High-precision Acoustic Positioning system

(HiPAP). This is a spherical sensor mounted on a beam lowered from the ship's keel that receives an acoustic signal sent by the towed body. The sensor employs beamforming to narrowly focus the receiving beam; by searching for and then pointing precisely to the signal, it can establish the angles at which the towed body can be "heard". Equipped with these angles, plus the length of the cable, the system then determines the precise position of the towed body relative to the ship.

Enterprise has two more sensors, both towed. One is a magnetometer. Comparing the local magnetic field to a database of the Earth's magnetic field allows it to identify distortions in that field resulting from large metallic bodies. The magnetometer is similar to the probes on the tail of aircraft like the P-3 Orion or the Atlantique, which use them primarily for pinpointing the large metallic bodies of submarines. In the context of mapping the sea floor, the magnetometer is especially useful for locating wrecks, even if buried under sediment. Because Enterprise is itself also a large metallic body, the magnetometer must be towed several hundred meters behind the ship.

The final towed body is the "sub-bottom profiler". Its task is to look "into" the seabed. It is fundamentally a seismic device whose acoustic noise penetrates the top layers of the seabed. The noise is reflected at layer boundaries, for example those between sediment and rock, or between rock and hollow spaces in the rock. A line of hydrophones "flies" behind the profiler. The later an echo returns, the further back in the line it is received, so the further it has penetrated the seabed. Oil and gas exploration relies on similar tools, but with louder acoustic pulses that are often created by explosions. The profiler used by Enterprise can penetrate maybe the top ten meters of the seabed, and it is primarily used to determine the thickness of the sediment.

Enterprise also carries a survey motor boat named "Spitfire" that can be deployed from the port side. Spitfire also carries echo-sounders and is optimised for shallow waters such as harbor basins. When Enterprise docks in Malta, Spitfire is scheduled to perform depth mapping of Malta's harbor.

DEPLOYMENT I witnessed the deployment of the side-scan sonar from the quarterdeck. Ships that deploy towed bodies are usually fitted with an A-frame. Imagine a capital "A" with its two feet mounted on the left and right side of the stern of the ship. The top of the "A" carries a roller over which a cable can be run. Typically the cable comes from one of the winches on Enterprise's quarterdeck. If the A-frame is stationary and vertical it acts a little like a crane, in the sense that you can lift or lower loads attached to the cable by taking in or letting out cable from the winch. In addition, however, the A-frame can pivot forward (over the quarterdeck) and aft (over the open water). To deploy a towed body, the crew first lean the A-frame forward, letting out the cable via the winch. They then attach the towed body to the cable and reel in the cable a little, lifting the towed body from the deck so that it hangs over the quarterdeck. They then start tilting the A-frame progressively aft, letting out cable at a rate that keeps the towed body at a constant height above the quarterdeck. Once the A-frame is tilted sufficiently far aft to be over the water, they slowly let out cable and lower the body into the water.

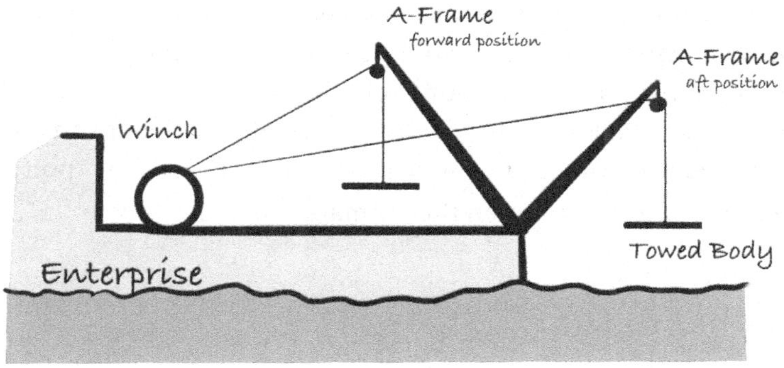

The whole procedure takes maybe ten minutes and is performed by five to seven crew. Before the deployment starts, there's a safety briefing:

> "Do not cross lines, or stand on lines, stand at least two meters clear of winches, do not stand in line of recoil, keep non-essential persons well clear. Wear boots and hardhats. The word STOP can be used at any time to stop the process if you see a dangerous situation."

I was a little surprised by the emphasis on safety I saw here and on many other occasions. After all, Enterprise is a naval ship, and working for the armed forces is, ultimately, dangerous. When I asked Phil about this, he replied that there is no point in taking risks for which there is no benefit. "Work accidents" are not one of the things for which it is worth taking a risk. "It's not that different from war—risk still needs to be balanced against benefit—perhaps more so, since the risks and benefits are both magnified."

PERMISSIONS There is one more interesting aspect of survey work: the circumstances under which it can actually be performed. Obviously there are weather limits; the system can compensate for the ship's heave or roll only up to a point. Working with towed bodies on the quarterdeck is not something you want to do when the deck is moving in heavy seas.

Another concern is that hydrography is only allowed in territorial waters with the permission of the respective country. For example, Spitfire has permission to map Valletta's port. Working with towed bodies, even in international waters, should only be done in areas that have been predeclared so as to avoid mishaps with other ships—and in particular, submarines.

Finally, it is well-known that loud sounds such as those from sonars can impact sea life, particularly marine mammals such as whales or dolphins. Whether these animals are permanently hurt by sonars is not completely clear, but there is a lot of evidence that such creatures are at least disturbed, in the sense that their navigation becomes confused or they stop hunting for food. This is not so surprising if you consider that these animals rely on sound to build a picture of the undersea world. However, Enterprise's sonars are not powerful enough, and on the wrong frequencies, to harm marine life of any sort, even if it swims through the sonar swath. Such problems are more critical for navigational or search sonars, which use higher sound energy and also insonify a greater volume of water, especially laterally. As Enterprise focuses the sound to below the ship, it is unlikely that animals would stray into the acoustic cone in the first place. However, whenever Enterprise works with her sonars, lookouts are positioned around the ship to watch out for sea mammals. If any are spotted, sonar work is halted—and everybody is eager to watch the animals.

4 | *Systems*

THE MCR Through a door, around a corner, one staircase down. Through more doors, then to the right. Or something—I don't quite remember. The couple of days that I was in Enterprise were not enough to really understand the structure of the decks below the one I was berthed on, and the bridge. Too many doors and staircases. Anyway. I arrive in the machinery control room, where I chat with marine engineers Ian Sutton and Lee Williams. They are ultimately responsible for propulsion and "hotel services", which is what they term the infrastructure that relates to living on the ship.

ENGINES The Enterprise's propulsion is diesel-electric: the engines in the Azipods that propel the ship are electrical, and the electrical power to run them is created by diesel generators. This has several advantages, which I discuss below.

The ship has a total of three generators, two main ones and a harbor generator. Depending on how much power is needed, only a subset of these engines may be running, which reduces fuel consumption. Enterprise can comfortably sail at around 12 knots with one main generator running at full power. Adding the second main generator gets the ship to 14 knots. The final, lower-power engine only adds a knot or so. There is no battery to buffer the energy created by the generators, so they have to produce the required power in real time: if the ship wants to go faster, more generators are brought online. This is done automatically by IPMS, the ship's management system, whenever the "spinning reserve"—the difference between the required power and the power available from the running generators—reaches a predefined minimum of 180 kW.

Relying on multiple engines also provides redundancy in the case of engine failure. More generally, it helps with availability: depending on the required maneuvers, maintenance can be performed on one of the generators while the other powers the ship. If only a single larger engine was installed Enterprise would have to go into dock for maintenance much more often.

IPMS keeps track of the maintenance intervals, and the engineers often choose to run the generator with more time available before its next service. Maintenance itself is also simpler compared to turbines, and, according to Phil, "If properly maintained, the generators run forever."

Diesel-electric propulsion also leads to a simpler mechanical infrastructure in the ship. There is no need for a shaft from the engine to the Azipod and no need for complex gearboxes, both simplifying maintenance. It also allows more flexible positioning of the diesels, as they don't have to be in a straight line ahead of the Azipods. Thus, one of the main generators are installed in the forward main engineering space together with the harbor generator, and one in the aft. This allows better management of the ship's trim, as well as resilience: in the case of flooding or a fire it is better if not all engines are in the same compartment.

A final benefit of diesel-electric propulsion is that the engines can be designed to run at their optimally efficient speed: they do not have to ramp up or down to adjust the speed of the propellers. In fact, similar to a power station, the generators run at a constant speed, because that speed directly translates into the frequency of the generated AC voltage, which must be kept stable. As the requested power changes, a governor adjusts the power output of engines—and the fuel burn—while keeping the RPM constant.

Like any engineering decision, this approach to ship propulsion also has drawbacks. I presume two engines with power P are more expensive to buy than one engine of power 2 x P. Regular maintenance, such as checking and refilling of oil, coolant and lubricants, also has to be done twice, and there are more moving parts, so I suspect there are potentially more parts to fail.

SWITCHBOARD While shafts and gearboxes are not required, Enterprise of course has to have an infrastructure for distributing the electrical power. Most of the ship's systems are supplied with two buses of 440 and 220 volts respectively. However, a bus of 690 volts is used to supply the up to 1,700 kW for each propeller, to reduce resistance losses. As power is the product of voltage and current ($P = U \cdot I$), for the same transmitted power you can increase either and correspondingly reduce the other. However, resistance loss is proportional to current ($W_R = I^2 \cdot R$), so it is generally better to use higher voltage and lower current. This is also demonstrated by power grids, which run at up to 380 kV in Europe.

As another example of resiliency, Enterprise has two main switchboards, one forward and one aft. The main buses are also installed on both sides of the ship. Each Azipod has two supplies, one from each switchboard. There is also an emergency switchboard with an emergency generator. If this detects any failure in the main systems it starts automatically to provide power to critical components such as steering motors, radars, water pumps, a compressor for starting air for the main engines, and some limited lighting. During the 45 seconds it might take for the emergency generator to come on line the ship goes dark, but important systems are buffered with their own, locally-installed uninterruptible power supplies and therefore stay "alive".

The fact that the emergency generator powers the steering motors is a requirement of the IMO, which specifies that a ship has to retain rudder power even if its propulsion system fails. As Enterprise has no actual rudders and the Azipods can only act as such when the propellers are turning, providing power to the steering motors alone does not provide any steering. Nevertheless, it's a requirement.

AZIPODS One more level down and (of course) all the way aft is the Azipod compartment. I was there while the crew was exercising the steering of the ship directly from here. The same throttle-like Azipod controllers that are available on the bridge can also be found here, right next to each pod. One of the many contingency procedures is to control the ship from various internal locations and on deck, where there are no actual engine controllers—just a telephone to talk to the guy down here, who then controls the pods.

Each 1,700 kW Azipod engine is installed in a teardrop-shaped gondola under the ship's (flattened) aft hull and drives its own propeller. They engines cannot be accessed from the Azipod compartment. Because the Azipods also serve as the rudder and must be rotatable through 360 degrees, they are mounted onto what looks like a tank turret ring that is driven by two electrical motors through a gearbox. There is also a converter to control the motors' speed and direction, as well as a brake to lock each pod into position. The day before, around 2 am if I remember correctly, there was a pipe through the ship that woke me up. It announced something about a "steering gear breakdown". I wasn't quite sure what to make of it and went back to sleep quickly. I asked Ian what this was about. "We lost steering control on one of the Azipods. One of the brakes got stuck. It was very hot, so we couldn't touch it for three hours." He continued, "Luckily it got stuck in the midships position. So we could still use the propeller, but we could only use the other Azipod to control steering." Basically, the Enterprise had a less effective rudder. Once the brakes had cooled down the next morning they were stripped down. "We found a broken disk. It has happened before. We had all the parts on board, so we could repair the problem quickly." Interestingly, just this scenario was practiced a day earlier as part of a steering gear breakdown exercise.

BALLAST A ship is designed to handle optimally at a particular weight and resulting draft. The total weight of the ship depends on the equipment loaded on deck, the size of the crew and the amount of consumables such as as fuel and drinking water. To compensate for varying loadings the Enterprise has ballast tanks that can be filled with seawater. More importantly, the distribution of the weight along the ship's long and lateral axes—its center of gravity—also impacts the ship's handling characteristics. This is why ballast water can be pumped between tanks at various locations in the hull. The pumping is controlled manually by the engineers and relies on calculations performed on a laptop that determine the optimal load and distribution. I assume that weight and balance is not such a huge concern for a relatively slow ship like the Enterprise. It's probably more important for faster and more maneuverable frigates. However, they usually pump fuel around instead of ballast water.

Managing the ballast is not as trivial as it might seem. An international treaty, the Ballast Water Management Convention, prohibits "transporting" water from one ocean to another, mainly to avoid introducing new organisms into foreign waters. "So taking on and dumping ballast requires quite some planning", Lee Williams explains.

STABILIZATION Another system that relies on water is the stabilisation system. Its purpose is to dampen any roll induced by wave action. Some ships use actively controlled fins for this purpose, but the Enterprise relies instead on a U-shaped tank installed in the hull that is partially filled with water. If the ship rolls, for example to the right, the water in the tank moves to the right, rising in the starboard part of the "U". Once the rolling motion has reached a maximum amplitude (on the starboard, in our example), valves in the bottom of the "U" close, trapping the water in its starboard side. This water, because it is higher in the tank and further away from the axis of rotation, has a higher inertia, and therefore slows the next roll to the port.

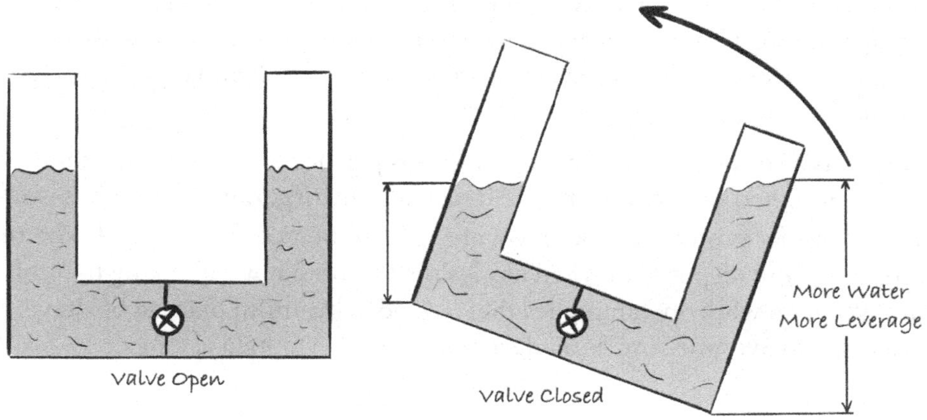

The bridge crew swear by the system. "It makes it a much more pleasant place to live," Kyle said. On the other hand, it is not so noticeable in the lower parts of the ship, for example in the engineering spaces. "The best place on the ship is low down and laterally in the middle. We've got a McDonalds there", Ian jokes.

FUEL Fuel is also interesting. Enterprise basically consumes "diesel oil that you could put into your car", not the heavy oil that is commonly used by large cargo ships. While Enterprise can in principle be resupplied at sea from a supply ship, she usually takes on fuel in port. As Phil says: "Setting up an underway replenishment requires a lot of planning. Considering that we have great endurance anyway, it is just so much easier to take on fuel in ports." However, different ports supply fuel with different levels of quality. Of particular importance is the "cloud point", the temperature below which fuel starts to "wax into sludgy stuff", because this makes it harder to be burned by the engines. This has two consequences. First, different quality fuel is stored in different tanks. Each fuel is tested for its cloud point and other quality attributes, and there's quite a bit of book-keeping around this. Second, fuel with a relatively higher cloud point has to be used before the Enterprise travels into colder climes.

Before fuel is injected into an engine it is cleaned in fuel separators, where water and particulates are filtered out. From the main storage tanks, fuel goes into satellite tanks, then into the separator, then into the service tanks from which the engines are fed. Unburned or spilled fuel is returned to the satellite tanks, and gets cleaned in the separators once more before it is used by the engines.

(WASTE)WATER Further down into the bowels of the Enterprise. Steep staircases are now replaced by vertical ladders that descend from higher levels through hatches in the floor. This is where we find a lot of the infrastructure that runs "Hotel Enterprise".

There are freshwater tanks, two holding 44 cubic meters at the stern and two with 51 cubic meters in the bow. "Far more than we can ever use. Hollywood showers for everyone every day", Ian quips, adding that average use is ten cubic meters per day. Enterprise can also make its own fresh water from salt water: reverse osmosis removes salt and other particles from water by pressing it through semipermeable membranes.

Of course there is also waste water treatment. This is performed by "huge machines that look like they were designed by a crazed man". Ian. These incorporate membranes, shakers and the like, and allegedly produce drinking water—which then goes overboard. The remaining sludge is collected and disposed of when alongside. Prompted by my asking that this must be the favorite system to work on, Ian replies that "It's not a beautiful place to be. It has exploded once or twice due to overpressure. It made a huge mess." I stopped asking.

HVAC Heating, ventilation and air conditioning—abbreviated to HVAC—is crucial, because a ship is basically a huge metal can. Many rooms have no windows, so fresh air has to be brought in by other means. Some of these windowless rooms contain equipment that produces heat that has to be moved out of the ship. In addition, humidity has to be maintained at a level that is both healthy for the crew and suitable for the electronics. Fresh air is created both by "scrubbing" used air and by replenishing with air from outside the ship.

Enterprise is divided into three HVAC zones. Each has a separate air-conditioning system, so that if one fails only a third of the ship suffers. In addition, for redundancy each zone's HVAC system has two compressors—the central components of an HVAC system. Rooms with computers have extra coolers, and equipment that produces especially high heat loads can also be cooled through a chilled-water loop.

There are a few additional and less obvious aspects to the HVAC system. First, in the event of a fire on board the HVAC must be shut down immediately to prevent it from spreading smoke throughout the ship. They don't just shut down the particular zone, because "The smoke is going to go outside anyway and the next zone will suck it back in", Ian explains. The diesel generators, however, require continued cooling to avoid overheating, posing an interesting tradeoff in the case of an emergency. In the worst case the ship must slow down or stop. In addition, special fans over-pressurise the central staircases, which act as the route for rescue operations. The inhabited parts of the ship are also designed so that they can be over-pressurised; this is to avoid the entry of potentially hazardous gases or contaminants, a precaution for chemical and biological warfare. Enterprise also has a "pre-wet" system that sprays the superstructure of the ship with water to prevent chemical, radiological and biological contaminants from sticking to it.

FIRE Fire is a grave danger for a ship. I mentioned the problem of smoke in the context of HVAC above, and I will talk more about fire exercises later. But there are more consequences for the ship's systems. For example, there are hundreds of fire and smoke detectors, and CO_2 and foam systems are available in the engine, Azipod and bow thruster spaces.

Dealing with water inside the ship is also crucial. First, water is needed to fight fires. Second, water must be pumped between the various ballast tanks, we discussed those earlier. But water must also be pumped out of the ship in the case of a leak. These three tasks are performed with a single system, the details of which I found particularly interesting. It uses "eductors", also known as jet pumps. These rely on the Venturi effect: if a gas or a liquid flow is choked (that is, its section is reduced), its flow speeds up. This, in turn, increases its dynamic pressure, but also reduces the static pressure at the same time. In other words, choking creates negative pressure. An eductor is a device that consists of two water flows: one is choked to create negative pressure, which then creates a second flow through suction. It is basically a pump without moving parts, which makes it very robust.

The catch is, of course, that you have to create that first flow, usually with "real" pumps. In the case of Enterprise the "first flow" is the high pressure saltwater main ring. The 6 bar of pressure that drives the flow in this ring is created by six electrical pumps. The main ring can be segregated into zones for resilience, and there are two pumps per zone. Eductors are spread throughout the ship, all driven by the flow in the main ring. By opening valves, the eductors can be activated and used for sucking water out of spaces, for moving water between ballast tanks, and to provide water for fighting fires.

IPMS Up the ladders and back into the machinery control room. I chatted with Ian and Lee while sitting in front of their IPMS (integrated platform management system) consoles. The IPMS manages all the technical systems of the ship. Users can switch between screens that show the status of the ship's systems. It collects, prioritises and reports alarms and status messages, and also automates some of the processes in the ship, such as starting up generators if the spinning reserve goes below the specified threshold.

Because of the automation provided by this system, fewer people are required in the engineering department: ordinarily only two crew are on duty during the day. In addition to working with IPMS, Ian and Lee also perform occasional walks through the various spaces to inspect the systems visually, although there are also cameras.

5 | *Life on the Ship*

CREW STRUCTURE With its crew of 50, HMS Enterprise is organised a little like a small company. The head of the company is Phil Harper. As Captain of Enterprise he "owns" the ship, whether he is on board or not. This is made very clear by the fact that the name on his jacket does not read "Harper", it reads "Enterprise".

The person who runs the ship at any given time is the "officer in command". If Phil is on board, he, as Captain, is also in command. If he is not on board, then his First Lieutenant or the Executive Officer are in command—but this will not make them the Captain. Both the CO and the XO are command-qualified by going through the respective courses and passing board exams. The 1LT is not command qualified, but is usually a very experienced senior Lieutenant or Lt. Commander. Two out of these three are always on board. The higher-ranked person will be in command, and the other one will run the executive and seamanship activities.

What makes the terminology slightly confusing are two oddities. One, the Captain is also known as the Commanding Officer. So, when Phil is on board, the Commanding Officer (aka Captain) is also the "officer in command". When he is not on board, then he is still the Commanding Officer, but he is not in command. Also, "Captain" and "Commander", in addition to identifying roles on a ship, are also ranks in the Royal Navy. Phil's rank is Commander, but his role is Captain of the Enterprise. If he were promoted one rank up, his rank would become Captain, independent of whether he "owns" a ship. "It's the military, so everything is as confusing and complicated as possible", Phil summarised jokingly when he explained this to me.

Next to the Captain, the crew can be grouped into officers, senior ratings and junior ratings. To stay with the analogy to a small company, a group of four or five officers make up the "board" of the ship. They include the First Lieutenant, the Executive Officer and the department heads for engineering, operations and logistics. In total there are 10 or 12 officers on the ship, so in addition to department heads you will find roles such as Navigation Officer ("Navs"), Meteorology Officer ("MetO") and Gunnery Officer ("Guns"). There are also usually some officers in training. Rumor has it that the Enterprise also has another officer called "TweetO" who is responsible for their Twitter account.

The senior ratings can be thought of as "middle management". As workshop foremen, transport managers or senior system experts, they manage and coordinate the day-to-day work. The junior ratings are the people who ultimately deliver the effect of commands.

The average age of the crew is mid-20s, which is relatively old for a Royal Navy ship, because of a relatively high number of senior systems experts for the survey equipment. Phil says that "It's quite a relief as a commanding officer to have more mature people on board." 20% to 30% of the crew are female.

THE SOCIETY ON BOARD As an outsider I tend to perceive armed forces as very hierarchical and authority-driven organisations. When I learned that officers, senior ratings and junior ratings each live in separate quarters and eat and socialise in their own messes, I took this as a confirmation of my bias. I asked Kyle about it. He replied that "It makes sense; you don't want to socialise with your bosses all the time." I admit I never saw it this way. When I talked about it with Phil, he seconded Kyle's point, but he also added that there is a lot of tradition to this arrangement, and at the time when that tradition developed the structure was probably much more hierarchical. He added: "Even I am not a member of the officers' mess. I have to be invited to enter. Similarly, the officers have to be invited to the messes of the enlisted personnel. These invitations happen routinely though; there are a lot of inter-mess activities."

Phil also elaborated on the relationship among the crew, especially cross-rank: "You don't get the results you want through a system based on authority and rank—teamwork is everything." This is confirmed by what I witnessed during the week. While some interactions are ritualised and showcase cross-rank respect (the ubiquitous "Sir" when enlisted men talk to officers, or the fact that officers address Phil as "Captain"), the general atmosphere is very collegial. There are many studies in a wide variety of domains, such as medicine or aviation, that show that outcomes are much better in less hierarchical structures (for example head physician and operating team, Captain and First Officer in civil aviation), and the military has learned from this as well.

However, Phil also points out a tension: "In contrast to commercial companies, I might have to give dangerous orders, for example ordering a man or woman onto the bridge wing to man the gun when fire is incoming. This requires a certain amount of 'friendly distance'." More generally, in a (time-)critical situation, a military force cannot function if every decision has to be taken democratically. Which is of course why armed forces have ranks, orders and discipline. A year later I had a similar conversation with Caleb Marheine, a pilot of an AH-64 Apache combat helicopter. I asked him who decides which missions he has to fly, especially if the weather is below a minimum, or when there is a lot of operational risk in a war zone. He replied that his commander will generally ask him for his opinion, and often this opinion is the deciding factor. On the other hand, there are situations in which the officer will override: he might decide that, for example, rescuing a downed aviator is worth "risking" an Apache and its crew. I don't envy officers, like those who command an Apache battalion, or commanding officers of warships like Phil. I am sure it is not easy to balance these two forces when leading a team of soldiers or seamen in dangerous situations.

ROTATION Of Enterprise's total crew of 75, 50 are on the ship at any one time, typically for a period of six weeks. On weekdays wakeup is at 0700, with breakfast in the various messes from 0730. The "work day" is from 0800 to 1600, often finishing earlier on Fridays. The work day is when exercises, training or operational activities are scheduled, at least during peacetime and, I guess, outside of major exercises.

While at sea the officers have to man the bridge in a rotating watch scheme. Kyle explains his watch rotation for the next few days: "Today I am on watch from 0800 to 1200 and then from 2000 to 2400. Tomorrow it is 1200 to 1600 and 0000 to 0400. Then again 1600 to 2000 and 0400 to 0800. Then I have a day off." I suggested that an ability to sleep at any time of the day might be a key qualification to live a happy life on a ship. He agreed. And his feather bed certainly helps.

Outside watches officers also have to take care of their admin chores. For example, Kyle is the crypto materials custodian and charge books officer, which involves lots of paperwork or working on the computer; all two-bed cabins have a computer workstation for this kind of work. He is also the Intelligence Officer. In addition, as a division officer, he has six junior sailors to look after regarding their career and make sure they are "happy, healthy and content".

I expected mandatory sports, such as an early morning exercise; after all, this is the military. But the Enterprise is too small for organised sports and the required physiologist, so sport is entirely voluntary. Good, I wasn't looking forward to sports anyway. Some people run across the deck or use the weights and other athletics equipment stored in the towed body workshop. I also saw a picture that showed a couple of people practicing morning yoga on the vertrep deck.

WHAT TO DO WHEN NOT WORKING So what does the crew do outside the working day when not on watch or working on their other duties? I saw people sleeping at all times of day, reading books or playing on their phones in their cabins. Watching a movie or talking to others in the mess is another pastime, sport is yet another. Internet is available for the crew to stay in touch with their loved ones at home, but only with fairly limited bandwidth. And I suspect a significant share of it is used by TweetO.

During her current mission Enterprise is often alongside for the weekend, which allows the crew to go ashore to visit a port city, go for extended walks or enjoy local restaurants. Alternatively the crew can of course stay on board and relax in the wardroom. When I joined Enterprise in Limassol I was surprised to see many of the crew wearing civilian clothes on board. I learned that this is common at weekends. Kyle was also out of uniform when he picked me up. As foreign sailors, going ashore in uniform is anywhere from impolite to outright prohibited, depending on the host country. In Cyprus Enterprise was alongside in a civilian harbor and serviced by a civilian logistics company. Sometimes, during exercises, she might use a military port facility, but these are often full, as Phil explains. “And the civilian ports are usually closer to where the fun things are”, Kyle adds.

I haven’t seen many folks stand outside watching the ocean. However there really wasn’t much to see on our trip: no dolphins, whales or impressive landscapes. I imagine people might be outside more when in the Arctic oceans or the Norwegian fjords, as during the 2018 Trident Juncture exercise. The pictures I saw on Twitter were certainly stunning. The crew tries to break the monotony that can arise on long cruises, for example through specially-themed food or the occasional open-ocean swimming session. For my part, I spent quite a bit of time editing interviews: by the time I disembarked I’d completed all the raw editing.

I found life in Enterprise quite comfortable, certainly for a week. The two-person cabins are really nice; on most ships cabins are shared between more people. The food was also really good. There was a light(er), usually warm meal in the afternoon and soup, a main dish and a desert every evening. I just read the book “Endurance” by NASA astronaut Mark Kelly, who spent a year on the International Space Station. He complained that NASA didn’t understand this fundamental law, according to which the chocolatey deserts are eaten first. “Nobody has a craving for vanilla or banana pudding.” The Royal Navy got that right with their desserts!

Two things I would have changed: a few degrees more warmth from the air condition system in the mess, and trading 20 centimeters of usable cabin volume for 20-centimeter-wider bunks.

MISSION PLANNING Let’s zoom out a bit to understand how the Enterprise decides what she does and where she goes in general. At the top level the demand on the Royal Navy comes from a cross-governmental body in the Ministry of Defence. They might specify an overall goal such as “Do more with NATO this year”, or “Deploy two units east of Suez while maintaining the ministerially directed tasks in the Gulf”. Navy Ops in Northwood then produce a long-term programme for the units that tries to meet these demands while also scheduling ships for maintenance and training. The “force-generating authority” in NATO terminology—the Devonport Flotilla in the case of the Enterprise—then directs the Enterprise to prepare for training and tasking. The crew of Enterprise then develops the detailed planning required to meet its directed tasks: logistics, extra equipment or people, or the daily detailed programme. The Devonport Flotilla, in addition to comprising the Royal Navy’s hydrographic survey fleet, also includes the assault ships HMS Albion and HMS Bulwark, as well as seven Type 23 frigates.

The specific plans for which ports to visit and when are then defined, either by the responsible operational control authority (MARCOM for NATO tasking and Navy Ops for national tasking) or by the relevant ship, depending on circumstances. For example, for the task of flagship for NATO’s mine countermeasures SNMCMG2, the programme is laid down a year in advance by MARCOM, including all the ports to be visited. When I arrived on Cyprus, however, Enterprise had been released from NATO tasking for four weeks (the so called “five days notice to move” period that happens twice a year), so she was free to choose her own programme.

Cyprus was chosen for its access to UK logistics, and Malta for maintenance in the Palumbo dockyard.

At the finest grained level, the planning an actual route is done by the navigator, then approved by the commanding officer. The crew rotation locations are determined by the programme, but the dates can be varied by the ship. For SNMCMG2, Enterprise changed its method of crew rotation, from a 28 days at sea followed by five days at home rotation to 18 days on task followed by three days at home, to fit in with the NATO programme. This required a lot of discussion with the tasking authorities, but in essence the crew can choose a way that works for them.

6 | *Safety*

MAN OVERBOARD The ship's siren sounds several times. The pipe: "Man overboard, man overboard, man overboard. Man overboard port side. Away sea-boat. Away swimmer of the watch." The engine indicators on the bridge show an increase in power, and the ship accelerates slightly and turns to port. A crew member stands on deck behind the bridge pointing an orange triangle towards the casualty in the water. He will continue to look at and point towards the casualty until the casualty is rescued—even in relatively quiet seas it is quite hard to see a head and shoulders, and losing visual contact must be avoided. The bridge and the upper deck fill with people, everyone helping with the rescue in one way or another.

A group of people start readying the sea-boat, a Pacific Mk 24 rigid-hull inflatable that is suspended from a crane on the port aft side of Enterprise. Three crew climb into the boat and a green flag appears. In the meantime Enterprise sails in a circle around the casualty in a maneuver called the "circling shark". This creates a visibly quieter region of water, thus helping with the rescue operation. A second green flag appears outside the bridge and the coxswain on the boat gives the command to launch. The crane immediately starts lowering the boat. After a few seconds it hits the water, is disconnected from the crane, and speeds away to the casualty. The eight-meter Mk 24 sea-boat is water-jet-powered and can reach an impressive 38 knots.

The Enterprise continues to circle while everyone watches the rescue operation. A couple of dozen seconds after launch the boat reaches the casualty. The boat crew lower a Jason's cradle into the water and pull the casualty on board. A Jason's cradle is a webbed stretcher that is designed to extract people from the water without risking spinal injury. The first-aider in the boat immediately takes care of the casualty and notifies the bridge about their medical condition as the boat makes its way back to Enterprise, which signals its readiness to recover via green flags on the bridge and the boat deck. The crane has already been lowered; the boat approaches, at first quite fast, then more slowly. They tenderly bump into Enterprise's port side, reconnect the crane and are hoisted up slowly.

THE DANGERS The whole MOBEX—man overboard exercise—took only a few minutes. After Phil's debriefing with the crews involved, I chatted with Phil and Kyle about the dangers of someone going overboard and the subsequent rescue activities.

A person going overboard is one of the gravest dangers on a ship. The risk is highest when the ship is moving in rough weather or at night, when it is easy to stumble over something on deck. This is why the upper deck is off-limits in such conditions, and may only be accessed by permission and with special precautions, such wearing a life jacket when working outside the guard rails. Rescue in such circumstances is dangerous and poses a potential risk to the lives of the rescuers. This is why "We perform the rescue with purpose, but we're not rushing it", as Phil says.

Luckily, a man overboard is exceedingly rare. Phil reports that he has seen it only once in his 27 years with the Royal Navy. However, the risk, and thus the need for training, is always a product of the probability (low) and the danger (high), so MOBEXes are performed regularly.

Falling into the sea is dangerous for many reasons. The casualty might die from hypothermia, from drowning when they run out of strength to swim, or from what is termed "secondary drowning": getting seawater into the lungs and preventing them from performing their job of oxygenating the blood. It is possible to die from secondary drowning even if only relatively small amounts of salt water enter the lungs, for example from spray. This is because osmotic pressure caused by salt in the seawater can force body fluids into the lungs, causing them to swell and preventing them from oxygenating the blood.

A rescue operation is triggered by a crew member noticing that a fellow sailor has gone overboard and raising a vocal alarm (that is, shouting). The faster the incident is noticed the better, of course, and working in pairs on the deck helps. Once the bridge crew become aware of the accident they drop a datum into WECDIS to record the current position. Because WECDIS knows (approximately) about ocean currents, the datum moves with the sea. In addition, a crew member starts pointing with the orange triangle, maintaining visual contact.

Even though the water was relatively calm when I witnessed the MOBEX, Enterprise's circling around the "casualty" leads to a visibly calmer area of water. The ship's wake also clearly indicates the location where the casualty will be found, easing the sea-boat's task of finding them. Finally, turning also creates a "slick", a quiet area of water adjacent to the hull on the inside of the turn directly beneath the sea-boat's crane. This simplifies launching and recovering the sea-boat in rough conditions.

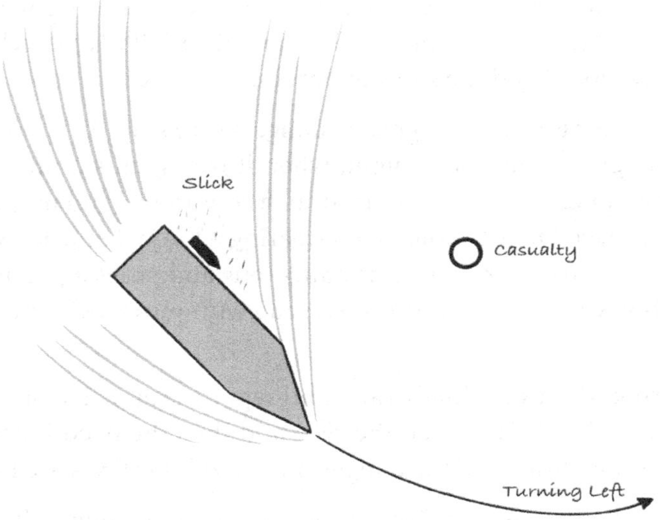

I was surprised by the use of red and green flags to indicate whether the boat and bridge crews were ready for launch and recovery, but if you think about it it makes a lot of sense. Conceptually speaking, a radio call is an event; it has a very short duration and can thus easily be missed. Flags, in contrast, indicate a state, and the indication continues as long as the state remains valid. People can always recheck the state by observing the flags. Flags provide lower bandwidth but are much more reliable—a tradeoff that makes a lot of sense for this kind of safety-critical information. Radios are of course used as well, for example to clarify why the boat deck might not yet be ready to launch.

It is interesting to think about this event-versus-state issue further. Flags, the indication of a state, have been and continue to be used a lot as a means of maritime communication. For example, there are flags that indicate a limited ability to maneuver, that radars are turned off (so it is safe to come close), or a man aboard. They can be seen by everyone, so no agreement on a common radio frequency is required. They are also an implicit broadcast—one "transmission" can be received by everybody in the vicinity. The big drawback is that their range is limited to line-of-sight, either visually or via binoculars. You also cannot communicate a change in state proactively: if observers don't recheck your flags, you cannot force them to, but a proactive radio call would accomplish this. The reason why flags work in the nautical environment, therefore, is because things usually happen (change) on relatively long timescales. You would never see this approach used for inter-aircraft communication in aviation, where things generally tend to happen much more quickly.

SWIMMER There is a secondary means of rescuing a man overboard: a rescue swimmer. On ships that do not have a boat, or where its launch might take too long, this is the only method.

Just as for the sea-boat, whose crew is predetermined for every watch, there is also a designated "swimmer of the watch". Once "piped away" they put on flippers and a drysuit and jump into the water from roughly seven meters. Secured by a 150-meter safety line, they swim to the casualty and pull them back to the ship, where they are both hoisted up: the casualty in a stretcher, to ensure blood isn't flowing into their lower extremities, lowering their core temperature, and the rescue swimmer in a strop. While I am sure some people enjoy the thrill ("Blake, today's swimmer, is half man, half fish"), it does take courage to jump from a height of seven meters into the (potentially rough) sea.

A swimmer-based rescue is a little different from rescue by boat, because Enterprise stops, ideally right next to the casualty; remember the line is only 150 meters long. It requires quite a bit of ship-handling expertise to do this precisely and quickly. They stops engines, because running the propellers or the thruster might "blow" the casualty away, suck them into the propeller, or create other dangers for the casualty and the swimmer. The ship stops upwind from the casualty, so that, if blown off by the wind, she is blown closer.

For both modes of rescue the maneuver is performed by the officer of the watch. The Captain stands back and maintains awareness of the big picture and commands by veto: as long as the Captain does not intervene, the officer of the watch runs the show.

I asked Phil which method of rescue he prefers. He says that using the boat is harder, because it requires more people and there are more things that can break or go wrong. In contrast, "The swimmer is just a bloke with some fins on." In terms of which method is faster, Phil says that, if you have a perfect crew and great ship handling, the swimmer might be faster. But in most cases, with average operating skills and average luck with the conditions, using the boat is faster. The casualty also gets into the hands of the rescuers more quickly, because the first aider comes to the casualty. On the other hand, when it is really rough, getting the boat launched can take a while, so the swimmer might be faster. Of course, it is then also more dangerous for the swimmer. The kinds of decisions a Captain has to make!

FIRE EXERCISE Safety-related exercises are ubiquitous on military ships. The MOBEX is one example, the steering gear breakdown exercise I mentioned before is another. Firefighting is trained for a lot as well. On (military) vessels everyone receives basic firefighter training, while some have advanced qualifications. CO_2 and foam extinguishers, as well as water hydrants supplied by high-pressure saltwater pipework are available throughout the ship.

Firefighting exercises come on various scales. Some train the "mechanics" of firefighting: where to find the gear, putting it on quickly, operating the eductor valves correctly to pump the water to the compartment where it is needed. On a larger scale an exercise might include most of the crew, where the focus is on coordination, reporting and decision-making.

The "internal battle", the fight against flooding, fire or other severe internal damage, is run from either the chart house adjacent to the bridge or the machinery control room, lower down in the ship. Both locations feature an incident board, a big wall-mounted schematic of the ship on which a crew member maintains the damage status or spread of a fire based on reports from crew members. During my stay the crew conducted several smaller exercises, a "fast-cruise" while in port, as well as a major fire exercise that lasted several hours and was preceded by a two-hour planning meeting. After the exercise there was ice cream for the whole crew—even for mere spectators like myself.

7 | *Conclusion*

MY LAST DAY My iPhone wakes me at 4 am on my last day of the trip. We are scheduled to sail into Malta early in the morning and I want to witness this from the bridge. On waking I notice a lot of pitch and roll, much more than on previous days. I get into the wet cell to take a shower. Showering while the world around you is moving in all directions is a really strange experience: as there is no visual external reference you can only see the shower's water moving around, and you have to rebalance continuously. I was told later that the instruments on the bridge had recorded a maximum of 22 degrees roll. The wind was gusting up to 40 knots. We would soon turn to a heading of 240 degrees, where the sea would come from the aft, reducing the roll.

I climb up the stairs to the bridge deck. I open the door of the bridge, close it, then peel away the curtain that prevents the light from the staircase spilling into the bridge. It is still quite dark outside; the bridge is illuminated in a dim, eerie red light. As one does when entering the bridge, I announce my presence quietly—only the officer of the watch is on the bridge: "Hi Markus!" Looking outside, I can barely distinguish the ocean surface from the sky.

TALKING TO MALTA We were around 15 miles off the north-east corner of Malta. Kyle suspected that, because of the rough seas, the Pilot will likely board inside the breakwater where it is quieter. In fact Valletta Port Control called us and asked for the Captain to be on the bridge. "Any watch officer could say anything. A Captain's word is almost like a verbal contract." Initial contact was established by a call from Enterprise to Valletta roughly two hours before the planned arrival. While almost every procedure in a port is specific to that particular port, it is common for ports to require ships to call in as they approach. Merchant ships typically report all kinds of details about the ship, the flag and their cargo. In contrast, military vessels typically just report that "Hey, we are here now." Kyle called Phil in his cabin, who was on the bridge two minutes later.

ENTERPRISE: Valletta Port Control this is Warship Enterprise, Channel 12 over.

VALLETTA: Warship Enterprise, good morning.

ENTERPRISE: Good morning, Captain speaking, over.

VALLETTA: Captain, good morning Sir. Captain, due to bad weather the Pilot will not be able to board outside Valletta. Are you willing to enter yourself and pick up the Pilot inside the breakwater and enter on your sole and full responsibility, over?

ENTERPRISE: This is Warship Enterprise. Affirm. Yes we are familiar with Valletta, very happy to do so and it's great to be back.

VALLETTA: Yes Captain, please bear with me, I have a few routine questions. Are your engines fully operational?

ENTERPRISE: This is Warship Enterprise. This is correct sir, all my engines work at 100% capacity.

VALLETTA: Message received. Is your steering gear fully operational?

ENTERPRISE: Yes Sir my steering gear is fully operational.

VALLETTA: Is your propeller fully submerged?

ENTERPRISE: Yes Sir, both my propellers are fully submerged.

VALLETTA: Are both anchors in good working order?

ENTERPRISE: Yes they are in good working order.

VALLETTA: Do you have twin propellers, over.

ENTERPRISE: I have twin Azipods, Sir.

VALLETTA: Twin Azipods, message received. Do you have operational bow thruster?

ENTERPRISE: Yes Sir I have a bow thruster and its operational.

VALLETTA: What is the horsepower of your thruster?

ENTERPRISE: 400 kW, over.

VALLETTA: Standby Channel 12 for instructions.

ENTERPRISE: This is Enterprise, standing by 12.

LIGHTER It was just beginning to get light outside, and several of the radar contacts resolved into the dark grey outlines of ships. The crew suspected that one of them was Enterprise's sister ship, HMS Echo. A contact eight miles out was a likely candidate because that particular ship did not use AIS. I asked whether there was no common UHF frequency on which we could raise them. Phil replied: "We could call her up, but the normal routine is to exchange identities with flashing light. We will do that when we get closer, so my seaman [the person who would operate the light] can sleep a bit longer." At just that moment we hear Echo call into Valletta Port Control. Phil: "I'm going to take a shower and get dressed."

A little while later the officer of the watch calls the Captain: "Captain, Sir, officer of the watch, shipping report, at green 30 range six nautical miles I have HMS Echo. I am a 140 degrees on her port bow, her bearing is drawing very slowly left. She has a CPA [closest point of approach] of half a nautical mile on my port bow. Sir, the time of CPA is 83 minutes, my intention is in 15 minutes time to conduct a turn to starboard to drive the SOA up before we make our turn to approach the pilot's station."

We are actually sailing on a little bit of a detour. Because the weather wasn't quite as bad as forecast we had arrived a bit early. Since we wanted to stick to our preplanned arrival time we'd have to go slower. However, a slow approach would annoy all the other ships entering the port through the fairway. Instead we decided to sail on a little circle further out, then approach the port at around 10 knots, "Looking smart and seaman-like", as Kyle puts it. Sailing more slowly in the current conditions would also lead to the ship heaving more significantly. Since most of the crew are still sleeping and will get up soon, it is more convenient to sail faster, reducing the motion.

It's seven o'clock sharp: time to wake up the crew. A sailor uses a traditional seaman's whistle (a "Bosun's call", or "pipe") and whistles into the ship's main broadcast to wake everybody. Immediately after the whistling they announce that "The ship is about to alter course and will roll heavily."

OOTW: Clear Port.

OFFICER: Clear Port Roger.

OOTW: Set both levers ahead 7.

OOTW: Starboard thirty, altering to 065.

And the Enterprise started to turn—and roll.

BRIEFING On the way towards the breakwater and Pilot pickup the crew got a weather report from the met officer (cloudy, windy, medium visibility) and agreed that they would raise flags only inside the breakwater, because it is too windy outside and thus unnecessarily risky for somebody to go outside and work on the flags. They also decided that they wouldn't call Valletta again because they had already been in contact. They briefed the approach into Valletta using a Powerpoint presentation. This included the navigational fixes they would use, which areas were off limits because the water was too shallow, how the wind would likely interact with the local landscape and buildings, (the lack of) problems with propulsion, navigation and other essential systems, the arrangements in the port, the readiness of the deck crew, the use of ropes and fenders, the status of the guns, as well as who would do what during which parts of the approach. This concluded with Kyle reminding everybody that "We were here exactly a year ago, on 21 November 2016"—"Kyle, it's December"—"Oh, ok, a year and a month." Phil: "Well I'm glad we've got that sorted out." Laughter all around.

THE PILOT ARRIVES Announcement from the bridge: "Special sea duty-men close up, cable party close up, assume damage control state three, condition Yankee, assume force protection posture Bravo, hands to harbor stations, hands to harbor stations. Close all upper deck screen doors and hatches. Rig on the upper deck will be foul weather jackets and baseball caps, rig on the bridge will be PCS and caps."

KYLE: I have the ship.

OOTW: N2 [navigation officer two] has the ship.

Other officers confirmed the status of their respective systems: WECDIS, sonar, generators, steering.

We approached our destination in a straight line towards the south-west. The entry of Valletta Grand Harbor, however, opens towards the south-east, so as we get closer we must sail around the tip of the breakwater, St. Elmo's Lighthouse. The closer we got, the more landmarks were fixed and their bearings entered into WECDIS to establish a precise position. Just as when we left Cyprus, a rating reports the depth measured by the sonar and the officer manning WECDIS confirms the depth according to the map.

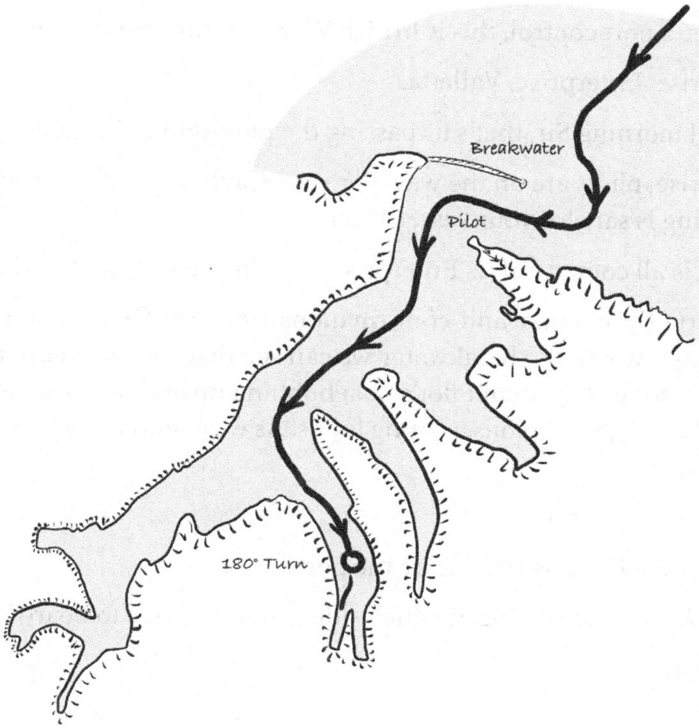

WECDIS: "Command, WECDIS, last fix, good fix, we are on track, and safe [far enough away from the danger line where the water gets shallow] with 1.5 nm to the pilot station and start of the pilotage leg in seven and a half minutes time."

NAV: Roger, which puts us one minute ahead of time.

SONAR: Depth 123 meters.

WECDIS: WECDIS concur depth.

I am once again astounded and impressed by how many people are involved in the process of operating the ship: Con, radar, WECDIS, sonar, the quartermaster who actually "drives" the ship, and the Captain in the background. I understand the rationale that Phil gave earlier, but I cannot help realising that this is only possible because the overall pace is really quite slow, very much unlike aviation (my default comparison). As we close in on the next waypoint where we turn and begin the leg on which we will take aboard the Pilot, Con confirms that the new course is clear visually.

CON (loud): Check quarters!

The lookouts look to their respective quarters and confirm they are clear of traffic.

CON: Both levers ahead 7, port 15.

The "voice control" of the ship through commands from the Con to the quartermaster intensifies as we approach the point where we'll turn onto the pilotage leg.

ENTERPRISE: Valletta port control, this is British Warship Enterprise, over.

VALLETTA: Enterprise, Enterprise, Valletta.

ENTERPRISE: Good morning Sir, that's us passing the pilot station inbound, over.

VALLETTA: Enterprise, pilots are on the way. For your Captain: keep proceeding to breakwater. Everything is safe for your vessel. Over.

ENTERPRISE: That is all copied, this is Enterprise listening on Channel 12, thanks.

More fixes, depth reports from sonar and confirmations from WECDIS that We are safe. While we are proceeding towards the breakwater we can see that the waves are much lower inside the harbor. I can also see the small Pilot's boat bobbing up and down right at the edge between the quiet and the rougher regions, waiting for us. As we come close to the breakwater we are called by the Pilot:

ENTERPRISE: This is Enterprise.

PILOT: I will board you when it is safe inside the port.

ENTERPRISE: This is Enterprise, Roger, which side would you like to board?

PILOT: Port side please.

The pilot boat makes its way towards the port aft side of the ship. Despite the lower wave action it still looks as if they have to work their way to the Enterprise. The boat leans onto Enterprise to prevent a gap from opening between the two vessels. The Pilot, wearing a life vest, climbs up a short rope ladder onto the quarterdeck. The lookout reports "Pilot on the ladder". A tug was standing by next to us, but, just like in Cyprus, we decided not to take it.

BACKING IN The Pilot arrives on the bridge and has a quick conversation with the Captain about the status of the ship and the planned route down the harbor. Phil takes over the Con: "I have the ship."

A little while later we reach another waypoint and turn roughly 90 degrees to port down the basin of Valletta's Grand Harbor. Despite the rather unpleasant weather, the city, with its historic sand-colored buildings, impresses. The harbor is actually a bay with several "fingers" running off it, some of them rather narrow. We are going to the far end of one of these fingers to berth at the Palumbo Malta shipyard. Our port turn from the pilotage leg puts us on track down the main bay. We have to turn to port once more to run into the finger, at the end of which we'll find the shipyard. While slowly sailing down the harbor we pass docked drill rigs, offshore supply vessels and, visible down another finger, a whole gaggle of super-yachts. The frequency of steering commands increases as the bridge crew works to stay precisely in the middle between the two danger lines. Phil regularly asks for reports on depth and the distance to danger lines.

When I later chatted with the Enterprise's crew about the approach via Twitter, they replied "We haven't forgotten how windy it was." When I suggested that, despite the wind, the entry into the harbor looked, to me, rather straightforward, they replied: "The agony of making it look easy—everyone thinks it actually is easy!" So it was a bit of a challenge after all.

There's one more interesting detail about our destination, the Palumbo dockyard: we have to go alongside with our port side to the dock, which means that we have to back into the dock. So we have to turn around at some point. The original plan was to do a J-turn: as we go down the main bay, we would basically turn our stern around, into the finger, then accelerate backwards. Kyle: "One pod will be at green 60, the other will be green 120. It's gonna be like a power slide. If there's a ship here [Kyle pointed to the chart], they'll think we'll hit them. But we absolutely won't."

However, there were several ships "here" and around it, so the maneuver was no longer possible and thus re-planned: we sailed down the finger bow-forward. This meant that we'd have to turn through 180 degrees in the small basin immediately to the north of the berth. The basin already looked rather small for the Enterprise's 90 meter length, but because of the shallow water the usable area was even smaller. Kyle briefed it as: "There's a 5.1 meter sounding. Plus tide 0.5 meters. We draw 5.1 aft. It is in a zone of confidence B, where the depths are plus or minus 1.6 meters. We really don't want to be there."

To cut a long story short, Phil decided that instead of attaching the tug (which was still doodling around us, standing by), he would use the dynamic positioning system to turn through 180 degrees and back into the dock. "Without DP it would be a real challenge in hand-eye coordination", says Phil.

DP allows the ship to turn on any point on the ship's long axis. When we arrived in the little basin ahead of the dock, we used it to turn, on the spot, around the tip of the ship. But only after Phil's confirming with the Pilot: "Pilot are you happy for us to turn on the spot here?" Slowly yet deliberately, the ship turned. You could follow the process on WECDIS, which plots not just the ship's position (as a point), but also renders the ship's shape on the chart. It clearly shows (and measures on the chart) the distance of the ship to the edges of the basin. Nonetheless, lookouts were positioned on the bridge wing to visually monitor these distances: "Distance to starboard 10 meters."

Once turned, Phil used the DP system's lateral mode to move the ship into the dock. "DP offset, bearing and range, 227, 85 yards." We ended up with the port side towards the berth and the aft end of the ship a mere three meters from the southern end of the dock. Really impressive work.

LEAVING The deck crew secured the Enterprise to the pier using ropes, and after maybe thirty minutes the gangway was deployed and the first crew members stepped onto land. Inside the ship the activity level ramped up, the central staircase buzzing with activity: one group was scheduled to leave for the airport soon and fly home for Christmas.

It was also time for me to leave. I had said goodbye to some of the crew before, but I did go back to Kyle and Phil, the two people I interacted with most. Luckily, Enterprise's fabulous Twitter account @HMSEnterprise allows anyone to continue to peek into life and work on board the Enterprise.

On the jetty I took a couple of final pictures. I walked towards the gate of the shipyard, where I asked the guards to call a taxi for me. I was staying in Malta for a short vacation: the weather was much nicer than at home. I never really formally entered Malta, I just walked off the ship. So I wondered during my vacation what would happen when I tried to leave Malta by air. I wasn't quite sure whether the immigration authorities at the airport would check against some kind of entry database, but they didn't and I flew back to Frankfurt without a problem.

Oh, and I never felt seasick on the whole trip.

4

Gliders and other Flying Machines

Ahead of me I can see beautiful clouds to the horizon. I'm flying along the "cloud street" at 2,600 meters and almost 200 km/h, racing down the Schwäbische Alb, a low mountain range in the south of Germany. Making very good progress towards my first turn point, Titisee-Neustadt, on today's cross-country flight, I can see the Black Forest from almost 100 kilometers away. Visibility is exceptionally good: on this June day the thermals are so powerful and perfectly distributed along the cloud street that there is no need to circle or even slow down to maintain altitude. I am busy, even a bit stressed, looking out of the cockpit to see and avoid small and hard-to-see oncoming gliders. Everybody is flying the Alb today because of the perfect conditions created by a convergence line. My FLARM, a collision-warning system based on peer-to-peer communication between gliders, is beeping regularly to warn me of some I might not be able to see yet. A small display indicates the bearing and elevation of the potentially most dangerous ones.

The cockpit is a strange place. It's a pilot's perch, his lookout over the hundreds of square kilometers that are a glider pilot's playing field. The view is magnificent—at least when atmospheric humidity is low and visibility is good, as they are today. On the other hand it's rather tight: the cockpit walls on the left and right are a few centimeters from my body and the wings grow directly from my shoulders, figuratively speaking. My head is two centimeters below the canopy and my legs from my knees down are tucked away under the instrument panel. It's almost like a cocoon. Someone once said "You don't get into a glider, you put one on". But to me it doesn't really feel tight or claustrophobic because of the vast open space beyond the acrylic of the canopy. The cockpit can be like a rollercoaster when you quickly pull up into a thermal, when you circle with 60 degrees bank and 2 g, or when a thermal doesn't "deliver" and you have to push down the glider to ensure it doesn't run out of airspeed. Primarily, of course, the cockpit is the command center of the aircraft: the soaring computer presents information about terrain, airspace, weather and how far you are able to glide, the radio connects you to fellow pilots and occasionally to air traffic control. An acoustic variometer fills the air with ever-important information about climb versus descent, which you take in subconsciously. The vario's beeping is mixed with the wind noise which, through its intensity, tells your subconscious mind about airspeed. The cockpit has the stick, rudder pedals, the flap handle and various other levers that you use, often also subconsciously, to control and guide the glider. Sometimes, the cockpit becomes a resting place and picnic area: while you cannot simply stop to rest, there are phases of long cross-country flights when you can rest your mind somewhat, maybe eat an apple, a bun or a cereal bar. Or pee into a bag. After thousands of hours of gliding, and many hundreds in this ASG-29, I really feel safe and sheltered in the cockpit even though, objectively, it's just a small cabin soaring through the heavens.

I am not the only one, by the way, to feel peculiar about cockpits. Steffen, my host on my A380 flight to Hong Kong, expressed awe at flying across half the world in this tiny cabin in the sky: "It's almost like a spaceship: you see the Earth scroll by down below, at night you see the stars above and you often experience beautiful sunrises at 38,000 feet." The cockpit of an airliner is the pilot's place of work and the command center for the aircraft, providing much more information about the state of the complex machinery that is the aircraft and its intercontinental flight. And they use it to eat and, occasionally, rest. The cockpit of the A380 is rather spacious though, and the crew has the comfort of using a real toilet.

1 | *Cross-Country Gliding*

BASIC ATMOSPHERIC LIFT Airplanes stay in the air because of aerodynamic lift produced by their wings. We will discuss the details of this later. Gliders, however don't have an engine, so there is another component to their staying aloft: atmospheric lift. Basically, a glider can only sustain a climb if the air through which it flies is rising. There are three mechanisms that produce rising air. First, the sun's energy might warm a bubble of air near the ground that, because warming reduces its density, will rise. This is called thermal lift. Glider pilots fly circles inside these bubbles and so climb with them. The second kind of lift is ridge lift, in which wind hits a mountain ridge and is forced up. Gliders can fly along the ridge on the windward side and climb using the upwardly deflected air. The third kind of lift is provided by mountain waves, atmospheric undulations on the lee side of mountains that can reach much higher than the mountain that causes the wave. While the first two forms of lift are often turbulent and require active control to remain in, mountain waves have a completely smooth, laminar flow. It's a surreal feeling that we will return to later.

Once a glider has been charged with potential energy (which is what the rising air effectively does), it can convert it into kinetic energy by flying on a slightly downward slope. A glider's flight alternates between climbing in rising air and then using its altitude in the cruise, making progress towards its destination or a turning point. The goal of a cross-country pilot is to fly as far as possible.

Below is an example where I flew 937 km from my home airfield, Donzdorf-Messelberg; I took the screenshot from skylines.aero. The solid black line is the actual flight path. I flew down the Schwäbische Alb, up the Black Forest, then towards Suhl, back down to the Ansbach area, over to Eichstätt and then a bit down the Alb again, trying (unsuccessfully) to complete the magical 1,000 km. I did this flight under the rules of the Online Contest, which means I can only fly 6 legs; the optimizer connects any six points from the flight in a way that leads to the longest net distance. The reason why I did not fly exactly along those straight leg lines is because I had to avoid airspace and fly where clouds guide me. If you'd zoom in, you'd also see me fly small circles to gain altitude regularly.

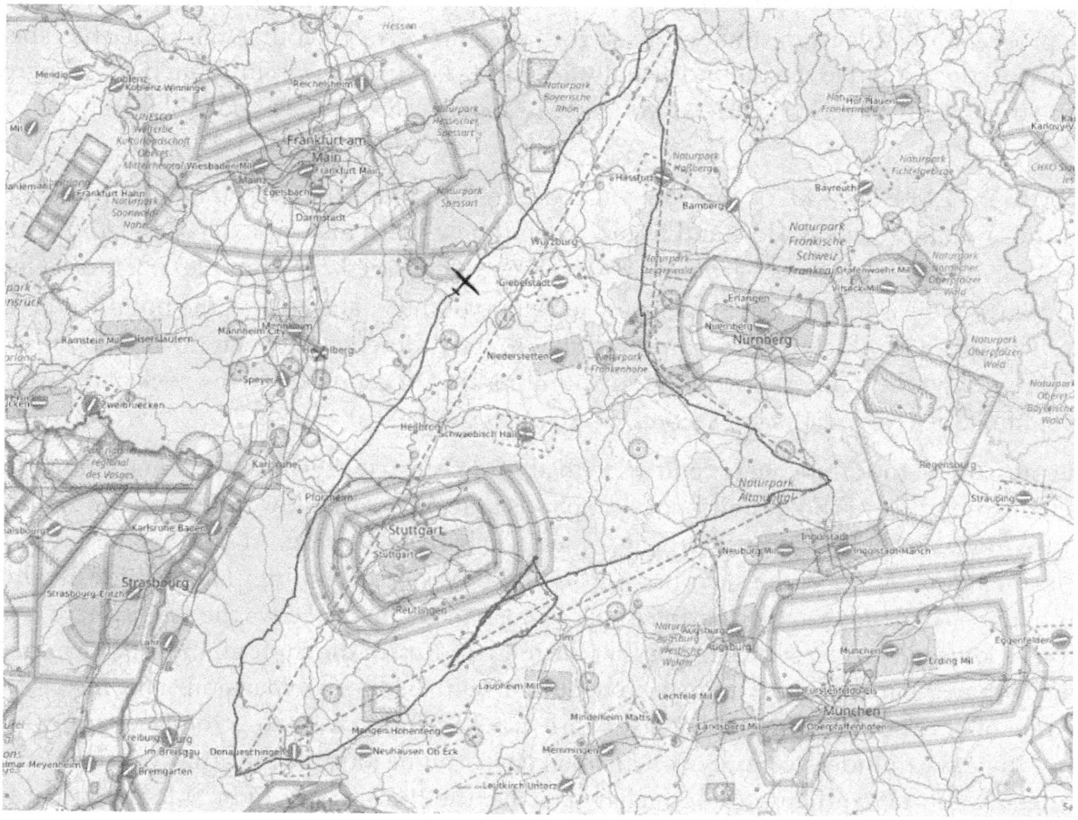

The world record distance—flown in mountain waves—is currently Klaus Ohlmann's 3,009 kilometer flight in January 2003. Triangular routes of over 1,000 kilometers are flown almost daily during the season in Africa, and can also be achieved on good days in regions with reasonable but not perfect soaring conditions, such as Southern Germany. Alexander Müller, an exceptionally skilled cross-country pilot, made nine flights of more than 1,000 kilometers from Bayreuth, Germany in 2014 alone—he does this more or less every year. I don't possess that kind of skill, but I do fly 600, 700 or 800 kilometers regularly. So how does this work? How can you fly so far using only the sun's energy?

GOING FAST The most important enabler for going far is to go fast. While this might sound unintuitive at first, it is quite logical: the total duration of a flight is limited by the availability of thermals. These only exist while the sun is at an angle that allows sufficient energy to be injected into the atmosphere to generate warm air bubbles. While the length of the "thermal day" varies with season, for any given day is it basically fixed. So to go as far as possible within a given time window, you have to go fast.

So what makes you fast? This is a question that has lots of answers. Let's explore them step by step. We have said that a glider climbs in bubbles of hot air and then uses the altitude gained to "glide down" along its planned track. While the glider is circling in a thermal it is not making progress towards its destination—its track speed is zero. Of course the glider has speed relative to the air and to the ground, usually around 95 km/h when circling, but its net progress along the track is zero. So, to be fast, you have to spend as little time as possible circling.

CLIMBING EFFECTIVELY How do you do that? Most importantly, you have to be lucky in terms of weather. Let's understand in more detail how thermals work. The sun supplies energy to a volume of air on the ground that is somehow confined, for example by a forest clearing. Warmer air is less dense—less heavy per unit volume—and so at some point it starts to rise. The speed at which it rises is proportional to the difference between the temperature of the bubble and the temperature of the surrounding air. On good days in my neck of the woods it might rise at 3 to 5 meters per second. In Africa, it can be 8 or more.

The pressure of the atmosphere reduces with altitude because there is less air mass above you to create the pressure. So our air bubble expands as it climbs into lower pressure layers. This in turn lowers its temperature, a process called adiabatic expansion. At some point it reaches the same temperature as the surrounding air and thus stops climbing. The altitude at which this happens depends on the layering of the atmosphere. On ideal days the atmosphere is unstable at lower altitudes, which means that the temperature of the static air mass reduces more quickly than the temperature reduction of our air bubble, due to expansion. This means that the rate of climb of the bubble increases with altitude. If this were to go on forever it would result in heavy thunderstorms. So, on an ideal day, at some altitude the atmospheric layering changes to a stable configuration, which means that the climb of our air bubble is stopped and overdevelopment is avoided. This kind of layering usually occurs with relatively fresh, cold air when it comes under the influence of high-pressure areas. Strong, fast thermals let you climb quickly, so you spend less time circling, which makes you fast.

This explanation of thermals based on temperature difference is not entirely correct, however. In the first few hundred meters above the ground one can measure relatively significant temperature differences between thermal bubbles and the surrounding air. Higher up, however, temperature differences are barely measurable. What drives thermals up there—and also contributes in lower layers—is water vapor. Air is almost never completely dry, and usually contains a certain amount of water. The ratio between the water vapor it contains and the maximum amount of vapor it can absorb is called relative humidity. It turns out that air with a higher relative humidity has a lower density. While temperature differences balance out as the air climbs, the water stays in the bubble and "keeps the air lighter". Water vapor has an additional and very important effect: cloud formation. Let's say our air bubble, when on the ground at some temperature T, has a relative humidity of 50%. Now it climbs, because it is heated. Because colder air has a lower capacity to hold water, the relative humidity increases as the rising air cools. At some point, with the absolute amount of water vapor staying approximately constant, relative humidity reaches 100%. At this point, a cloud forms. This is good and bad. It is good, because a cumulus cloud that forms in this way is a visible indicator of a thermal below it. It is bad, because as a glider you typically cannot fly into clouds, so the bottom of the cloud is the maximum altitude you can climb to. Days with very dry air, or where the atmospheric layering does not allow the rising air to reach 100% relative humidity, are called blue days. Glider pilots don't like blue days very much, because the thermals are invisible and also often weaker, because there's not much driving effect from water vapor in the air bubbles. Glider pilots want days when between 40% and 60% of the sky is populated by nicely-shaped but not too big cumulus clouds with dark undersides. Such days make you fast, because you can confidently approach locations where you can expect strong lift. Of course it never works out quite as perfectly in practice, but clouds do help a lot.

Even on a good day not all thermals are equally strong, because of course conditions for each particular lift are different between locations. Thus one way of becoming fast is to climb only in very strong thermals. This is easier said than done. First of all, you have to maintain a moving average of the day's typical climb rates so you can recognize the strong ones. Pilots do this intuitively, but your soaring computer also maintains this kind of statistic for you. Second, while you will stumble into good thermals by accident, you might want to increase the odds. Experienced glider pilots can tell from the way clouds look, from the structure of the ground below, or just from local experience, how good a thermal might be under a particular cloud—or at least they think they can. It's an art, not a science. You can also listen to pilot chatter on the radio and maybe hear about particularly good thermals ("Wow, five meters above Oberkochen station!"), ask your buddy who is down track a few kilometers, or use the FLARM anti-collision system in a mode that shows the climb rates of gliders in the surrounding five to eight kilometers.

More generally, flight strategy and tactics are crucial for flying fast. Strategy refers to how you plan your route. Weather changes over the course of a day, and changes in different ways in different places. Instability moves in, leading to local overdevelopment, cirrus layers might show up inhibiting the sun, or the air might dry up, changing a nice cloudy day to a blue one. Planning your route in a way that allows you to spend as much time as possible in good weather is crucial. Remember that you might cover an area of 300 × 300 kilometers, so you actually do fly into different weather zones. Tactics refers to the smaller-scale decisions of specifically where you fly. You can't just fly a straight line between your turn points because you have to consider where the lift is. Deciding which cloud to fly to, which thermal to use and which to ignore is very important for a fast flight. Is it worth taking a small detour to exploit the lift under a cloud that is slightly off course?

One factor that influences your tactics is the maximum altitude you can use: the bottom of the clouds. The higher they are, the faster you can typically go. Why is this? Gliders don't have engines, obviously, so when they get very low they have to find an airfield to land, if none is available land on a suitable field, or if they have a retractable engine, extend it and climb out. Each of these options effectively ends your cross-country flight. So the lower you get the more carefully you must fly. In practice this means that as long as you are high you only use the strong thermals. The lower you get, the more willing you are to take the weaker ones to avoid landing or having to use the engine. Of course relying on weak thermals makes you slow. So on days with a high cloud base you can cruise for a long time—and make lots of progress along your track—without getting into low altitudes where you have to think of landing or using the engine, and therefore potentially use weak thermals. Of course you can gamble... I'm sure the killer thermal will come, so I will just press on even if I am low. Many ambitious flights include such gambles. If they don't work out, the flights end up somewhere down in the rankings and nobody sees them. If you gamble correctly, you're the hero of the day.

There is a final ingredient to good and effective climbing: the mechanics of actually finding the center of lift and staying there. Air bubbles are not always exactly circular (glider pilots will tell you about the weirdest shaped thermals: "Man, this one, I tell you, it was a triangle!"), they might become wider or narrower with altitude, and on windy days they are displaced downwind with altitude. If you have problems finding or staying in the center of lift you don't climb optimally. Often, on days with strong thermals, the bubbles are also relatively narrow, so you might have to circle at 45 or more degrees of bank. And on good days you are usually not alone in the air, and through cloud optics, radio chatter, FLARM and general group dynamics, several gliders will often end up in the same thermal. Circling with four gliders in the same circle or with 15 stacked above each other doesn't always make it easier to agree on exactly where the center actually is, and part of your mental capacity is diverted from staying in the center to avoiding a collision.

CROSS-COUNTRY FLYING So we have successfully climbed, only in strong thermals, mostly in the center, and we are good at finding those reliably. Now we have to optimize the cruise parts of the flight in which we don't climb, but instead fly straight to make progress along the track. What can we do there?

You want to lose as little altitude as possible during cruise, because the altitude you don't lose is altitude you don't have to regain through "zero speed" climbs. So—getting back to tactics—we try to fly where the air doesn't sink as much. Remember that, if the air goes up in some places, it has to go down in others, and we try to avoid those areas. So you have to decide: do you fly the shortest path to the next cloud, potentially going through a stronger sink, or do you take a slight diversion (which takes longer) through air with less sink, or through air that might even climb a bit. So we try to fly under "whiskers of cloud", or monitor nearby gliders to see if we are climbing or descending relative to them, then adjust our track laterally. Similarly, because the rules only allow a limited number of optimizer-produced legs, flying directly from turnpoint to turnpoint in a straight line is better, you avoid detours. On the other hand, the thermals along that "detour" might be stronger, making the longer track a good trade-off. You have to make such decisions all the time, which is one reason why serious cross-country flying can be really tiring.

You might have noticed that I haven't talked much about wind yet. As you probably know, gliders are also called "sailplanes", but unlike their terminological cousins, sailboats, gliders are not directly propelled by wind. We have seen that they use thermals instead, or exploit indirect effects of wind: air rising because of ridges or waves. But wind does play a role in classical thermal-based cross-country flight. First, if the wind is at your tail it makes your speed over ground faster for a particular airspeed. As we will discuss in more detail later, airspeed is what governs airplane performance, but ground speed is what counts in terms of cross-country distance. So you make faster progress if you fly with a tailwind. This is why most competition scoring schemes give more points for routes that bring you back to the takeoff location. In these cases you have to fly both with the wind and against the wind, and the impact of airspeed versus groundspeed over the day is reduced. But it is often better to fly out with a headwind so that the way back is made easier by exploiting the speed-up from the tailwind. On the other hand, you might want to exploit the "push" of a tailwind in the morning when thermals are still weak. If the wind direction is expected to change during the day, such decisions can be non-trivial.

Wind also has another effect. Above wind speeds of 20 to 25 km/h clouds will often align along the direction of wind, leading to the phenomenon of cloud streets I mentioned in the introduction. If your track is aligned, at least roughly, with the direction of the street, you hardly ever have to fly diversions to find good lift: there's always a cloud straight ahead and it's often quite close. If the street is solid enough, you don't actually have to circle at all to maintain or increase altitude, you just fly along the street: when the air around you rises, you slow down to climb with it; if the air sinks, you speed up to get through the sink as quickly as possible.

Your soaring computer tells you about the state of the air in which you are flying, both visually and audibly. On the June day I mentioned at the beginning the thermals were so strong that I didn't even slow down to climb. If I did, the thermals would push me against—and then into—the cloud: not good. Instead I just raced ahead as fast as I could, as fast as the mechanical structure of the airplane would allow in turbulent conditions, around 200 km/h in the case of my ASG-29E.

So let's say the weather is great and you make no mistakes in terms of strategy or tactics, and you can do this consistently for hours on end to exploit the full "thermal day", not needing to take too many of the rest phases I mentioned in the introduction. I would say you've probably done 90% of what you can do, you will be really quite fast and you will go far. But there is still the question how you can optimize the actual straight-ahead glide performance. This is the point at which we have to talk about aerodynamics.

2 | *Glider Physics*

AIRPLANE PERFORMANCE The performance of a glider can be characterized in three ways. The number everyone primarily cites is glide ratio. This describes the distance a glider can travel from a particular altitude in undisturbed air with no thermals or sink. For example, a glider that can travel 50 kilometers from one kilometer up has a glide ratio of 1:50, or just 50 for short. My ASG-29, a modern glider with an 18-meter wingspan, has a glide ratio of 52—at least on paper. Open-class gliders of more than 25-meter wingspan can get above 60, with a few touching 70. For comparison, the Ka-6, designed in the 1950s, achieved a glide ratio of 32 with a 15-meter wingspan. The 1930s Grunau Baby had a glide ratio of a mere 17. Quite an impressive evolution.

The second relevant number is the minimum sink: the minimum vertical descend rate the glider can maintain, again in undisturbed air. My ASG-29 has a minimum sink of 0.47 m/s. You might ask what the difference is between the minimum sink speed and the best glide ratio. Why are both numbers relevant? The fact that one is expressed in terms of distance (meters distance per meter altitude) and one is expressed in terms of time (meters per second) suggests that there is a difference, and we will explore this below.

The third relevant characteristic cannot be captured by a single number—and actually, the numbers above are also slightly misleading, or at least incomplete, as we will see below. A glider's handling characteristics are also relevant. How agile is it; how quickly can it turn into a thermal? How stiff are the wings; how rough is the ride in bumpy air? How well does it maintain a circle on its own—do you have to continuously control it? How well does it handle at low speeds; what is its tendency to stall, or even spin? These and other characteristics are hugely important, because they affect how much of a pilot's mental capacity has to be spent on actually flying the airplane, as opposed to making all those tactical decisions and looking outside, scanning for other traffic. The Zacher protocol, named after Hans Zacher who pioneered the academic and engineering aspects of test-flying gliders in the 1930s, attempts to capture these characteristics objectively: it quantifies low-speed handling, the effectiveness of the controls, the distances and forces required to move them, as well as various aspects of static and dynamic stability. Idaflieg, the association of Germany's academic flying groups, evaluates new gliders according to this protocol every year at their summer camp in Aalen-Heidenheim. Handling qualities are to some degree a matter of personal taste, but I guess everybody agrees on the importance of graceful low speed handling and agility, especially in roll.

HOW A WING WORKS Now to the physics. We all know that as a body moves through the air it experiences drag. We also know that the faster the body goes, the more drag it experiences. Furthermore, it is intuitively understandable that a body experiences more drag if the medium through which is moves is denser. And it is also clear that a larger body, one with more frontal area, experiences a larger drag force. Finally, we know that the shape of the body influences drag. In fact, the word "streamlined" literally describes the fact that the (imaginary) "stream lines" of a gas flowing around a body are aligned; if they are not, this is a sign of turbulence induced by the body, which consumes energy and thus leads to drag. Based on these intuitive observations we can write a formula for the force that affects a body due to aerodynamic drag:

$$D = \frac{1}{2}\ \rho\ v^2\ A\ c_D$$

Here, ρ represents the density of the air, c_D is the drag coefficient, a dimensionless (experimentally determined) factor that represents the shape of the body, v is speed and enters the formula squared, A is the frontal area, and there is an additional factor of 1/2. The product 1/2 · ρ· v^2 represents the dynamic pressure and is abbreviated to q. This is the pressure due to oncoming air, and is different than the static pressure p that results from the weight of the air mass above us in the atmosphere, the one mentioned in weather forecasts. Both are measured in Pascals.

The characteristic cross-sectional shape of the wings produces an additional force: aerodynamic lift. This force is calculated in exactly the same way as drag, but the dimensionless number that represents the shape is a different one: c_L, the lift coefficient. The performance of a wing—its ability to convert speed into lift—is given by the ratio of lift and drag. Note that this ratio does not depend on speed, area or air density, because all of these are cancelled out once we establish the ratio; it is captured by c_L and c_D, it is purely a geometric property. Actually, that's not quite true: as the speed approaches the speed of sound, beyond .4 Mach, we can no longer treat air as a incompressible gas, and everything changes. We also cannot completely ignore the size of the wing. As we scale up or down, the Reynolds number, another dimensionless number that characterizes the development of turbulence in airflows, changes. This is why an aircraft wing doesn't perform quite as well on model airplanes or when it is scaled down for testing in wind tunnels. For the purpose of understanding the performance of gliders, however, we can ignore these factors and stick with the claim that a wing's glide ratio is purely a matter of shape.

This suggests that a glider's performance remains the same as it flies faster. This is not true, however, for various reasons. The dominant one is that a glider has lots of parts that create drag but not lift: the fuselage with its relatively bulky cockpit at the front, its long tail, the horizontal stabilizer (which, in conventional aircraft generates a downward force to maintain overall balance of moments), as well as the interference of aircraft parts such the wings' connection to the fuselage. As the airplane accelerates, therefore, the overall drag becomes greater with the square of the airspeed while the wings' ratio between lift and drag remains constant. Performance decreases, which means that the glider sinks faster. We will return to this later.

If drag is the force that slows you down, what then is the force that counterbalances it and keeps you going? Let's look at the following diagram, which shows the forces acting on a wing in a steady-state glide at some glide angle β.

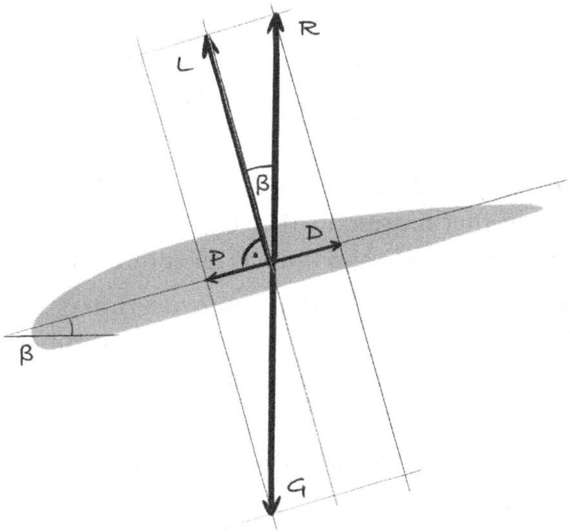

I emphasize "steady-state" because this implies that there is no acceleration, and thus no overall resulting force acting on the mass. In other words, all the forces shown above are balanced. Let's look at each of them. We start with G, the gravitational force m · g, where m is the mass of the glider and g is the well-known 9.81 m/s^2. G points straight down towards the center of the Earth. L is the aerodynamic lift produced by the wing, perpendicular to the wing's chord line. D is drag, and acts in the opposite direction to the velocity vector, which is oriented "down" the slope β. The force that drives the glider forward is marked as P in the diagram. This is in balance with the drag (otherwise the glider would slow down or accelerate) and is a component of G; in fact, it is $G \cdot \sin(\beta)$. Remember the downslope force in the context of inclined planes in physics class?

So, what can we learn from this? First, a glider must fly on a downward slope, because if β was zero the forward force P would be zero. Sloping downwards is a glider's way of transforming the potential energy inherent in its altitude into forward motion. The angle β is of course directly related to the glider's glide ratio: $\cot(\beta) = L/D$, where cot is the cotangent function.

A quick interlude: the angle β represents the (normally downward) glide slope, not the angle of attack α, the angle at which the airstream hits the wing relative to its chord line. When you watch an airliner landing you can clearly see that its nose is up—a positive angle of attack—while it descends towards the runway. We don't need α in our discussion yet: we'll cover it later.

MCCREADY'S THEORY Let's recap. For a wing c_L and c_D are always in the same ratio and characterize the wing's performance. In practice, however, because a glider has components that produce drag but not lift, the drag force D increases more quickly with speed than the lift force L. In other words, the glider's glide angle β becomes larger the faster it flies; its speed of descend increases. This is represented in a glider's polar curve:

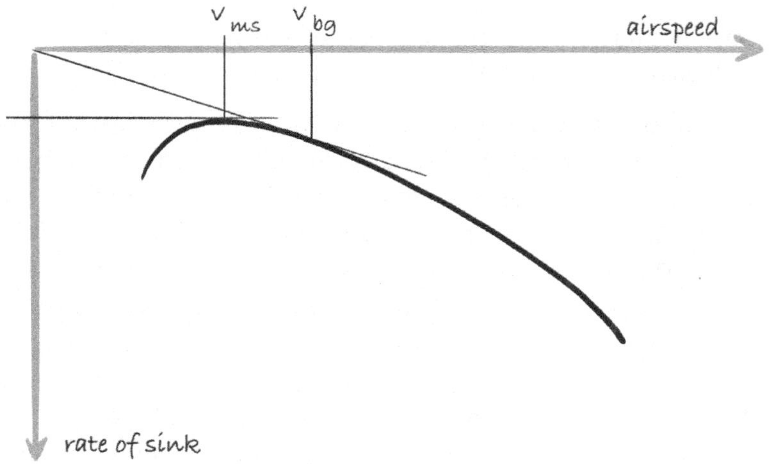

The topmost point of the curve is the minimum achievable sink rate, corresponding to v_{ms}, the speed of minimum sink (the 0.47 m/s for my ASG-29 that I mentioned earlier). This is the speed you fly when you either want to sink as slowly as possible, for example on a winter's day with no thermals and you are just gliding down from a tow to enjoy the scenery, or when you want to climb efficiently, for example when circling. This speed is usually relatively close to the stall speed, which is why it is important that a glider has benign stalling characteristics: when circling in a thermal you really don't want to stall and maybe drop onto a glider circling below you. The speed of best glide v_{bg} can be found by projecting the tangent from the diagram's origin. This represents the speed you want to maintain in straight flight, because it is where the glider has its best L/D ratio.

Based on this diagram we can refine our statement about the performance characteristics of a glider. It is not just important that the glider has a good glide ratio and a low minimum sink speed. It is also important that v_{bg} is as high as possible, to allow you to go fast with little penalty. Much progress has been made in this area over the decades. Comparing the previously mentioned Grunau Baby or a Ka-6 to an ASG-29 in this respect is pointless, the speeds are too different. But even compared to the ASW-20, which is only two generations older than the ASG-29, the v_{bg} has increased from around 90 km/h to around 105 km/h—that is, you can fly 15 km/h faster without any performance penalty! For the 28 meter EB-29R it's another 10 to km/h higher.

We said above that v_{bg} is the speed at which you fly in cruise because it gives you the best L/D. In fact this is also not quite correct. Let's say you go faster. This means that you arrive at the next cloud sooner, which is good. But you are also lower, so you have to spend more time in this (or the next) thermal—remember, effectively at zero track speed. So which alternative is better? Arrive early and lower, or later but higher? As usual, looking at extreme values helps: if your average climb rate in thermals was zero (no thermals that day), then it's obviously crucial to stay up, so you fly slower. In fact this is what you do when struggle to get home in the evening. In contrast, if climb speed was infinite, then of course you would want to fly ahead as fast as possible, because it would take you no time to get back up.

So the trends are clear: the stronger the expected thermal, the faster you fly between clouds. If you try to work out the ideal speed exactly, however, things get more complicated: it involves several parameters, a few simplifying assumptions and about 30 lines of (not too sophisticated) math. Paul McCready did this math first in the 1950s and wrote down a formula that calculates the ideal speed to fly. He won the 1956 world gliding championship, presumably because of these insights. Today the formula is programmed into the soaring computers that are installed in most gliders. It's used as follows: as you fly, you keep track of the average climb, either in your head or by relying on the computer. You then enter this average climb rate into the computer as your "McCready value". During cruise the computer tells you the speed at which to fly for ideal performance, depending on the relative movement of the surrounding air at that particular point in time. The indication is visual and acoustic, and as a routine pilot you subconsciously control the speed to follow the McCready recommendation. It's an imperfect science, of course: you must still decide which thermals to take and which to pass on, as the computer doesn't know the distance to the next thermal and doesn't consider your altitude or your willingness to accept the risk of landing out. But at least in the care-free zone high up most pilots fly according to the speed-to-fly indicated by the "McCready in a box".

We can now once more refine our notion of what it means for a glider to have good performance: you don't just want a high L/D ratio at high speed, but you also want it not to degrade too much at even higher speeds. Looking at the polar curve above, you want the curve to be flat, which will allow you to go faster without too much of a penalty. The speed-to-fly computation in your soaring computer takes this polar curve into account, by the way, so it recommends different speeds depending on aircraft type. Again this is where a lot of progress has been made over the years.

WATER BALLAST Are there more ways we can improve the speed at which we can fly? There are. Let's revisit the diagram that shows the forces on a wing and see what happens if we increase G, which we can do by making the glider heavier:

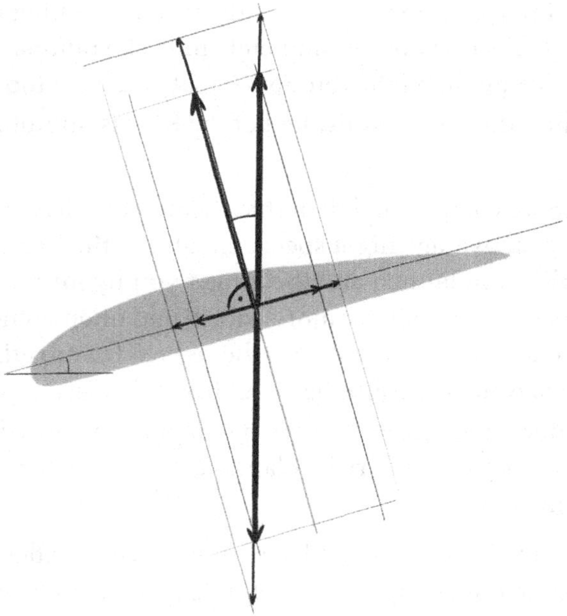

Remember that the angle of descent, or the L/D ratio, is purely an aerodynamic property: it is only a matter of the two shape-related factors c_L and c_D. If G is increased it has no impact on L/D, so we have to keep β the same. However, you can see that the arrows representing drag and propulsive force become longer. In other words, for the same L/D ratio you can fly faster by making a glider heavier. This is why gliders carry water ballast—it moves the best glide speed v_{bg} to a higher speed, making the glider's polar curve flatter. You tell your soaring computer how much ballast you have loaded and it will then take this into account when computing the speed-to-fly, depending on the sink rate of the surrounding air.

The effect can be striking. On that June day I launched without water because the day was not forecast to be that good, and I was also just too lazy. Above the south of the Black Forest I met a famous cross-country pilot in his ASG-29. We met underneath a cloud and both started racing towards the next cloud along the track, initially next to each other at the same altitude and at the same speed. We were not next to each other for long: I literally fell out of the sky compared to him. I am sure his glider was full of water. After maybe ten kilometers at 180 km/h I was several hundred meters lower—a very impressive illustration of my mistake of not taking water!

There is no free lunch, however. Water ballast obviously makes the glider heavier, so your minimum sink rate becomes higher, moving the polar curve downwards, which in turn means that you climb more slowly in a thermal of a given strength. However, this is a tradeoff similar to the one involved in McCready's theory: how much water should you take on to achieve the optimum balance between increased cruise speed and reduced climbing efficiency? The better the weather, the more water you want to take on. Most pilots "guesstimate" their ideal water load based on (their interpretation of) the weather forecast. Others, such as the aforementioned Alexander Müller, always launch completely full. The rationale for this is that you can always release water if during the flight you notice that you were too optimistic, but you can never add water in flight—there's no equivalent of the KC-135 Stratotanker to supply water to slow gliders.

As before, though, it's more complicated than that. Remember that the total distance you can fly is not just related to your average flight speed, but also to the fraction of the "thermal day" you can use. So the earlier you launch and the longer you fly for the better. But early in the morning thermals are weaker, spread out more widely and often do not reach so high. Flying in these circumstances chock-full of water, with the reduced climb efficiency that this causes, can be a challenge. A heavy airplane also has a higher stall speed, is often not quite so agile, and can generally feel more sluggish—although not in pitch, because it builds up speed much more quickly. It certainly increases the risk of landing out and potentially wasting a perfectly fine flying day early in the morning.

So we can now refine our idea of good glider performance: a glider should cope well with water: its handling characteristics should not get much worse if it's fully loaded. Lots of progress has been made with this over generations of gliders. In 2017 I had an opportunity to try out the then brand-new Ventus 3. A good day was forecast, and the guys from Schempp-Hirth who briefed me that morning basically forced me to take on a significant amount of water, around two-thirds the Ventus' maximum capacity. "You need it, it's too light otherwise, and you really won't feel it in handling" is what they said. I was skeptical, but they were right. The airplane handled beautifully, and even though it was my first flight with the type, I flew 650 kilometers with an average track speed of 105 km/h. To put this into context, this was the fastest flight that day in that part of Germany, although not the farthest—a few people flew as far as 800 kilometers.

3 | *The Future of Gliders*

STRUCTURAL IMPROVEMENTS When I mention water ballast to non-pilots they are surprised. Most people think that a glider should be as light as possible to be able to soar like a leaf. They don't immediately see the benefits of a heavier airplane, and the above discussion suggests why: it does take a bit of physics to see the benefits of greater weight.

On the other hand, on weak days, or in the evenings of good days when you have dumped your water, you really do want a light airplane, one that has a very low minimum sink rate. Overall you want the best of both worlds: a light airplane for when the weather is weak, but one that can take on lots of water for when it's strong. Glider pilots put this in terms of wing loading, the mass per unit area of wing. Most modern gliders can adjust wing loading from around 35 to around 60 kg/m^2.

This concern leads directly to structural strength. The design regulations for gliders require them to be able to ....withstand 6.5 g positive and 3.5 g negative plus a safety factor of at least 1.5. To be competitive, a glider has to be able to fly at around 200 km/h, even in turbulent air that might stress the airframe. The structure of gliders is built mostly from carbon-fiber reinforced plastic. Over the years the materials themselves, as well as design and manufacturing techniques, have become more advanced, so less structural mass is needed to achieve the required structural strength. This allows construction of gliders that are light for weak days while at the same time being able to support a high water load for really strong days.

CONCORDIA I chatted about this with Dick Butler, the designer of Concordia, a 28-meter one-of-a-kind glider. He built it for himself, so he did not necessarily optimize it for a broader market in terms of flight characteristics or ease of production. For example, it's not rated for winch launch because its elevator is not effective enough. However, it has an L/D of over 70, and complies with those regulations that are relevant for admission to competitions, such as a maximum takeoff weight of 850 kg. Butler joined forces with Gerhard Waibel, a legendary sailplane designer who worked all his life for Alexander Schleicher, one of the world-leading glider manufacturers, and Johannes Dillinger, a young engineer who at the time worked with Loek Boermans at TU Delft, one of the main research outfits that design high-performance airfoils for gliders. The group took the fuselage of an ASW-27 and attached a custom-built 28-meter wing. To reduce technical risk they relied on basically the same airfoil as the ASW-27, but extended the wing span by a factor of almost two.

One problem with long wings and heavy weight is that the wings flex upwards under the influence of positive g. Search YouTube for "eta sailplane landing" and you'll find a video that demonstrates this bending nicely, and it's not even in a high-g situation. You can also find videos that document load tests for sailplane wings; they can be bent up to around 45 degrees before they break! One problem with wing bend is that it cross-couples into torsion loads, which then lead to wing twist. This in turn changes the angle of attack of the outer wing, the angle between the oncoming air and the chord, reducing performance. You don't want to necessarily eliminate bending, because this would require much thicker components and thus add structural weight, and as we have learned, this is not desirable, because it makes the glider heavy without water. Dick recounts: "Dillinger spent a year with aeroelastic tailoring, optimizing the way the carbon fibers are laid out, oriented in such a way that wing twist when it bends under load is eliminated. With a conventional approach to designing and building the wing we'd have gotten almost 1.5 degrees of twist at 200 km/h." This was the first time that aeroelastic tailoring was used for a glider, and it demonstrates the lengths to which designers go these days to optimize overall performance. Another example: because of the high loads induced by the large wingspan, the spar at the center became so thick that it didn't fit into the tiny ASW-27 fuselage. At least this would have been the case if they'd used the traditional way of building the spar. Instead they involved Christian Streifeneder, who pioneered the use of resin infusion molding in the construction of the eta, another experimental sailplane with a 31-meter wing. With this type of construction resin is sucked into previously laid out carbon fabric by vacuum, instead of manually applying it to each layer of carbon with a brush. The resulting wing spar is much stronger using the same material for two reasons. First, it makes it possible to achieve a higher fiber-volume fraction, meaning that there are more fibers per unit volume, increasing strength (mostly tensile). Second, it leads to more uniform saturation of resin in the carbon fabric, increasing specific strength. This technique allowed the spar to be 30% thinner, allowing it to fit into the fuselage.

AUTOMATION In the future I expect performance improvements from various directions. Continued research into airfoils will lead to improvements, as has taken place over the last 30 years. But those are likely to be marginal, maybe two points of L/D ratio at 5 km/h higher speeds every 15 years or so. Other factors will be more relevant for increased overall cross-country performance.

For example, if significantly stronger and stiffer materials become available (carbon nanotubes, anyone?) then one could build thinner, higher-aspect ratio wings that would still have the required strength; aspect ratio is the ratio between the wingspan and the mean "width", the chord length of the wing. Thinner sleeker wings are generally better from an aerodynamic perspective. For future materials to be suitable, they also have to be a good fit with the kind of manual processing and craftsman-style production methods that are common with today's glider manufacturers. The highly automated processes that companies such as Airbus use to build structural members of much higher-strength for a given size and weight have not yet been adopted by glider manufacturers for this reason, and also maybe because of the required financial investment and the "We've always done it like this, why change now?" factor.

The second direction is automation of some aspects of flying itself. Something we have not talked about yet is that high-performance gliders have flapped wings, which means that they can move the trailing edge of the wing up or down. Moving it down leads to a thicker, higher-cambered wing that is suitable for low speeds. Moving the flap up results in a thinner, less cambered airfoil optimized for high speeds. Today, the pilot operates the flaps manually with a lever in the cockpit, potentially guided by recommendations provided by the soaring computer. Doing this optimally over long periods of time is tedious and pilots make mistakes. This task could easily be automated, as could flying at the optimal speed-to-fly recommended by the McCready theory. The Nixus Project is experimenting with these ideas, controlling the flaps automatically using servos installed in the wing. Another factor that limits how thin you can build a wing is that the rods and cranks necessary to drive the flaps and ailerons have to fit inside it, while a bunch of electric cables and servos require much less space. This is why Nixus builds a new, thinner wing for an ASH-30 fuselage. As I write this in early 2019 they have just had their first flight; I hope an omega tau podcast episode will be forthcoming.

The third possible direction for faster cross-country flight is data. There is a huge treasure of unused data hidden at the Online Contest, a website where almost everybody uploads their flights. Analyzing this data could identify thermal hotspots as a function of weather. This hotspot data could then be made available to pilots to help with flight planing, or even to guide pilots actively to possible thermals during a flight. Today, pilots rely on their own experience to "know" where "it will be good". The introduction of Online Contest in 2000 has boosted average cross-country performance, because it allows pilots to see where everyone else flies and how well they did. But you could take the use of data even further. Imagine if you could see, on your soaring computer, realtime data about cloud base and cloud cover for, say, all of Southern Germany, and exactly where other gliders are climbing, up to 50 kilometers away. Maybe even averaged over the last hour. You could adjust your route in real time based on live data instead of on the morning's forecast or anecdotal stories on the radio. The computer could make optimal tradeoffs between many of the parameters we have discussed in this chapter.

Of course some will say "Well, if everything is automated, that's not really flying any more." But they said this about the introduction of GPS and soaring computers, and for sustainer engines. And probably about radio ("If you can simply ask your buddy where the thermal is..."). And about "these new plastic airplanes". It's not an argument!

LANDING We're back in the cockpit on this exceptional June Saturday. After five hours of flight I'm approaching my home airfield. "Donzdorf, 2M is five minutes out, do you still have runway 27 in use?" I ask on the radio. "Yep, 27, report downwind" they confirm and ask me to report when I begin my approach. I have lots of altitude, so I can "waste" it by flying fast, at 235 km/h. I don't want to go to the glider's maximum allowed speed of 285 km/h because it is still relatively turbulent out there. As I arrive at Donzdorf I kill the energy through a couple of steep turns and then report "2M, downwind" as I begin the approach about 200 meters above the ground. The air pressure has changed during the flight so I judge altitude visually, as the altimeter is probably off by a few dozen meters. I reduce to around 100 km/h, extend the landing gear, move my flaps to position 5 (slightly positive) and tighten my safety belt. After 20 seconds or so I turn right onto base, slightly extending my air brakes to adjust my altitude. It's important that the first landing attempt works out because, in contrast to airplanes with engines, I cannot go around! I turn right once more onto final approach and extend my flaps to fully positive, 47 degrees down. I have to push the nose down quite a bit to keep my airspeed. I use the air brakes to control my rate of descent so that I will hit the start of the runway precisely. As I come close to the ground I carefully pull the stick back to flare... and touch down gently. I roll out and leave the runway to the left side.

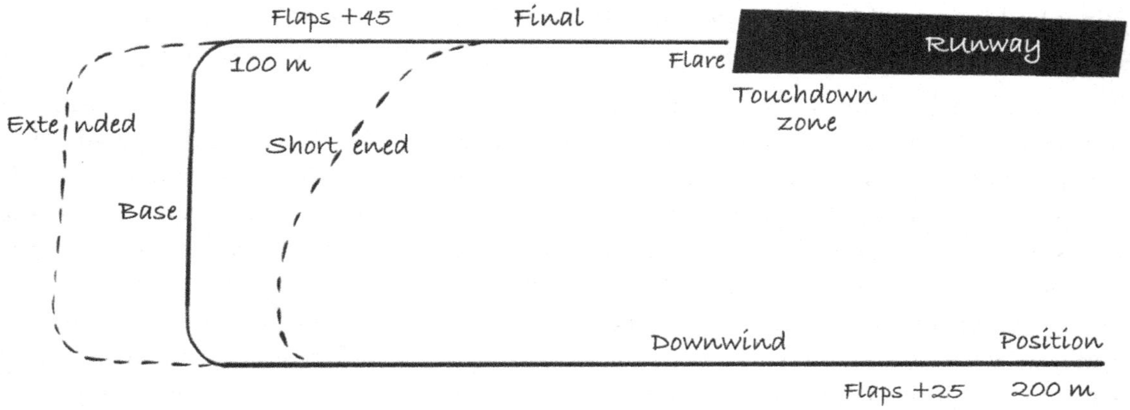

Satisfied with the day, I open the canopy. I've covered 600 kilometers with an average speed of 123 km/h. The fastest leg, the one down the Alb along the convection line, averaged 135 km/h, making it my fastest 600 kilometers ever. And then came Sunday. Unlikely as it sounds, Sunday was even better. I flew 810 kilometers at an average of 149 km/h, my fastest long flight ever and the second fastest of the day in my region. My buddy Max almost cracked the magical 1,000 kilometers and ended up with 979. What was even crazier was that the first 600 kilometers were at an average of 160 km/h, a full 30 km/h faster than the flight the day before. What a weekend!

4 | *Wave Flying*

IN THE WAVE It is 2004 and I am flying from Samedan, Switzerland, with my ASW-27, registration D-6642. I launched on the winch and climbed on the local ridge. The wind is quite strong today. High mountains, between 3,000 and 4,000 meters, result in very good ridge lift—I can get quite high even without thermals.

On the radio I hear the other pilots who fly from Samedan starting to mention waves. There were rumors in the morning while we were all standing around on the airfield, waiting for the weather and chatting, that this might be a wave day. But people weren't really sure; nobody was betting on it. Now, as we fly back and forth on the ridge at around 3,000 meters, my buddies start talking about it again. I look up, and indeed, I can see a glider that is much higher than I am. This is my first time in Samedan, so I am not familiar with the local conditions, and in contrast to thermals, wave lift is extremely specific to the local geography and wind conditions. I do know the wind direction, of course, and I do understand the fundamentals of how waves form in the lee of mountains. The following illustration is adapted from OSTIV's Met-Panel; OSTIV is the Internation Scientific and Technical Soaring Organization, where glider enthusiasts from all over the world look into the technical, safety-related and meteorological aspects of gliding.

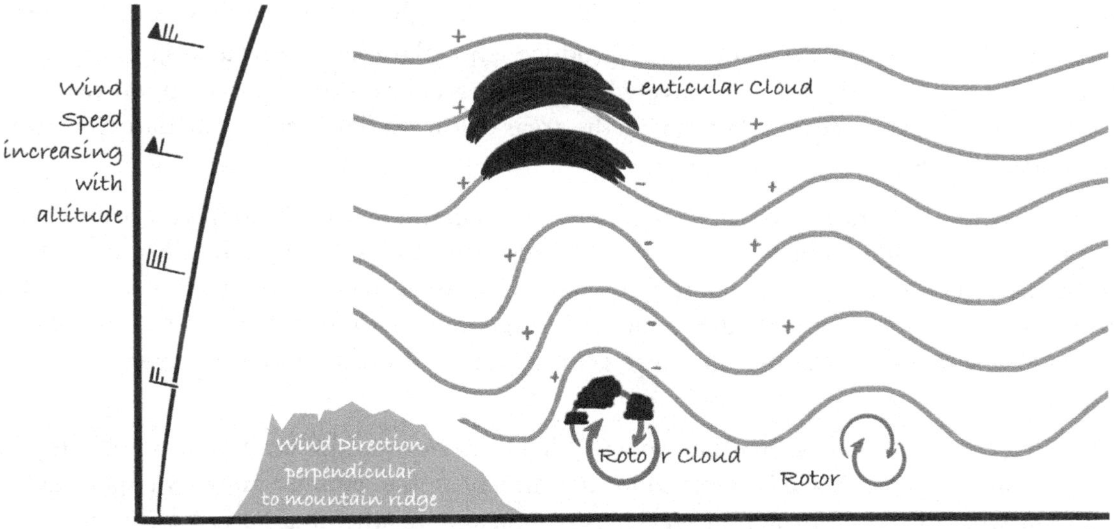

I can see the other guy who is higher, so I decide to carefully fly out into the valley, away from the ridge and towards the position of the higher glider. As expected, the reliable ridge lift stops and I start to sink. Not too badly though, and I can still see the other guy in front and above me. I press on at about 150 km/h, always in safe gliding distance of Samedan airfield so that I can get back and land if I don't find any more lift. As I approach the center of the valley, approaching the higher guy, I can feel (and hear, via the vario beeper) the sink become less... and less... and less... zero... a little bit of plus... I seem to have indeed stumbled into a wave. The climb becomes stronger, about a meter per second now.

Once above the very turbulent rotor zone the wave has a completely laminar flow; there is no noticeable turbulence at all. The glider is totally stable, perfectly trimmed, with no need to touch the controls. You can imagine it as every "air particle" traveling on its own track in parallel to others. If you inject smoke into a wind tunnel these tracks become clearly visible. Turbulent flow is the opposite, where you can see vortices and mixing. To form vortices air must be accelerated continuously (in the sense of changing its direction). This acceleration requires energy, and this energy is taken out of the flow. Turbulence created by the shape of an aircraft is a major reason for drag: the turbulence "sucks out" energy from the flow, which has to be replenished through engine power, or by sacrificing potential energy in the case of a glider.

One of the factors for successfully exploiting a wave is to make sure you find and then stay in its center, and prevent strong winds from blowing you away. Nowadays, with soaring computers that have a moving map that plots your flight track and color-codes the track by climb and sink, it is very easy to explore a wave area and effectively record your own little map of where the lift is strongest, so you can return to areas where the track shows strong lift. Back in 2004 computers were more limited. Mine certainly did not have this feature, so I used geographic landmarks and the higher glider as a reference to stay roughly in the same location despite 80 km/h wind. Days with higher humidity produce lenticular clouds—clouds that look like a lens—which form at the top of a wave where the air reaches 100% relative humidity. These clouds also help to locate and then stay in the areas of lift. On that day in Samedan there were no lentis, however.

As I fly higher the climb becomes stronger—very exciting! The view becomes more and more impressive. At around 4,000 meters I put on oxygen. With increasing altitude, atmospheric pressure lowers, and so does the partial pressure of the oxygen in the air, making it harder for your lungs to extract the necessary amount of oxygen. Regulations require the use of oxygen if you fly for more than 30 minutes over 3,600 meters; it is prudent to use it even lower and earlier.

Climbing further, I realize that the wave will likely go beyond the limit of 4,500 meters: the airspace above this altitude is reserved for instrument traffic. Hmm. I grab my paper map to find out which flight information service is responsible for the area here. It's Zürich Information. I decide to give them a call and see if they are willing to give me permission to climb higher. Not very likely, since I have no transponder, and my small plastic airplane is effectively invisible on air traffic control's primary radar.

ME: "Zürich Information D-6642, Hello."

ZÜRICH INFORMATION: "D-6642, Zürich Information, grüezi, go ahead."

ME: "Zürich, D-42, glider 7 kilometers north of Samedan airport, at 4,400 meters in a wave. Is it possible to climb higher than 4,500 meters?"

ZÜRICH: "D-42, do you have a transponder?"

ME: "Zürich, D-42, negative."

ZÜRICH: "Hmm, then it will be problematic, but I will ask."

She presumably calls her colleagues at Zürich Radar or Center to see if they can deal with one (or more) gliders above Samedan intruding into IFR airspace. A little while later:

ZÜRICH: "D-42, how high do want to climb?"

ME: "Not sure how high the wave will go. Maybe 6,000 meters?"

ZÜRICH: "And what is your heading and destination?"

ME: "Zürich, 42, nowhere, just trying to get higher here in the wave."

ZÜRICH: "So you are stationary?"

ME: "Yes, within a few kilometers."

ZÜRICH: "D-42, standby."

This sounds promising! I have arrested my climb by extending the air brakes to compensate for the now 1.5 m/s climb produced by the wave. That's a bit painful!

ZÜRICH: "D-6642, Zürich?"

ME: "D-42"

ZÜRICH: "You have permission to climb up to flight level 200 as long as you stay within 10 kilometers around Samedan. Please confirm."

ME: "Zürich, 42, flight level 200, within 10 kilometers of Samedan. Thank you very much!"

ZÜRICH: "You're welcome. Please stay on the frequency, we will call you back if we need the airspace."

ME: 42, "Wilco, thanks."

Very cool! Retract air brakes, continue to climb. Flight level 200 is 20,000 feet above the standard atmospheric pressure of 1013.2 hPa. Depending on the actual atmospheric pressure, this is slightly above or below 20,000 feet above mean sea level. Flight levels are used to make sure everyone defines their altitude relative to the same pressure level, otherwise the same numeric altitude would be physically different and aircraft might collide despite their altimeters showing different altitudes. Anyway, on that day flight level 200 is almost exactly 6,000 meters, just as I had hoped. The limit to 10 kilometers around Samedan is to ensure we don't conflict with instrument traffic; Zürich is basically routing the IFR traffic around a 10 kilometer cylinder centered at Samedan so gliders can safely "play" in this area.

I continue to stay in the wave with a few other gliders. Either they all called into Zürich, or everyone thinks that someone else got permission, and "One more glider doesn't hurt". To cut a long story short, that day in Samedan I climbed to 7,300 meters, with permission of course, where the wave finally subsided. The climb increased to over 2 m/s and remained completely smooth. Since I was trying to fly sufficiently slowly to be almost stationary above the ground, the wind noise was low. I switched off the vario beeper as I didn't need it—there was continuous lift anyway. It was almost completely quiet. The mountains there are up to 4,000 meters, but the valley bottoms are 1,500 meters as well. So I was more than twice as high as the mountains compared to the "visual bottom". What a wonderful view and feeling. My first time in a wave, and immediately in such a wonderful one. This, of course, was the only day I did not bring my camera, and before the days where a phone served as one. So no pictures, unfortunately.

Since this remarkable premiere in Samedan I have been in waves over Lienz, Austria (5,000 meters), over Gap, France (6,000 meters) and at home over the Alb (2,500 meters). I'm not much of a wave expert—I pretty much stay in one place and enjoy the view and the sensation. But others do much more.

MOUNTAIN WAVE PROJECT One of these is Klaus Ohlmann. He holds the world distance record for cross-country gliding of 3,009 kilometers that I mentioned before. Soaring is not just his passion, it is also his profession. He's an expert in mountain flying, especially in the southern French Alps, where they regularly get the Mistral, a strong northerly wind that streams down the Rhone valley, partially accelerated by a Venturi effect created by the Alps and the Massif Central. This leads to strong wave systems in the Southern Alps that Klaus has been exploiting for years, with flights of well over 1,000 kilometers.

At some point he must have become bored and started looking for other areas where strong geographically extended wave systems occur. Together with the meteorologist and pilot René Heise, he formed the OSTIV Mountain Wave Project to explore advanced wave flying all over the world. Their goal was to find the meteorological conditions for a straight distance flight of 2,000 kilometers in waves and thus win the 2,000 kilometer OSTIV Kuettner Prize. They considered the Pyrenees and the Caucasus, but then settled on South America and the Argentinian Andes in particular. René commented: "Usually, good favorable thermal and wave conditions are created via atmospheric processes with a maximum horizontal length scale of 1,400 to 1,700 kilometers and a timescale of 12 to 14 hours. The challenge was to find an optimum of all atmospheric processes coupled to a long mountain ridge and a maximum of daylight. The occurrence of two jetstreams over the longest mountain ridge of the world, the Andes, with the favorable orthogonal upper airflow, seemed to be the solution for the Kuettner 2,000 kilometer challenge." Klaus added: "For me, the primary challenge was to fly in an area that is completely new and unknown to me. And though the basic approach to mountain flying is similar everywhere, the local peculiarities really have to be explored." In addition to pioneering glider flights over the Andes and Himalaya, the project also emphasized research: how do wave/rotor systems work, how does their strength relate to atmospheric conditions and how can we improve forecasting? One reason why this is interesting is that mountain waves can reach up to the altitudes where airliners fly. Waves are one of several causes of clear-air-turbulence, which can be a problem for airliners, because they are hard to detect and thus hard to avoid. Flying into clear-air-turbulence can affect passenger comfort and, in the extreme case of breaking waves, can impact flight safety. Strong updrafts produced by waves can also inject so much energy into an airplane that they are pushed into overspeed; we'll see why this can happen very quickly later when we discuss the "coffin corner". The mountain wave project pilots collaborate with scientists and regularly take instruments on their flights to the edge of the stratosphere. For example, in 2006 they carried one that measures turbulence. It was developed by Jörg Hacker, and combines GPS and inertial sensors to establish a precise position-referenced wind vector. These were the first ever flights to measure atmospheric turbulence over the Andes.

The flying they do is really extreme. Klaus: "During a speed task over a 500 kilometer out-and-return track, I was able to achieve a 307 km/h average." Remember, this is the average speed over ground, so the waves had to be so strong they never really had to slow down at all to climb: "On this flight I also experienced the strongest lift ever, with 18 m/s. I mostly flew between 4,000 and 8,000 meters." That's a very long way from the quiet, stationary kind of wave flying enjoyed by beginners like myself!

TRUE AIRSPEED Flying at these speeds and altitudes is interesting. We'll have to add a little bit to our understanding of aerodynamics to see why. How fast exactly are we going? Earlier we distinguished between the airspeed (the speed of the aircraft relative to the air), the ground speed (how fast we are over ground, which is basically air speed plus/minus wind speed) and the track speed (which is the speed at which we progress along our track, which is zero when we circle).

Let's understand how we measure airspeed. The speed indicator in an aircraft measures the dynamic pressure, which we have previously defined as $q = 1/2 \cdot \rho \cdot v^2$. Technically, it does this by measuring the total pressure of the oncoming air p_t with a pitot tube, a tube that is open at the front, then mechanically subtracting the static pressure p (with $p_t = q + p$). So in some sense the formula above contains two values for speed: v, the speed at which the aircraft truly moves through the air, and the indicated speed, proportional to the square root of q. Notice that q depends on the density ρ, which becomes lower with altitude: as we climb, the indicated airspeed becomes lower and lower compared to the true airspeed. In effect, the airspeed indicator becomes increasingly incorrect with altitude. It is calibrated so that true and indicated airspeed match at zero altitude, standard pressure and standard temperature. The error with altitude is significant: at 3,000 meters, you have to add 16%, at 5,000 meters 25% and at 7,500 meters 50%.

What determines the maximum allowable speed of an aircraft? Drag is one answer. If a glider were to go straight down it would accelerate to the point at which the drag (which now points straight up) is as large as its weight. For modern gliders, which have very low specific drag, this would be a very high speed; most can't even go straight down without exceeding maximum allowed speed with air brakes fully extended. What governs this maximum speed is "flutter".

Aerodynamic flows can induce flutter: flags "fluttering" in the wind demonstrate this phenomenon. The different parts of an aircraft can be induced to flutter as well, just like a flag. The wing of a glider, because of its long thin shape and flexible structure, is especially prone to aerodynamic flutter. Different flow speeds induce oscillations at different frequencies. Everything is fine as long as the natural frequency of the wing (or any other part of the airplane) is sufficiently far from the frequency induced by the airflow at the particular speed. If some kind of kick, turbulence, or the pilot accidentally hitting the stick induces, say, a bending moment, the bending is damped and nothing critical happens: the system is stable. If the two frequencies become close, however, the oscillation can diverge, which means it increases in amplitude. This instability is known as flutter. It can get to the point where the wing or the entire aircraft literally disintegrates. An aircraft is therefore limited to speeds significantly below its critical flutter speed; this has to be demonstrated during flight testing. Importantly, flutter depends on the true airspeed, so you have to be careful with your airspeed indicator when it gets progressively incorrect with altitude!

You can now start to see a potential problem: as you get higher and faster, like for example Klaus' 307 km/h average at 4,000 to 8,000 meters, you have to be very aware of your glider's maximum true airspeed. You don't want to fly faster than that, especially since the less dense air has less of a damping effect on a potential flutter event. Klaus flew the record with a Nimbus 4DM, a 26.5-meter two-seater that has a maximum speed of 285 km/h ... less than Klaus' 307. So what is going on here? Is the guy crazy? Well, maybe a little bit, but not in this way. Schempp-Hirth, the manufacturer of the Nimbus, offers a modification kit that increases the "never exceed" speed by installing lead weights into the wingtips to change its natural frequency; this avoids flutter at speeds up to 320 km/h.

PERLAN PROJECT In addition to the science, Klaus' goal has always been to fly far and fast and use waves to achieve this. While they achieved altitudes of up to 12,500 meters in Argentina, they never explicitly tried to beat altitude records. The Perlan project, in contrast, has this as its stated goal. Of course it also does science, in particular regarding climate change and the ozone layer. But if you look at its communications, the effort that goes into the "flying" part and the fact it is sponsored by Airbus, you cannot avoid the impression that it's really about the flying.

Perlan's members have set their sights on 90,000 feet—around 27 kilometers—in a glider. If they were to pull this off they would have reached a greater altitude than all aircraft with air-breathing engines, in particular the U-2 and the SR-71. The SR-71 holds the record for sustained horizontal flight, 85,000 feet, although there have been parabolic zooms to higher altitudes by jet-powered aircraft. The F-104 Starfighter holds that record officially with 103,000 feet and unofficially with 111,000 feet. Rocket-powered vehicles have of course "flown" higher: the X-15's record is 107.8 kilometers (almost three times as high as the Starfighter), SpaceshipOne reached 112 kilometers (proudly beating the X-15) and of course the Space Shuttle, if you want to count it, has been up to 611 kilometers during STS-61 when it deployed the Hubble space telescope.

I mentioned before that above about 4,000 meters pilots use additional oxygen. If they didn't do this they would suffer from hypoxia, whose symptoms range from headaches through fatigue to nausea: not good when you are in charge of flying an aircraft. As you go higher, much more severe problems arise from expanding gas enclosed in the body, leading to compression sickness and fluids in the throat and lungs boiling away. Above 15 kilometers the lungs stop doing their job entirely, because the pressure at which they excrete CO_2 is higher than the atmospheric pressure, so nothing goes "in". To prevent this the pressure around the body has to be increased, either by wearing a pressure suit or by pressurizing the aircraft's cabin.

Perlan tried pressure suits first. Their Perlan I glider was a modified off-the-shelf aircraft, a two-seat DG-505M. They removed the engine and installed oxygen tanks and scientific instrumentation. Because the cabin was not pressurized, the crew had to wear bulky pressure suits that NASA was kind enough to lend them. You may remember my telling you about how tight my ASG-29 cockpit is. The 505 is a bit roomier, but it is still so tight that full movements of the stick while wearing a pressure suit were a challenge. The plan was to reach 50,000 feet with the DG, and on August 29, 2006 project founder and former NASA test pilot Einar Enevoldson, together with copilot Steve Fossett, indeed reached around 50,700 feet.

On to 90,000 feet then—but not with this airplane. Two things would have to be different with the new one: first, they wanted a pressurized cabin, so they wouldn't have to fly with inflated and therefore quite rigid pressure suits. They also wanted to use rebreathers, to avoid the need to carry huge amounts of oxygen for the very long flights anticipated. You can't buy a pressurized glider off-the-shelf, and adding a pressurized cabin to an existing model is not feasible because traditional fuselages and canopies cannot withstand the pressure difference between the 14,000 feet of (pressure) altitude inside the cabin and the potentially 90,000-foot pressure outside. So they had Windward Performance build the Perlan II, "Basically a spacecraft with sailplane wings", as chief pilot Jim Payne said. It has a unique forward fuselage without the large canopy known from traditional gliders. Instead it has several small round porthole-style windows, because those are easier to seal and leave more carbon structure to withstand the pressure. Of course visibility is not that great, and I asked Jim whether this is a problem, especially during takeoff and landing: "It's surprising how much you can see if you move your head side-to-side." Generally, a major challenge was to make the fuselage rigid enough to withstand the pressure but still keep the weight down—we'll see why this is crucial in a moment.

The second difference relates to flutter. As we have seen, high-altitude flight leads to very high true airspeeds: the Perlan II glider is rated to 700 km/h of true airspeed! Achieving this demands a very stiff fuselage and wing structure. One way to achieve these is to use a wing with a low aspect ratio. Perlan's 25-meter wing has an aspect ratio of 27, whereas an ASH-25 cross-country glider, also with a 25-meter wingspan, has an aspect ratio of 38, making it 40% "sleeker". The ASH-25 bends a lot, whereas the Perlan II seems to exhibit no wing bend at all.

Let's discuss the challenge of high-altitude flight a little more. Remember our original formula for lift: $L = 1/2 \cdot \rho \cdot v^2 \cdot A \cdot c_L$. We can directly see that lift decreases with lower air density, ρ. At 30 kilometers—the 90,000 ft design limit of Perlan II—air density is only one hundredth of that at sea level. The weight of the aircraft remains the same as it climbs, so we have to somehow produce the same lift despite the lower density. What can we do?

First we want to build a very light airplane, to ensure that the amount of lift needed to keep it aloft is relatively low—this is why they used the very weight-efficient design of the forward fuselage mentioned above. Second, you want to give the airplane lots of wing per unit weight—the wing loading we mentioned before. Perlan II has a wing loading of 27 kg/m^2 when empty and around 35 kg/m^2 at its maximum takeoff weight. In comparison, a modern cross-country glider like my ASG-29E starts at 41 kg/m^2 with me in the cockpit and can go up to 57 kg/m^2 when full of water. From our discussion of aircraft weight, water ballast and L/D at high speeds, we can see that this is clearly not what Perlan II is optimized for—on their flights they generally stay within 100 kilometers of their launch site, and fast and far cross-country flying is not their goal. Calculations predict that Perlan's L/D ratio is somewhere between 40 and 50 at 60,000 feet. For comparison, the ASH-25 has a ratio of around 60 with the same wingspan—but at lower altitudes, of course.

The other thing we can do to produce more lift is to fly with a higher true airspeed. Since speed goes into the formula with a power of two this a very effective way to increase lift: to compensate for 100 times less dense air we have only to increase speed by a factor of 10. If we calculate back from the 700 km/h true airspeed, then a factor of 10 would be 70 km/h. So we have to design a wing that produces its optimum lift at 70 km/h at sea level in a standard atmosphere. A few other factors also come into play here, so it's not quite a factor of 10, but it's clear that the wing has to be optimized to produce lots of lift at low (indicated) speeds. This means that, compared to a "classical" glider, Perlan II has a wing with a relatively thick and cambered cross-section. It's a low Reynolds number wing. How does this affect its handling qualities? "She actually handles much better at 60,000 feet. She doesn't roll very fast at low altitudes, but then your Nimbus 3D doesn't either", Jim says, comparing Perlan II to a traditional open-class glider from the 1980s.

Flutter really is a major concern, though, and the team uses a systematic approach to ensure there are no nasty surprises as they make their way to higher and higher altitudes and true airspeeds. They started by putting weights into the moving surfaces (aileron, elevator, rudder) to ensure that their natural frequencies are far enough away from those of either the wing or those induced by the air. They then performed ground vibration tests, in which the structure is excited with electrical motors to verify the absence of (divergent) oscillations. They even installed servos inside the wing to induce oscillations during flight, and as they stepped up the altitude, repeatedly performed in-flight tests to validate the absence of dangerous flutter phenomena. Every 5,000 feet they ran a 30-second sweep of induced frequencies between 2 and

20 Hz and recorded the wing's response with accelerometers. A computer in the cockpit applied a Fourier transform to resolve the frequency content and analyze whether the response was decaying (as required) or not. They also telemetered the data to "mission control" on the ground to perform a more in-depth analysis. This approach allowed them to see trends in the flutter behavior before it became catastrophic. This careful, systematic approach meant that they made relatively slow progress in terms of going higher, but it also meant that they would not risk their lives and their US$ 2M one-of-a-kind glider. The fact that Einar and Jim are both military test pilots certainly helped here in terms of mindset and the necessary patience.

The Perlan project has not yet reached their target altitude of 90,000 feet, but on 3 September 2017 Jim Payne and Morgan Sandercock achieved a new altitude record for gliders of 52,000 feet. I asked Jim about how it felt to fly at this altitude, hypothesizing that as a military test pilot, he might be used to it? "I was at 50,000 feet a couple of times while flying the F-16, but we couldn't go higher because we'd have been required to wear pressure suits." In a tweet I saw him quoted as saying "The biggest impression is, it's a long ways down from up here." I'm pretty sure that, as a test pilot and a life-long glider pilot, he wouldn't have expected that his highest flight would be in a glider.

I also asked Jim how he judged the likelihood that they will make it to 90,000 he replied: "It looks like if we go a bit further south from El Calafate we may have more opportunities to get high. To do that safely you'd probably have to tow a bit higher than we've been towing [to get back to the airport if they cannot catch the wave, because Perlan II is not good at penetrating strong headwinds at low altitudes], so we're working on different tow planes that might be able to give us higher tows." He didn't get more specific as to which kind of tow plane this might be and I joked about Airbus, their sponsor, being willing to spare an A320. A few months later I read on Twitter that the Perlan project had obtained a Grob 520, also known as Egrett, a high-altitude turbine-powered reconnaissance airplane built in the 80s. Wow, what a deal! In September 2018 they towed to around 13,500 meters in 62 minutes. That is an average (!) climb rate of 3.6 m/s up to 13.5 kilometers. I'm very jealous! It's also a sight to be seen, because the tow plane has a much greater wingspan (33 meters) than the glider (25 meters): ordinarily it's the other way round. I could have suspected the Egrett deal, by the way, because in the interview with the Perlan guys I learned that Einar , the project founder, was chief test pilot for the Egrett at Grob in the late 1980s. But I wasn't aware that there were still airworthy Egretts; after the fall of the Berlin Wall the need for the aircraft diminished and only six were built. It was never used operationally. Its piston-powered successor, the Strato 2C, wastes away in front of Grob's Mindelheim factory: I see it whenever I fly towards the Alps from Donzdorf.

With this new towing option the team delivered once more. Jim Payne and Tim Gardner reached 76,000 feet on September 2, 2018, breaking their own world record of 52,000 feet from the year before. In addition to beating their own altitude record, they also beat all other altitude records for subsonic aircraft. They were 2,500 feet higher than the previous record holder, Jerry Hoyt, who achieved his record in a U-2 spy plane. We'll look at this remarkable airplane in a moment.

5 | *Higher and Faster*

POLAR DIAGRAM Before we do this, though, we have to add a bit more to our understanding of wings and lift, and in particular the angle of attack, abbreviated AoA and α. This is the angle between the chord line of the wing and the direction of the oncoming air.

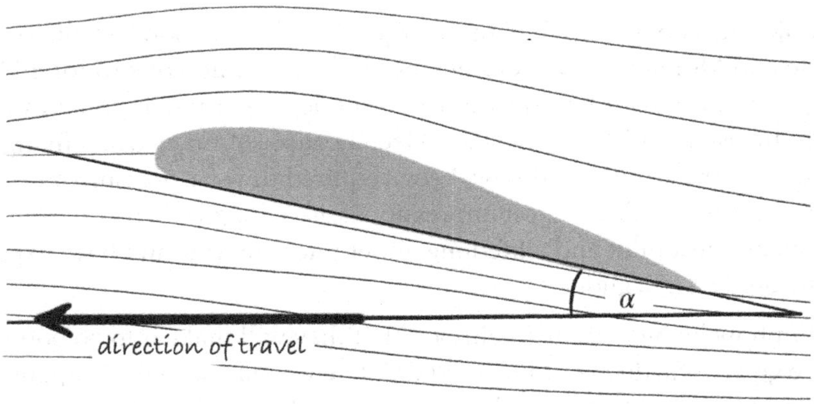

The AoA is relevant because a wing's performance, c_D and c_L, depend on it. It's quite intuitive: effectively the air "sees" a different shape, so the numbers that characterize this shape change with α. But how? This is different for every wing profile and is generally determined experimentally, or with numeric computational fluid dynamics simulations. The results are then captured in the polar diagram of a profile. It is a parametric curve, which means that it relates two quantities (c_D, c_L) to a third (α). Graphically it is shown with c_D on the horizontal axis and the c_L on the vertical axis, and the points in the diagram are labelled with the α at which the particular c_D and c_L were measured. Note that this is not related to a glider's polar curve, introduced earlier—that relates forward speed and rate of descent.

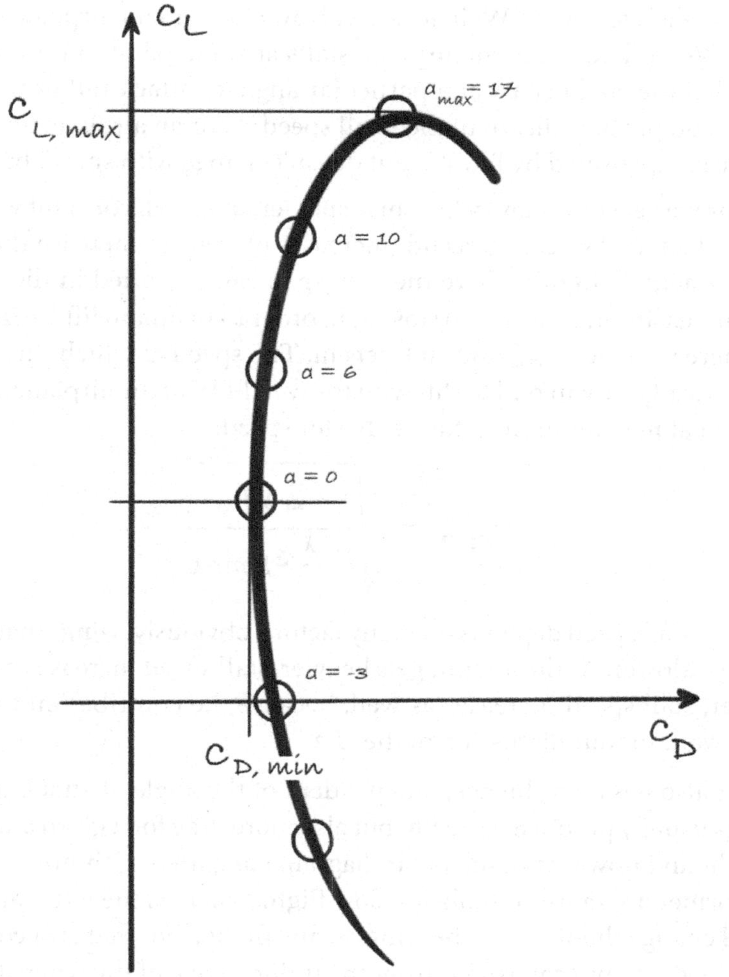

The diagram above shows a typical shape for a wing profile. We can observe a couple of characteristic features. With increasing α, both c_D and c_L increase. However, at some value of α, 17 degrees in this example, c_L drops off quickly: the wing stalls. Stall describes the condition in which the air can no longer "follow" the wing's upper surface, it doesn't have enough energy to overcome the increasing static pressure towards the trailing edge; the flow becomes turbulent and ultimately separates from the wing. Modern wing cross-sections used in gliders can cope with α up to around 15 to 20 degrees. The profile usually has its lowest drag at α= 0, but notice that it still produces lift! This is the reason why wings are cambered airfoils: a simple plate, or a symmetrical airfoil, would have no lift for α = 0. Modern profiles exhibit the "laminar bubble", a range of α values at which lift increases but drag does not (or only very slightly)—a bit of a free lunch. Discovering such wing profiles has been a major breakthrough.

Does any of this relate to speed? Well, let's see. How slow can an airplane go without falling out of the sky? We all know that an airplane stalls at some point, but in the polar diagram above we saw that the stall occurs at a particular angle of attack rather than at a particular speed. Why then do people talk about the "stall speed"? For an airplane to continue flying its weight has to be compensated by lift. Weight doesn't change with speed, but lift does: $L = 1/2 \cdot \rho \cdot v^2 \cdot A \cdot c_L$. So as we get slower and v becomes smaller, to keep L constant we have to increase one of the other factors. We can descend (increases ρ), we can increase the size of the wing (there are experimental aircraft where the wingspan can be varied in flight), but mostly we increase c_L by increasing α. So as we go slower, in order to maintain lift we increase α, right up to the point where we reach α_{max} and stall sets in. The speed at which this occurs is the stall speed. Mathematically we can replace lift with the weight W of the airplane, and cL with $c_{L,max}$, the lift coefficient at maximum AoA, then solve for speed:

$$v_{min} = \sqrt{\frac{2\,W}{\rho\, A\, c_{L,max}}}$$

As we can see, the stall speed depends on many factors: obviously, wings that continue to work at higher α can go slower. As the aircraft gets heavier, stall speed increases (we need more lift). As we go higher, stall speed increases as well, because the contribution to lift of ρ reduces. This will be relevant for our discussion of the U-2.

The shape itself also has an influence, independent of the angle of attack. Thicker and more cambered wings usually produce more lift, but also more drag for a given angle of attack. They move to the right and upwards in the polar diagram compared to thinner and less cambered wings. So the former are more suitable for slow flight, whereas the latter are better for going fast. If we could change the shape of the wing during flight... but that is of course exactly what flapped wings do: the flaps that extend from the trailing edge of the wing effectively increase the wing's camber and thus allow the aircraft to fly slower. Not all gliders have flaps, but those that do can even extend the flap upwards. Because the flap is then "in the shadow of" the wing, this reduces not just camber, but also thickness, optimizing for fast flight. My ASG-29, as well as Ohlmann's Nimbus, are designed to go fast, so they have flaps. Perlan II, in contrast, does not.

PARABOLIC FLIGHTS Before we move on to higher- and faster-flying airplanes, let's briefly look at parabolic flight because they are an interesting case that illustrates some of the issues around speed, AoA and weight. The purpose of such flights to produce weightlessness for whatever activity is going on in the cabin. To this end, the aircraft flies on a parabolic trajectory in which the centrifugal force of the arc compensates for gravity to create weightlessness. Is a parabolic flight really true weightlessness? Well, the Earth is right there, a few kilometers below the aircraft, so its gravitational force is certainly present. But Einstein's equivalence principle says that it is not (easily) possible to distinguish gravity-induced acceleration from acceleration due to inertial forces; effectively, if the vector sum of the accelerations is zero, then it's indistinguishable from feeling no force at all. This is what a parabolic flight tries to achieve, by flying a path where the centrifugal force is opposite and equal to gravity. It is exactly the same effect as on the ISS, just with a smaller radius and a shorter duration: in some sense the aircraft flies a segment of a mini orbit.

You probably know about parabolic flight from astronaut training: in these the airplane's cabin is mostly empty, allowing trainee astronauts to learn how to move around in weightlessness. Today, however, the majority of parabolic flights are used for research, and the cabin is filled with experiment racks from life sciences, materials science or physics, to name a few. The experimenters are held down by straps, there's only limited opportunity to "fly around". Novespace is the European operator of such flights, and in 2019 I was able to participate in a science campaign with the DLR. Novespace also offers these flights commercially under their brand AirZeroG; you can buy a ticket for around 6,000 EUR.

In the case of the A310 used by Novespace the parabola begins by pulling up to 50 degrees pitch. The crew then push over the top at zero g until 45 degrees nose down, then recover back to level flight. The whole maneuver lasts around a 72 seconds, with 22 seconds of weightlessness between injection and pull out. A science flight has 31 of these parabolas, separated by breaks of various lengths; the whole sequence takes around two hours.

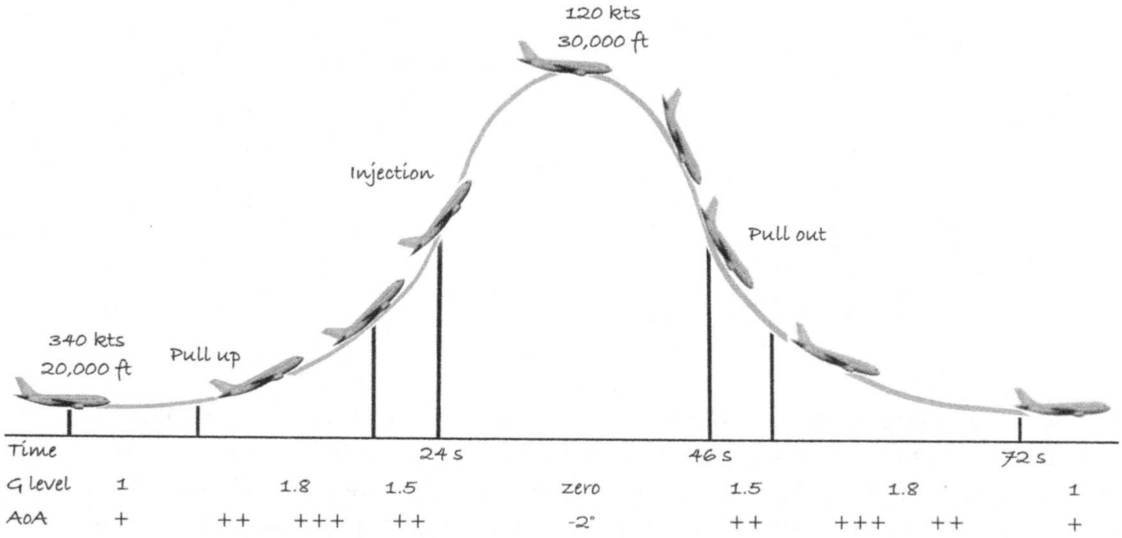

Before we look at how to actually fly such parabolas, what other approaches exist for creating weightlessness on Earth? An obvious one is to just let something fall freely in a vacuum, so that aerodynamic drag does not affect the 1 g acceleration. Drop towers do exactly that; the one operated by the University of Bremen creates up to 4.5 seconds of weightlessness. This is much shorter than a parabolic flight, and there's a huge deceleration at the end of the fall when the capsule is "caught", which an experiment has to survive. You can also shoot something vertically upwards in a vacuum, let it decelerate naturally at 1 g and then have it plummet back down. Sounding rockets use this approach, delivering up to six minutes of weightlessness. The Bremen drop tower can do that too: it has a catapult at the base of the tower that shoots an experiment straight up and thereby doubles the time of weightlessness to around nine seconds. However, neither of these approaches allow humans to participate; in an aircraft, in contrast, scientists can sit next to their experiment and monitor, operate and optimize it.

Assuming that you want to retain this benefit, what other maneuvers could an aircraft perform to create temporary weightlessness? It could go straight up and then fall straight down again (part A in the illustration below). This has two problems. First, drag will change the acceleration, so it won't be exactly 1 g, although maybe you could do something clever with the engines to compensate. Second, flying vertically up, stopping and then dropping down again is something you can do with aerobatic aircraft (in which it's a lot of fun), but not with aircraft that have a large cabin for hosting experiments, like the A310. The second option for creating weightlessness in an aircraft is to fly straight and level, then push down into an arc with zero g (part B below). In some senses you're flying the second half of the parabola but wasting the first half. You will get really fast in the vertical section because you have to start out with at least the minimum at level flight, which 180 knots for an A310, so this is not a great idea either.

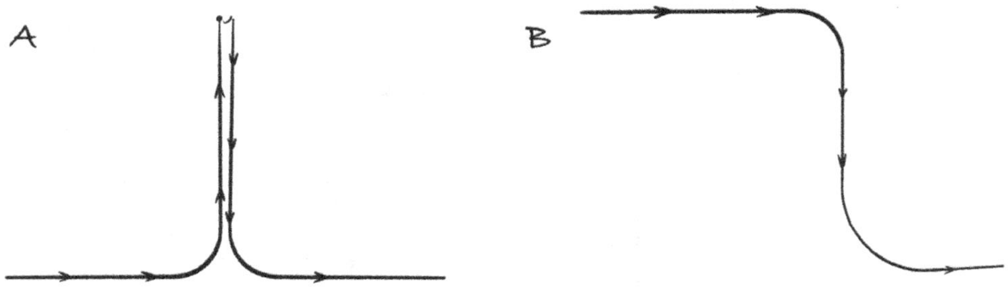

Parabolas are therefore the best option. Let's look a bit more closely at how to fly them and how this relates to some of the aerodynamics we've discussed before. The maneuver starts with the A310 flying at 340 knots indicated air speed, which corresponds to around Mach 0.84 at 20,000 feet. Since the parabolic trajectory essentially converts kinetic energy into potential energy (on the way up), it makes sense to fly as fast as possible. The throttles are first set to sustain this speed, then, to start the parabola, the aircraft pulls up with 1.8 g while the throttles are moved to full power so as to lose as little speed as possible "up the hill". In principle, you want to pull up as quickly as possible, because the longer it takes to get to 50 degrees, the more airspeed you lose. 1.8 g is a compromise between several factors. First and most obviously, you don't want to over-stress the airplane. Transport aircraft must be designed and certificated up to 2.5 g. While there is some margin left, it's advisable not to go up to the limit, because atmospheric turbulence might add additional loads while you're pulling the gs. Angle of attack is another issue. If you pull up too hard, the power provided by the engines won't be able to overcome the increasing drag and eventually the aircraft will stall. In fact the last 10 degrees of pitch are flown only at 1.5 g, because at the corresponding speed and altitude the wing would enter buffeting at the AoA required for 1.8 g. The final issue is passenger comfort: up to 40 scientists have to sustain the g focus without getting sick, 31 times a flight.

The precise values of initial speed and altitude, g level and pitch angle at injection depend on the aircraft. Jean-Francois Clervoy, founding engineer of Novespace and three-time Shuttle astronaut, wrote a program as part of his test pilot school thesis that figures out the optimal parameters for a given airplane based on its performance characteristics. For example, the A300 they used before the current A310 would only pitch up to 47 degrees, while the Caravelle, the airplane used initially, started at 25,000 ft instead of 20,000 for the A310.

When the aircraft reaches 50 degree pitch the crew push the nose over until the aircraft is at zero g. You don't want the wing to produce lift at this point, so zero g corresponds to minus two degrees AoA. The thrust is reduced to almost idle, just enough to overcome the drag of the aircraft, which decreases as it slows. Up to this point the crew use a (human) control loop taking g level and pitch angle as the input. During the zero g phase, the crew controls exclusively for g level—pitch is only relevant in the sense that you need to pull out at some point. Novespace guarantees between −0.05 and +0.05 g during that phase, with 95% of the time spent between −0.02 and 0.02 g on average. The quality of the weightlessness is critical for the success of the onboard science: it should be as close as possible to zero, and any fluctuations should be minimal. Experiments are bolted to the floor of the airplane, so every deviation from a perfectly parabolic trajectory will affect them directly. In contrast, people floating freely in the cabin will not notice the aircraft "wobbling" around them with a few hundredths of a g. This is why pilot proficiency training is done during public "fun" flights. As an aside, Novespace did trials with experiments that floated freely in the cabin, using a camera to show their position to the pilot so that he could control around this "sensor". This worked well for the one floating experiment, but the potentially more abrupt control inputs were suboptimal for those experiments bolted to the floor. Novespace then decided on good quality for all experiments instead of perfect quality for one and degraded quality for the rest.

At the top of the parabola, at which the aircraft flies horizontally for a short moment, the speed is only around 130 knots indicated. This is 50 knots below the above-mentioned stall speed of the aircraft in its clean configuration! Why is this possible and not dangerous? Because the aircraft is on a parabolic trajectory, it is effectively weightless, the wing does not have to "carry" the aircraft. The wing is at −2 degrees AoA, the angle at which it does not produce any lift. No lift means no pressure differential and no air failing to "climb up" the pressure gradient at the trailing edge on top of the wing. Hence there's no stall: the aircraft just "falls" over the top of the parabola. This is the difference to the alternative trajectory B shown above, in which the aircraft would have to fly above 180 knots because it is not weightless during the horizontal portion. What then determines the minimum speed at the top, and why is it not zero? Remember that, because of drag, the aircraft is not actually on a ballistic trajectory: it has to control its parabolic path actively. So effectiveness of the control surfaces has to be retained, in particular the elevator, as it does have to produce lift! This requires an aircraft-specific minimum speed.

Once the aircraft reaches −45 degrees pitch the crew start the pull-out, first at 1.5 g and then 1.8, for the reasons explained above. If they get too fast during pull-out they might also pull harder and use up their 0.7 g margin to the design limit of 2.5 g. It's generally better to over-g than to overspeed, as they only have a 20 knot margin for speed after pull-out. There are also speed checks as the nose drops through −30 degree pitch, so they can pull out earlier if the speed is higher than planned at this point.

To be able to fly a precise parabola, the cockpit is fitted with an additional instrument, a screen that shows two g scales, one between −1 and 2.5 and a higher-resolution scale between −0.1 and 0.1 g. The indicator also shows the AoA and highlights the approaching buffet zone so that the crew can reduce to 1.5 g. The pilots focus on this screen and the pitch indicator during the maneuver. As a beneficial side effect, staring at the screen means that the pilots don't move their heads during the high-g phases, which is the most important trick for avoiding motion sickness.

The parabolas are flown by a crew of three, with all three making control inputs. One pilot exclusively operates the elevator, and is ultimately responsible for the quality of the parabola by precisely following a zero-g trajectory as indicated by the high-resolution g scale. He also handles ATC communication, because this happens only between parabolas. The second pilot keeps the airplane's wings level. Because this is considerably less work than controlling the elevator, he's also in charge of the overall flight path and communication with the cabin. The third pilot manages the engines. As the aircraft reduces its speed along the parabolic trajectory, the elevator must be deflected more to produce the same pitching moment and keep the parabola going. This, in turn, increases the elevator's drag, so the engines must spool up a little: active adjustment of the throttles is required throughout the parabola.

The procedure is flown at neutral trim. Together with the position of the center of gravity used in these flights, which is restricted compared to that allowed for an A310 in regular service, this means that the pilot has to pull on the yoke quite hard during pull-up and pull-out. On the other hand, neutral trim ensures that not much control force is required in the zero g phase, which makes it easier to control precisely around zero. The A310 has a mechanical flight control system with hydraulic actuators, so any resistance to inputs the pilot feels is artificial. The crew turns off the artificial feel system for pitch, because you couldn't pull to 50-degree pitch while working against the feel system. But this means that over-controlling the aircraft becomes easy, so very deliberate flying is crucial. Remember that the crew is also exposed to repeated changes of g between 1.8 and zero. Eric Delesalle, chief pilot of Novespace, told me that "It's actually quite physically exhausting, you can feel it at the end of a week of three flight days." To reduce the strain the crew of three rotate their roles during the sequence of 31 parabolas. A fourth pilot is on board so that everyone can take a break.

There are several additional peculiarities. The pilot who controls the elevator attaches a special extension to the yoke that isn't connected to the ailerons; he can only push and pull, making it impossible for him to roll the aircraft either intentionally or accidentally. This contraption is shown in part A of the next illustration. Conversely, the pilot who controls the ailerons uses two rubber bands attached to the roll handles of the yoke so as not influence pitch (part B). The reason why this works is that the yoke's pitch axis is at its base, but the axis for roll is at the top. By attaching a special set of levers to the stem and thus bypassing the roll axis, it becomes possible to isolate pitch actuation from roll actuation.

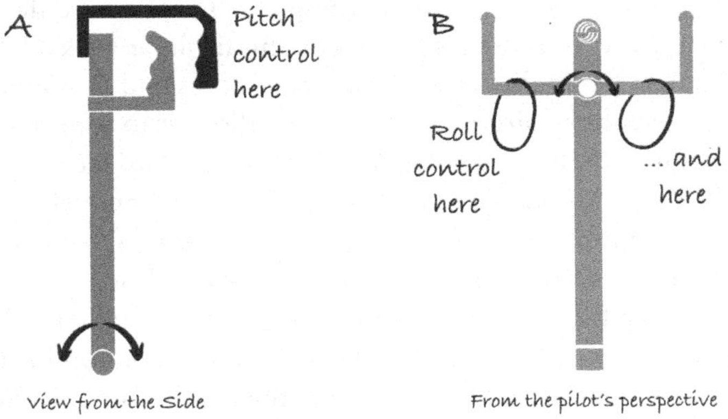

The crew darkens the side windows of the cockpit to make sure no reflections or other disturbances distract their focus from performing a perfect parabola. They all wear baseball caps and align their visors with the top of the instrument panel. Effectively the crew tries to isolate themselves from any external visual impressions. A dedicated controller on the ground ensures their separation from other traffic.

Flying precise parabolas is not an easy task—I tried many times with my glider. Although the Airbus has a much higher wing loading, and is thus less susceptible to turbulence, Novespace cockpit videos clearly show all three pilots concentrating intensely on the precision-g instrument and making very deliberate control inputs. Most of the pilots are former or current test pilots. For example, Eric Delesalle started his career as a military pilot flying the Mirage F-1. He eventually graduated from the French EPNER test pilot school and flew most of the other French military jets of the time, including the prototype of the Rafale. Since leaving the military, his main job has been chief test pilot for ATR, the joint venture between Airbus and Leonardo that builds the ATR 42 and ATR 72 turboprop regional airliners.

Novespace's A310, registration F-WNOV, was the German government's former VIP transport "Konrad Adenauer". It is unmodified apart from the special g indicator in the cockpit and a couple of recorders that track the quality of the parabolas: all components even of the regular A310 can sustain zero g for up to a minute. It nevertheless took six years for studies, paperwork and ops procedures to mature, carried out jointly with Airbus and Lufthansa Technik, before it was ready for its special role. Compared to a routine commercial flight from A to B, the aircraft is stressed more heavily during a flight that includes 31 parabolas: changing g levels, rapid change of altitude and the corresponding pressure and temperature differential between 20,000 and 30,000 feet, engines spooling up and down. A correspondingly tighter maintenance program is therefore required. Novespace treats a single parabola as equivalent to one entire regular airline-style flight. They fly around 30 flights per year, a total of around 800 parabolas (flights for the general public only have 15). The rest of the year is effectively used for preemptive maintenance because of the high "flight" rate.

You're probably waiting for my excited description of how cool it is to be weightless, and of course the plan was for me to fly as part of the campaign. On the third day, Thursday, we got our anti-sickness meds, put on our flight suits, got on the airplane, took off, flew to the airspace reserved for us ... and then heard this announcement: "Ladies and gentlemen, we have discovered a mechanical issue with the aircraft that does not allow us to continue this flight. It is not dangerous, but we cannot fly the parabolas. We'll return to Bordeaux. We are very sorry for the inconvenience." Everybody was disappointed and it was completely quiet in the cabin on the way back, in stark contrast to the journey out, when everyone was audibly excited. It turned out that the windscreen had cracked. Novespace was able to track down a replacement within a few hours and the flight was rescheduled for the following Sunday. Most people rebooked their flights home and extended their hotel bookings, but on Friday morning we all received the dreaded SMS that said that the Sunday flight was also cancelled: more problems had been discovered during the removal of the defective windscreen. This is only the second time in 20 years that Novespace has had to cancel a science flight. The scientists and I were hit really hard by Murphy this time. I hope I'll get another chance to fly with Novespace in the future.

THE U-2 In the early 1950s the CIA wanted an airplane that could overfly the Soviet Union to collect photographic images of their military installations and "scoop up" electronic signals. Remember that this was before the dawn of the satellite era—Sputnik launched only years later, in 1957. Clarence "Kelly" Johnson, chief engineer and all-around guru at the Skunk Works, Lockheed's boutique factory for special and often secret aircraft programmes, proposed an aircraft for this mission in 1953, and in 1955 the U-2A made its maiden flight. The key requirement was to be able to fly so high that Soviet surface-to-air missiles could not shoot it down. The U-2 delivered for years, until its luck finally ran out and one was shot down in 1960, with pilot Francis Gary Powers at the stick (actually, strangely, a yoke).

The U-2's altitude record stands at 73,000 feet. To be able to fly at this altitude it very much resembles a glider with a jet engine. The current U-2R/S model has a wingspan of almost 32 meters and an aspect ratio of 11, not quite as sleek as a glider. It also operates at a much higher wing loading: between 78 and 193 kg/m^2. Depending on the pods and antennae attached, it does however have an L/D ratio of 25, almost as good as the L/D of 27 of the ASK-13 training glider in which I learned to fly. Despite the advent of satellites and unmanned aerial vehicles such as the RQ-4 Global Hawk and the MQ-9 Predator, the U-2 is still going strong today. With its air-to-air refueling capability and minimal ground support infrastructure it can reach any crisis point on earth, even though today they are only used as stand-off sensors, or above areas where no ground-to-air threat exists. Flying at high altitude is still very useful, because it allows the airplane's sensors to look over the horizon. Over the years the U-2 has acquired larger wings, stronger engines and a modern glass cockpit. All kinds of sensors and antennae have been installed in its replaceable nose and wing pods and at various other locations on the fuselage. U-2s were never armed in any way: "We had a patch that said 'alone, unarmed and unafraid'", former U-2 pilot Carl LaRue told me in our interview.

So what determines the maximum altitude of the U-2? As it turns out, it's all about speed. We have seen that the minimum speed at which an aircraft can fly is governed by the maximum angle of attack its wing can tolerate. We've also seen in the case of the Nimbus 4DM glider that top speed is limited by the increased risk of flutter. Are there other limitations?

We've not really discussed how a wing produces lift. There are, in fact, several explanations, some simpler than others—and the simpler ones are also mostly wrong. However, what they all have in common is this: the air that flows over the top of the wing is accelerated, which leads to a reduction of the static pressure above the wing, creating lift. The particular shape of the wing determines by how much, but the speed of the airflow on top of the wing is certainly significantly faster than the overall airspeed of the aircraft. If and when the accelerated air approaches the speed of sound, we have a problem.

Let's understand how supersonic flight is fundamentally different from subsonic flight. In subsonic flight, as the aircraft moves through the air it "pushes away" the air particles (yes, I know there are no actual air particles, but the model works for now). The aircraft essentially creates a bow wave, just like a ship. Sound is composed of waves too, and the notion of "speed of sound" means the speed at which sound waves propagate through a medium. It is constant for a particular medium, although, as we saw in the previous chapter when we discussed sonars, it varies slightly with parameters such as temperature or pressure. For air the speed of sound effectively only depends on the temperature. The Mach number represents the ratio between the aircraft's current speed and the speed of sound: $M = v / c$.

So the speed of sound is the maximum speed at which the bow wave can be pushed away by a moving aircraft. If the aircraft goes faster than the speed of sound, the particles cannot move away fast enough, and are instead compressed. This has many consequences. For example, it takes more energy to push through this increased pressure, causing friction and thus heating to increase. Also, the bow wave morphs into a shock wave, which extends conically from whatever edge moves through the air. When the shockwave reaches the ground and the sudden increase in pressure reaches our ears, we perceive it as a sonic boom.

Back to our wing and its acceleration of the air. The aircraft speed at which the wing accelerates the air above its upper surface to supersonic speeds is called the critical Mach number, M_{crit}. Various smaller shockwaves form on the wing. If they hit other parts of the wing it leads to vibration of the aircraft; if they hit control surfaces they can lead to a controllability issue. The shockwaves can also disturb the delicate boundary layer on the wing and cause it to separate, creating a condition similar to a stall; this is why the phenomenon is also called high-speed stall. In any case, the shockwave-induced turbulence results in increased drag. For aircraft like Perlan II that drag is a problem. Jim Payne: "We are limited to Mach 0.62, because this is when we get supersonic flow over parts of the wing. The shockwave basically causes the drag to increase, and unless lift is very strong, you stop climbing." The fact that at Mach 0.62 parts of the wing's airflow is supersonic means that the wing accelerates the air quite significantly. This is not so surprising: the wing is intended to create high lift at relatively low speeds. High lift means a high pressure difference between the wing's upper and lower surfaces, which in turn requires that the air is accelerated a lot. For the U-2 the engine might help to overcome the drag, but the Mach buffet still leads to issues with structural stability and controllability and might also induce flutter. You therefore want to avoid supersonic flow over the wing even if you have an engine that could push you faster. So an aircraft's "never exceed" speed is given either by the speed up to which it has been demonstrated not to flutter, or by the critical Mach number, whichever is lower.

We've already seen that the tendency for flutter increases with altitude, because the reduced density provides less damping of oscillations. How does the Mach number change with altitude? As we have already said, the Mach number is affected mostly by temperature, plus a few chemical constants that characterize air: the colder, the slower. Since the atmosphere (usually) becomes colder with altitude, $M = 1.0$ occurs at lower true airspeeds higher up. To summarize: as we climb, our true airspeed increases for a given indicated speed, and at the same time the speed of sound decreases. At some point the true airspeed hits M_{crit} and we have to stop accelerating to avoid Mach buffet. So that's our maximum speed. What about the minimum speed? We saw in the previous section that, in terms of true airspeed, the minimum speed increases with altitude because density becomes lower, so, assuming we are at α_{max}, we have to compensate with a higher (true) airspeed. So the maximum speed becomes lower and the minimum speed increases; where the two meet, that's the end of our climb.

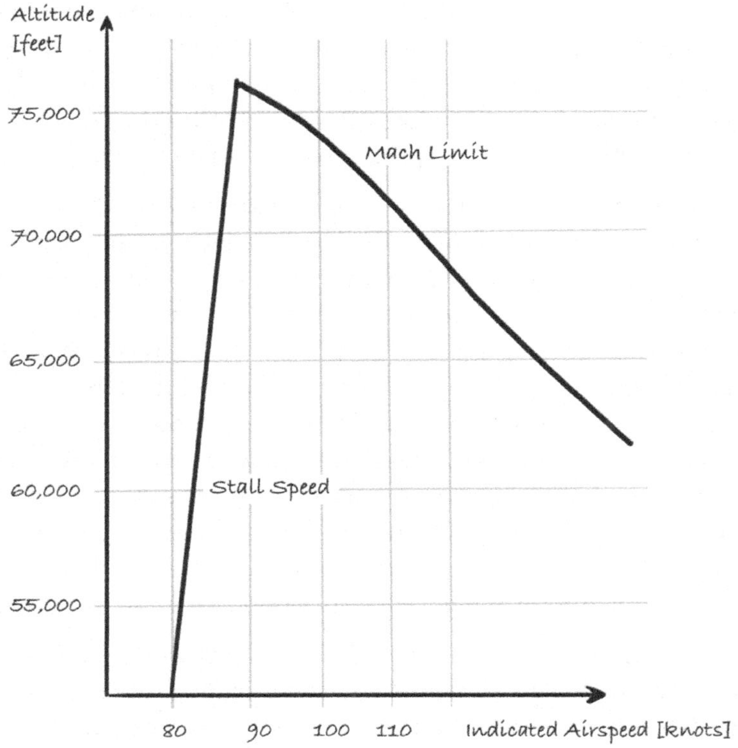

The corner created by the two lines in this U-2-specific diagram is called the "coffin corner". To fly as high as possible, the pilots move into that corner as snugly as possible. Carl: "In the C model, we only had 5 knots between v_{min} and v_{max}." This was only feasible because there was no turbulence at all at the altitudes they flew. Turbulence of course leads to spontaneous acceleration or deceleration, potentially making the aircraft penetrating either limit. "Nothing—absolutely smooth", Carl remembers. At 21 kilometers they were above the weather, which plays out primarily in the troposphere, the lower 11 kilometers of the atmosphere. What made flying in the U-2's coffin corner particularly tricky was that "Both [low speed and high speed] buffets felt the same, so you didn't know which way to correct." The U-2 relied on its autopilot: "It was impossible to fly that precisely at altitude manually. If we ever lost the autopilot for any reason, it was a mission abort." What the autopilot did at high altitude was to maintain 0.7 Mach, just below the critical Mach number for the U-2. As the aircraft burned fuel and became lighter (but produced the same lift) it would "cruise climb" very slightly over time. But because the autopilot controlled for Mach it automatically stayed right in the middle of the admissible speed band. The altitude is the free parameter and automatically adjusted accordingly. A cruise climb is the most efficient way to cruise. Airliners do something similar, but they climb in steps, to ease aircraft-to-aircraft separation managed by air traffic control. This is not a concern at the U-2's altitude, of course.

You must also be careful with turns and other maneuvers, anything that loads the airplane. As we know, when you fly a turn the airplane feels a centrifugal force; at 30 degrees bank angle a coordinated turn leads to 1.2 g, 45 degrees to 1.4 g and 60 to 2 g. The stall speed increases along with the g-load, because effectively the airplane gets heavier, so you need more lift. If you are only a few knots above your stall speed, you need to be careful with pulling g.

The five-knot coffin corner on typical U-2 missions is extreme, an attempt to tease out every inch of altitude. But other aircraft also make do with relatively narrow bands. SOFIA, for example, also needs to fly high to get over the atmosphere's water vapor. Ace Beall reports that "At the beginning of a mission when we are still heavy [and stall speed is higher], we often have a coffin corner of only around 10 knots." And even Perlan II is designed to operate with about a 15-knot margin; the lack of an autopilot requires a slightly larger tolerance. The phenomenon is also relevant for passenger aircraft. Steffen, our A380 pilot and former 737 captain, reports: "With the not-so-sophisticated wings of the 737, when you crossed the yellow bar you noticed after 5 knots or so how it got louder and the aircraft started to shake slightly. Fuel consumption increased significantly." The reason why airliners often descend a little when they enter turbulence can be explained by the coffin corner as well: as they get lower, the speed band becomes wider, and turbulence does not lead to high or low speed buffet.

So what can we do to improve the situation and allow the U-2 or a potential successor to fly higher? What about a more powerful engine? Let's first understand what an engine does. In a glider, we have to overcome the drag by sacrificing potential energy, so we always fly on a downward slope; a sustained climb is only possible if the air in which we fly is rising. An aircraft with an engine can use the thrust produced by the engine to overcome drag, so it can fly in a straight line and no downward slope is necessary. If we have excess power, we can use it to compensate for part of the weight of the airplane and climb. Importantly, we can also use the thrust to overcome the increasing drag that comes with a higher angle of attack. So, going back to our discussion about whether a more powerful engine can help us to fly higher: yes, to the degree that we can use the additional thrust to sustain flight at a higher AoA, with its increased drag. In the discussion above we assumed though that we're limited by the maximum α of our wing, so more thrust doesn't do any good, because we cannot increase α any further. If the angle of attack we can sustain were limited by drag (and the wing could do a higher α), then additional engine power would help.

A more powerful engine of course helps at lower altitudes, because it allows us to climb faster. When I was in Palmdale for my flights with SOFIA I saw one of NASA's ER-2s, a variant of the U-2, climb out at a 45-degree angle. One is used to such performance from fighters, but for the U-2, with its long wings, it was a very impressive sight. Carl: "At Davis Monthan Air Force Base once, with a clean airplane [no external payload], I've been at 20,000 feet at the end of the [4,150 meter long] runway."

What about a higher-performance wing, one that produces more lift for a given speed? This helps in principle, assuming you can find suitable wing profiles. But you have to make sure that it does not accelerate the air more on its upper side, because then you just lower the critical Mach number, meaning that you run into buffet earlier. Higher performance wings also often come at the cost of a higher stall speed; this moves the stall line in the altitude/airspeed graph to the right and thus lowers the altitude of the coffin corner as well, so is not very useful either.

What about a bigger wing, so that you can produce more absolute lift from the same aerodynamic profile properties? Yes, that helps, but you have to take structural considerations into account: the wing must still be robust enough to deal with the expected turbulence, and it must exhibit the necessary bending and torsion stiffness to not run into flutter issues. Increasing stiffness adds to aircraft weight, which in the context of going high is also counterproductive, because it requires more lift.

And the U-2s wing is already pretty impressive. At the end of its mission, when its fuel is depleted and the aircraft is comparatively light, even with the engine in idle it does not descend! Pilots have to extend the speed brakes and the landing gear to produce enough drag to convince the airplane to descend. It takes around 30 minutes to get down from 70,000 feet, a descent rate of 11 m/s.

The high-lift wing, together with the ground effect, also makes landing a challenge. As Carl explained, "The trick was to land with both wheels simultaneously, and as smoothly as possible, to avoid bounce. It was easy to get into a pilot-induced oscillation, especially if you landed on the main wheel first." Basically you fly the U-2 down to the runway at just above the stall speed, "and at two feet you'd add a little back pressure and hold it off and stall it right onto the runway." Rumor has it that the U-2's nickname, "Dragon Lady", comes from the landing experience. Because it is so hard to judge height over the runway precisely from the cockpit, U-2 pilots use what's called the Mobile. Another pilot waits at the end of the runway in a high-power sports car, and as the landing U-2 crosses the threshold they accelerate to catch up with the airplane. The pilot in the car then uses a radio to call out the precise height to ensure a safe landing. I saw the 2006 Dodge Charger that NASA uses for chasing their ER-2 when I was in Palmdale. It had little vertical tails attached. The guys said this improves high-speed stability. I suspected that their primary purpose was to look cool.

Because of the U-2's tandem landing gear, pilots had to ensure that any remaining fuel was balanced between the wings, otherwise they would not be able to keep the wings level as they slowed down on the runway. I heard that ground loops weren't uncommon because if, after touchdown, the pilot didn't keep the wings completely level, the fuel moved to one side, further lowering that wing. As they come to a standstill, one wing drops just like with a glider, and ground crew attach two little pogo wheels to allow the U-2 to taxi off the runway without dragging one wingtip on the ground. I've seen pictures of only one pogo wheel attached, with a crew member lying on the wing to keep the other one off the ground when the taxiways were too narrow for both wheels.

There is one thing you might do to lessen the problem of Mach buffet and allow an aircraft to fly higher at higher Mach numbers: a supercritical wing. Al Bowers, NASA Armstrong's former chief scientist, was involved in a NASA experiment in the 1960s that attached a transport aircraft-style supercritical wing to an F-8 Crusader fighter jet. "We chose the F-8 because one was available, and because it had a shoulder-mounted wing that made it relatively easy to replace." The key insight by NASA aerodynamicist Dick Whitcomb was that if you make the wing flatter at the top, the air is not accelerated as much and Mach buffet is reduced.

The experiments with the modified F-8 showed that it works; Mach buffet onset occurred only at Mach 0.98! In fact, with its new wing the aircraft could go supersonic without an afterburner, something known as "supercruising" and only ever achieved in a mass-produced military aircraft in the late 1990s by the F-22. Today essentially all transport-category aircraft use this kind of wing design. When I asked Al about potential drawbacks, and in particular whether these profiles produce as much lift as classical ones, he said "There is a slight penalty in lift from the flatter top, but the increased curvature at the leading and trailing edge compensates for most of it." However, he added, "They have a lower maximum AoA, but that is not a problem for transport-category aircraft." The U-2 does not have a supercritical wing. For one thing it was designed in the 1950s, a decade before the NASA experiments with the F-8. It's also not absolutely clear (to me at least) whether it would be useful. While you might be able to go faster because of delayed Mach buffet, the reduced total lift and the reduced maximum angle of attack might thwart the benefits. In other words, supercritical wings are useful for fast, lower-level flight (like the 30,000 or 40,000 feet of airliners), but not necessarily for the U-2's high altitude requirement.

As you have seen, wing and aircraft performance depends on many factors, interacting in intricate and often nonlinear ways. Aircraft design is a complex process, with lots of tradeoffs and lots of experiments or numerical simulations.

I listened to the episode with Carl again when I wrote this and came across this passage: "We had a configuration, partial flaps if I remember correctly, where we could simulate a dead engine. And we experimented with this in the ridge and wave lift in Arizona. We did indeed climb." And a bit later: "I was at a record ceremony once with Steve Fossett [who later became involved with Perlan], and I talked to him about my experiences as a U-2 pilot, and the mountain waves in the mid-70,000s both in the US and on the lee side of the South American Andes." QED Perlan!

THE SR-71 It became clear, particularly because of the shooting down of Gary Powers' U-2, that another approach was required for signals intelligence. Satellites were becoming available, but they had their own challenges regarding photographic reconnaissance. To get high-resolution pictures, physics dictates large optics, which in turn requires large and heavy satellites, but they could not be launched with the boosters of the time. In addition, making sure that a satellite is at the right place at the right time is cumbersome. There are two options: you either have to wait until the satellite's high-inclination orbit naturally brings it over your area of interest, which could take days or weeks (and then only for a brief time). Of course this implies predictable overflight times, so your targets might be hidden; during the "black" operations in the 1980s, the F-117 were always moved into shelters during known Soviet satellite passes. Alternatively, you can also change the satellite's orbit. However, changing the orbit requires lots of fuel because of the huge velocities involved, which in turn requires bigger satellites. Finally, this was a time before you could easily transmit collected images down by datalink. Early satellites used wet film photography, then dropped little capsules with exposed film into the atmosphere, to be snatched out of the air by military transports such as the C-119 and the C-130. So spy satellites were not able to replace spy aircraft in the late 1950s and early 1960s.

On the other hand, the Soviet Union was building surface-to-air missiles that could reach higher. To avoid the threat of being shot down you had to be not just high, but also fast enough to outrun a missile. Fast the U-2 was not, so a new approach was required. Kelly Johnson got the job again, and he delivered once more with an aircraft that ultimately became the SR-71. It remains the fastest and highest-flying air-breathing aircraft to this day, cruising at around 85,000 feet and Mach 3.2. The SR-71 first flew on 22 December 1964 and the type was in service from 1968 to 1990, then from 1995 to 1997 at the request of the US Congress, and again until 1999 as a high-speed test bed for NASA. The last Habu, as it is called by its crews, flew on October 9 1999. You might well wonder why the SR-71 has been retired while the much older U-2 is flying to this day. One factor certainly was that the SR-71 cost US$ 38,000 per flight hour, compared with around US$ 7,000 for the U-2. And I am sure that this figure doesn't even take into account the tankers required for a useful SR-71 mission.

Let's take a brief look at supersonic aerodynamics. We've already mentioned that a supersonic aircraft moves through the air so fast that air particles cannot get out of its way and thus get compressed, forming shockwaves. This has consequences for the wing. In supersonic flight, wings should be as thin as possible and razor sharp at the leading edge to "cut through" the air instead of pushing it into compression. The problem here of course is that very thin wings do not create a lot of lift at low speed (and can be a challenge structurally). So you end up either with extremely high takeoff and landing speeds, or you have to use large flaps to add camber for slow flight. For example, the F-104 Starfighter, an aircraft that was radically designed for supersonic flight, also by Kelly Johnson, had a landing speed of 170 knots with flaps. If they didn't work (a contingency one has to design for) it was well over 200 knots!

The angle at which the shockwaves extend from their point of origin depends on the speed. One way to avoid them disturbing the flow over the wing is to sweep it backwards: if you make the angle of the wing sweep greater than that at which the shockwave expands, the wing never "sees" the shockwave and essentially lives in a subsonic regime. However, highly swept wings can lead to stability problems, because of how the moments move over the speed envelope; we'll discuss this later. One way to overcome this issue is through the use of variable-sweep wings, which are moved forward for slow flight and swept back for fast flight. The F-111, the Tornado and of course the famous F-14 use this design. But it has its own challenges: building a wing pivot that is strong enough to cope with the high g-loads generated by fighter jets is really challenging—and heavy.

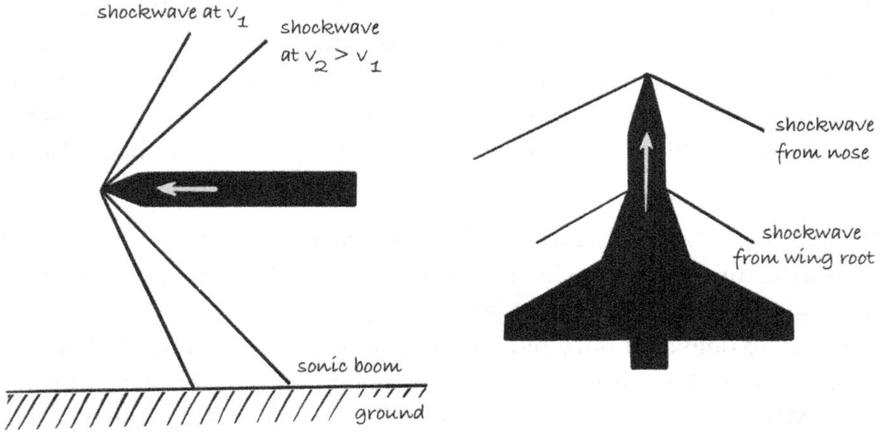

Delta wings solve some of these problems. In high supersonic flight, the shock cone from the leading edge creates an attached shockwave under the wing on which the aircraft can ride, which creates lift without additional drag. This really is key: thin air reduces drag (remember our formula), but also lift. Riding on the shockwave increases lift, enabling suitably designed aircraft to fly fast relatively efficiently. However, delta wings have a limited AoA range and develop lots of drag at high AoA, requiring a lot of thrust. Forward-fuselage strakes, or "chines", can help with this. If designed correctly, these can create vortices that inject energy into the flow over the wing at high AoA. They also improve directional stability. After a number of design studies, Kelly Johnson and his team settled on a large double-delta wing for the SR-71; since it would be optimized for sustained straight and level high-altitude flight, the limited AoA range was not a problem. The engines would have to be strong enough anyway, so using them for slow flight during landing would not be a problem.

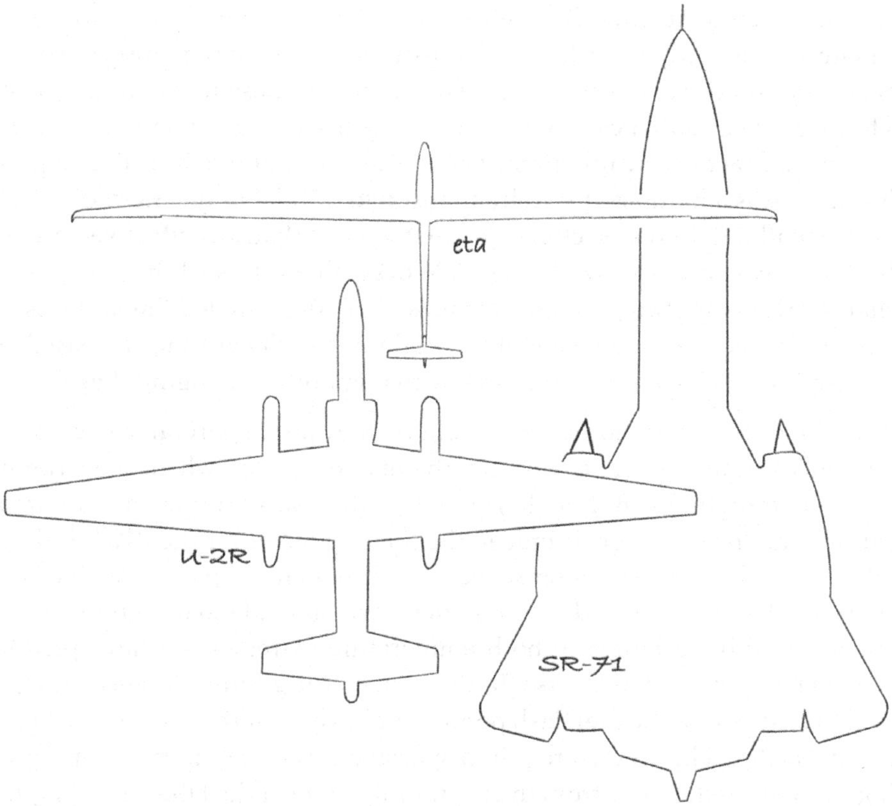

Remember Dick Whitcomb? He was the guy who did research on supersonic flight at NASA in the 1950s and came up with the supercritical airfoil we discussed before. He also discovered the area rule: for a body to have minimum supersonic (wave) drag, it should have the same frontal area along its length, or at least change that area very slowly and continuously. A long thin tube fulfills this requirement. But aircraft have wings, engines and other things protruding from the fuselage (a long thin tube), which naturally changes their frontal area. Once the rule was discovered, however, aircraft design was adapted: early area-ruled designs had a "thin waist" where the wings were attached, a shape that resembles a classic coke bottle. In later aircraft the area rule was applied more subtly. For example, it turned out that the 747 would fly faster with a longer upper deck, because the upper deck would taper out where the wings began, keeping the frontal area approximately constant. The SR-71 is area-ruled as well: the fuselage becomes thinner as the wing increases in span along the delta.

The CIA, the initial customer for the SR-71, also had a requirement for stealth. Stealth refers to reducing the radar cross-section of an aircraft: as little as possible of the incoming radar energy should be reflected back. This ensures that surveillance or targeting radars will detect the aircraft later, which is obviously advantageous in a military context. Two strategies can be used to make an aircraft stealthy. One relies on covering the airplane's surface with radar-absorbent material, basically a fine honeycomb-structured paint that absorbs some of the radar wave's energy and dissipates it as heat. The other is to design the surface in such a way that it has few right angles that can reflect the energy back to its origin (effectively forming a cat's eye reflector). Practical stealth aircraft combine both approaches. The requirement for the stealthy shape was a hard one for Kelly Johnson to fulfill, but the SR-71 did end up with a comparatively small radar cross-section. The first aircraft that was effectively invisible to radar was the F-117, also designed by the Skunk Works, albeit under Johnson's successor, Ben Rich. Its characteristically strange shape, composed of oddly-angled flat surfaces, is a consequence of the strict requirement to avoid any right angles. Thanks to increased computing power, Northrop's B-2 achieves the same goal with a smoothly changing shape.

Supersonic flight also creates another constraint on how fast an aircraft can go: friction heating. The aircraft's surface can get so hot that the materials from which the structure is built start to lose their strength. Even though 92% of the SR-71 was built from titanium and other advanced alloys, the "heat barrier" is indeed the limiting factor for the Blackbird's maximum speed. At Mach 3 the heating is so intense that its components expand, which is why its wing surfaces are beaded and corrugated. The engineers weren't able to design the fuel tank seals such that they would be leak tight at both temperature extremes, so they opted for tight at high speed, which is why the SR-71 leaks fluids when on the ground. A special fuel, JP7, had to be developed to ensure that the fuel flash point was above the temperature to which the tanks were heated at Mach 3. This in turn required special chemicals to initiate combustion. Since air refueling was part of the story from the beginning, some of the USAF's KC-135 tanker fleet had to be modified to the Q variant in order to be able to carry JP7. One thing led to another... you get the point.

Let's talk about engines. A jet engine works by the "suck, squeeze, bang, blow" principle. It ingests air at the front, compresses it (a lot) through multiple successive compressor stages, and then, when compressed and thus hot enough, injects fuel. The fuel ignites thanks to the heat, then expands out of the back of the engine which—action and reaction—propels the aircraft forward. In addition, the hot air that streams out of the combustion chamber passes through fans, making them rotate. These are connected to the compressor fans at the front of the engine through a shaft. The whole system is self-sustaining, propelled by the energy supplied by the fuel. The shaft can also drive additional fans with a much larger diameter than the compressor and engine core. These also accelerate air, but that air never enters the engine's core; there's no contact with fuel. The fan is basically a propeller with a very high rotational speed. Modern high-bypass engines such as the GE90 installed on the 777 produce 80 to 90% of their total thrust from the bypass. They aren't suitable as fighter engines because they take up too much volume and the large fans take too long to spool up, limiting agility.

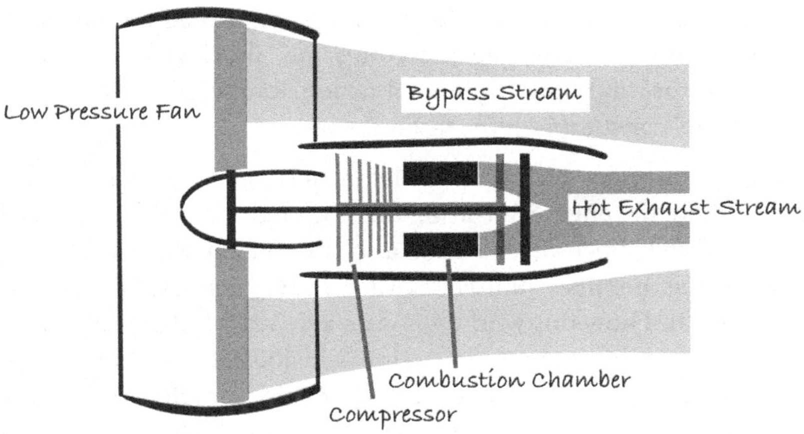

Some turbojet engines also have afterburners. These inject additional fuel into the hot engine exhaust, where, because of the heat, it also ignites and combusts. This increases thrust significantly. For example, the General Electric F110 engine installed in some variants of the F-14, F-15 and F-16 roughly doubles its thrust in full afterburner. Of course this also increases fuel consumption significantly, so most military jets are not designed to use afterburner continuously. As Nick Anderson, our Phantom guest, told me, the F-4 would run out of fuel after 20 minutes in full afterburner at low level, and a bit more at higher altitude, even with additional external fuel tanks. An afterburner is a bit like a rocket engine, in the sense that you just inject fuel into a volume where it combusts and then expands. Just as in rockets, that expansion has to be controlled to be effective. This is the job of the nozzle; ideally its diameter should increase at the same rate as the natural expansion rate of the exhaust gas. Since this rate of expansion changes with atmospheric pressure, engine performance and the amount of fuel injected into afterburner, the nozzle can be opened and closed. The nozzle is also important in non-afterburning mode, as it regulate the back-pressure seen by the engine.

The key to a jet engine therefore is that air is compressed and heated to a point at which, if you inject fuel, the whole mixture "explodes", propelling the airplane forward. The compression is normally performed by the sequence of compressor stages, but can also be achieved by other means. Enter the ramjet. Let's assume that something else propels the aircraft to above Mach 3 or so. The compressed air in front of the aircraft has a huge dynamic pressure, and therefore is hot enough to ignite fuel if you inject some into it. This is how a ramjet works: it is a jet engine with no moving parts, except possibly a movable ramp that manages pressures and shockwaves, plus a turbo pump that injects fuel into the high-pressure combustion volume. A drawback, of course, is that you need the "something" that first accelerates the aircraft to speeds at which the dynamic pressure becomes high enough for the ramjet to work.

Now we have all the information to understand the Blackbird's propulsion system: the key to making the SR-71 fast is that it combines a classical jet engine, the J58, with a ramjet. At its cruising speed of Mach 3.2 only 20% of the total thrust is created by the engine—the rest is produced by the inlet and the nozzle: the inlet routes the "shockwave" air to the nozzle, where the afterburner injects fuel for ramjet-style combustion. Remarkably, the SR-71 becomes more efficient in terms of fuel consumption the faster it goes!

No jet engine can ingest supersonic air. The inlet must shape the shockwave in such a way as to deflected it away from the engine, ensuring that the engine's compressor face only sees the subsonic air "behind" the shockwave. This is also true for the SR-71's J58 engine. However, instead of just deflecting the supersonic shockwave away, the inlets "capture" it and use it for the ramjet. To understand how this works it is crucial to realize that the two cylinder-shaped pods on the wings are not the engines; an engine is inside each pod, but the pod is larger in diameter than the engine itself. The gap between the engine and the outer diameter of the pod is where the air from the captured shockwave travels—around the engine itself—to the nozzle to supply the ramjet. So in some sense the J58 has a bypass, but it is not a classical bypass engine in which a huge fan accelerates air; instead it supplies the ramjet. The pod is also much longer than the engine. At its aft maybe 10% of the pod is the nozzle, while the forward 35% is the inlet. Its job is to capture the shockwave and route it around the engine, but inside the pod, for ramjetting. How this works is really interesting.

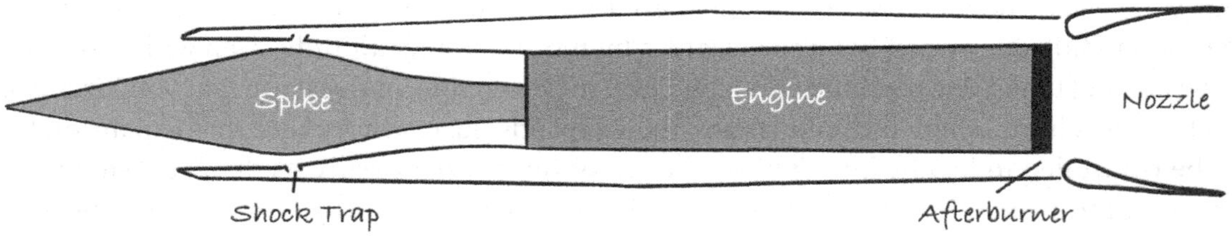

We have already learned that a supersonic shockwave extends conically from a sharp edge, and the angle at which it expands depends on the speed: the faster, the more pointed. This is why each inlet has a conical spike that projects forward: a conical supersonic shockwave expands from its tip. To capture this shockwave, it must strike the inlet just inside its lip. Over a speed range of Mach 1.6 to Mach 3.2 the spikes move backwards and forwards to achieve this capture. An onboard computer maintains the spikes at a position that ensures that the shockwave always impacts the forward edge of the inlet lip's inner surface; the Habu community calls this situation a "started" inlet. This word makes sense: we have seen that the inlet, plus the ramjet, produce around 80% of total thrust at high speed. So, just like starting an engine, you "start" the inlet.

Remember that jet engines cannot work with supersonic air; the purpose of an inlet is to ensure that engines "see" only subsonic air. Here's how the SR-71's inlet does that: within the engine pods, the spikes are shaped in such a way that, together with the shape of the inner surface of the pod, they reflect the shockwave multiple times to a point inside the inlet, where doors (called shock traps) remove the supersonic flow from the path that leads to the engines. The doors redirect the captured shockwave around the engines and directly to the nozzle through the radial gap between the engines and the pods. The subsonic air "behind" the shockwave continues down the inlets, where further expansion slows it to around Mach 0.6 so that it can be consumed by the engines.

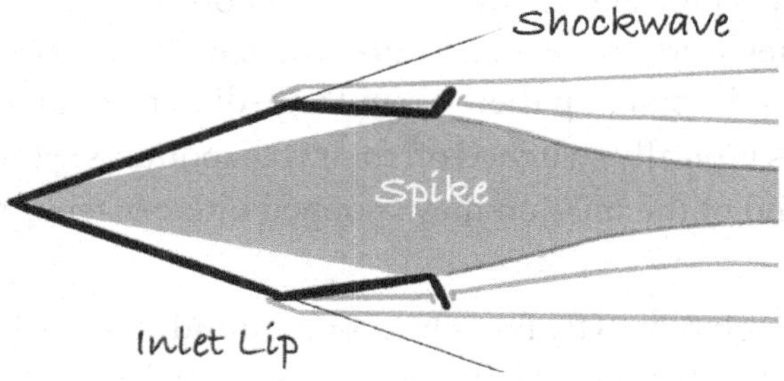

For this to work correctly, the pressure inside the inlets must be at a specific value. To fine-tune it, the inlets have two sets of doors, forward and aft (both are in the inlet, in front of the engine). Opening the front door lowers the pressure ahead of the shockwave, while opening the aft door lowers the pressure behind it. This allows the shockwave to be positioned precisely inside the inlet to hit the shock traps. The doors are operated continuously by the engine controller and sometimes manually by the pilot. There are also additional doors behind each engine, just in front of the nozzles, additional suck-in doors for operation on the ground and during takeoff, as well as boundary layer suction grids on the inside of the spikes, but we will ignore these for our discussion.

By now it should be abundantly clear that this was a very complicated system, and keeping the inlets started was a delicate process in which lots of parameters had to be precisely tuned relative to each other. If these parameters got out of whack this was called an "unstart". When I interviewed former SR-71 pilot Richard Graham he recalled: "An unstart resulted in a very violent maneuver with an extreme rolling and pitching movement in the direction of the offending engine. It feels like you're hitting a barn door. Bang!" This isn't surprising, considering that 80% of the thrust would be lost instantly and drag would immediately increase. The reason for an unstart was always some kind of glitch in the control of the spike or the doors. Turbulence could lead to an unstart as well, but luckily there is none at 80,000 feet. The engine controller usually detected an unstart and initiated an automatic restart programme that took 10–12 seconds, during which the pilot worked frantically to keep the plane aloft and on track. Sometimes the automatic procedure didn't work and one unstart followed another. In this case the SR -71 had to fly lower and slower, effectively aborting the mission. With the Blackbird's initial analog engine computer an unstart occurred every fourth or fifth mission. They became rarer with the later digital engine controller.

SR-71 missions lasted between one and eleven hours and must have been a surreal experience. Richard remembers: "The air is all yours and there's no one talking to you. You are very lonely with your back-seater there. It's all yours and you're flying halfway around the world, not talking to a soul the entire time. There's no sensation of speed; you get a great view, you see the curvature of the earth. If you look about 45 degrees above the horizon all the stars are out, even during daytime. Taking the airplane [from Beale Air Force Base, California] over to RAF Mildenhall [in the UK] via the North Pole I saw some northern lights, they are really great up there at 80,000 feet, they were so clear and pure, you feel like everything's dancing right outside your cockpit. Even the stars up there seem three-dimensional. I can't explain it but I always felt like it." Habus typically switched off their transponders—there was nobody up there anyway—and at the end of the mission they "popped into existence somewhere", reconnecting with ATC.

While the U-2 had enough fuel to fly for 12 hours or around 11,000 kilometers, the SR-71's J58's afterburners' thirst meant that most Habu missions required a lot of tanker support. Multiple refuelings were the norm, which required the Blackbird to "come down and slow down" to meet the tankers in their preplanned refueling orbits. They usually refueled directly after takeoff as well, because this allowed them to launch with lower gross weight, making a rejected takeoff less risky. There was an additional reason: to reduce the risk of fuel igniting from frictional heat, the tanks were filled with nitrogen as they emptied of fuel. Post-takeoff refueling filled the tanks completely; from then on pressurized nitrogen was injected into the tanks as fuel was consumed. If they'd done that on the ground, the pressure would just exacerbate the leaking due to loose (cold) seals.

While at least officially the SR-71 never flew over "enemy" territory, missions were often flown close to not-so-friendly areas to give side-looking cameras, radars or other sensors a good view. To avoid jeopardizing the airplane's partial stealth characteristics by actively transmitting, missions were routinely flown in complete radio silence—including during refueling. Optical signaling was routinely used for close-in coordination between a tanker and an SR-71. Missions were planned precisely in terms of waypoints, tracks and timing, and tankers loitered in agreed refueling orbits. Encrypted TACAN-style radio beacons were used to guide a Habu in from about 300 kilometer range. The tankers were visible to radar anyway, so they might as well transmit. Richard: "I can honestly tell you that through the whole history of the 22-year program that we flew the SR-71, we've never had a missed air refueling because the tankers weren't there. They've always been there, every single time. In fact a lot of them have come out of their orbits when they didn't have to. When they heard over the radio that the SR was out of gas or had an emergency, they came and got us. My hat goes off to the tanker guys."

There were no ground-based radio beacons to help with navigation, and of course no GPS at the time. The Blackbird instead relied on three techniques. First, like most airplanes and spacecraft, they carried an inertial measurement unit that measured accelerations in all three directions. Integrating twice from these values and knowing your point of origin allows you to compute your current location. This kind of integration drifts, however, because errors are integrated over time as well. This is why the Habu's primary means of navigation during the high-speed portion of a flight was a star tracker. Richard explains: "There was a compartment on top of the airplane that contained a computer about the size of half of a refrigerator. We nicknamed it R2D2 because it was smart. It controlled everything on the airplane. Inside of R2D2 there was a chronometer with an accuracy of a hundredth of a second. It also had a star catalog of the 64 brightest stars in the northern hemisphere. On top of R2D2 there was a telescope on a gimbaled platform that looked through a quartz glass window at the very top. The telescope went through an expanding rectangular search pattern looking for three of the brightest 64 stars to lock on to. With three locked stars we were good to go as far as accuracy, it kept the aircraft always within no greater than a half-mile off course the whole time at Mach 3." The third means, and ultimate backup, was manual dead-reckoning, in which pilots determine a position by keeping track of which directions they travel, for how long, and at what speed. For general aviation, at low altitudes, with wind, weather, airspace and the need to avoid other aircraft, this could be cumbersome and imprecise. At 80,000 feet, however, where none of these were a concern, Richard reported that it was pretty accurate.

The SR-71 was a remarkable airplane in terms of altitude and speed, but it was not at all agile. At Mach 3 it had rather large control forces and was very sensitive to changes in AoA or g-loading. An autopilot did the flying, just like in the U-2. "It was like airliner flying." In fact the AoA had to be at a very particular value for maximum efficiency, and this had to be achieved without deflection of the SR's elevons, to avoid drag. Fuel was pumped around to maintain the center of gravity, and thus the angle of attack, within the required narrow range. Finally, Kelly Johnson paid a price for the airplane's low radar signature: it had limited directional stability, which is why a stability augmentation system was installed to remedy the flaw.

We will look at stability and control in the next section. But before we do that, do yourself a favor, put the book away, open Youtube and search for "SR-71 speed check brian shul" and get a final impression of the SR-71 and its crews.

6 | *Stability And Control*

STATIC AND DYNAMIC STABILITY Let's move on to stability and control. To get started, we have to establish the concept of stability in an aerodynamic sense. You've probably all seen these pictures that illustrate the notion of stability:

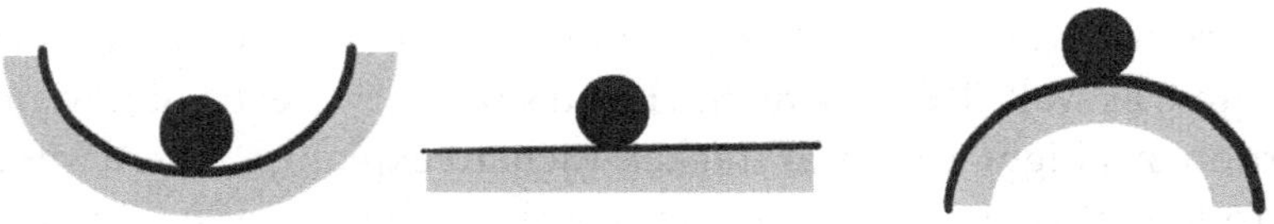

Starting on the right, if we move the ball very slightly in either direction it will accelerate (forever). A disturbance amplifies. In the indifferent (also called neutral) case, the ball, once accelerated, will continue at that speed forever as long as we ignore friction. In the stable case the disturbance will dampen; it will reduce to zero, in this particular case after oscillating in either direction a couple of times. Stability is not black and white: the degree of instability of a system varies; if you change the radius of the surfaces, the "acceleration" changes accordingly. Similarly, damping can be different. If you increase friction, the unstable case will accelerate more slowly, while the stable case will oscillate less.

This notion of stability can be applied to different aspects of an aircraft. Static stability applies it to the aircraft's attitude: a statically stable aircraft returns to its neutral attitude (zero pitch, wings level) after a disturbance. Indifferent means that if, say, a turbulence tilts the nose slightly up, the aircraft will remain in this new attitude (until it runs out of airspeed and stalls). Unstable means that the pitch angle increases further after the nose is forced up. Similarly for the other two axes. Many training aircraft, such as the ASK-13 glider I learned to fly on, are statically stable. My current ASG-29 is more neutral.

Dynamic stability applies to oscillations about either of the three axis. Stable means that oscillations are damped, neutral means that they continue unchanged, and unstable (or negative) dynamic stability means that they increase in amplitude. The SR-71's directional instability is a form of dynamic instability; it means that it yawed slowly from left to right and back about its vertical axis. This was annoying for the pilots, because they had to correct continuously with the rudder. Depending on the gimbaling of photographic sensors, this might also be a problem for observations. So how can you improve directional stability? You add more vertical surface area. A larger vertical tail, winglets or the kind of strakes you can see on the aft end of the F-16 help. Of course larger surfaces create more drag, so you might not want to use this approach too much. Another alternative is to make the existing vertical tail more effective by lengthening its moment arm by making the fuselage longer, but that creates additional drag and has structural consequences. In the case of the SR-71, Kelly Johnson was certainly limited by other constraints, such as the requirement for stealth, or a particular wing sweep because of the high speeds. But let's start with simpler planes first: gliders.

GLIDER HANDLING Remember the situation in the wave? I let go of all controls and the glider would just sit there at 90 km/h, flying straight as if on rails. Because of the laminar flow there were really no disturbances at all. Under such conditions a glider will fly perfectly straight. In particular, the ASG-29, and its predecessor the ASW-27, have really good directional stability—no yawing from left to right.

It becomes more interesting when you fly in turbulent air, for example during a good thermal day. There are lots of "disturbances" from thermals—in quotes because these are what keep a glider in the air they are not really disturbances—and the question is: how does the glider react? Let's say you are flying at 140 km/h and a thermal bubble moves your right wing up, you roll left. What happens? If you don't do anything the roll rate will return to zero. You might end up in a 20-degree left bank, but the bank angle does not continue to increase. So you just use some right aileron and fly straight and level again. The same is true for the other two axes, pitch and yaw: a disturbance accelerates a rotation around that axis, but the rate decays quickly to zero. All gliders I have flown behave in this way, although with different decay rates. For example, the Ventus C requires a bit more active control, and aerobatic gliders like the Fox even more.

Now let's say we are in a 45-degree left bank, circling and climbing in a thermal, and we briefly get stronger lift under our right wing, initiating a left roll and increasing our bank angle: many gliders will not return to a zero roll rate. In fact many gliders require some degree of right aileron to maintain a stable relatively steeply banked turn. Despite this, gliders are stable enough to be controlled with the human in the cockpit providing the control algorithm. Once you acquire some degree of routine with gliders in general, and with a particular type, the control algorithm runs mostly in your spinal cord: you don't think about it consciously.

Before we move on to other aircraft for which a human control algorithm is not good enough, let's quickly revisit how you control an airplane in the first place. Let's look at the following illustration with our knowledge about how wings work:

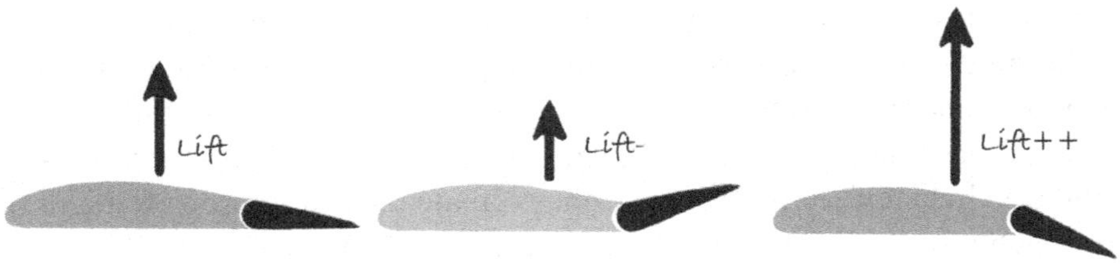

If we put a flap at the trailing edge of the wing and move it upwards, the wing's cross-section becomes thinner and less cambered. This reduces lift, all other things being equal. Conversely, if we move the flap down, we get increased lift. This is the basic principle of how you control an airplane. If you put such a flap at the very aft end of an airplane and move it upwards, the tail moves down and the nose moves up. The opposite is true if you move the flap downwards. You've invented the elevator. Airplanes also have such flaps on the outer part of wings. These are the ailerons, which rotate the airplane around its long axis. They work in opposite directions, so that one wing moves upwards and the other moves downwards. Finally, the rudder, which yaws the plane around its vertical axis, works this way as well, except that the airfoil is symmetrical, so does not create any lift when neutral, which would act as yaw for the rudder.

These control surfaces are connected to the cockpit by cables, rods cranks and ball bearings to ensure that they are light and smooth to operate. The stick actuates elevator and ailerons. Pushing the stick forwards moves the nose down; pulling it backward moves the nose up. Moving it left and right rolls the aircraft in the respective directions. The pedals actuate the rudder. So to descend (and speed up—the two are connected because you have no engine) you push the stick forwards. To climb and slow down, you pull it back. To turn left you move the stick to the left and press the left pedal.

It is instructive to compare this with how you steer a car—notice how nobody calls that "controlling a car". When you want turn left you turn the steering wheel to the left, and keep it there as long as you want to stay in the turn to change direction. When you reach your intended direction you move the wheel back to its center position. In aircraft it is different, because the stick controls the roll rate. So if you want to fly a left turn, you move the stick to the left and press the left pedal. The aircraft starts rolling to the left and also starts turning to the left. Once you have enough bank, say, 40 degrees, you move stick and pedals back to neutral. The aircraft is now in a new steady state, and will stay in this state and continue to turn forever if we ignore the disturbance and stability considerations discussed above—it basically flies a circle. If you only want to turn, say, 90 degrees left, once the nose is pointing in the desired direction you move the stick and rudder right and start rolling to the right. When your bank angle reaches zero, you move both back to neutral. In some senses you control the first derivative of the turn rate, not the turn rate itself as you do in a car.

It's similar for up and down. Let's say you fly straight ahead at 100 km/h (slightly sloping downwards, of course) and you want to accelerate to 200 km/h. You push the stick forwards and the nose moves down, let's say 25 degrees. The aircraft accelerates until the drag and the downward force from the slope are in balance. With my glider this will be way more than 200 km/h at 25 degrees, so when I reach 200 km/h I pull the stick back slightly until the acceleration stops. To stay at 200 km/h my stick remains a little further forward than when I was traveling at 100 km/h. If I now pull the stick back and keep it pulled back, the aircraft will fly a loop, because the elevator's position is proportional to the rate of rotation around the lateral axis.

So does the stick control rate of descend and climb, or speed? Technically both, because in a glider the two are linked. In terms of a pilot's intentions the stick is most often used to control speed, and climb/descend is more or less a consequence. During approach, for example, when I want to arrive precisely at the beginning of the runway, I use the stick to control speed and the air brakes, which basically destroy some fraction of the lift, to control the sink rate. During launch, conversely, when I am towed by an aircraft with an engine, my speed is determined by the towplane and I use the stick to ensure that I am at the same altitude. In some situations you want to control speed and altitude. For example, let's say you want to impress your buddies on the ground by flying along the runway at 270 km/h in 20 meters (which is technically not allowed, but you see it done all the time). In this case you have to begin your push-over at an altitude where, once you've descended to 20 meters, you've accelerated to around 270 km/h. This requires a bit of practice.

STABILITY AUGMENTATION Let's return to the SR-71, which was slightly unstable in all three axes and so used a stability augmentation system for all of them. It used accelerometers to sense uncommanded accelerations around each axis, and controlled hydraulic valves that actuated the elevons (a combination of elevator and aileron common on delta winged-aircraft) and the rudders. The channel for the yaw axis took some tuning. In the context of unstarts, the system initially seems to have overcompensated for drag on one side, causing pilots to think that the unstart was on the other side, thus overcompensating the overcompensation and worsening the unstart. One SR-71 crashed because the system was somehow mis-wired. This hints at something we will revisit later: if you add automation and computers to the control of your airplane, you had better make sure their magic works reliably!

When looking at the shape of the SR-71, you would probably not be very surprised that its flight characteristics weren't that great and that it required a stability augmentation system. However, such systems are common even on airplanes that look more normal: almost every passenger aircraft has a yaw damper to reduce left/right oscillations of the nose. One reason for this is that the tail is relatively close to the wing, so the moment arm of the vertical tail is not very large and its effect on directional stability is therefore limited. Second, aircraft with a swept wing are susceptible to what's known as "Dutch rolls" at frequencies of between a few seconds and a minute. Let's say the aircraft yaws right. Because of the wing sweep, the right wing increases its sweep as seen by the air while the left wing reduces it. Thus the left wing produces more lift, causing the aircraft to slowly rotate into the direction of the turn. But it also increases drag, which is why the aircraft eventually starts to yaw to the left. Repeat. A yaw damper, just like in the SR-71, detects the onset of yaw and corrects by applying rudder before Dutch rolling can develop.

For some aircraft a yaw damper is mandatory; they are not allowed to take off if it is not working. For others, altitude or speed is limited; based on what we have already learned, it's not surprising that higher up, with less dense air, the natural damping is reduced and yaw oscillations are exacerbated. However, none of these aircraft will fall out of the sky if their yaw damper fails. A couple of passengers might want to, though, since the oscillations probably induces motion sickness for those who are susceptible. For other aircraft, however, this is different: they will definitely fall out of the sky if the flight control computers fail.

7 | *Flight Control Systems*

GENERAL DYNAMICS F-16 Let's talk about the F-16, one of my all-time favorite aircraft. The Viper was "born" in the same year as I was, 1974, and today, with over 2,000 flying and continued production, is still the world's most widely used fighter jet. Single engine, single seat, single tail, and just overall absolutely beautiful. In 2017 I had the opportunity to get to know one personally at Nellis Air Force Base. A few months earlier I received an email from one of my listeners, Lt Col Jan Stahl, which basically said "Hey Markus, I am working for the US Air Force, flying the F-16 as an aggressor at Nellis. If you ever happen to be in the area, you are welcome to visit the base, check out the airplane, and maybe record a podcast." That's the kind of listener initiative I love, and of course I made sure that I was "in the area" a few months later. Let me tell you: it's quite cool to sit in the cockpit of an F-16 while an F-15 taxis past majestically a few meters away. While I was in the US I also visited NASA Armstrong Flight Research Center at Edwards Air Force Base, the place where much of the military flight research that I was so fascinated by as a child and teenager was conducted. Hallowed ground for an "avgeek": a wonderful week. But I digress.

The F-16 is a relaxed-stability airplane: it is intentionally designed to be unstable. More specifically, it has slightly negative static stability and negative dynamic stability. This is useful for a fighter aircraft because you want it to be highly maneuverable. Say you want to fly a turn. Because of negative dynamic stability, you basically just have to start a roll "oscillation", and it will accelerate, making the roll very quick. Once you get to the desired roll angle you actively stop the oscillation. Because of its relaxed static stability, there is no natural tendency to roll back to zero degrees bank; it will stay in whatever attitude you've put it in. As we will see, the flight control system helps with this as well.

How do you make an aircraft unstable? We previously saw that a short tail leads to directional instability. Making an aircraft "tail-heavy" also helps.

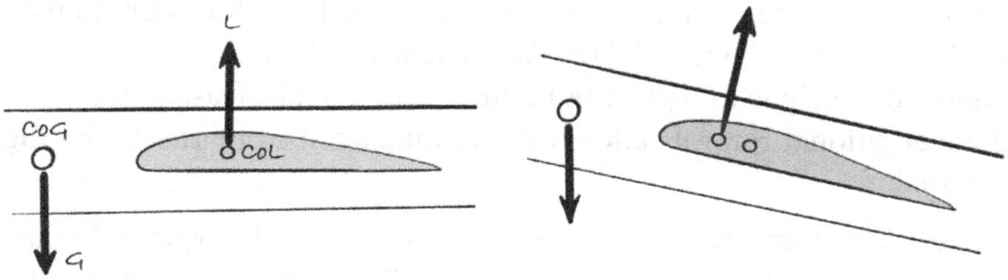

In the illustration above the left part shows a stable aircraft: the center of gravity is in front of the center of lift, the (imaginary) point at which the wing's lift connects to the aircraft. This keeps the nose down, slightly accelerating the aircraft. As the AoA increases, the center of lift moves forward, which is a fundamental characteristic of lift-generating airfoils. It's not a problem if your center of gravity is sufficiently forward, as it will remain in front of the centre of lift and work to move the nose down; this arrangement of CoG and CoL is stable: it counteracts a pitch-up movement. But if the CoG is even slightly behind the CoL, increasing AoA means that the CoL will move further forward, which, because of the increasing moment arm, leads to an even more increasing AoA. This positive feedback causes instability.

An unstable airplane, especially one that is seriously unstable, is not manually controllable by a pilot, especially at high speeds. And the real job of a fighter pilot is to focus on the tactical situation and the mission rather than on the mechanics of flying. This is where flight control computers come into the picture. Even if the pilot keeps the stick centered, the controller actively actuates the various control surfaces to prevent divergent oscillations. I read somewhere—I can't find the reference anymore, so take this with a pinch of salt—that the F-16 self-destructs from these oscillations within one second if the flight control system fails. To avoid this fate the flight control system has lots of redundancy in its computers, and is supplied by five different power sources, including batteries, the emergency power unit and the F-16's regular electrical buses.

The F-16's flight control system processes two sets of inputs: data about the current state—attitude, speed, plus various data about the airstream, such as angle of attack, side-slip angle, total pressure and total temperature—as well as the pilot's input about how he wants this state to change, as expressed through stick and rudder. It then computes appropriate deflections of the elevator, aileron, rudder and flaps to effect this change. There is no direct connection between the pilot's controls and the control surfaces—it is all mediated by a computer. A computer is fast enough to create artificial stability in an unstable aircraft.

This kind of flight control system is called fly-by-wire, because electrical wires are used instead of cables, rods and cranks to pass signals to the hydraulic actuators that drive the surfaces. But maybe it should be called fly-by-computer, because it's probably more important that there's a computer between the inputs and the surfaces to interpret the pilot's commands in light of the current flight state. The F-16 was the first operational fighter that used fly-by-wire, which contributed to its exceptional maneuverability. Today all fighters use this approach. Many of the newer ones have additional control surfaces: for example, the Eurofighter has canards, little wings at the front of the fuselage, while the F-22 has thrust-vectoring nozzles, both intended to help with high-AoA flight. It would not be feasible for the pilot to manually use these additional controls effectively: an integrated computer-based flight-control-system is essential.

The fly-by-wire system ensures that the pilot experiences a stable airplane. In fact, they experience an airplane that is much more stable than a traditional one would be, while at the same time being extremely agile. Jan recalls: "One thing that hit me about the F-16 from the get-go is how easy it is to learn and how easy it is to fly compared to any other platform that I flew previously." Jan transitioned to the F-16 from the non-fly-by-wire F-15. Of course, even in non-fly-by-wire aircraft like the Eagle, the stick and pedals are not connected directly to the control surfaces by wires, rods and cranks, because the aerodynamic forces are much too great for a pilot to be able to move the surfaces manually. As in the F-16, they are hydraulically actuated, but the valves that control the hydraulics are themselves controlled by cables, making the deflection of the control surfaces directly proportional to the pilot's input. No electrical signals and computers are involved in the primary flight controls, although there is one involved in the automatic trim system of the F-15.

In traditionally-controlled aircraft like my ASG-29 the pilot can feel the state of the aircraft to some degree. For example, when I fly faster the stick becomes heavier to move, because the aerodynamic forces required to move the surfaces become greater. As I slow down towards the minimum speed, the stick becomes "wobbly"; when I approach stall the stick starts shaking a bit, because the buffet created by the wing's separating airflow strikes the elevator. My reptile brain uses these cues a lot when controlling the aircraft, but if you have no direct "connection to the air" this feedback loop is not available. Aircraft like the F-15 therefore have artificial feedback systems, essentially springs on the stick that are preloaded with a force that is proportional to airspeed. Transport aircraft have stick shakers that indicate an approaching stall. The F-16 has no such feedback at all. Jan: "In the F-16 the buffet cues are mostly absent, so you grow very reliant on displays that precisely indicate the state the jet is in." In fact the stick did not move at all in the original F-16, it only sensed the force that the pilot applied to it. Pilots didn't like this, because, in addition to stick force, the stick's position also provides important feedback for a pilot's subconscious flight control algorithm. General Dynamics subsequently changed it and the stick now moves a few millimeters, but there is still no force feedback.

Since the computer has a say in the translation of the pilot's inputs to what the control surfaces do, it might as well prevent the pilot from doing stupid stuff, such as trying to increase AoA over what the wing is capable of, or stressing the aircraft beyond the g-limit it can tolerate. Such protections are common in all fly-by-wire aircraft. Jan: "It is much easier to perform the F-16 to the maximum capability of the flight envelope than it is a jet like the F-15, because I really don't need to think about it. All I need to do is to pull on the stick: if the computer calculates that in the current state I can get to 9 g it will get me to 9 g and it will stay there, as long as the aerodynamics permit. In the F-15 it was a much more complex process in the sense that I had to check my airspeed, and based on the airspeed I had to essentially forecast how much g is available, and if I know that g-available is higher than the maximum g the airframe can permit, I have to make sure that I'm not pulling my stick all the way back. I have to pull a little bit slowly to a more restrictive position, until my g-available becomes less than my g-allowable, at which point I can bring the stick all the way back." To help the pilot, the F-15 has an Overload Warning System, which sounds beepers in the pilot's headset to indicate what proportion of g-allowable the pilot is pulling: 91–94% is denoted by a single-rate beeper, 95–97% by a double-rate beeper, and 98–100% by a solid tone. Beyond 100% a voice annunciation system (traditionally called "Bitchin' Betty") tells the pilot "Over-G, Over-G". In my glider there are no protections: if I fly at 250 km/h and pull the stick fully back, something will break—probably the wing. So you just don't do that, just as you don't jerk the steering wheel all the way to the right at 200 km/h on a German autobahn.

We will look at the workings of a fly-by-wire system in more detail below, but before we do I want to differentiate the flight control system from other automated systems in the cockpit. It's important to realize that in fly-by-wire aircraft you still manually control the aircraft's speed, altitude, heading, bank angle or g-level. It's just that you don't fly the unstable and hard to control physical airplane, but instead a more stable, more care-free "artificial" airplane that is simulated by the flight control system.

Autopilots are different, and they have existed since well before fly-by-wire systems. The simplest forms just maintain aspects of the current attitude: altitude hold, airspeed hold, heading hold. Once you have manually put the aircraft into a particular state, the autopilot holds it for you. This is very useful if you have to fly long straight legs. The SR-71, on its Mach 3 cruise, used basically such a system. More advanced autopilots can also make changes to the flight state. For example, in modern airliners you can dial in headings, speeds, altitudes and rates of climb and descend and the aircraft will "get you there". Today's autopilots can also fly towards a radio beacon or towards a GPS location, or fly "down" an instrument approach. Since we mentioned Bitchin' Betty above, I should probably mention that the autopilot is traditionally called George.

Airliners also use what is called a flight management system. This allows pilots to plan their complete route, with multiple waypoints, speeds and altitudes, taking weather and fuel consumption into consideration. The aircraft will then fly the route automatically. The autopilot is part of the flight management system. Today, flight management systems also have various forms of datalink, such as ACARS, to allow for remote monitoring by an airline's maintenance department. Pilots interface with all this through a set of displays, keyboard and the like. In Airbus aircraft, for example, this system is called the Multi-function Control and Display Unit. It also provides access to "systems pages" that allow pilots to monitor and control many of the other systems on a large aircraft, such as fuel—which might have to be pumped between tanks during a flight to manage the center of gravity—the electrical systems or the hydraulics.

Military aircraft have systems that are similar to a flight management system, usually called a Mission Management System. This also takes care of weapons and other stores, integrates the various sensors (typically radar or infrared) and supports the pilot in aiming and releasing weapons. For data exchange, NATO aircraft use Link16, a networking system that allows fighters to exchange sensor data, or an AWACS to "push" the big picture to the fighters. However, these are separate systems and are not technically part of the mission management system.

None of this is related to whether or not an aircraft uses fly-by-wire. Traditionally the autopilot used servo motors to control the wires, rods and cranks that operated the surfaces. For example, a 747-400 has a modern flight management system with all the "fly-me-there" automation on top of an aircraft that uses conventional cables to connect the cockpit controls to the hydraulic actuators that deflect the control surfaces. Similarly, an aircraft can have a glass cockpit (screens instead of "steam gauges") even if it does not feature a fly-by-wire control system.

A final word about automation. There is an ongoing discussion in the commercial pilot community about how much automation is too much. There are several arguments here. One is that if pilots only "dial in" altitudes, headings and speeds, and routinely let the aircraft land automatically, they might lose their manual flying skills. This is certainly true—practice makes perfect, and lack of practice is bad. On the other hand, it is probably a good thing if pilots get help from automation if they have to land an aircraft in adverse weather after a long and fatiguing transatlantic flight. A second argument, however, concerns the transparency of the system. As long as everything works, "magic" is fine. But when things start to fail, it is often crucial for pilots to understand the interdependencies between systems to be able to diagnose problems correctly and react appropriately. If you read the training manuals, there is indeed a lot of mode- or fault-dependent behavior: "If X happens, then Y is no longer done automatically", where X and Y are functionalities that are not obviously related. Instead, the dependency stems from details of the underlying system. For example, two functionalities might run on the same computer. Understanding these dependencies certainly takes a while. In the end, an accident is always a combination of several causes—the famous Swiss cheese model in which all the holes have to align to lead to a catastrophic outcome. One of the holes can certainly be lack of practice, or the lack of understanding of a complex system. We will return to this problem after we have discussed how fly-by-wire systems work.

AIRBUS A320 The Airbus A320 is one of the most successful airliners in history; only the Boeing 737 has sold better than the A320 and its brethren, the shorter A318 and A319 and the longer A321. The A320 family was also the first commercial fly-by-wire airliner. The A320 is not dynamically unstable, so its reasons for employing fly-by-wire are different. First, a fly-by-wire system is lighter than cables and cranks, which saves fuel over the lifetime of the aircraft.

Second, if you have ever looked into the wheel well of a traditionally controlled airliner, you will probably agree that it is a benefit if you can get rid of the mechanical complexity of the control cables; they produce quite a bit of maintenance effort. The primary driver for fly-by-wire systems in airliners is safety: assuming that you can build a reliable fly-by-wire system, it can provide lots of protection mechanisms and benign flight characteristics, similar to the F-16's just-pull-and-I-will-give-you-appropriate-g-levels feature Jan mentioned earlier. A secondary reason is family commonality: because the system simulates a virtual aircraft, you can simulate the same virtual aircraft for several real aircraft types. In other words, the A318, A319, A320 and A321, and to some degree the A330 and A340, all handle very similarly, simplifying pilot training and currency. The system can presumably also fly more precisely than a pilot, which may lead to increased efficiency because of fewer unnecessary control surface deflections or imperfect trim. Finally, you might be able to build a lighter structure, because the system can actively compensate for gusts and turbulence, reducing the load on the physical airframe—search for "aileron waltz" on YouTube for a nice demonstration of this on the A380. In 2014 I chatted with Reinhard Reichel, professor for aircraft systems at the University of Stuttgart and an expert in fly-by-wire systems, who has a deep knowledge of the A320 flight control system. In the following I will give you a sense of how the A320 flight system works and how it has been engineered to be reliable and safe.

The design of a fly-by-wire system is an exercise in systems engineering, involving hardware, electronics and software. Engineers start out by capturing the requirements that describe the intended functionality of the system, as well as its quality attributes. A particularly important quality attribute is reliability, which in turn can be broken down into safety and availability. Safety means that the system does not produce any erroneous, dangerous behaviors (such as fully deflecting the rudder during high-speed flight). Availability means that it is extremely unlikely that the system fails to provide its functionality. For commercial airliners, the European Aviation Safety Agency's CS-25 certification specification requires a reliability of 10^6 for an aircraft as a whole. This means that a fatal accident is acceptable once per million flying hours. This number governs the design of the flight control system. Reinhard Reichel summarized the challenge as follows. CS-25 requires 10^6. The pilot is statistically responsible for 80% of accidents, the structure for 10%, so there are around 10% remaining for all other critical systems. Taken together, they can therefore only fail once in every 10^7 flight hours. The aircraft has around 100 critical systems: one of them is the flight control system. This means that the flight control system has to achieve a reliability of no more than one failure in 10^9 flying hours. However, many of the components, such as the computers, have a much lower reliability: 10^4 is typical. So the challenge of engineering the flight control system is to assemble a system that achieves a reliability of one failure in 10^9 hours from components that individually have a much lower reliability. To understand how this is achieved, we have to zoom out a little bit.

System engineers work a lot with models. A model is an abstracted, simplified, less-detailed representation of a system that supports analytic methods that make reliable predictions about the real system: we will discuss models much more in the next chapter. Maybe the most important model is the functional model, in which a system is hierarchically broken down into subsystems and the interactions between these subsystems are captured. The following representation resembles block diagrams that are found in many systems engineering modeling languages, and in particular SysML.

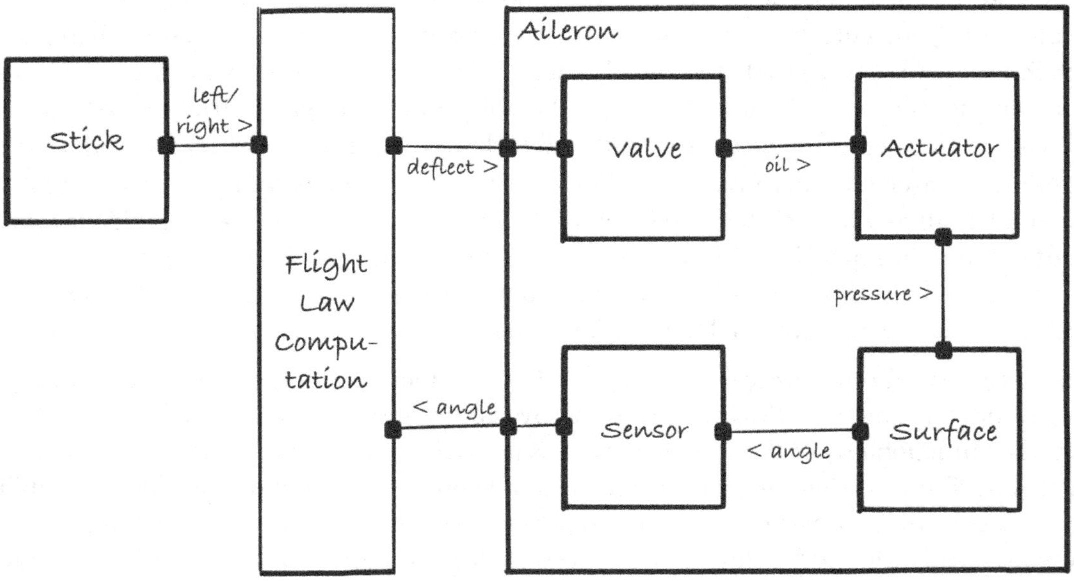

The system shows three functions: the stick, operated by the pilot, the flight control law computation, which takes the pilot's input and determines which control surfaces to deflect and by how much, as well as one of these control surfaces, an aileron. The flight law computation tells the aileron by how much to deflect and receives feedback from the aileron about its actual deflection. It then runs a control loop to keep the aileron at the angle that it wants it to be based on the stick position (and lots of other sensor inputs that we will ignore here). The flight law considers the aileron as a black box: commanded deflection in, actual angle out. But the aileron system is itself built from several subsystems: the electrical deflection request arrives at a valve that opens or closes according to the received command. This allows oil into the actuator, which then puts pressure on the actual surface, which then moves. An independent sensor measures the angle of the surface and reports it back to the flight law (indirectly—the actuator law is a separate system). For the engineer who designs the flight law, the aileron is a black box. But for the engineer who designs the aileron, it is a whole (potentially complex) subsystem in its own right. This hierarchical decomposition is crucial to being able to understand and design complex systems.

The various subsystems exchange data. For example, the flight law computation sends a deflection request to the aileron and the aileron sends the measured angle back. In early stages of system design this level of precision is enough to allocate basic responsibilities to subsystems (and recursively, to their sub-subsystems). But as we make progress in the design we have to become much more precise: is the angle in degrees or radians? What is the valid range? How many decimal digits? What is the error in each value? Once we think about representing these values in software we have to decide how to encode them. Let's say the angle is a value between −30.00 and +30.00 degrees. Should we encode this as a floating point number or as 16-bit integer, multiplied by 100, as a range between −3,000 and +3,000? Finally, we will have to define how often a new value must be sent. This is very important: if the rate is too low the control algorithm might not be able to run at the required precision, and if the rate is unexpectedly high the receiver might be overloaded, for example because data buffers could fill up too quickly. The documents that capture all of this are called interface descriptions. They might be just words initially, but as the system design progresses they are expressed precisely in various modeling or programming languages. Interfaces are agreed on by the producer(s) and consumer(s) of signals. Conceptually they are then attached to ports: ports are the "connection points" of the boxes that represent the functions in the diagram above. Two ports can only be connected if they have the same interface, to ensure that both data producer and consumer have the same understanding about the data they must exchange. The notion of an interface also extends to the physical world; in this case they do not specify the exchange of data, but the exchange/flow of heat or materials or forces.

Clear, unambiguous and agreed interfaces are the first ingredient for building a reliable system. They avoid any number of errors from "misunderstandings", and they also help with finding problems during integration testing. For example, if we specify the rate at which the flight law is able to receive and process the angle signal, we can monitor this rate as the system runs. If the rate is too high or too low, we can log an error that specifically indicates this problem. If we didn't do this we'd eventually see the flight control become less accurate due to reduced algorithmic precision because the signal is not received at the minimum rate, or fail entirely because the input buffer overflows and the program crashes. The more specific log entries are about their root cause, the easier it is to fix the problem. A crash, and even an overflowing buffer, can result from all kinds of problems.

We can also detect flaws in the functioning of the aileron; if the aileron reports an angle that is outside the −30/+30 degree range, something must be wrong. Here it is not obvious what the root cause is, but we know that something is wrong with the aileron. This approach is known as plausibility checking. The flight law computation could detect this, because it is a consumer of the angle signal produced by the aileron. There are also more advanced forms of plausibility checking. For example, one could say that, if the valve is opened and the sensor does not detect a change in the angle of the surface after some specified time delay, then something must be wrong. In fact the A320 flight control system runs a model of how the surface is supposed to behave as a consequence of the valve opening, and if the surface behaves differently it constitutes a fault. If this kind of plausibility checking were to run inside the aileron function, then the aileron would have to report an additional signal back to the flight law computation that reports its "valid" status to notify the flight law computation of the failure of the aileron subsystem.

The flight control system in the F-16 works similarly. Jan explains: "If the four channels of the flight control computer don't return the same flight control command for a given set of inputs, it will filter out the offending channel and pick the control deflections the other three channels calculated. If the flight control system is unable to do that, then it will lock out the affected servo actuator and give me reduced flight control authority around that axis until I can verify that the control system isn't somehow getting invalid data from the air data computer. If I can isolate the sensor that's feeding incorrect data to the air data computer, I can then reset the appropriate actuator and continue my mission. If not, I may have to terminate the mission and try to land with degraded flight controls around that axis, which the flight control system can partly compensate for with the remaining flight control surfaces and the trim."

Now let's look at different kinds of faults. A fault might happen spuriously and then disappear: a bit flip caused by solar radiation is an example. Such faults do not necessarily indicate a problem with the design or construction of the system, but the system has to be able to cope with them nevertheless. The classical approach to detecting such errors is to use a checksum: for a four-digit number XXXX you compute an additional value Y from the digits and append it to the number, XXXXY. The receiver performs the same computation from the received digits X'X'X'X', and if one has changed (because of a bit flip or any other reason) the checksum will be different and won't match the one received with the number. To "fix" the fault you just try again, and very likely the problem will disappear.

Then there are faults that result from wear or because a mechanical part is not strong enough. Importantly, these kinds of faults become more likely as the system ages. For a system to be able to function in the face of such faults, critical components are either significantly overdesigned, closely and regularly inspected over the lifetime of the system, or they are installed more than once. For example, in the A320 there are in fact two actuators for each aileron, so that if one fails the other can take over. A third kind of fault is systematic, which means that it always occurs if a particular condition is met. Such faults often occur in software and are usually a design or implementation error. For example, the internal model of the control surface might expect the aileron to react faster to an opening valve than it actually does and report an (invalid) error, for example at high speed. Finally, there are logic errors where the system correctly does what it was programmed to do, but that program specified inappropriate or unsafe behaviors. Such a logic error was partially to blame for the accident of Lufthansa Flight 2904 when landing in Warsaw, which overshot the runway in poor weather on 14 September 1993. Consider this paragraph from the accident report:

> The aircraft automatics comprises, for basic landing configuration if the aircraft (i.e. with flaps extended to FULL position), the programme which subjects actuation of all braking

> devices to some specific conditions. Ground spoilers, when selected, will extend provided that either shock absorbers are compressed at both main landing gears (the minimum load to compress one shock absorber being 6300 kgs), or wheel speed are above 72 kts at both main landing gears. Engine reversers, when selected, will deploy provided that shock absorbers are compressed at both main landing gears. Such a logic results in the lack of possibility of immediate actuation of two mentioned above aircraft's braking devices without meeting the conditions described.

While mistakes by the crew also contributed to the accident, the aircraft's logic that determines when it is considered to be on the ground, and that therefore brakes and ground spoilers are allowed, was partially to blame: basically, a bumpy landing and hydroplaning "confused" the system. As a consequence of this accident, Airbus issued a software update that changed the required weight from 6.3 tons to only 2 tons. Preventing these kinds of faults is extremely hard: essentially they are a result of failure to anticipate the conditions during landing correctly. Testing helps, but you still have to test in conditions where the fault reveals itself. In the case of Flight 2904 the system did exactly what it was designed to do—it's just that it was designed to do the wrong thing.

We can now return to the CS-25 requirement of a reliability of 10^7. Systems engineers will assign reliability factors to each of the sub(sub(sub))systems. The respective factors have to be justified either by experiment, long-term experience or through a fault-tree analysis that mathematically derives the reliability. Let's look at our aileron again. Say the actuator fails. The rest of the aileron cannot compensate for this; there's nothing it can do to keep the surface moving. To make it more reliable, we can use redundancy and add another actuator that can take over if the first one fails. Assuming an actuator fails with some probability P_A, then the probability of both failing is $P_A \cdot P_A$. Since probabilities are numbers between zero and one, this product is smaller than the original P_A: that is, the probability of two actuators failing at the same time is lower. However, for this to be true, the failures of the two actuators must be independent. For mechanical systems this assumption is usually valid, but if they for example get their hydraulic pressure from a single supply, they suddenly have a common failure mode if the hydraulic system fails. Finally, the assumption requires that a failing actuator does not prevent the other one from performing its job by blocking the control surface.

So how does the A320 do it? It indeed has two aileron actuators, but they are supplied by two independent hydraulic systems. Each actuator has an override valve that connects the two chambers of the cylinder to each other, ensuring that a failing actuator cannot block the other from taking over. That valve is commanded through a separate signal. It is similar in the F-16. Jan: "Two channels of the flight control computer control one set of actuators, and two channels control the other. At the same time, the actuator is actually comparing the inputs from both opposing channels, and locking out the actuator as described before. Getting a servo actuator warning light is actually relatively common when 'assaulting the aerodynamic limiters', such as in a visual fight at high g."

We can now see how redundancy in the aileron subsystem helps us build a reliable system (the aileron) from components that do not have the required reliability (the actuators). In this case it was local component redundancy: we simply duplicated a component. When designing an aircraft, however, redundancy must be seen relative to the capabilities of the system and its overall reliability. This is beneficial, because it has the potential to increase reliability without too many additional components (which add weight, complexity and maintenance effort). This can be nicely illustrated with roll control. Roll control is a capability provided by the flight control system, but it cannot just be realized through ailerons; it can also be realized using the spoilers. These are a bit less efficient, but the A320 can tolerate the failure of an aileron because it can compensate with the spoilers. The flight law computation, when detecting a fault in an aileron, will automatically fail-over to the spoilers. They are mechanically independent and have no common mechanical failure modes. As we will see, ailerons and spoilers are also controlled by different computers, another ingredient to overall system reliability.

Software does not only have logic errors such as those that lead to the crash of Lufthansa Flight 2904: the underlying computer hardware might have its own low-level logic errors, or the compilers that generate the machine code from high-level programming languages might have errors. To avoid fatal effects from such errors, dissimilarity is used: redundant computers that have the same responsibilities don't use the same hardware architecture, programming language or compiler. This avoids common failure modes that might otherwise result from low-level logic errors in the hardware or compiler. This principle is used in all the computers in the A320's fly-by-wire system. In total the system has five full-authority computers that control pitch, roll and yaw, all of them active all the time. They perform both the flight law computation as well as control of the actuators; there are no dedicated actuator control electronics. There is enough redundancy in the system to provide full functionality if one computer fails, and safe operation of the aircraft is possible even with only one computer working. However, the system degrades in steps between the two extremes of full and minimum functionality, as we will see below. In the A320 there is also mechanical backup for trimming the horizontal stabilizer and controlling the rudder; it has since been replaced by an independent electrical system in the newer A330/A340 family. However, the intention here is only to stabilize the aircraft while the computers are restarted in an emergency; landing with only the mechanical systems is not part of the requirements, even though some airlines try it at least once as part of the type rating syllabus—in a simulator, of course.

The five computers are comprised of two ELACs (elevator and aileron computers) and three SECs (spoiler and elevator computers). Notice how ELAC and SEC overlap in the sense that they can both control the elevators. However, they are distinct in control of ailerons and spoilers but, as we have seen above, these back each other up to some degree. For dissimilarity the ELACs are based on 68010 processors and the SECs on 80186s. These are produced by different companies, and a total of four different architectures are used for the software running on the computers. There are three independent hydraulic systems and two independent electrical buses, as well as three identical air data computers that measure AoA, sideslip and speed, plus a number of attitude parameters. The number three is not coincidental: the five central computers consume the data from all three air data computers. If the data from one of them differs from that from the other two, it is deemed unreliable and ignored. This kind of majority voting is a common strategy for deciding on the correct data values among multiply redundant data sources.

To increase the reliability of each of the five computers, each consists internally of two channels. These are implemented with two identical computers that perform the same computations in parallel: one commanding, the other monitoring. The monitoring computer compares the results of each computation with the other channel's results, and if they diverge beyond a predefined threshold this constitutes a failure of the overall computer. It would then report the problem, stop its work and electrically disconnect itself from the system. The other ELAC, or eventually, one of the SECs, takes over. The two channels use the same hardware architecture but different software.

Let's look at the degradation in functionality that I mentioned above. The flight control system provides three different flight control laws. In accordance with the two main reasons for using fly-by-wire systems, each law provides a particular set of flight characteristics and protections. If all computers are operational, the A320 operates in normal law. The aircraft is flight path-stabilized: the system provides stability in the sense that, with the stick at neutral, the aircraft retains its current flight path; it controls for 1 g, or the absence of any acceleration (this is true only up to a point: if a divergence is large enough pilot input will be required—it's not an autopilot attitude-hold mode). The pilot's stick input governs the first derivative of the flight path. Pulling the stick requests a particular load (g-level). Deflecting the stick to the right rolls the aircraft to the right with a roll rate that is proportional to the deflection; full deflection results in a 15 degree/second roll rate. Importantly, the mapping from stick deflection to load factor or roll rate is independent of speed, which makes the A320 quite different than a conventional aircraft. The roll behavior changes at 33 degrees bank, however: instead of controlling the roll rate, the stick now controls the roll angle. As with the steering wheel of a car, the stick has to be kept deflected to continue a turn of more than 33 degrees. The intention behind this behavioral change is to make the pilot "force" the aircraft into such a tight turn. Bank angles of more than 67 degrees are not possible—the law protects against overly steep turns. Pilots can also ignore the rudder pedals in normal law, because the law automatically applies the right amount of rudder to fly a coordinated turn for the commanded bank angle. It also does not allow flying with more than 30 degrees of nose-up pitch or 15 degrees nose-down, protects against overspeed and limits the AoA to what the wings can do. The system is quite invasive: it will automatically add power and/or push the nose down to keep the aircraft safe. Finally, it also protects against over-stressing the airframe: it won't permit more than +2.5 g or less than −1.0 g.

If subsystems start to fail, the flight control system performs a stepwise degradation to simpler laws. Let's say that both ELACs fail and the SECs take over. Since these have less computing power, simpler control laws are used. If the air data computers fail to provide data reliably, then the automatic flight path stabilization can no longer work and the system degrades. The specific rules that govern when which law will be engaged depend on the specific failures, but generally at least two failures are required in either computers, hydraulics or sensor systems for normal law to become unavailable.

Let's briefly look at the two simpler laws. The first is called alternate law. In pitch, the stick continues to express a load factor demand, but in roll there is a (virtual) direct stick-to-surface connection; it basically commands a non-stabilized roll rate. There's no more turn coordination and pilots have to use their feet to operate the rudder pedals. There are also fewer protections: over-stress protection remains active but AoA protection is gone, and so is the 67-degree maximum bank angle limit. "Direct" law is the final degradation step: the computers do not modify any of the pilot's commands, there are no more protections, and it's back to manual flying in every respect.

There are many more details. For example, normal law has a ground mode that provides a direct stick-to-surface connection. This is to allow pilots to perform checks of the controls and rotate the aircraft during takeoff. There's also a landing mode: at 50 feet above ground the ELACs memorize the pitch attitude; at 30 feet they add a little nose-down to the memorized pitch attitude, which the pilot then compensates by pulling on the stick. This is supposed to make the aircraft feel more conventional during flare, the final part of a landing, when pilots are used to pulling on the stick. Finally there is also an abnormal attitude law: if the aircraft leaves its protected performance envelope, for example because of severe turbulence, abnormal law stabilizes its attitude but without most of the protections. This is designed to make sure that the computers never prevent the pilots from recovering from an abnormal attitude.

Remember our brief discussion of parabolic flight earlier? After injection, the pilots control the aircraft manually to maintain exactly 0 g. I asked chief pilot Eric what they plan to do once they have to replace their A310 with something newer, probably something that has a fly-by-wire control system. He said that they tried parabolic flights with an A380 together with Airbus, "and it worked". I assume they used basic law, because the others would no allow such a maneuver. He also said that, in principle, because Airbus' flight control system works to maintain 1 g when there are no pilot inputs, one would just have to replace 1 g with 0 g in the control loop: "You'd pull up to 50 degrees pitch and then just press a button that engages the 0 g law and the system would automatically control for a 0 g trajectory." He added that one would have to remove and adapt all kinds of protections, but in principle, he suggested, the architecture of the flight control system should allow such a modification with acceptable effort.

So what exactly is a "law"? We have already said that a law governs how the pilot's inputs are translated into attitude changes and which protections are available. But how does it do that? The best way to summarize it is that the flight law contains a mathematical model of the ideal airplane. When the pilot makes a control input, the attitude of the model is changed based on how the aircraft should behave, and the initial deflections of the control surfaces are determined by predictions based on that model. There is also a representation of the actual attitude of the aircraft. When the two differ (as they will initially when the pilot makes a control input), the control algorithm tries to align the real aircraft with the model. It uses the available control surfaces to do so—for example, aileron and spoilers if everything works, or just the spoilers if an aileron has failed. This also explains why the system degrades to simpler laws if air data computers or attitude sensors fail: the computer does not have a trusted representation of the aircraft's current state, so it cannot reliably use the control surfaces to "align" the real world with the model.

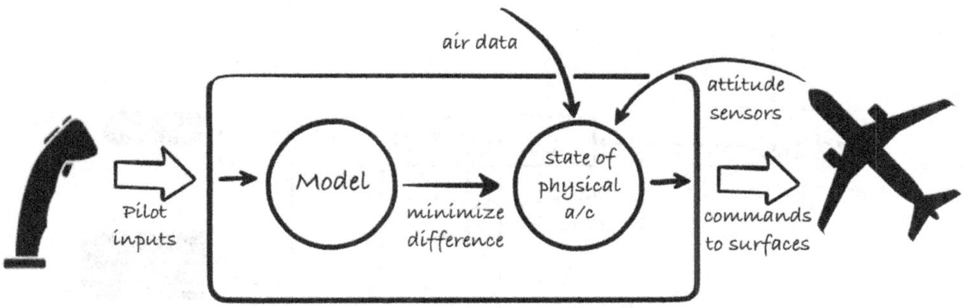

The aircraft has many more computers than ELACs and SECs. There are the flight control data concentrators (FCDCs) that acquire data from the primary computers so that they can be logged and shown to the pilots. There is also the electronic centralized aircraft monitor, or ECAM, a system that monitors the health of aircraft functions and communicates them to pilots; an independent monitoring infrastructure is an additional means of achieving reliability. There is a whole set of computers in the flight management and guidance system (FMGS) which, among other functions, is also home to the autopilots. Two more systems, the flight augmentation computers (FACs), deal with the flying itself: they control yaw damping, turn coordination and rudder trim, and also compute minimum and maximum speeds and handle AoA protection. Because all of their functionality is optional, however, they are not part of the core reliability architecture.

To summarize, the diagram below shows the final architecture. Stick and pedals feed into the ELACs, which control the ailerons and elevators. They also uses the spoilers to help with roll, which is why they send roll orders to the SEC. To fly nicely coordinated turns, the ELACs send yaw orders to the FACs. To build an accurate picture of the aircraft's attitude and flight path that is crucial for the flight laws, the ELACs receive air data and data from the inertial reference unit (ADIRU), slats and flaps (SFCC), accelerometers (ACC), landing gear (LGICU) and the radio altimeter (R-ALT). If the ELACs fail, the SECs take over. The ailerons are then dead, but the aircraft can roll with its spoilers. Note that the SECs cannot pass yaw orders to the FACs, so the rudder becomes inactive. However, spoilers also provide a yawing moment, so the missing rudder is not a huge problem; also, the mechanical linkage is still there. Both the ELACs and the SECs can talk to the elevator and the trimmable horizontal stabilizer (THS).

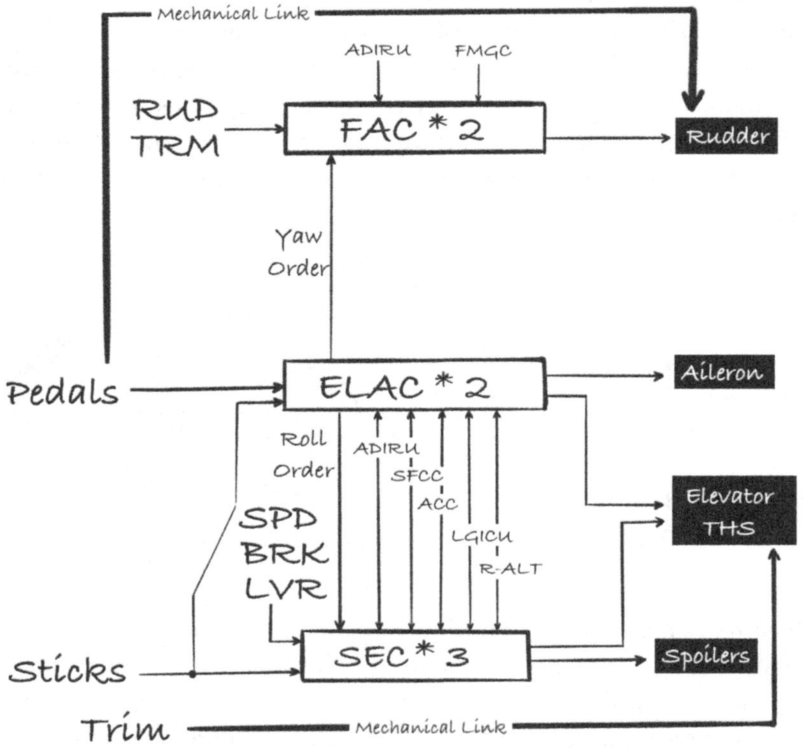

Even though they are not essential, a failure of both FACs has significant consequences and the flight law will degrade to alternate. The AoA protections are lost and speed is limited due to loss of high-speed protections and the yaw damper. The rudder travel limiter may be stuck in the wrong place, so the rudder must be used with care. During landing, control automatically reverts into direct law, because alternate law does not provide the landing mode mentioned above. Here's what the ECAM will show the pilots; the abbreviated words stand for

> Flight Control Alternate Law (Protections Lost). Maximum Speed 320 knots. Use rudders with care above 140 knots.

```
F/CTL ALTN LAW
(PROT LOST)

MAX SPEED 320 KT
RUD WITH CARE ABV 140 KT
```

Let me very briefly revisit the discussion about pilots not getting enough manual flying practice. As I said earlier, this discussion concerns higher-level automation such as autopilots and flight management systems more. These are also available in traditionally controlled aircraft. However, there is a connection to fly-by-wire control. The "virtual aircraft" represented by the normal law of the A320 is simpler to fly manually than the unprotected, direct-law physical aircraft. If a pilot who is used to lots of automation and at most flies the virtual aircraft is suddenly forced by system failures to fly manually in direct law, I do think that this might introduce a risk.

Let's revisit the Swiss cheese model. The loss of an A330 performing flight Air France 447 over the Atlantic, in which 228 people lost their lives in June 2009, can be attributed to (1) the icing of pitot tubes, which invalidated the airspeed signal, (2) the autopilot switching off and the flight control system switching to alternate law, (3) a misunderstanding of the situation by the pilots, who basically pulled the stick back until the aircraft stalled, then ignored the stall warning, (4) the pilots not sticking to the predefined procedures for loss of airspeed indication, and (5) the problem that the two pilot and copilot's sticks don't move together, so it was hard for the pilot who was not in control of the aircraft to realize that the pilot who was flying it was pulling continuously on his stick. This account of the causes is slightly simplified, but the resulting accident was definitely a combination of strange, unintuitive and suboptimal system behavior together with the pilots making mistakes. An initially similar problem occurred with an A320 over Spain where two of the three angle-of-attack sensors reported faulty values because of icing—the flight control system followed its algorithm and voted 2:1 against the remaining good one! Because the two sensors were stuck at around 4.5 degrees, the angle-of-attack protection kicked in and tried to lower the nose. Since the sensors still reported an unchanging high AoA, the flight control system pushed more and more, forcing the aircraft into a descent. The pilots, together with technicians on the ground, understood the problem, switched off the air data computers and flew the aircraft home manually in alternate law. The holes in the cheese didn't align in this case.

SAFE SOFTWARE Let's conclude our discussion of how to make a fly-by-wire system reliable with a brief look at how one builds reliable software. Many ingredients are important, but verification and validation are especially critical. Verification is about ensuring that the software behaves exactly as specified (nothing less, nothing more). For example, ensuring that buffers never overrun if a sender of data stays below its maximum sending rate and the receiver takes data out of the buffer at its minimum guaranteed processing rate is an example of verification. This is very different from validation, which is about ensuring that those behaviors are appropriate with regard to the real world. The problem of detecting the "on ground" state by Lufthansa Flight 2904 was a validation failure.

Consider the following diagram, which represents what software engineers call a "state machine". This captures various states of the aircraft (the ellipses), as well as the conditions under which the aircraft transitions from one state to another (the arrows). The boxes also contain actions, denoting things that happen when the system enters that particular state. Note that this is just a simple example, not at all an attempt at replicating actual logic from the A320. However, state machines and the verification and validation techniques we discuss below are used in the development of flight control software.

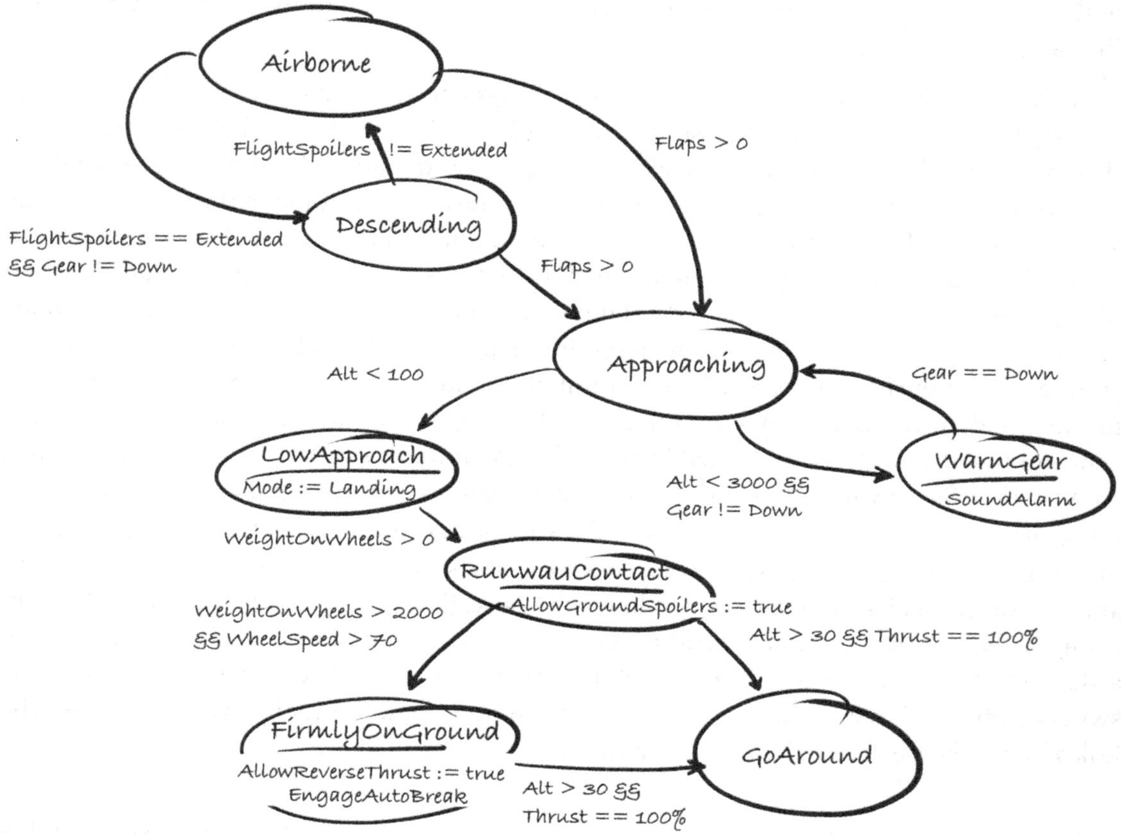

Let's walk through the diagram. Initially we are in the Airborne state, flying straight and level. If we now extend the flight spoilers while keeping the gear up, we go to the Descending state. If the spoilers are retracted (no longer extended) we are back to Airborne. If we extend the flaps in either of these two states, we move to the Approaching state. If we get lower than 3,000 feet above ground with the landing gear not down, we go into a GearWarning state and sound an alarm. The alarm stops when we extend the gear. Below 100 feet we enter the LowApproach state, in which the flight law's Landing mode is activated. Once we detect weight on wheels we enter the state RunwayContact; this allows the pilots to extend the ground spoilers. When we detect a load of more than 2 tons on the wheels and they have spun up to at least 70 knots, we decide we are FirmlyOnGround, engage auto braking and allow the thrust reversers to be used. Finally, we detect a GoAround if altitude increases and the engines are in full thrust.

How can we ensure that this state machine is correct? We essentially have two tasks. We first have to validate that the state machine expresses the correct behavior—that we actually want the aircraft to behave as the state machine prescribes. Then we have to verify that the software program that implements this machine behaves exactly as the model says. Let's start with validation. How can we do this?

First we can manually review it. For example, by looking closely at the diagram we might notice that if we do not extend the flaps (maybe they have failed and we cannot extend them) we never enter the Approaching state, so we will not get a gear warning below 3,000 feet! Or we might notice that extending the spoilers might not result in descent, but just in a slowdown (if we pull on the stick enough). However, this manual inspection is really tough once this sort of model becomes realistically complex.

The next best thing we can do is test it. We can "exercise" the model by scripting an execution, checking that the model reacts correctly based on the requirements provided by people who have an understanding of how the system should behave in particular scenarios. The following is an example that assures that we get a gear warning at 2,500 feet.

```
assert state == Airborne
set    Spoilers := 50%
assert state == Descending
set    Alt    := 4000 ft
set    Flaps  := 25°
set    Alt    := 2500 ft
assert SoundAlarm == true
```

Ideally tests are scripted once and then run automatically whenever we change something in the state machine model; this is to ensure we don't accidentally break previously valid behaviors. But we still have to "manually" think of all relevant test scenarios, and it is easy to forget corner cases. Think of LH 2904: to catch the fault condition, we would have had to write a test that used less than 6 tons of wheel loading. That might be hard to think of, especially if you are the engineer who has just decided that you will always have at least 7 tons, and so a threshold of 6 tons is correct.

Can we do better than test? Yes. For example, we can write down invariants. These are Boolean conditions that must always be true. Here are two examples; the X => Y arrow represents implication, which says that if X is true, then Y must also be true, otherwise we don't care.

```
invariant !WeightOnBothWheels     => !AllowReverseThrust
invariant Alt < 3000 && !GearDown => SoundAlarm
invariant ReverseThrustEngaged    => Alt == 0 && AF[Speed == 0]
```

The first invariant expresses the fact that, unless we have weight on both wheels, we are not allowed to use reverse thrust. The second says that we will get an alarm whenever the altitude is less than 3,000 feet and the gear is up. The third is especially interesting. It says that if reverse thrust is active, the altitude must be 0 (that is, it is only allowed on the ground). It also says that the speed must become zero eventually: AF(c) implies that for all (A) possible executions and branchings of the machine, in some future (F) state the condition c becomes true. AF is an example of a temporal operator, one that lets us express invariants that involve the execution of the state machine over time. There are many others; for example, we could express that, once the reversers are deployed, for all future states they have to stay deployed until (U) speed is zero:

```
ReverseThrustEngaged => A[ReverseThrustEngaged] U[Speed == 0]
```

Now that we have stated these invariants, what can we do with them? One thing we can do is to check them during tests. So whatever test scenario we run, we also check the invariants; if any fail, the test fails. In this way we can hopefully detect unexpected behaviors even though we didn't write a test that specifically checked for them.

However, we can do better than this by using a technique called model checking, which uses software tools to analyze the state machine model and tries to to prove that the invariants hold for every possible execution of the machine. If the tool cannot find that proof it shows us an example execution of the machine where an invariant does not hold; this will hopefully help us to fix the machine. Model checking is interesting, because it frees us from having to manually write tests that cover all possible (and/or relevant) executions of the system. On the other hand, we still have to come up with the invariants. Both testing and model checking rely on complete requirements and on the engineers' correct understanding of them. In practice a combination of reviews, testing and model checking is used, and more. For example, ensuring that tests are written by people other than those who designed the state machine in the first place, to ensure that they throw "nasty" tests at the machine.

Now that we have validated the model, how can we ensure that the actual software, written in C/C++, Ada or Rust, implements this model faithfully? The mapping from the model to the code is often non-trivial because, in addition to behaving identically in terms of the logic, the code also has to be fast enough to fit into the constraints demanded by a realtime control system. Developers also have to care about memory management, buffer sizes, integer overflows, concurrent executions of parts of the program and many other low-level issues that are not relevant for the behavior expressed in the model.

So what can we do? We can for example generate code from the model automatically. Assuming that the model is correct, and assuming the code generator does not make any mistakes, we will get a faithful implementation. But the assumption that the code generator will not make mistakes is a big one. We have moved our problem from verifying a particular model to validating and verifying the code generator. There are tools, however, that have been proven to be correct; Esterel's SCADE is an example. As far as I know SCADE is used for flight control software at Airbus. Another approach is to run the set of tests written for the model against the code as well, to check that the code behaves in the same way. We can also use software tools to extract tests from the model automatically and run them against the code. Finally, there are tools that allow the equivalent of model checking on the level of the program code. We will look further at the relationships of model and implementation code in the Models chapter.

BOEING 737 MAX As I finalize this chapter, the Boeing 737 MAX 8 has just been grounded worldwide due to two fatal accidents that are likely to have a common contributing cause: the Maneuvering Characteristics Augmentation System (MCAS) and/or the pilots' unfamiliarity with the operation of this system. I think it makes sense to look at it from the perspective of what we have discussed so far in this chapter.

The story starts when Airbus updated its A320 to the A320neo, which added new and more economical engines. Boeing had to react with a more economical airplane: they decided to update the venerable 737 to the MAX variant by installing more fuel-efficient engines, the CFM International LEAP 1B. I don't know whether Boeing ever considered developing a completely new aircraft, but clearly this would have taken much longer and would have been more expensive. So an update it would have to be.

An important factor in the increased efficiency of the new engines is their higher bypass-ratio. To achieve this they have much larger fans, which in turn increase the diameter of the engine housing. The 737 has always sat relatively low on the ground, which is why the -300/-400/-500 variants' CFM-56 engine cowlings have a slightly flattened bottom. The larger engines could not be fitted under the wings, and higher landing gear would have required lots of other modifications and was (probably) considered too expensive. The solution used was to install the engines much further forward, almost in front of the wing, and higher up. They also extended the nose gear a little. Together, these measures provided the necessary ground clearance.

This decision had an important consequence: because the engines in the MAX variant are further forward of the center of gravity and the center of lift, increasing engine thrust leads to a nose-up moment. In addition, at high AoA part of the wing is "behind" the engines, reducing lift. If the increasing AoA remains uncorrected by the pilots it could lead to a stall. This is potentially catastrophic for an airliner, especially if it happens at low altitude, for example during a go-around when the pilots increase thrust quickly and AoA is already high.

So what can you do about such a change in flight characteristics? One option is to describe it in the flight manual and train pilots accordingly. Many aircraft have "unpleasant" corners of their flight envelope, and instruction and training is a valid way to deal with them. However, the requirement to (re-)train pilots when they transition from the older 737 variants to the MAX leads to costs for the operators, and so (again, I am speculating here) this is something that Boeing probably wanted to avoid. In addition, when the unpleasant behavior becomes too bad—and dynamic instability when adding power might very well qualify as such—this might prevent certification and mandate correction. A hardware fix was too expensive, so Boeing decided to make the unpleasant nose-up moment go away by installing MCAS, a system that counteracts this behavior by pushing the nose down using the horizontal tail. Such augmentation systems are nothing new, as we know from the discussions in this chapter. The SR-71 has one, and almost every airliner, including the 737, has a yaw-damper and various auto-trim systems.

The horizontal tail of the 737 has two moving parts. The tail end of the stabilizer moves up and down to form the elevator, but the whole stabilizer can also move up and down to trim the aircraft. In our discussion of the A320 flight control system it was called the THS (trimmable horizontal stabilizer). In Boeing airplanes the THS is moved by means of a worm gear, colloquially known as a "corkscrew mechanism". It can be actuated manually through the trim wheel in the cockpit and with a rocker switch on the yoke. It is also moved automatically by the autopilot and by trim systems. For example, the Speed Trim System becomes active at low speed, aft center of gravity and certain flap settings to help make the aircraft "feel" right. Another system, the Elevator Feel Shift, increases forward stick pressure as the aircraft approaches high AoA, forcing the pilot to pull even harder to move to even higher AoA.

The MCAS is a new system in the 737 MAX that uses this THS trimming/corkscrew mechanism to push the nose of the aircraft down when the AoA gets close to critical for whatever reason, but in particular due to a nose-up moment resulting from the forward location of the engines. Assuming this system works reliably, the "unpleasant" aspect of the flight characteristics is taken care of. There is no need to train pilots, with a resulting simplified transition to the 737 MAX. As far as I know, the transition to the MAX can be done through a few hours of computer-based training on a iPad. No actual flying or even simulator training is required. So far so good.

From here on the story becomes interesting. I have no inside information, and the investigations of the two accidents are still ongoing, so by the time you read this things might have changed. Which is why I will just put out a couple of facts and my thoughts about them; you can then form your own opinions about the cause of the accidents, or—better—wait for the reports.

The original paperwork for the MAX is said to have claimed that the MCAS can/will only move the stabilizer by 0.6 degrees within 10 seconds. Because of this very limited authority, the criticality of the system was classified in a way that required no crew training. In actual fact the MCAS has up to 2.6 degrees of travel in 10 seconds; this would have put the system into a category that would have required a more stringent review during certification. How exactly this was communicated to the FAA, and which reviews the FAA did or did not perform during certification certainly raises questions, but this is not a technical issue and I will not discuss it further.

In addition to the 0.6 versus 2.6 degrees of authority, the system resets itself after five seconds of inactivity, thereby gaining an additional 2.6 degrees of authority after each reset. This effectively gives the MCAS unlimited authority to push the nose down, which is one of the suspected contributing causes for the two accidents.

One reason for a malfunction of MCAS that might lead to it pushing the nose down over and over again is faulty AoA input. If this input was "stuck" near the critical AoA, MCAS might continuously push the aircraft's nose down even after the pilots or the MCAS lowered the actual AoA, because the measured AoA, as seen by MCAS, remained critically high. As this AoA signal is so important, it was really astonishing for me to learn (from an Airworthiness Directive issued by the FAA) that the MCAS only uses the data from a single AoA sensor as its input!

MCAS is a software function in the Flight Control Computer that also manages the other trim-related functions mentioned above. The 737 has two of these: one is active at any one time and they alternate with each power-up of the aircraft. The computers do not cross-check each other during the flight! And each computer is connected to one AoA sensor. So even though the 737 MAX has two of these sensors (and two computers), only one is used at any given time. A divergence of the two, indicative of an error with one of them, cannot therefore be detected by MCAS. Worse, a warning/indication for pilots in the cockpit was only available for a surcharge—it was not installed by default. In the Lion Air accident the two sensors disagreed by 20 degrees even before takeoff where, depending on speed, either no AoA is measurable at all or it's zero!

After our earlier discussion about redundancy in flight control systems, this lack of redundancy should surprise you as much as it surprises me. I suspect that the only reason why Boeing could justify using only one AoA sensor to itself and the FAA is that MCAS is not a critical system: the airplane can be flown without it, it (theoretically) has only 0.6 degrees of authority, and can therefore produce only limited harm. As we have seen above, though, this 0.6-degree claim is not really true. And while you can fly the 737 MAX without MCAS, the pilots have to be aware of the system and turn it off explicitly in the event of a problem. Which brings us to the role of the crews in the accidents.

Boeing initially did not communicate the presence and functionality of this system to the crews. This puzzles me: even if the system is only meant to "rescue" the aircraft from a dangerous situation (critically high AoA), I do not understand why its presence on the aircraft was not communicated. As a result there was also no discussion or training regarding a malfunctioning MCAS, and when the system behaved erroneously on Lion Air 610 the crew did not really know what to do. In the wake of this crash the presence of the system became known and was actively communicated by Boeing to the crews. One can argue that the pilots on Ethiopian Airlines 302, the second of the two accidents, should have known about the system and how to deactivate it.

To deactivate MCAS pilots have to cut the power to the THS. They may then use the manual mechanical trim wheel to correct the nose-down THS position. Depending on how far the MCAS had mis-trimmed the aircraft, manual adjustment of the trim may take a while. Alternatively (and in addition) they can also just pull on the yoke; it is a certification requirement that the elevator is powerful enough to control the aircraft even if the THS is deflected to either extreme. However, there is a catch: depending on how late the pilots decide to do this, the aircraft may have accelerated so much that the elevator experiences blowback—that is, the speed-induced aerodynamic forces on the elevator become so large that the hydraulics are no longer powerful enough to move the elevator. At this point there is probably no way to save the aircraft. Even before this fatal situation is reached it can be non-trivial to trim the aircraft correctly. An elevator in the up position (because of pilots pulling on the yoke to keep the nose up while trim is nose-down) creates a force on the stabilizer trim screw that makes it very hard to move the trim wheel manually. With earlier 737 versions Boeing recommends the "roller coaster method": pitch the airplane up using the elevator, then let it nose over and pitch down by stopping pulling on the yoke. This relieves the air loads on the horizontal stabilizer, allowing the stabilizer and the trim wheel to be moved manually. Boeing suggested to repeat the maneuver until the aircraft is trimmed correctly. However, you need enough altitude to achieve this, plus you have to have the idea in the first place. As far as I was able to discover, this maneuver was not trained, at least in some airlines ("They only tell us about the trim wheel to make sure the handle is stowed and doesn't hit our knees when it rotates automatically"). Exactly how all of these factors contributed to the crashes, and exactly what the pilots did, will hopefully be discovered by the investigations.

Cutting off the power to the THS should be a familiar procedure for crew, because it is also used if the THS malfunctions for another reason: a "stab trim runaway" is trained for regularly. The button to completely turn off the THS has been located in (almost) the same place on the center panel of all Boeing airliners since the 707. On the other hand, I suspect that a (continuous) trim runaway feels different than the repeated stepwise push down of the nose induced by a malfunctioning MCAS. And there is an important difference between the previously installed trim systems (such as the speed trim) and MCAS: when the pilots didn't want the system to trim the aircraft automatically, they could stop the process by clicking the trim rocker switch on the yoke. It is slightly different for the MCAS: when the pilot clicks the rocker switch, the MCAS will stop, but only for five seconds. It will then automatically restart another ten-second nose-down trim run. To disable MCAS pilots have to cut off trim power completely. If they misidentify an MCAS intervention as a stab runaway, they might lose valuable time trying to "fix" the issue using the stab runaway checklist, on which the trim cutoff is only step #6.

Another way of looking at the situation is as follows. The earlier systems that affected trim, the speed trim system and the elevator feel shift system, were considered assistance or convenience functions. Because of their non-essential nature they could be switched off easily, so the requirements for safety and reliability in the underlying system architecture were relatively low. MCAS is intended as a protection system—one that protects against the effects of an unfortunate aerodynamic design. It is justifiable that a protection system is more "insisting", that it is harder to switch off. However, the underlying system architecture was not adequate for the increased responsibility of the system. It was a departure from Boeing's longstanding philosophy that the pilot is always ultimately in command.

So where does this leave us? Let me repeat: we should hold back accusations and blame until the authorities release a report about the cause of the accidents. However, it is certainly fair to say that this is yet another example of the Swiss cheese model in action, in the sense that several factors had to "conspire" to lead to a fatal accident: (1) the (at least) unfortunate design of the MCAS regarding only one AoA sensor; (2) the fact that it resets and thereby gets unlimited authority without plausibility checking with other data; (3) the lack of communication to and training of crews regarding the system and the subtly different behavior regarding how to turn it off; (4) the pilot's failure to identify the problem correctly and cut off the THS when MCAS misbehaved (other incidents have since been reported where pilots did cut off the THS); and (5) in case of Ethiopian 302, the MCAS malfunction at low altitude, leaving little time to diagnose the problem and react.

Boeing is now developing and testing an updated version of MCAS which only works once for 10 seconds (thus limiting authority) and will use the signals from both AoA sensors; if they disagree by more than five degrees MCAS switches off. I have also read that they will rework the software architecture of the two computers so that they are both active during a flight, cross-checking each other. The cockpit indication/warning of divergent AoA will also be made available as standard equipment. However, for how long do the two AoA sensors have to disagree to disable MCAS? Can a transient disagreement lead to a (erroneous?) disabling of MCAS? And might not knowing that MCAS has switched off then become a contributing factor to a crash? I am not suggesting Boeing will get this wrong, but I want to drive home the point that remedies to one problem might come with their own risks and lead to other problems. And in fact, it was reported that the more complex logic required to "safely" operate MCAS overloads the processors installed in the computers! The more distinct behaviors a system has, the harder it is to make sure all combinations of them are safe and reliable.

Can we learn something about fly-by-wire systems from these accidents? The 737 MAX is not a fly-by-wire aircraft: it has conventional controls, with the auto-trim, MCAS and yaw damper overlaid. This is why AoA sensors had to be connected to MCAS explicitly. On a fly-by-wire aircraft, MCAS would very likely not have been a separate system. Instead, it would just be part of a (modified) control law that uses the THS or elevator to cancel out the effect of the different engine position. This would be roughly similar to the A320 family, where the 318, 319,

320 and 321 are quite different in length and therefore the elevator has different authority (and stabilization effect). However, the control law (at least those above the direct law) abstracts over this difference and "simulates" very similar characteristics for all types. As we have seen in this chapter, the control law takes into account a whole set of data: the values from all AoA sensors (maybe even after voting) would have been available to an MCAS-like control law feature. In addition, signals such as speed, change in altitude and attitude, as sensed by an inertial platform, would also have been available. These could have been used for plausibility checking of the AoA value. For example, if, after pushing down the nose several times to reduce AoA, the AoA signal remained high because of an error in the AoA measurement, but the attitude suggested nose down, speed increases and altitude decreases, the aircraft would likely be not near a stall, so pushing down the nose any further would make no sense.

More generally, I think the risks introduced by fly-by-wire systems are likely to be more than compensated by the fact that in the vast majority of situations the pilots have a simpler virtual airplane to fly. This makes all the situations in which normal law is available much safer (or makes flying the aircraft possible in the first place, such as the F-16). I am sure this has prevented many accidents; as far as I know there is no statistical evidence that fly-by-wire aircraft are more accident-prone. Finally, protections really are the last resort, because operationally pilots should never reach and thus rely on them. A bank of more than 30 degrees requires a callout by the pilot flying, operational overspeed starts below the aircraft's limit, and if you can't hold +/– 100 feet of altitude or +/– 5 degrees of heading manually in A320 direct law you might well fail a simulator check!

Can we learn something about automation in general? One thing that is completely unacceptable is that crews are kept in the dark about systems that affect the flight path of an aircraft. I talked to several pilots when I wrote this text, and all of them very clearly expressed the sentiment "I want to know what the hell these systems are doing!" On the other hand, Rick Ruiz, my guest on Flying the Big Boeings, said on Twitter "Long story short: airliners aren't made to be hand-flown in [instrument flying conditions], they aren't made to be hand-flown in high traffic density areas, they aren't made to be hand-flown for passenger comfort. Turn the damn autopilot on and focus on your flight path."

I also read this statement by John Hamilton, chief engineer of Boeing Commercial Airplanes, defending the fact that by default there is no AoA indication in the cockpit: "There are no pilot actions or procedures during flight which require knowledge of angle of attack." Considering how important AoA is for the aerodynamics of the aircraft, I find this attitude surprising, even if, as he then continues, the procedures pilots use to operate the airplane are "focused around airplane roll and pitch attitude, altitude, heading and vertical speed." A 737 pilot I asked about this said "Indeed, I don't know of any situation where I'd use the AoA sensor routinely. On the other hand it might be useful for tracking down problems if other sensors fail—if you have the time." Exactly: if the aircraft knows about such essential data, show it to the pilots, especially when it becomes unreliable. While there is of course a tradeoff with overwhelming pilots with too much information, I think it is justified to display such a fundamental characteristic of the aerodynamic state of the aircraft. A dialog in the Airline Pilot Guy podcast, where a group of airline pilots chat about their profession, brings it home nicely:

> **NICK:** And of course the one gauge we are often missing from modern airliners that we would all love to have is AoA.
>
> **EVERYBODY:** Yes!!
>
> **NICK:** Which actually tells us directly, and very simply, exactly what the airplane is doing.
>
> **JEFF:** And guess what: all of the airplanes that we are flying have AoA sensors that are feeding various systems [...] but in most airplanes we don't have the display of AoA, but that's critical information that the airplane should display to the pilots.

So: is the malfunction of the MCAS a software problem? If you think that an unfortunate hardware design, such as the engine positioning on the 737 MAX, must never be "fixed" in software (see Gregory Travis' IEEE Spectrum article "How the Boeing 737 MAX Disaster Looks to a Software Developer") then this is a software problem, simply due to the fact that Boeing has used software. I do not agree with this position; it would rule out building any aircraft that rely on computers to fly, and all modern military aircraft, as well as all Airbus aircraft since the A320, do so, very successfully. A more useful question to ask is whether the software that implements MCAS on the Flight Control Computer malfunctioned. Let's revisit our distinction between verification and validation. I think we can say with certainty that this is not a verification problem. If we look at our five Swiss cheese ingredients cited earlier, it is obvious that only number two, the fact that it "restarts" with another 2.6 degrees after a five second break, could potentially be due to an internal bug in the software or a developer's misunderstanding of the requirements. Much more likely, however, is that this is a problem with validation: the software in the Flight Control Computer did what it was programmed to do, but this behavior was wrong. Somebody had specified exactly this behavior, and everybody who reviewed the specification (or tested a model, verified temporal properties or experimented with the real system) considered this behavior to be acceptable. It is probably most useful to attribute the problem to systems engineering: the use of only one AoA sensor at a time is clearly a systems engineering problem, and so is the issue of human-machine interaction that does not communicate AoA sensor disagreement to the pilots.

A final thought: in this section I tend to lay the blame on the engineering, and not the pilots. In an article called "What Really Brought Down the Boeing 737 MAX", William Langewiesche looks at the irregularities around the maintenance of the Lion Air aircraft and the lack of proficiency and airmanship of both accident crews. While acknowledging Boeing's imperfect engineering, he very clearly attributes the primary cause of the accident to the crews. In the end, every aircraft designer has a fundamental question to answer: do you expect the flight control system to never make mistakes and therefore give it the last word (as Airbus does) or do you let pilots have the final say regarding the trajectory of the aircraft, relying on their proficiency (which is Boeing's philosophy)? It's not a black and white answer: flight control systems are never 100% bullet proof, and pilots must be reasonably competent. And as I have said in the introduction to this section: let's wait for the investigations to determine the cause of the accident. However, in the context of our discussion about flight control systems in this chapter, we can certainly conclude that the design of the MCAS system was not a masterpiece of engineering!

8 | *(Flying) Simulators*

FLIGHT TESTING Flight testing is a very important part of the development of an aircraft. I visited Airbus in 2016 and talked with Peter Chandler, a former chief test pilot at Airbus. Peter started his career with the Royal Air Force flying Hunters and Tornados. He then attended the USAF test pilot school at Edwards Air Force Base and worked as a test pilot for the RAF, then spent six years as captain on A340s with Virgin Atlantic, before joining Airbus.

There are two kinds of test pilots at Airbus. Production test pilots check every newly built airplane for conformance with specifications and help iron out any kinks before its delivery to customers, while experimental test pilots work on the initial test and certification program of new types. Peter mostly works with the latter but also helps out with the former. He regularly flies the A320, A330, A340, A350 and A380. When I asked him about how it is possible for him and his colleagues to have more than two type ratings at the same time, he said that, because they do not fly passengers, airline regulations do not apply. He doesn't really have a "type rating" for all of these—type ratings are concepts from the airline world. Peter: "And initially, when we first flight test a new type, there isn't even a curriculum for a type rating." Good point!

An important part of the early phases of flight test is the validation of the flight control system. They "fly" the aircraft in the simulator for months before committing to the first actual flight. The simulator uses exactly the same flight control system with exactly the same software. There is an "iron bird" test rig that contains the systems relevant to flight control—actuators, hydraulics, electrical—which are all exercised as the pilots fly the simulator. The pilots learn to fly the virtual aircraft represented by the flight control system, hoping that the real aircraft behaves just like it. Peter: "As I remember, there are hardly any big surprises, certainly with the 380 and 350 there were none."

During the first actual flight they perform takeoff and landing in direct law, and engage the more advanced laws only during the less critical phases of the flight. In contrast to production aircraft, the prototypes have switches in the cockpit to change the flight laws. The first few flights are known as "identification flights"; their purpose is to identify where the real airplane behaves differently than the model implemented in the flight control system. Next come envelope expansion flights, during which the flight envelope—speed, altitude, attitude—is explored systematically. This also involves the extremes of the flight envelope, where flights are often flown in direct law to make sure the protection mechanisms don't prevent the crew from exploring these regions. For example, they determine the minimum control speed, the lowest speed at which the aircraft is still controllable if an engine fails. This is relevant because the thrust is then asymmetric and has to be compensated by the rudder. They verify the minimum speeds at various weights and flap settings, and actually stall the airplane: "As you get slow you'll start to get a little bit of burble, a little bit of buffet, which then increases and gets really quite strong just before the stall. [When it finally stalls] Sometimes you get a wing drop, sometimes you get a build-up of sideslip, sometimes you get a little bit of pitch up, depending on how the wing has stalled."

Of course they never spin the airplane; that would be unsurvivable for the structure. They do, however, go well beyond the speed up to which the aircraft will be certificated. A major reason for this is to validate the absence of flutter. They induce oscillations, just like Perlan, and the flight test engineer in the cabin, as well as the crew on the ground, verify that they are properly damped. Like Perlan, the aircraft has strain gauges and other sensors installed throughout to measure the loads on critical components. These sensors are so tightly integrated with the structure that they cannot be easily removed, which is why early flight test airplanes are not subsequently sold to airlines. Peter: "We still have the very first A320 here, from 1987."

During Farnborough 2016 I talked to Tony Flynn, who works for Airbus Defense and Space, testing the A400M, the four-engine turboprop tactical transport. The A400M has basically the same cockpit philosophy and control laws as the Airbus jets, so I wondered how he could fly an air show display where he would pull up to 50 degrees nose-up, then roll the aircraft to 120 degrees: why don't the protections prevent these maneuvers? Do they fly in direct law? His reply: "It's the standard normal protections, we are fully protected during the maneuver. It is very easy to fly, you could teach it to your grandmother. We actually use the protections to help with the maneuver: I pull the stick full back, and the flight control laws give me exactly 3 g. When we get 50 degrees nose up, I just apply full aileron, stick in one corner, and the protections stop me at 120 degrees of bank, and I just wait until we come out the other side, and I recover." The crucial piece of information here is that the protections let them go to 3g, 50 degrees pitch and 120 degrees of bank. "Obviously the military wants more maneuverability, and 120 degrees seemed reasonable." He was generally quite happy with his role of testing the A400M: "We have much more maneuverability than the jets, twice the roll rate [30 degrees per second] and we can pull 3 g. It feels like a much smaller aeroplane. Feels light and maneuverable. You can throw it round the pattern like a trainer, you can fly down a valley at low level, feeling very very comfortable. For a pilot it's a pleasure to fly."

As we continued to chat he mentioned that he had spent the last two years developing the receiver role for air-to-air refueling. The challenge here is to maneuver a relatively small probe at the front of the A400M into a 3-foot receiver basket that streams behind the tanker on a hose. It's challenging to get the probe in because (a) it's a small target area, and (b) the receiver airplane, especially a large one like the A400M, tends to push the basket away with its own bow wave. This kind of flying is very different than, for example, racing down a valley with lots of yanking and banking. The beauty of a fly-by-wire aircraft is that you can provide a tailored control law for different parts of the flight. This is exactly what Airbus did: they added a flight control law for the air-to-air receiver role. The law includes a feature called "direct lift control" to adjust altitude very precisely: "When in receiver mode, the spoilers are automatically deployed by a little amount. So then if you need to come up 5 centimeters, we can just bring in the spoilers a tiny amount." And while Tony didn't elaborate on that, I suspect the law also uses the control surfaces cleverly to provide very precise translation in forward/backward and left/right directions. I could even imagine that the stick, maybe just for one of the two pilots, is reconfigured so that "pitch" means forward/backward and "aileron" means left/right.

This idea of changing the flight laws during flight can be taken much further, as illustrated by ATTAS, an experimental airplane operated by Germany's DLR until 2012. It was based on the Fokker VFW-614, and as such had a traditional mechanical control system. However, a fly-by-wire system was overlaid on that and connected to the mechanical controls (so strictly speaking it was not fly-by-wire because the mechanics were still there; rather it was computer-mediated). The point of this system was to fly with different laws that simulated different aircraft. ATTAS was basically a flying simulator. For example, DLR had experimented with a blended-wing transport, an airplane basically shaped like a ray (the fish, not light). Because the fly-by-wire system could always be mechanically disconnected, allowing the pilot to continue flying with the undisturbed mechanical system, the requirements for safety and reliability were much lower, which is why it was feasible to develop new flight laws relatively quickly. Of course the VFW-614 could only simulate aircraft whose flight characteristics were a subset of its own—you couldn't go faster, climb higher or roll more quickly than the actual aircraft could. Which I suspect is one reason why many research projects were not so much about using radically different flight laws, but instead exploited the fact that other, higher-level systems could be integrated with the flight control system easily: remote control from the ground, steep approaches, or energy-optimized continuous descend approaches to an airport.

Since the retirement of ATTAS, DLR operates ATRA, a modified A320, but with the regular flight control system. The focus has shifted from experimenting with flight laws to working on improving aircraft efficiency or reducing noise during approach. I guess one reason for this change in focus is that computer-based simulations have improved to the point where a flying simulator is no longer critically important. On the other hand, DLR operates a relatively new EC-135 with a flexible flight control system; the same idea as the ATTAS, but for helicopters. The USAF test pilot school operates the F-16 VISTA, which stands for Variable stability In-flight Simulator Test Aircraft. It also has a similar adaptable flight control system; because the base aircraft is more maneuverable, it can simulate a wider range of aircraft types, such as the C-17. The simulation profile can be changed while flying. This is exploited in the syllabus of the USAF Test Pilot School, as Marija Jovanovich, a recent graduate, told me in our interview. For example, the instructor in the backseat may change the gain of the yaw damper, demonstrating an aircraft that is laterally stable up to some particular sideslip angle, and then becomes unstable. They can also make a random change, asking the student to identify what it was they changed. Noticing and pinpointing changes in flight characteristics is a very useful skill for future experimental test pilots who will have to evaluate flight characteristics and then communicate with engineers about problematic aspects of how an aircraft handles.

THE SPACE SHUTTLE The Shuttle Training Aircraft was another aircraft modified to replicate "foreign" flight characteristics. A fleet of four Gulfstream IIs were modified to simulate the Space Shuttle during its landing approach. Remember that the Shuttle landed like a glider, except that it had a L/D ratio of 1 at hypersonic speeds and between 2 and 4 subsonically. The technical term for this kind of a glider is "brick". Shuttle pilots and commanders practiced the final 6,000 meters of altitude, racing towards the runway at 20 degrees nose-down and approximately 30–50 m/s rate of descent! With the Gulfstreams, the main gear was extended to allow for a 20-degree nose-down approach without over-speeding, and the thrust reversers were engaged during flight with up to full thrust. A computer-controlled flight control system was installed that provided a very high fidelity approximation of the characteristics of the Shuttle in approach. Custom-fitted direct lift control flaps were used by the computer to fine-tune the flight characteristics. For example, the Shuttle's center of gravity, the point around which it pitched, was much further aft than the Gulfstream's, and the point around which it rolled was much lower. The direct lift control was used to simulate this difference: when the pilot pulled on the stick the whole aircraft moved up instead of just pitching up. Originally the STA also had vertical surfaces mounted under the fuselage to simulate the side forces the pilot feels because the center of yawing is further aft as well; these were subsequently removed because they didn't add too much useful realism. During a training mission the training pilot sat in the left seat, which had Shuttle-style controls. He used a center stick with no feedback and an implicit attitude-hold, not unlike the Airbus. His side also featured the Shuttle's head-up display and screens, and the windows were masked to resemble the smaller windows of the Orbiter. The right seat had the Gulfstream's regular controls and was occupied by a safety pilot.

When I interviewed Shuttle pilot Duane Carey, he told me that he, like any other Shuttle pilot or commander, did around 1,000 landings in the training aircraft, until it was "what a fighter pilot would call part of your reptile brain, part of your instinct, you just know how to do it."

Because the Shuttle had no jet engines it could not go around—the first approach had to work out. It's all about energy management: you had to carry enough energy, either potential or kinetic, to be able to reach the runway, but not too much, because then you would overshoot. This is basically similar to me flying my glider, but the big difference is that the Shuttle operates at very different distances, speeds and altitudes. This fact was perfectly illustrated by Duane: "When you're getting closer to the runway, in other words you're getting over Texas, you're getting over Alabama, and approaching Florida..." In my glider I can see the airfield when I land and I know the local wind. I can basically estimate all the parameters relevant for the approach: how far out can I fly in the downwind, do I need spoilers on base, do I have to shorten the base, do I need to go faster on final because of headwind, do I use full flaps or not, and how should I employ the spoilers during flare? You can't really fly by the seat of your pants if your approach takes you all across the US—you need instrumentation and automation to help.

Let's look at how the Shuttle returned to Earth. There are basically four phases. The first is the entry phase, which starts at around 400,000 feet at what NASA calls the "entry interface". During this phase the following issues are relevant: if you enter the atmosphere at too shallow an angle, you skip off and escape into space again. On the other hand, if you enter too steeply, you get to dense air with too much airspeed and the Shuttle runs into design limits for heat load and dynamic pressure. So at the entry interface an open-loop guidance system keeps the Shuttle at 40 degrees AoA until the increasing atmospheric drag produces around 0.1 g deceleration. From then on the computer maintains a constant heat shield temperature down to around Mach 20. At this point it transitions smoothly to a constant drag phase of around 1 g deceleration down to Mach 10. The limits are shown in the following diagram. It is similar to that for the U-2 in the sense that the designers of the flight control system have to make sure that the Shuttle stays within specific boundaries defined by physics.

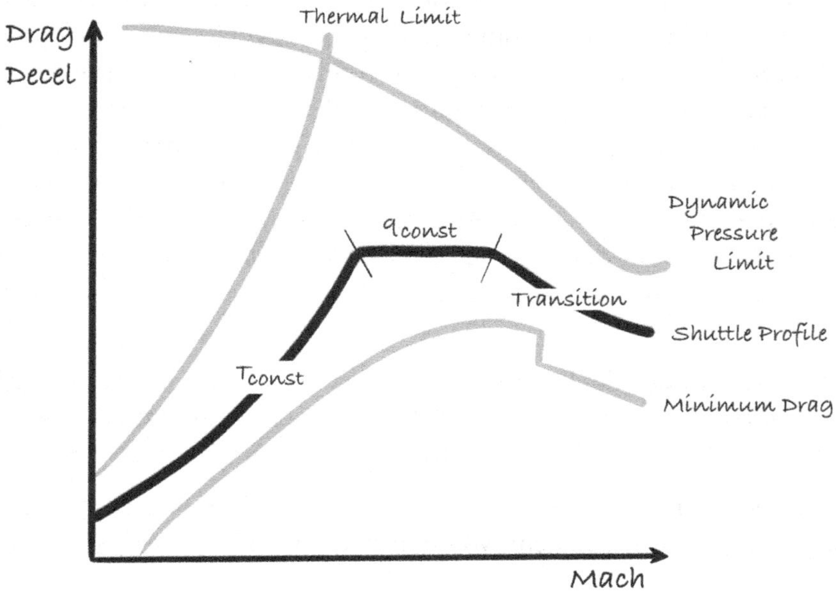

The high heat load phase especially has a very narrow range of AoA, plus/minus 3 degrees are required to ensure that the Shuttle stays within its design parameters. To still be able to influence the flight path, the crew use banking to maintain the right balance between heat and pressure. They don't really call it "turning" because it is not commanded aerodynamically—the atmosphere is still too thin; instead the attitude control system initiates the bank. Banking makes the lift vector, weak though it is, points sideway, causing the Shuttle to descend more quickly into thicker atmosphere. But it also makes the Shuttle deviate from its course, which is why at some point they reverse the direction of bank. This is called the roll reversal; you might have heard this term mentioned by the commentator when watching Shuttle landings.

As an aside, you might wonder whether they couldn't have designed the Shuttle with smaller wings so that they didn't need the banking. You don't really need the big wings during the flare for landing, as NASA had demonstrated (before they designed the Shuttle) with the X-24A, M2-F3 and HL-10 experimental aircraft. Like the X-37B "mini shuttle" and Sierra Nevada's Dream Chaser, these are lifitng bodies and have much smaller wings, and all of them also land airplane-like. The reason for the large wings on the Shuttle was a requirement by the military during the early phases of its development: they wanted it to be able to launch, quickly perform its mission and then, after only one orbit around the planet, land back at the launch site. Because this was envisioned to be a polar orbit, during the roughly 90 minutes the launch site would have moved east by around 2,000 kilometers. The Shuttle would have had to glide back that distance during re-entry, which required relatively effective wings. However, the military subsequently dropped out of the Shuttle program; but NASA was stuck with the big wings.

Back to the landing: as the Shuttle progressed into thicker parts of the atmosphere it slowed down and started to behave much more like an airplane; this second phase of the approach was called transition. Because of the lower speeds, heat was no longer a problem, and the adjustment of the glide angle could be done through AoA. On a nominal approach they would be at Mach 2.5, 80,000 feet, 60 nautical miles from touchdown when TAEM began, the Terminal Area Energy Management. Here the goal was to fine-tune the approach to make sure that the Orbiter arrived at exactly the right point, the right speed and on the runway's heading. If, because of high-energy orbits and high return mass, they could not get rid of enough energy through banking during the earlier phases of the entry, the autopilot commanded additional S-turns; the Shuttle's split-rudder air brake couldn't be used at these speeds. On this leg the Shuttle eventually slowed down below Mach 1, at which point the commander took over manual control: still fly-by-wire controlled, and computer-guided by what would be called a flight director in a transport aircraft. This was the phase practiced in the Shuttle training aircraft. The Shuttle tangentially intercepted the heading alignment cone, a roughly 360-degree circle (or less, depending on how the flight path and runway align) that was used to adjust heading and glide slope. If the circle was flown more tightly, with more g and more of the lift vector required to "pull" the g, the Shuttle would descend more rapidly. As they exited the circle they intercepted a microwave landing system that provided ILS-like guidance for the final phase, approach and landing. The speed brake was used to maintain 300 knots indicated airspeed, which, because it was a glider, was another means of shortening or elongating the glide. At around 2,000 feet they pulled up from the 20 degree "plunge" into a 1.5-degree approach that was used to bleed off speed and approach the runway with a reasonable rate of descend at 8 degree AoA. Gear down at 300 feet, flare between 170 and 30 feet, touchdown at around 200 knots, brake chute deployment and finally wheel brakes. Then the well-known dialog: "Atlantis, wheels stop." "Houston copies, wheels stop. Welcome home Atlantis." I was always deeply impressed by Shuttle landings; I found it the most elegant and majestic part of a mission, maybe because of its similarity to gliders. Unfortunately I never got to see one live.

MY EXPERIENCE WITH SIMULATORS Remember Tony's claim that my grandma could fly the A400M? I never got my grandma to try, but I had the opportunity to fly several airliners in the simulator. I flew a Dassault Falcon 50, a three-engine 10-person business jet, for about 45 minutes with an instructor, and spent 15 minutes each flying a Boeing 767 and an Airbus 380.

So how was it? We have to distinguish the actual flying (stick, rudder, flaps, engines) from managing the avionics (autopilot, flight director, tuning an ILS). For the latter, you need extensive training. These are non-trivial systems with a particular user interface philosophy, and if you aren't trained in using them there's no chance of doing so effectively when you fly a simulator for an hour or less. So I didn't even try. On the other hand, you can practice these systems with the respective add-ons for the X-Plane, Prepare3D and DCS simulators, to a degree of realism that is probably not much different from the large, professional simulators that airlines and training centers use.

Let's talk about the flying itself. The Falcon was reasonably straightforward. With my "copilot" telling me the correct speeds and flap settings I was able to land it in visual conditions (I could see the runway, didn't need an ILS), even with quite a bit of crosswind. I then tried to land the A380 at San Francisco. I initially over-controlled quite a bit, because my glider pilot's brain was not quite adapted to the flight path stabilization provided by the flight control system. Once I got the hang of it, I managed to land the Airbus as well. A less elegant landing, though. The 767 was the hardest. It is much bigger and therefore much more sluggish than the Falcon, and it does not have the fly-by-wire system of the Airbus to make it "easy". We were "flying" around Innsbruck, constricted by the narrow valley, and somehow the visualization system felt strange to me. I got the thing down, but don't ask!

If you ever decide to pay for an hour or two flying one of the pseudo-commercial simulators that are popping up in cities around the world, I suggest you spend a few hours in advance familiarizing yourself with the cockpit: find the relevant levers (flaps, slats, gear) and make sure you know where essential data (speed, altitude, rate of descend/climb, heading) is displayed. You will get much more out of the simulator; you don't want to learn these basics in a simulator that costs hundreds of Euros an hour.

"This is your purser speaking. The pilots are incapacitated. Are there any pilots among the passengers who can land this airplane?" There is this question that everybody asks all the time: would a glider or general aviation pilot be able to land one of the big jets in an emergency? Under the right conditions, maybe: enough fuel to loiter and familiarize yourself with the aircraft, visual conditions so you don't need automation to land, acceptable wind, and somebody on the ground who can talk you down and tell you the respective speeds and so on. Then it could work. In practice, of course, the proverbial stars won't align perfectly, and don't forget: you would be nervous as hell! So in a practical, real-world emergency, I think it's unlikely.

SIMULATOR MOTION CONTROL Explaining how simulators work would be a chapter on its own; there are models of the aircraft in terms of shape, mass distributions and moments, models of the environment, visualization systems, drivers for stick forces and instruments, as well as the motion system. The quality of all but the latter is purely dependent on computers: more computing power allows more detailed models, which result in higher fidelity in the simulated aircraft physics, the "cleverness" of the environment, or the details of the visuals. But the motion system has some very interesting tradeoffs that are governed by more fundamental constraints, so let's look briefly at this in more detail.

The motion system is crucial in all simulations that concern the handling of an aircraft or other vehicle, because the human in the control loop takes the forces they feel as an implicit input into the human control algorithm. So to develop intuitive flight control systems, to find out if pilots react correctly and quickly to attitude changes, or whether the driver of a car can safely avoid an obstacle that suddenly appears in front of them, an accurate replication of the motions of the real system is crucial.

As we know, force is proportional to acceleration: $F = m \cdot a$. Acceleration integrates to speed, and speed to position. So whenever we want to produce a force to make simulated flight feel realistic, it means that the simulator must move. Ultimately this requires space. There is only one exception to this rule, one way to get a force without needing movement: gravity. Gravity is there all the time, albeit acting in a fixed direction. But we might be able to exploit it cleverly to simulate other forces.

Let's say you want to simulate an airplane accelerating on the runway. The acceleration is in +x direction (towards the nose of the aircraft); the pilot feels a "push" in their back. The simulator could create this force by actually moving the simulator cab forward with the respective accelerations and resulting speeds. But then you'd need a physical space that is as long as the real runway. This is clearly not feasible. What else can we do? One idea is to scale down the acceleration, which means that you produce a force in the correct direction, for the right amount of time, but with lower intensity. It turns out that humans are not very good at relating a visual cue to the magnitude of a force. By scaling the force down by, say, a factor of three, one can save some space—but a third of the length of a real runway is still way too much for a simulator, of course.

A better alternative, and the way it is done in today's moving-base simulators, is to tilt the cabin backwards: a component of gravity now acts towards the subject's +x direction, it feels like a push in the back. Because the visualization system does not show the backwards tilt, and because the subject has no outside visual reference, the body "believes" that it is accelerating. However, there are several problems with this approach. First, the maximum acceleration that can be simulated is obviously 1 g. And when you use that, all of gravity acts in the pilot's +x direction, leaving no –z component. In other words, your butt no longer feels any weight, which is unrealistic: it is as as if you're in a weightless spacecraft accelerating forwards at 1 g.

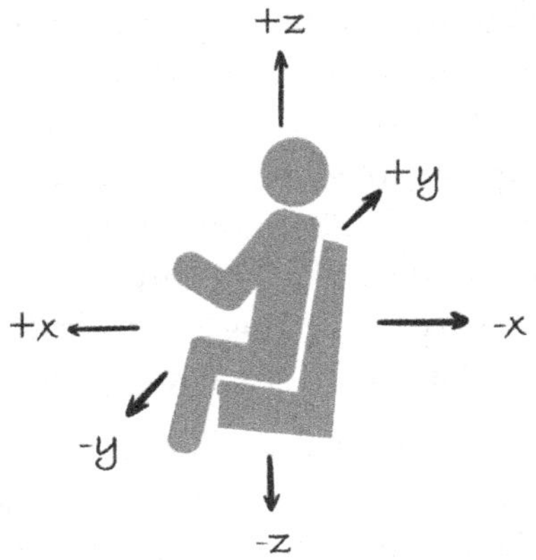

This is particularly ironic when trying to simulate acceleration in a turning aircraft. In reality you get heavier in your seat: you feel accelerated in the –z direction. But if you use backward tilt to simulate this, the force you feel is actually in the –x plane; you get lighter in z! In my experience in simulators you do feel "accelerated", but it also feels a bit strange.

A second problem of using tilt instead of actual linear motion is that, in order to get to the tilted position, the cab has to be rotated. If this rotation exceeds around 3 degrees per second people can feel the rotation itself, which of course is unrealistic. In practice, this means that you can only build up the acceleration at a relatively slow rate. A physical quantity that represents a change in acceleration is known as a jerk. When I talked with Friedrich Hoffmeyer about the driving simulator at Daimler, he said that trying to simulate stop-go traffic like this makes almost everybody sick, because the cabin sways back and forth.

So what are typical accelerations, and is tilting useful to simulate them? An airliner accelerates down the runway at around 1/3 g, a sports car at 1 g, an F-16's initial take-off roll is around 2 g, and a catapult launch on a carrier is 3 g. The sine of 30 degrees is 0.5, so you can get to around half a g of simulated acceleration by tilting the cab. This is good enough for realistic simulation of an airliner, and possibly for a scaled-back sports car, but it isn't realistic for military jets. Thirty degrees is also the maximum bank angle used by an airliner, so the acceleration from the turn can be reproduced realistically. They also roll in and out of a turn relatively slowly, and the jerk is small. A fighter jet's "yankin' and bankin'", with up to 9 g in an F-16, is clearly infeasible. Centrifuges can get up to such high g loads, but of course they build up really slowly because of the time needed to spin up. This is not useful for simulators.

Today's mainstream simulators mount the cab on a hexapod, also known as a Stewart platform. This allows whatever is mounted on top of it to move in six degrees of freedom: the three linear directions, plus rotation around all three axes. Hexapods use six hydraulic (traditional) or electric (modern) actuators arranged in a way that allows each to move independent of the others, which means that there is no risk of mutual blocking due to imperfect control.

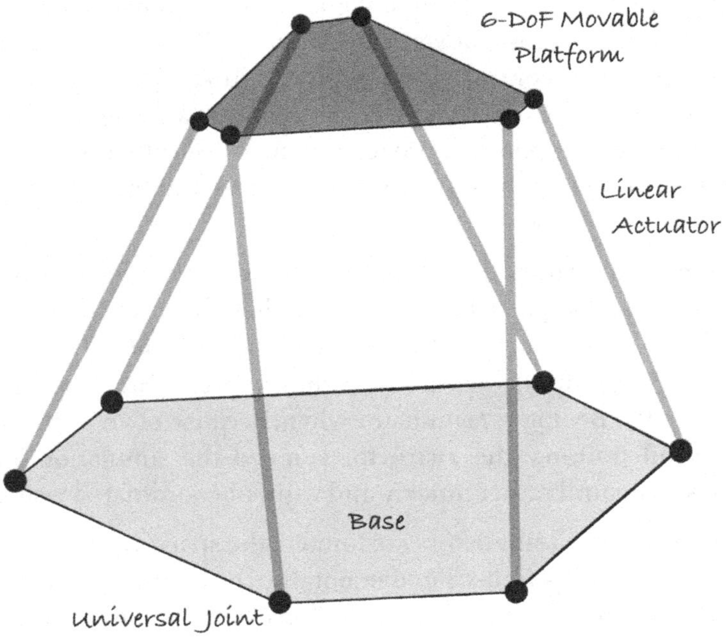

It is fairly obvious how a hexapod can produce the tilt angles. It is similarly obvious that it can produce a little bit of linear up/down movement. For example, the rumble of a runway can easily be simulated in this way (if the runway at Düsseldorf airport is as bad in reality as it was in the simulator, it is in dire need of repair). But it can also move left or right a little by cleverly controlling all six actuators. However, there are limits because all the axes are connected: imagine your hexapod is fully extended upwards (for example if you have just lifted off vertically in your simulated helicopter) and you now want to tilt backwards or move linearly sideways. The platform can't produce this motion because some of the actuators would have to extend further, which they can't.

One way to reduce all these problems is to add an actual linear axis and mount the hexapod onto a platform on this axis. This allows accelerations in the direction of the linear axis to be replicated more precisely. For example, Daimler's car simulator has a 10-meter linear axis. The simulator is used mostly for evaluating driving characteristics, in which the linear axis is used to simulate side forces realistically, for example when switching lanes on a highway. NASA's vertical motion simulator is even more sophisticated. Like Daimler's simulator, it has a roughly 10-meter lateral axis. But in addition it has a 2.5-meter linear axis for forward/backward motion, a dedicated axis for 25 degrees of roll, pitch and yaw, and also a 20-meter vertical axis (which defines its name). This simulator can faithfully reproduce accelerations in every direction. As Bimal Aponso told me, NASA uses it mostly for the design of flight control systems in which, as I have said above, the accurate reaction to a pilot's input is crucial to give the pilot a "feel" for the airplane. Examples include the challenging task of landing a helicopter on a ship's moving deck, the development of flight control systems that compensate for loads slung under a helicopter, and those for vertical-lift aircraft like the V-22 or the F-35. It has also been used to design the flight control system of the Shuttle, especially for the last few dozen meters during the landing flare and rollout. In addition to its use as a research tool, it is also used routinely to train Shuttle pilots: approaches with the Shuttle training aircraft end a few meters above the runway, while the remaining few meters and the rollout are trained in this simulator.

Bimal said that in their simulators "People usually do not get motion sick because the visuals and the forces are so closely aligned. People get sick when the forces do not correspond to the visuals." I felt this misalignment issue in my stomach when I was driving a visually very immersive fixed-base driving simulator—none of the expected forces occurred! I felt similar stomach confusion in the Boeing 757 simulator when, because of an impending crash during one-engine-out takeoff training, the instructor stopped the simulation and the simulator moved the cab back to neutral rather quickly and without coordinated visuals.

Of course even these advanced simulators are limited: the space available for movement, and hence to produce the forces, is finite. Fundamentally, the challenge for the designer of the motion simulation algorithms is to reproduce the forces that occur in the (usually) unlimited reality within the finite space available in a simulator. The algorithms used here are called "wash-out algorithms". Consider the example where the real-world requires a continuous linear motion which the simulator simulates with a finite linear axis. The simulator will eventually run into a hard stop. A wash-out algorithm might slowly reduce the motion to zero, hoping that the human in the loop will not notice. Or it might combine different axes: for example, an acceleration might initially be created with a linear axis because it can create a bigger jerk, but then, as it slows down because it runs out of space, a corresponding tilt might be phased in to compensate and produce a force that can last forever.

There are other quite clever approaches to better exploit the available space or degrees of freedom. For a nice example let's go back to Daimler's driving simulator. Suppose you are driving in the right-hand lane of a three-lane autobahn and the goal of the simulation is to find out how reliably the driver can avoid a sudden obstacle. The scenario will probably start by asking the driver to follow a truck for a while. This has two purposes. First, the "suddenly occurring" obstacle can be created by the truck moving to the left-hand lane because the truck driver sees a stationary obstacle. Driving behind the truck, the car driver can't see this, so that when the truck moves away the obstacle is suddenly there. But more importantly, while driving behind the truck, the simulator can know, based on the predefined scenario, that the car will soon accelerate to the left, so it can move the cab to the right of the linear axis, very slowly, so the subject in the simulator does not notice. The full length of the axis is now available for use in the evasion maneuver. This pre-positioning of the actuators to maximize the travel for an upcoming maneuver is a clever idea, but it only works if the scenario is closely controlled. If movements are unconstrained, the best the simulator can do is to move slowly to the center of its axes while the simulated movement requires no acceleration. There's another requirement for this trick to work convincingly: the actuators, hexapod or linear, must not produce vibrations or noise when they move, otherwise the driver might be tipped off that something is going on. Daimler's linear axis rides on frictionless air bearings. The simulators used by Deutsche Bahn for training ICE train drivers apparently are not quite so well designed. My friend Tobi tells me: "You can hear the actuators clearly. They always remind me that I'm in a sim."

A final limitation for the fidelity of the motion simulation is of course the speed of the actuators. The total mass that is moved by the VMS' vertical axis is around 70 metric tons; it can be accelerated at roughly 3/4 g. It's quite impressive to see this huge structure move at such accelerations and the resulting speeds. Eight 150 horsepower motors drive the vertical platform, with additional motors for the less demanding other axes. The reason why 1,200 hp is sufficient is because the weight is neutralized by an equilibrator, consisting of nitrogen-pressurized columns that take the static weight of the platform. Similarly, Daimler's air-pressurized, electrically driven hexapod "feet" can only simulate motions of up to around 1 Hz. Daimler has a separate ride simulator that can generate up to 50 Hz, to simulate the damping of a car's shock absorbers on bumpy roads.

9 | *Helicopters*

FLYING THE ECUREUIL Let's conclude this chapter with an aircraft that is unstable yet still does not have a fly-by-wire system: the helicopter. In early 2018 I got to fly an AS 350 Ecureuil, a light single-engine turbine helicopter that is used for logging, dusting forests, checking high-voltage lines and the occasional sight-seeing flight. Its most powerful variant, the AS 350 B3, holds the record for the highest landing: 8,848 meters on top of Mount Everest. I visited Helix, located in Neuenstein, Germany, and spent an hour flying a "Squirrel" with Achim Widmann, operations manager and flight instructor.

We pushed the helicopter out of its hangar where it had spent the night with its fellow Ecureuils. Achim did a preflight check and verified that there was enough fuel in the tanks for an hour of flight. I indicated that I knew, in principle, how to fly a helicopter, so he kept the theory part brief. We strapped in and Achim started the engine. As soon as it had spooled up and ignited, he switched on the avionics and the intercom so we could talk again. We waited until the engines reached their operating temperature, and Achim then gently lifted the helicopter off the ground, flying at a meter or two towards the helipad a few hundred meters beyond the hangar. The helipad was surrounded by what he called the hover square: a marked area of a few dozen meters in length that is used as an exercise area for hovering. He set the helicopter down, then told me: "Ok, I will lift the helo a few meters and then hand it over to you. Just hover, and try to stay within the square." Ok, I thought, he's starting me off with the hardest part, hovering, but let's give this a try. Once I took over it took only a few seconds I had involuntarily left the square, with Achim taking over control, laughing compassionately. Holy cow the thing was sensitive! Let's understand why, and how you control it.

Instead of a wing, a helicopter has a rotor. The blades, three in the case of the Ecureuil, are basically little wings that have an airfoil similar to an aircraft's wings. Because the blades are rotated through the air by the engine, a helicopter does not require forward airspeed to be able to stay airborne. This is of course the point of a helicopter: it can hover, fly in any direction as slowly as required, and take off and land vertically in tight spots. There are drawbacks, of course. It's relatively slow compared to aircraft, relatively inefficient in terms of fuel consumption and more expensive to maintain because of its gearboxes and lots of other moving (and shaking) parts.

The rotating blades create torque: the engine, mounted in the helicopter's fuselage, turns the rotor, and the rotor tries to turn the fuselage. This is basically Newton's action/reaction principle for rotational movements. To prevent the fuselage from rotating, the helicopter has a tail rotor. It works fundamentally like the main rotor but is mounted sideways so it can create a force that counteracts the main rotor's torque. We can already see a necessary condition for the helicopter to be in a stable hover: the force created by the tail rotor must be equal to the main rotor's torque, otherwise the helicopter will rotate around its vertical axis.

So how does a helicopter climb or descend? Like any other aircraft, the "wings" have to produce more lift. The only way to achieve this with the (non-flapped) airfoils of the rotor is to increase the angle of attack collectively for all blades. As we know, this does not just produce more lift, but also more drag, which the engine has to overcome by applying more power. This of course also results in more torque, so the tail rotor has to compensate more. The necessary force is created by increasing AoA of the tail rotor blades collectively.

To fly forward, a fraction of the thrust produced by the rotor has to be directed aft. This is achieved by changing the angle of attack of the blades cyclicly: whenever a blade passes the aft hemisphere of the rotor circle its angle of attack is increased. Depending on the specific design of the helicopter's rotor, this either tilts the rotor plane forward or tilts the whole helicopter—eventually the helicopter will pitch down in either design. Because a fraction of the rotor's power is now used for acceleration instead of compensating for gravity, the overall force created by the rotor has to be increased, by collectively increasing the AoA a bit. Which leads to more torque, which requires a change in the AoA of the tail rotor.

The job of the pilot is to control all three mechanisms. He has a collective lever on the left of his seat. It looks like an old-school handbrake in a car, and by pulling it up the rotor AoA is increased collectively. Moving it down does the opposite. The pilot's right hand operates the stick, which changes the rotor AoA cyclicly, and the pedals are used to change the AoA of the tail rotor.

As we have seen, the three are connected through physics: changing one requires changing all others. This is what makes flying a helicopter so difficult, at least initially: getting your subconscious brain to intuitively understand these interactions and coordinate them correctly takes some time. Experience from fixed-wing aircraft doesn't really help much, apart maybe from being sensitive to accelerations in any direction. Because none of the dimensions of control has any inherent stability, the controls are very, very sensitive. When I finally got the hover more or less under control I had to move the stick only a millimeter to stabilize the helicopter! Glider pilots pride themselves at how delicate their controls are and that they can control the glider with two fingers and stuff. But there's no comparison to a helicopter: you basically just think of the control movement you want to perform! When Achim instructed me to hover the Ecureuil after takeoff I completely over-controlled it and we jumped out of the hover square—but at least stayed "in the country" as Achim dryly observed.

The faster you fly, the more stabilization and damping is provided by the airstream, and control sensitivity becomes much more conventional. Above about 80 knots the Ecureuil flies basically like an airplane and I felt much more at home. But the space between the two extremes is also interesting. Let's say we want to pull the helicopter off the ground a few meters, then accelerate slowly, and at maybe 50 knots lift it up into a climb. I start out by pulling the collective, which lifts the helicopter up, and stabilizing it with the pedals. I then carefully push the stick forward to accelerate, adding a bit more collective, and compensating with the pedals some more. As you accelerate past 25 knots, transitional lift becomes noticeable. This is when the rotor gains efficiency because the airstream becomes less turbulent and more horizontal. If you want to remain at the same altitude, you have to reduce rotor pitch a bit by pushing down the collective slightly. After accelerating further, if you want to climb higher you have to pull the collective, and you have to pull quite a bit once you leave the ground effect. The ground effect refers to the air cushion that builds up under a helicopter or airplane when it flies very close to the ground; a helicopter's rotor has more lifting power while it is in ground effect. So the technique I've just described—lift off a few meters, accelerate beyond transitional lift, then climb—is an energy-efficient means of launching the helicopter, because it cleverly exploits ground effect and transitional lift. It's also the technique of last resort to get airborne if your helicopter is heavy or you fly in hot-and-high conditions. If your helicopter has wheels rather than skids you can also do a rolling takeoff to get some of the same benefits. I saw the tilt-rotor V-22 do that several times at Bayreuth.

You also have to consider the ground effect during landing. Let's say you are flying around the pattern 300 meters above ground at 100 knots. When you turn onto base you'll want to descend to around 200 meters. You do this by pushing the collective down a bit. At the same time you want to start slowing down, so you pull the stick back. Essentially you're descending with the nose slightly high. When you turn onto final maybe a kilometer away from the helipad you want to be at around 50 knots. You keep the collective down and the nose up to slow down some more. When you get below 25 knots, transitional lift ceases and you have to pull the collective a bit to compensate. As you approach the ground effect the drag slows you down and you have to push the stick forward to keep going. For the last few meters, hopefully at close to zero forward airspeed, you have to push the collective down significantly to "push through" the ground effect. After practicing a few takeoffs, patterns and landings I did manage a safe although not very elegant landing myself. At the end of my one-hour flying experiment we tried the hover square again: not perfect by any means, but I managed to stay inside it.

I wondered whether there wasn't some kind of automatic hover mode. Achim told me that yes, these do exist, but he said it wouldn't be useful, because their use case is very dynamic and hovering for long periods isn't typical. I guess such a system would also make the helicopter way more expensive, because it involves a lot of safety-critical electronics. In 2018 I visited the US Army's Apache helicopters in Illesheim. Pilot Caleb Marheine told me that the Apache indeed has an automatic hover system. During their ground support missions they often hide behind a ridge, monitoring ground activity, coordinating with ground troops or aiming their weapons. Not having to hover manually makes a big difference there.

EMERGENCIES If the engine fails in an aircraft you can continue flying as a glider. Depending on the type of aircraft, you'll have an L/D ratio of 15 to 19 in an airliner and between 8 and 10 for light aircraft such as Cessnas. You will therefore have some chance of gliding to a place where you can "land" without killing everyone or destroying the airplane. How does this work in a helicopter? Very similarly, actually. You also convert potential into kinetic energy, but it's the kinetic energy stored in the rotor. The maneuver is called "autorotation": if the engine fails, you immediately push down the collective. This reduces the drag on the rotor, preserving rotor speed. The helicopter descends rapidly with the airflow turning the rotor. A helicopter "glides" with an L/D ratio of about 4, so a suitable landing spot had better be right beneath you. Some helicopters, such as the Apache, put a symbol onto the pilot's helmet-mounted sight that represents the point where the helicopter will "land" given the current altitude and forward speed. When you get close to the ground you pull the collective to produce lift; you basically use the rotational energy of the rotor to create one last bit of lift to arrest your rate of descent. At the same time you pull the stick to slow down your forward speed, pushing the stick forward at the last moment to make sure the tail rotor does not hit the ground. If you time it correctly—and you can try only once—this will give a nice, soft touchdown and look rather unspectacular. Achim demonstrated it a couple of times, it works. Every pilot practices this a lot!

What you need, of course, is altitude, otherwise you can't "fall" to speed up the rotor. Alternatively, if you are only a few meters high, you can just push the helicopter down and the gear or skids will absorb the energy; you'll likely not damage anything. If you are fast, you can use the forward speed to either climb (and then autorotate) or directly accelerate the rotor (nose up, slow down). There is a critical range of altitude and speed, the "avoid region", maybe between 5 and 200 meters, where neither approach works well. In a civilian helicopter you try to avoid this altitude band if you can.

The reason why we did not launch from the slightly cramped area in front of the hangar relates to engine failure as well: in that case we would not have been able to perform an autorotation with its required slight forward speed—there would have been no suitable place to land. Instead we lifted off to about two meters, an altitude where we could safely "push down" if the engine failed, then flew at two meters to the helipad. The area around the helipad was clear from obstacles and autorotation from 150 or 200 meters would have been possible. A similar approach is used by the London Air Ambulance helicopters when they lift off from their helipad on top of the Royal London Hospital, 100 meters above London: as they climb away from the pad, they move slowly rearward up to around 100 meters above the pad. If one of their two engines were to fail during this critical phase (they are too heavy to hover on one engine), they can always "glide forward" back to the pad. As you can imagine, there are not too many other places to land in a crowded city, especially when you are in an engine-failure-induced hurry.

What do you do if the tail rotor fails? You autorotate in just the same way. Since the main rotor is no longer turned by the engine, there is no reaction, no torque that the tail rotor must compensate. I'm sure that's not so easy in practice, however. "Many have tried, and many have failed", or so I've heard.

Let's focus back onto the Apache. They routinely operate inside this critical altitude band of five to two hundred meters during their close air-support missions. What do they do if an engine fails? The Apache has two of them: can it hover with only one? In ground effect it is possible, but only in a very low density altitude. A pilot told me: "The only time I hovered single-engine was on maintenance test flights, in ground effect, without wing pylons installed on a cold January day in Germany." Out of ground effect the Apache cannot do it, a good demonstration of the significant impact ground effect has on the performance of a helicopter. They could at least lower themselves slowly with only one surviving engine, or push the nose down, accelerate, exploit transitional lift and fly away.

What if both engines fail? The Apache is designed to make the resulting crash survivable. Some of the energy is absorbed by the collapsing gun cradle mounted under the nose, some is absorbed by the landing gear, which also transforms some of the vertical speed into forward speed to lighten the load on the crews' spines. The collapsing seats make a contribution as well, and the helmet-mounted sight system breaks away at a particular g-load to make the head+helmet+sight lighter and thus reduce the load on the pilot's neck. What if the tail rotor fails at this critical height? The Apache has a tail wheel that is sturdy enough to land at 100 knots, a speed at which the airflow provides enough stabilization even without a tail rotor. In particular, the vertical tail is not a symmetrical airfoil; it creates a sideways force that helps to compensate torque at speed. Pragmatism for the win.

The right reaction to a particular in-flight emergency is crucial. It's usually also time-critical—there really is no time to think. Pilots have to diagnose the problem and then react instinctively, which is why such things are practiced a lot.

FLY-BY-WIRE If helicopters are so work-intensive to fly, how is it that fly-by-wire is not used as widely in helicopters as it is in airplanes? I don't really know, but the situation is changing slowly. The first fully fly-by-wire controlled military helicopter is the NH-90, a medium transport helicopter developed in the 1990s by one of these purpose-built European multi-company consortia, NH Industries, and introduced into service starting 2007. In the same timeframe a Boeing-Sikorsky team developed the RAH-66 Comanche, a helicopter that was supposed to replace the OH-58 Kiowa and work together with the Apache. It also had fly-by-wire controls but the programme was cancelled.

I visited the German army's helicopter transport regiment 30 in Niederstetten to learn about how the fly-by-wire system helps the pilots to fly the NH-90. Like the Airbus, they can use different laws to control the helicopter. In the most basic law, the collective, cyclic and pedals work just like on a traditional helicopter. However, the crew usually elects to fly with attitude hold enabled: when the pilot lets go of the stick after bringing the helicopter into, say, a bank, the helo will stay in that bank. Heading, speed and altitude hold modes can be activated as well.

Perhaps more importantly, the NH-90 has a hover mode that stabilizes the helicopter around four axes, the fourth being power: it can deal with weight or balance changes during hover. This is relevant in the naval environment when refuelling from a hose while hovering over the deck of a ship. Tom, my guest says: "We have this joke around here that we'll just park the helo in hover mode over a café, winch ourselves down, get a coffee, and come back after a few minutes to continue the flight." When the helicopter is in hover mode, it actively keeps its position (as long as it is technically able to) in the face of gusts. However, the pilot can still use the stick to slowly move the helicopter without exiting hover mode; more importantly, he can hand over (fine-grained) control to the guy on the which who might have a better view of whatever object they are hovering over; he can also use the winch trim unit to dial in a little bit of forward speed to, for example, precisely follow a ship. A number of specific procedures are also built in, such as a precision approach to a particular location (useful for a landing in environments with reduced visibility) and flying a programmable pattern (useful for search and rescue, especially over the sea where visual references are scarce).

The system relies on quad-redundant computers and data paths, two hydraulic systems and two engines. The main and tail rotors and their gearboxes are of course still single points of failure. In terms of sensors, the system relies on GPS, inertial platforms and downward-looking doppler radar to tracks the ground as it moves under the helicopter as a fine-grained reference to detect movement. There are no sensors that track a slung load, so dealing with its movements and oscillations is still left to the pilot.

The US Army is currently working on the Future Vertical Lift programme, a project to develop a common platform that will eventually spawn replacements for "everything": the UH-60 Black Hawk, the CH-47 Chinook, as well as the Apache and the long-dead Kiowa. Two teams are still contending for the contract at the time of writing: Bell is developing the V-280 Valor and Boeing-Sikorsky the SB-1 Defiant. Both have in common the use of two contra-rotating rotors, thus avoiding the need for a tail rotor. Both will also be much faster than traditional helicopters, but in all other respects the designs are quite different. Bell is proposing a tilt-rotor configuration similar to its V-22 Osprey. The Valor, as well as the V-22, has a fly-by-wire system, which is especially important for the transition from helicopter mode (with the proprotors axes vertical) to airplane mode (axes horizontal). The angle of the engine nacelles can be changed by a little thumbwheel on the collective. Like the NH-90, the Osprey can hover in low wind with no pilot input.

The Defiant, in contrast, has vertically-stacked coaxial rotors. To increase speed over that of regular helicopters it also has a pusher-prop at the tail. This design had been researched with Sikorsky's X2, which achieved an unofficial helicopter speed record of 400 km/h. A key to making this possible is to slow down the rotor at such high speeds to avoid the forward-moving tips going supersonic, one of the limitations for helicopter speed. Its fly-by-wire control supports Airbus-like attitude stabilization. It has a thumbwheel, too, but this is used to control the pusher prop. Using the prop, the Defiant can accelerate forward without lowering its nose—a very uncommon sight for a helicopter. I saw a video in which the test pilot, while flying in a simulator, quipped that, "It is as easy to fly as an airplane." There you have it: finally, helicopters will be just like airplanes :-)

10 | *Flying the F-16*

A MIRACLE This chapter was supposed to end with the helicopters; I had written it over Christmas 2018. But then something unexpected happened. On January 6, 2019 I sent an email to the USAF Thunderbirds, applying for a media flight. I had wanted to do this for a long time, and I finally decided that omega tau was now "big enough" to give it a try and have a non-zero chance of getting accepted. I told them about the podcast and its audience, my flying experience and my personal dream of flying in an F-16. Two days later I got a reply from Staff Sergeant Ned Johnston, one of their aerial photographers and media people. It started with this:

> Thank you for your interest in aviation and the Thunderbirds!

Sounds like a polite "no", right? It continued:

> Would you be able to share with me some analytics from your podcast? In order to get your flight staffed and on the route to approval, I have to write up a justification to Air Force headquarters. Any statistics you might have to sell yourself to them will make it easier for me to get you approved and started on this journey.

I almost fell off my chair when I read this! This might actually really work out! Holy smokes! Five months later, after providing the requested analytics, carefully reading their operations manual, measuring in detail my physical dimensions, signing an air force transportation waiver (you can't sue us if you die during the adventure) and getting a medical from my physician, I am on a Delta flight from Stuttgart through Atlanta to Fort Wayne. I am scheduled to fly with the Thunderbirds on June 7, the Friday before their Fort Wayne Airshow appearance.

Between January and June, and in particular after we had agreed on the June date in early April, I couldn't stop thinking about the ride whenever my brain was in idle. Flying in an F-16 has been a dream forever, and now the dream would become real.

THE BIG DAY I was supposed to show up at Fort Wayne Air National Guard Base at 0630; no problem, I was still jet-lagged because I'd arrived in the US only two days earlier. I was up at 0400. Arriving at the gate, base security checked my passport and searched my car. I was then allowed to drive onto the base, parking in front of the headquarters building. I was walked over to the ops building by a media person of the 122nd Fighter Wing, the host unit at Fort Wayne. There, after a short wait, we met Cory Bush. He is a member of the Thunderbirds' media team, and he and Ned had been my primary contacts before the flight.

I also met another "flyer", Rod Bradtmueller, a policeman from Fort Wayne. The Thunderbirds typically do two "guest flights" at each air show weekend. These guests are either non-pilot team members, celebrities, members of the media like me, or "local heroes", people who have stood out in their community. Rod fell in this last category: he's a police officer who thwarted a bank robbery! Rod was scheduled to fly the previous afternoon, but because the Thunderbirds arrived late—their C-17 transport had come in late from its previous mission—his flight was postponed to Friday. Luckily the team brought both of their two seaters, so we could fly at the same time.

MEDICAL BRIEFING The day began with a conversation with Jafet "Doc" Ojeda, the team's flight surgeon. We talked primarily about how to avoid motion sickness and how to handle the g forces. But even before this briefing, I had to get a certificate from my physician that declared me generally healthy. We also got an email a few days before the flight that read as follows:

> There are a few things you can do to minimize the possibility [of motion sickness]: hydrate, sleep and eat. Starting right now, hydrate! Drink more water than you normally do, then drink some more. Hydration combats motion sickness, so this step is key. Please do this only until the evening [before the flight] to avoid the need to use the restroom every few hours during the night, as sleep is important as well. Try to get a good night's rest, at least seven hours if possible. The day of your flight, we want you to have food in your stomach, but nothing greasy and nothing spicy. We recommend fruit, bread, bagels, oatmeal, or a light sandwich.

We complied, of course, and both Rod and I walked around with a water bottle all the time. Doc then told us what to do during the flight to minimize the risk of motion sickness: move the seat as high as possible to make sure you can see out as well as possible, look outside and don't stare at the instruments, and make sure you don't move your head during high-g maneuvers. Make sure you open the ventilation, as feeling hot increases the risk of motion sickness. And of course talk to the pilot; let them know how you feel and tell them to stop maneuvering if you start to feel sick. They can tell from the tone of your voice if things get critical. These are all things, by the way, that I also tell my passengers in the glider. I usually also tell my passengers to (lightly) grab the stick, so they can anticipate from the stick's movement that the aircraft will soon move, reducing the surprise factor. This cue is not available in the F-16 because the rear stick does not move in sync with the front stick. Instead there's a last trick you can use: switch to 100% oxygen. This works miracles. As a last resort there's a barf bag. We were reminded to remove the mask completely before we used it. The briefing ended with this slightly funny exchange:

> **Rod:** How many people get sick that do this?
>
> **Doc:** I don't know if you wanna know that.
>
> (Laughs)
>
> **Rod:** I do wanna know.
>
> **Doc:** (Pause) Hmm... not everybody gets sick.
>
> (Laughs)

More seriously, Doc said that getting sick is not uncommon. He added "If you need something after landing let me know—I have my bag of meds." I was optimistic that I would not feel sick; after all, I had never ever felt sick in an airplane no matter what the airplane did. And indeed I felt fine for all of the flight.

I wasn't quite so confident about how I would handle the g. In the glider during a normal flight 2 g is common. I often do steep turns for fun in the evening: those might get up to 3 g, maybe 3.5. And I've experienced up to 6 g during glider aerobatics. However, the F-16 can pull up to 9 g! Maybe more importantly, it can sustain these high g forces: it's not a problem to pull 5 g for half a minute. So, if you will, the g integral—strength multiplied by time—will be very much higher in the F-16 than in the glider, even higher than a comparison of the pure numbers (6 vs. 9) suggests.

The problem with positive g forces is that they drive the blood towards your lower body and away from your head. The organs in your head that need most oxygen are your eyes, so as g increases your field of view starts to shrink—you get tunnel vision. In the extreme case you might become temporarily blind. It is also possible that you could lose consciousness for a moment, a phenomenon known as g-LOC, or g-induced loss of consciousness. It's not dangerous per se, but after a g-LOC your body is quite relaxed and has reduced g tolerance, so it's quite likely that a g-LOC will end the maneuvering part of the flight. Negative g is not much of a concern by the way, since fighters try to avoid it as much as possible. For example, instead of pushing the nose over to get to level flight after a climb, a fighter will typically roll to inverted, pull to level, and roll again. A roll rate of 270 degrees per second makes this feasible.

To avoid all these problems you have to keep your blood above your abdomen. To help with this fighter pilots (and their lucky passengers) wear a G-suit. This is basically a pair of pants that inflate with air and compress your abdomen, thighs and calves. A G-suit increases your g tolerance by about 1.5 g. As a corollary, if you pull between 7 and 9 g, you still have to deal with 5.5 to 7.5 g yourself. People have a natural tolerance for around 4 g. The remaining gap has to be filled with what's known as the anti-g strain maneuver. This is what Doc taught us in the rest of the briefing. The idea is to strain your lower body. Specifically, you pull up your toes towards your knees to strain the calves, press your thighs outside (against the pressure of the G-suit), tighten your butt and push out your belly. At the same time you try to keep your upper body relaxed so you can talk and breathe.

Breathing is also hard at high g; your lungs are compressed by the g force because your chest wall becomes heavy as well. This is why the anti-g strain maneuver has two more ingredients. Before the g builds you take a preparatory breath: you breathe in once, normally, not a deep inhale. The pilot will announce oncoming g by saying "Here come the geeeeeeeees". You're supposed to prep-breathe during the "here come", because the g will build quickly during "geeeeeees". And then, while the g forces are present, you replace the air in your lungs every three seconds with a purposeful exhale/inhale. Doc taught us to forcibly exhale by mouthing a pronounced K-sound and then quickly re-inhaling the expelled air. We practiced the maneuver for a bit; while it's not hard, it's easy to forget about it and fall behind the gs during the flight when you are excited and overwhelmed by the experience. Doc also warned us that the body wants to breathe much more often than every three seconds. Indeed, during the flight, I had to control my breathing rate consciously, I breathed too fast initially.

Doc suggested we keep our hands in our laps. While there are "towel rails" for the hands below the canopy, putting your hands there risks them dropping down under heavy g, which might hurt your hands or inadvertently press buttons. He also helpfully suggested that we don't use the ejection handle if we want to hold on to something. Doc closed with the suggestion that we should treat the flight as a workout. And indeed, I felt quite sweaty after landing from "fighting" the gs. He then signed the all-important form that declared us medically fit to fly. "Somebody will want to see this later", he said. Nobody cared.

EQUIPMENT Suit up was next. Let's read again from this email we got a few days before the flight:

> Wear or bring a cotton t-shirt, cotton underwear, and a pair of cotton athletic socks [...]. We will provide you with the flight suit, harness, helmet, G-suit and all other gear necessary to make your flight as comfortable as possible. All you will wear under your flight suit is your socks, t-shirt and underwear.

Since I had told them various body and shoe sizes in advance of the flight, everything was prepared and neatly laid out when I arrived in the room used by the AFE specialists Jasper Roberts and Kyle Boddie, the people who deal with aircrew flight equipment. They handed us an olive flight suit made out of Nomex, a "flame-resistant meta-aramid material", as Wikipedia says. It was surprisingly light. "Go to the restroom and put this on." So we did. We also put on comfy boots.

When we returned to the AFE room, there was what felt like a little ceremony: we were patched up. The guys attached to our flight suit a patch of the Air Combat Command, the part of the USAF that operates fighter and bomber aircraft, one for the 57th Wing, the parent unit of the Thunderbirds at Nellis AFB that also owns the Weapons School and the Aggressors I had visited in 2017, one for the Thunderbirds themselves, and finally a name tag that said "Thunderbird VIP". For the rest of the day we walked around in this flight suit feeling very cool.

The G-suit was next. You put it on like a (slightly heavy and bulky) pair of pants that are open on the side of the legs. You close the legs with a zipper and a set of latches—these are intended as a backup should the zipper fail. So far so good. However, in order for the pressurizing air hoses in the pants to do their job, the G-suit has to fit tightly in the first place. This is why we spent the next 15 minutes standing still while the AFE specialists used ten or so laces built into the G-suit to adjust its shape specifically to our bodies. Once done you have your own "personal" G-suit, flush with your particular body shape.

Next was the harness. This is like a parachute, in the sense that it has a strap across your chest and two straps around your butt/crotch. But instead of being part of a parachute, as in a glider, the harness is attached to the parachute in the ejection seat once you enter the cockpit. AFE explained to us how to connect and release riser lines; we'll have to do the former when we enter the aircraft and the latter, in a hurry, in the unlikely case that we have to eject and the wind pulls us over the ground after a parachute landing. We also learned how to attach and disconnect the seat pack, a bag with survival equipment that is part of the ejection seat and will dangle below us when we descend on the chute and the seat has been released. Finally, they told us how to disconnect the oxygen pressure reducer that is attached to the harness; we'd have to do that if we have to exit the aircraft in an emergency on the ground. We then took off the harness and G-suit; we'd see it again at the aircraft.

On to the helmet. We learned how to put it on without messing up our ears, how to close the chin strap that holds the helmet in place and how to lower and raise the visor. The helmet was surprisingly light; this makes sense, because you still have to be able to move your head under g. The final piece of equipment is the oxygen mask. It's connected to the helmet, fixed at one side and releasable at the other. During cruise phases you can remove it, which is good because it fits quite tightly and isn't very comfortable. I left it on all the time.

We spent the last 15 minutes with the AFE guys watching a Powerpoint presentation about ejection: the body position to take in preparation for ejection (all limbs tied back to the seat, head against the headrest), how to control the chute, how to control it if it has a hole, how to control it if the riser lines are tangled, and how to land—on ground, in trees and hopefully not on power lines. It was very obvious that this was considered a necessary formality and not a realistic possibility: "The Thunderbirds have never ejected a passenger", they told us. Thank God.

FLIGHT BRIEFING I then met Major Jason "Flack" Markzon, my pilot. As Thunderbird Eight he narrates during the demonstration programme and acts as the advance pilot; which means that he arrives at a show site a day early and sorts out any remaining organizational issues. He typically uses "his" F-16D to get to the show site; he also uses this two-seater to fly civilians like me on the Thursday and/or Friday before a show.

Jason started out by showing on a map where we'd be flying. After take-off from Fort Wayne, we'd cruise via Huntington and Wabash towards Logansport, around 100 km west of Fort Wayne. The airspace where the Twelve Mile and Hill Top military operating areas overlap is ours to play in, from a few hundred feet above the ground up to 18,000 feet MSL.

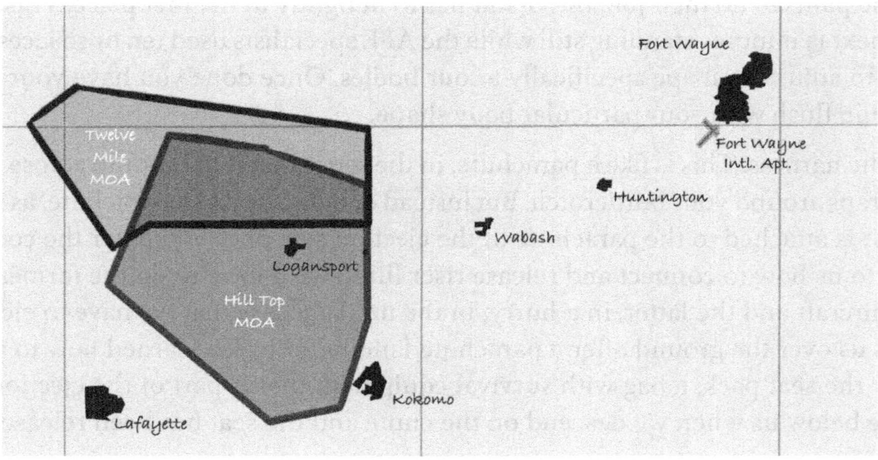

Apart from getting to know each other, the main part of the briefing was to talk through the maneuvers we would be flying; I won't spoil this for you, as we'll see those later. We also discussed how to get into the jet without breaking switches in the cockpit, how to strap in, how to arm the ejection seat, how to operate the intercom, how to switch to 100% oxygen and what I'll see on the HUD repeater screen: primarily indicated airspeed, mach number, altitude, pitch angle, g and heading. It turns out I wouldn't look much at anything except speed, being too busy taking in the experience.

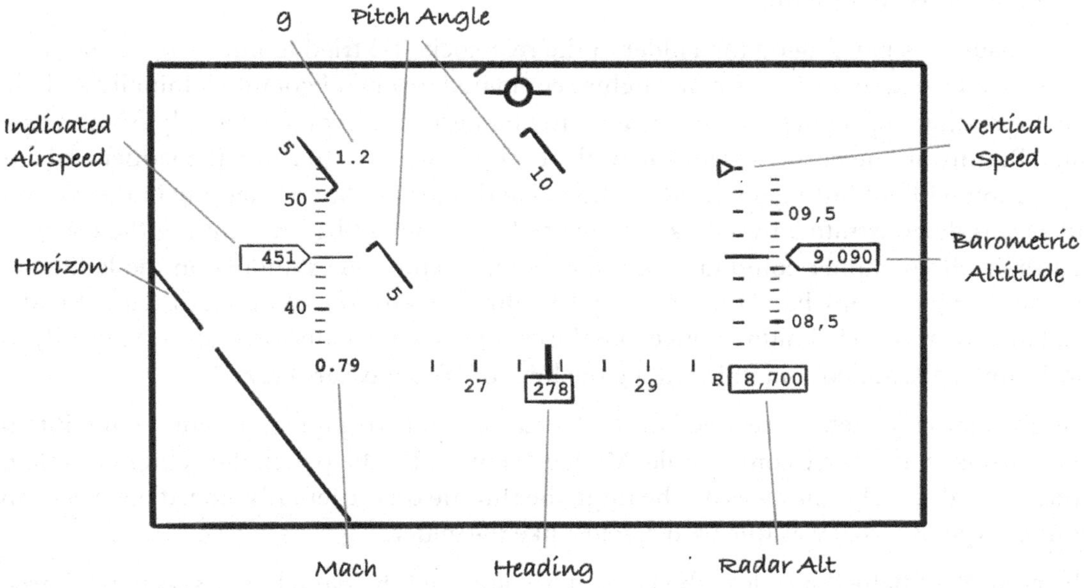

Jason pointed out that it will be relatively cramped in the back seat of the F-16 and warned me to be careful about bumping into the stick when I moved around. After all, everything is live in the rear cockpit!

He briefed how he will announce an ejection ("Bail out, bail out, bail out") and also explained how this works if the intercom has failed and we cannot talk: "I'll wave this paper manual above my head to announce an ejection. A few moments later I will pitch up the nose and pull the throttle back (the two are connected, in contrast to the stick). You can veto the ejection once by immediately pushing the throttle back forward. But only once. Then we'll go." More generally, if the intercom fails before takeoff we'd postpone takeoff and fix it. If it fails after we're airborne, "then enjoy the ride, and I'll talk to you after we get back," he said, laughing.

We concluded the briefing by talking about how we'd coordinate when I will fly. We'd pass control with the usual "You have the aircraft/I have control" exchange. He also said that he'll ghost the controls and make sure we stayed within the bounds of the airspace that is reserved for us to play in. Apart from that, "Do whatever you want."

STRAPPING IN It was time for the flight. Interestingly, I was no longer nervous. I had been nervous in the weeks and days before the flight, mainly because of my fear that something could go wrong last minute. Now that I was well into the big day I was confident the flight would happen and I felt fairly relaxed. Jason and I hitched a ride on a golf cart and made our way to the line of six beautifully polished F-16s, two of them D models. We grabbed the one marked as 8; it would be re-decalled 2 or 3 later to fly as part of the formation. Crew chiefs were out there prepping the plane. In line with the Thunderbirds' emphasis on teamwork there was lots of high-fiving.

"My" G-suit was hanging on the ladder of the rear cockpit. I tried to put it on, but because it was now adjusted to me it was much tighter compared to when I got into it initially and I had trouble closing the latches. Maybe I was a little bit nervous after all. I got help from the crew chief. Putting on the harness went smoothly. I was shown how to climb the ladder and enter the cockpit without hitting my head on the rail and/or destroying something in the cockpit. I did as was demonstrated. I was asked to pause for a moment before entering the cockpit so that their photographer could take the "hero shot" picture: me standing on the ladder, one foot on the aircraft, my hand on the canopy, smiling towards the camera. Although I had my head directly next to it, I didn't notice that the canopy had my name on it until somebody told me. I think I exclaimed something along the lines of "You guys are crazy."

I climbed into the seat. It's reclined thirty degrees to improve g tolerance and to fit it into the small fuselage. It's fairly comfortable. My feet went under the panel, the center console between my calves. The stick was on the right, the throttle on the left. It's indeed not a very spacious cockpit. But that was fine with me: just like the glider.

The crew chief helped me close the lap belt, we attached the parachute risers to the harness and connected the seat survival pack. I thanked him as he left. Then Jason came up and re-emphasized that I should re- and re- and re-tighten the lap belt: "It's the only thing that keeps you in the airplane if we go inverted." He didn't have to tell me twice: I know from glider aerobatics that it's important for feeling "safe" in inverted flight that your butt continues to feel the seat even with minus 1 g or more. I was surprised to realize that the shoulder straps that are attached to the parachute are not taut; just like the safety belt in a car, they are flexible and only tighten upon ejection. "It's so your upper body remains flexible, so you can turn around." This makes sense—still, I was surprised. It's different in a glider and everything else that does aerobatics. Put in earplugs, put on the helmet, lower the visor and attach the mask. Thunderbird Eight went back down the ladder and entered his own cockpit. We were good to go. Thunderbird Seven and Rod had intercom issues and they would take off a bit later. We'd take off separately, in contrast to what was briefed.

TAXI AND TAKEOFF Jason went through the preflight checklist and requested taxi clearance. He pushed the throttle forward ever so slightly, and the Viper started to move. Right after we started moving, he pressed both toes forward to test the brakes—the F-16D made this slight bow that I always loved to see when aircraft taxied past. We rolled down the taxiway to the holding point off runway 14. A few more checks, a request for take-off clearance and we entered the runway.

Jason: "Alright, runnin'er up... it's on the roll... here comes the afterburner... here it is!" A pronounced push in the back starts us down the runway, accelerating to 160 knots in 10 seconds, when Jason lifts the jet off the runway, retracting the gear immediately to avoid over-speeding it. He keeps the jet a few meters off the ground, accelerating further. Strong acceleration is always impressive, but it's even more impressive when it just doesn't stop! After another 10 seconds we reach 350 knots at the end of the runway. "Here come the geeeeees", Jason announces, and pulls 5 g into a 60-degree climb. After a few seconds: "Chicago Center, Thunderbird Eight, passing 5,000 for 16,000." 25 seconds after pull-up we reach 10,000 feet, and Jason pulls over the top for around 12,000 feet. 45 seconds after brake release we're cruising at 14,000 feet, climbing towards 16,000, on the way to "our" military operations area. We just performed an "unrestricted takeoff", and it is damn impressive! If you watch the video you can hear an assortment of enthusiastic exclamations by yours truly.

We then cruised for 10 minutes or so to our airspace. 14,000 feet, rock solid, like an airliner, just with a better view. And it's really quiet in the airplane: you can hear the engine a little bit, but otherwise there's only the radio and intercom. The earplug, helmet, and canopy do a really good job of isolating your ears from the engine and the air rushing around the fuselage.

THE PROGRAMME Once we reached our military operating area, we start with the aerobatics. We begin with two g familiarization turns, one with 4 to 5 g and one with 6 to 7. Apart from my breathing too fast, everything is fine. We roll to 120 degrees and pull down to a few hundred feet to fly low and fast; at this altitude and 400 knots, the ground races past! We see a silo, follow a road, and wonder about the fact that we are allowed so low over farmland. "Let's pop up to 5,000 again", Jason says, as he pulls to 25 degrees pitch, climbs for a few seconds, rolls to inverted, pulls to level flight, and rolls back upright. This kind of flying is so 3D, flying gliders or general aviation should be redesignated 2.5D. Next up is the knife-edge: level flight at 90-degree bank with lots of rudder to keep the nose high.

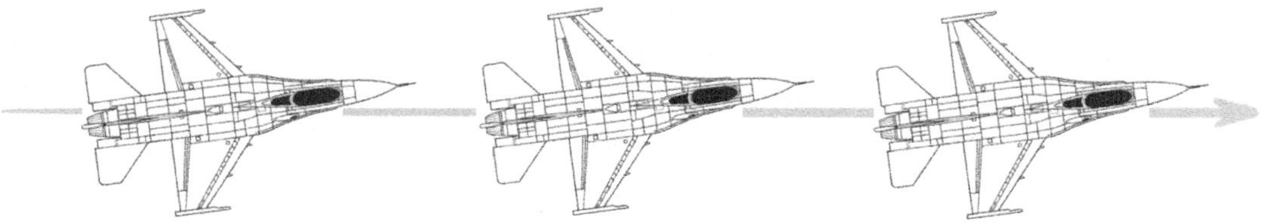

We do all of this with smoke on, performing a little airshow for the folks on the ground who pay for our pleasures with their tax dollars. An eight-point roll follows. I try to keep my head against the headrest to make sure it's not bounced around for each of the quite pronounced stops. I count in German, and Jason is glad to learn that it was indeed eight stops.

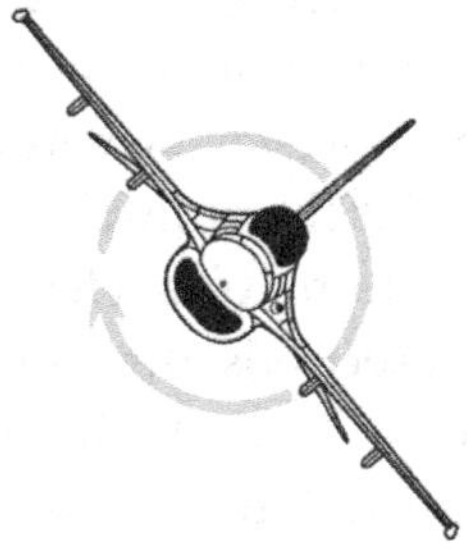

Now a four-point roll. Here the inverted flight lasts slightly longer and I actually feel the negative 1 g—I didn't really in the eight-point roll. Staying with inverted flight, we perform a single aileron roll from inverted to inverted. Nice! We have reached the airspace boundary, so we turn to the opposite heading by flying an Immelman: half a loop, then half a roll.

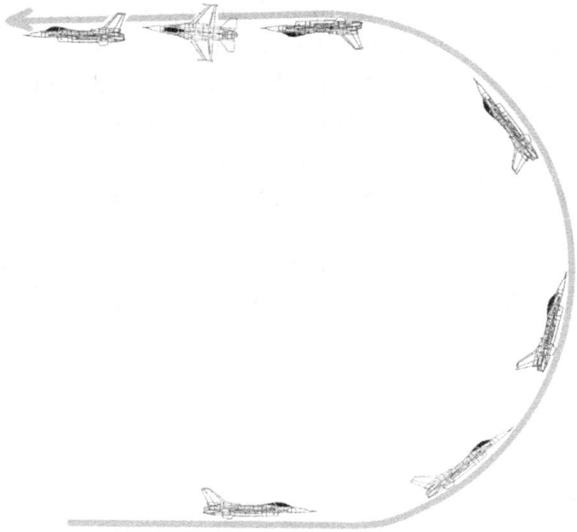

"Alright, this concludes our warm-up", Jason jokes. I know what's coming: the 9 g turn. It's a bit of a rite of passage: the Air Force is kinda proud that their F-16 can do 9 g while the F-18, used by the Navy and the Blue Angels, is limited to "only" 7.5. I switch to 100% oxygen just to be sure. Jason kicks in the afterburner, accelerates to 450 knots, rolls to 90 degrees and pulls the stick back: the maneuver lasts for about 10 seconds, 3 of them at 9 g. Tough work straining against the load, but I survived and didn't g-LOC. After cruising straight and level for a little while to recover, the barrel roll is next on our list. While not easy to fly well, it's undramatic in terms of gs. It's also a good approximation of what the non-center members of a formation roll have to do: "Now imagine an aircraft left and right!", Jason says.

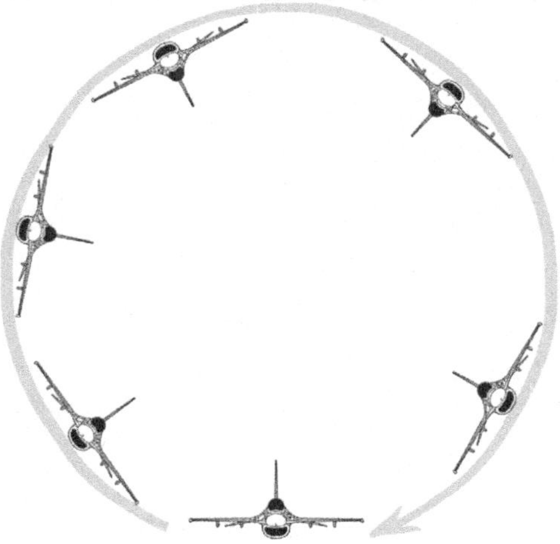

Next up are the four consecutive aileron rolls; we complete them within six seconds! Immediately afterwards we pull the nose up slightly and add a single, slower roll in the other direction, "to unwind", as Jason explains. This is the first time I feel slightly queazy, but after a few seconds of 100% oxygen the feeling dissipates. In a further demonstration of the 3D-ness of this kind of flying, we now fly three consecutive rolls while going straight up, pulling over the top as gravity slows us down to to 200 knots and we approach the 18,000 ft ceiling of our airspace. The F-16 rumbles slightly.

FLYING MYSELF This ends the "official" programme and it's time for me to take control. "At 7,300 ft you have the aircraft my friend," Jason says. I reply with "Ok, I have control." I start by trying to get a feel for the aircraft in terms of roll and pitch. I roll left and right and fly a couple of turns with 4 to 5 g—if I want to finish the turn in a reasonable time and without leaving our playground I have to pull some g, otherwise the turn radius is large. Flying turns in a fighter jet is a pitch maneuver! Then I try an aileron roll. It's really easy: fly straight and level, put the stick to the right, wait for a second or two, done. The only challenge is to stop the roll at precisely level flight (which didn't quite work out). Contrast this with the ASK-21 aerobatic training glider in which an aileron roll takes around 12 seconds: you first pull the nose up to 10 degrees, put full right aileron, as you approach 90 degrees bank you use some left rudder to keep the nose up, during the inverted part you push quite a bit to keep the nose over the horizon and create enough negative AoA to produce lift, and in the last quarter you again use rudder and the elevator to somehow force the beast to finish the roll at the precise heading you started. Admittedly, with more advanced gliders such as the Fox it's much simpler: its roll rate approaches 90 deg/s. For comparison, the Edge 540 that is used in Red Bull air racing has a roll rate of 420 deg/s. I am sure aileron rolls are quite simple there as well. Next I try a half Cuban Eight, another maneuver I know from glider aerobatics: three fifths of a loop and then, as the nose points down again, half a roll to wings level.

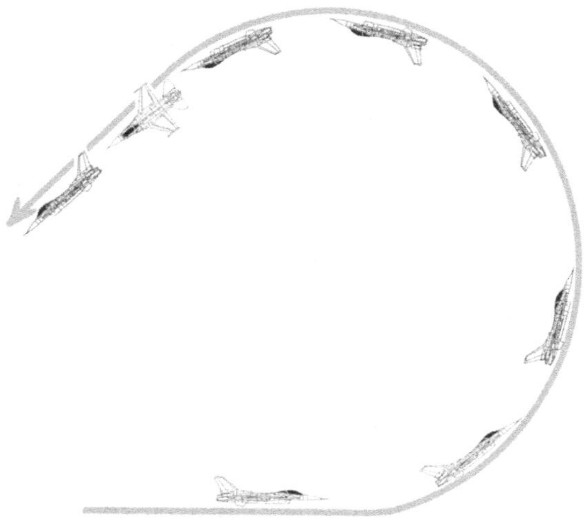

I notice that I have to pull on the stick with several kilograms of force to get 5 g for the loop. Somehow, because of the fly-by-wire, I expected the stick to be light all the time: there's no aerodynamic speed-induced feedback, and feel, to the degree that it exists, is artificial anyway. On the other hand it makes sense, of course, that you have to tell the airplane in no uncertain terms that you want a bunch of gs; you don't want to get 7 g whenever you accidentally bump lightly into the stick. I flew a Chandelle or two before Jason took control again.

As suspected, the mere flying of the aircraft (in the air, not during landing) felt relatively straightforward, thanks to the digital flight control system and my general flying experience. Understanding the systems and flying missions is an entirely different matter, of course. In retrospect, I should have flown a little bit more aggressively, less inspired by glider maneuvers. I guess I have to do this again.

FORMATION FLYING "We are about to do our 9 g turn and then we are ready to rejoin.", reports Thunderbird Seven over the UHF radio. "Ok, ready to join, timber sweet as well", Jason replies. We're about to meet the other F-16 to do some formation work. Bitchin' Betty tells us that we are Bingo (enough remaining fuel to head back home), so we turn towards Fort Wayne. I spot the smoke trail of Thunderbird Seven to the left of us. He takes the lead, and Jason joins on his wing at around 1.5 meters separation. I can take pictures of the other F-16 with my (slightly wide-angle) iPhone, a testament to how close we are!

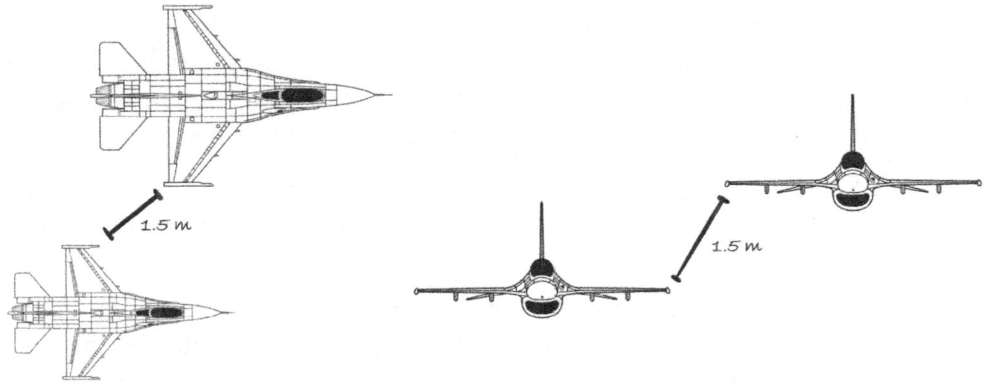

Thunderbird Seven announces changes in flight path and attitude and Jason keeps station with lots of small corrections with the stick and throttle. Very impressive, and really really cool "bonus content" for the flight. It also reinforces my awe for Thunderbird-style formation aerobatics!

HEADING BACK As we get close to Fort Wayne, the airshow air boss instructs us to hold for a bit until a display finishes and a historic B-17 completes its landing. Betty blurts another "Bingo, Bingo" into our headphones. "Not a bad perspective to watch an airshow," Jason quips. I agree. Finally, the air boss grants access to the pattern. We perform a pass over the runway, smoke on, in trailing formation and then break to the left. The relatively hard turn helps reduce speed for the pattern. We fly a downwind, then a descending left turn onto final. Jason has to work because there's a fair bit of crosswind and a sink just before the threshold. Touchdown. He keeps the nose up for aerodynamic braking (and I'm sure to make the mighty Falcon look cool). After an hour and a few minutes the nose wheel touches down as well and we slow to a speed that allows us to leave the runway. "Alright, well, Markus, it was a pleasure flying with you, I hope you enjoyed it as much as I did." Jason says as we taxi back towards the gaps between the other four F-16s on the ramp. I reply with a bunch of stuff that makes clear that I very much did.

AFTER LANDING Once the engine was shut down and the canopy opened, I took off my mask, removed the helmet and put it back onto the rail. Helped by the crew chief who came up the ladder, I opened the lap belt and removed the hooks for the seat pack. I got out carefully, making sure I didn't fall down the six-foot ladder. I took off the harness and G-suit.

We all assembled in front of Thunderbird Seven's aircraft. About 20 of the Thunderbirds were there. Lots of smiles everywhere. Rod and I were presented with a nice framed photo of the Thunderbirds flying over Half Dome, with signatures from all 12 officers. Thunderbird Seven talked a bit about Rod, why he was selected, and thanked him for his service. Then Eight talked a bit about me and omega tau, and thanked me for reaching out to the STEM community. It felt a bit weird for them to thank us—so I thanked everybody and briefly narrated my 30-year dream of flying on an F-16. Both of us thanked all present with a shake of hands.

Jason, Cory and I found another golf cart that took us back to the Ops building. I changed back into my regular clothes, and, just like that, the experience of a lifetime came to an end.

THE MIRACLE, REVISITED If you are an omega tau listener, or read through this aviation chapter, you should have a sense by now that I am a huge avgeek. If you read the How I Got Here section of the book you might remember this letter to General Dynamics which the 12-year-old Markus wrote, asking for pictures of the F-16. I have been in love with this airplane since I was 12, and flying in one has been a dream ever since. You might also remember from that section that one of the triggers that prompted Nora and I to start omega tau was the Thunderbirds flight of Steve Tupper, the host of the Airspeed podcast. We always joked that, once I got to fly in an F-16, we'd stop the podcast because then we'd have "played it through". We won't stop. But from all of this you can probably tell how big a deal this flight was for me. It was literally the biggest dream of my life. So, Dear Thunderbirds, thank you very much for turning my Big Dream into reality. Thank you Jason Markzon for flying me, thank you to Ned Johnston and Cory Bush for coordination in advance and afterwards, and thanks to everybody else in the team for what you do and for letting me take a sneak peek into your world.

5

Detecting Gravitational Waves

If you read the introduction to this book, you will know that I had a relatively untypical time at school. In particular, the physics teachers let us do lots of cool stuff. At one point we were supposed to measure the speed of light. How do you do that? You let light travel a known distance, as far as possible in the school building, and measure the time it takes. Obviously. Our specific experiment involved a laser, a mirror at the end of a long corridor, a photodiode to detect the reflected laser, plus a way to precisely measure time. We used an oscilloscope: when the laser is triggered, the beam of the oscilloscope starts, and the detection of the laser by the diode showed as a peak on the screen. You need very high resolution, because the time would be in the order of 0.3 microseconds. The school had one oscilloscope that could do this and we were allowed to use it. Not a very good idea, as it turned out. We cleverly plugged the power cord into an outlet that was right outside the door of a classroom, and it was the slot right before lunch. To be at the front of the lunch queue and thus maximize the usable time during lunch break, everybody sprinted through the school building to the cafeteria when the bell rang. It was actually dangerous to be in the corridors at that time. A tiny little power cord wouldn't stop anybody. And it didn't. The oscilloscope ended up on the ground, and then sent for repair. My reputation among the teachers took a hit that day. Also, we didn't get to measure the speed of light. But no worries, we knew it anyway: c = 299,792,458 m/s in vacuum, and for our purposes also in air.

I | *Relativity and Gravitational Waves*

C IS A CONSTANT Measuring the speed of light was also pointless in another, more fundamental way. The value of c is defined to be 299,792,458 m/s in the SI system of units. Historically, of course, the value was measured, with increased precision over the years and decades. Back then it was defined relative to the international prototype meter, a platinum bar stored in Sèvres, just outside Paris. In 1983, at the 17th Conférence Générale des Poids et Mesures, the relationship was reversed: the numerical value of the speed of light was fixed at 299,792,458 m/s, and the meter was instead defined as "the length of the path traveled by light in vacuum during a time interval of 1/299792458 of a second." The length of a second is also fixed, based on stable, fundamental atomic properties. Future experiments would thus, by definition, not change the value of the speed of light, but the physical length of what we call a meter.

The rationale for this reversal was that the SI system should be based less on prototypes and more on physical constants that are inherent—and thus stable—properties of our universe. In 2019 the SI system was further changed to get rid of the last remaining prototype, the one for the kilogram. Now the kilogram is defined based on the Ampere and the Kelvin, two of the seven fundamental units in the SI system.

But beyond mere definitions, the speed of light is even more constant: it is constant independent of the reference frame. Let's recap a famous Gedankenexperiment (German for "thought experiment", one of these German words that somehow seeped into English). You stand next to a track in a train station. A train approaches. There's a guy on the train who throws a ball at you. Let's say he throws the ball at 10 m/s and the train is going at 25 m/s. If you measure the speed of the ball as it passes you, the stationary (no pun intended) observer, what will you measure, assuming we ignore drag?

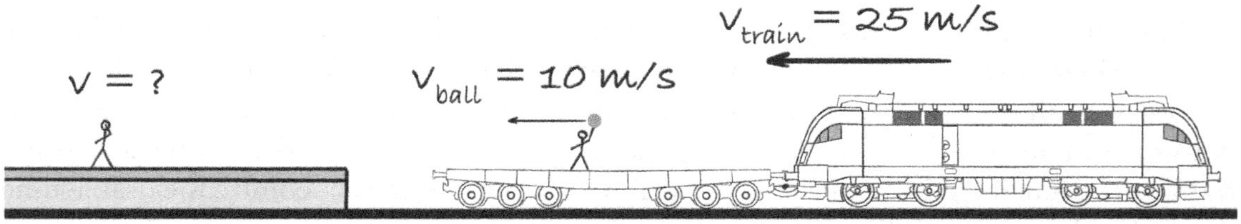

We measure 35 m/s, of course. And if some other guy on the train catches the ball, what will he measure? 10 m/s. This is all very intuitive and consistent with our experiential world: the measured speed depends on whether your reference frame is the moving train or the stationary platform.

But what if the guy on the train fires a laser at you instead of throwing the ball, and you measure the speed of light (I wonder what sort of creative way we might have found to destroy the oscilloscope once trains got involved)? You would measure the same speed in both cases, for the guy on the platform and for the other guy on the train, because c is a constant independent of the reference frame! You would also measure a slight blue shift in frequency because the light waves are "compressed"—just as the pitch of a police car's siren is raised while the car approaches—but the speed would be constant. The absolute constant nature of the speed of light is one of the central postulates of Einstein's 1905 Relativity Theory, where he (successfully) attempted to get rid of contradictions between Newtonian mechanics and Maxwell's equations for electromagnetism. If you think about it, this is quite a courageous postulate; it's really quite unintuitive.

SPECIAL RELATIVITY The speed of light is not just constant, it is also finite. This has consequences for the notion of simultaneity, of events happening at the same time. The reason is basically that, to establish the notion of simultaneity, one has to transmit the occurrence of an event to an observer somehow, for example through a flash of light. But that signal isn't transmitted instantaneously: it travels to the observer with the speed of light. Consider again our trains. Let's say there are two clocks at either end of a railcar. Whenever they show a particular time, say, twelve o'clock, they emit a flash of light that is detected by an observer who is in the middle of said car. In a stationary railcar, the observer sees both flashes at the same time, and therefore he considers the clocks to be synchronized. If the system is moving, the situation does not change, because all non-accelerated reference frames are equal in terms of their laws of physics—another postulate of relativity. Let's now reintroduce the guy on the platform. From his perspective (frame of reference), because the train is moving, the observer between the two clocks travels towards one and away from the other, while the flashes are moving at the speed of light towards that observer. So the distances the two flashes have to travel are not the same. Since they both travel at a fixed speed, c, they must arrive at a different time. According to the observer on the train, they arrive at the same time. How do you resolve this paradox? The only way is to say that they did not start at the same time, in other words, the clocks are not synchronized. The consequence of this? There can be no global notion of simultaneity for spatially distant events.

Spacetime is a mathematical model that helps with resolving and visualizing those apparent contradictions. Let's understand why. Consider a vector: it can be described by two points, its start and end. Each of the points can be defined by its three spacial coordinates. But these coordinates are relative to a particular coordinate system: I could define any number of other cartesian coordinate systems by translating or rotating my original coordinate system. Reinhard Prix, in our episode, had a nice example. Let's say you enter your office and sit down at the exact same position today and tomorrow. From the perspective of the local coordinate systems, you were at the same location. But for an observer who watches from outside the

Earth, the location is quite different because the Earth, and with it the office and its local coordinate system, have moved on. The Galileo transformation is a set of formulas to transform one set of coordinates into the coordinates in another coordinate system, if the coordinate systems differ only in terms of translation, rotation, speed or time. The fact that coordinates are relative to a coordinate system means that the start and end coordinates of the vector cannot be "real" physical properties. However, it's length and its direction relative to another vector are: they are unaffected by any such coordinate transformation—they are invariant.

Relativity introduces a similar idea for two different events that happen "at the same time" at two different places: simultaneity is no longer independent of the reference frame of the observer. In a relativistic world one has to use the Lorentz transformation to transform between coordinate systems instead of the Galileo transformation, as it changes the time coordinate as well, in such a way as to keep the speed of light constant. It treats space and time together: hence spacetime. The invariant in the transformation is the "distance" in spacetime. I am putting the term distance in quotes because it is not the traditional Euclidian distance, but instead a Minkowsky distance. In a Euclidian space you measure distance by summing the deltas of the distances in each dimension. For two dimensions and a Cartesian coordinate system, this results in the familiar Pythagorean theorem, but it's obvious how to generalize it to more dimensions:

$$\sqrt{(x_1 - x_2)^2 + (y_1 - y_2)^2} \qquad \sqrt{\sum_{i=1}^{n} (x_i - x_i)^2}$$

Minkowski distance is like a four-dimensional Euclidian space (three space plus one time dimension), but time has a negative sign:

$$\sqrt{dx^2 + dx^2 + dz^2 - dt^2}$$

The lack of a notion of global simultaneity has more consequences, for example for the length of an object. Length is basically defined as the start coordinate and end coordinate of an object measured at the same time. If the notion of "same time" is no longer globally agreed, it is not surprising that the observers in different reference frames cannot agree on a length! The factor is by which the length changes is:

$$\frac{1}{\sqrt{1 - \frac{v^2}{c^2}}}$$

This is the so-called Lorentz (or gamma) factor; it is only measurable for really high speeds. The SR-71 at a cruising speed of 3,200 km/h would shorten by about 0.14 nm—so this effect certainly doesn't contribute to it leaking fuel :-) Objects become heavier as they become faster by the same factor; in fact their mass grows without bounds as v approaches c. We'd need infinite energy to accelerate further, which is why bodies that have mass cannot reach the speed of light. There is also time dilation: time moves slower for a fast moving object. Science-fiction authors have exploited this idea to allow people in very fast spaceships to live longer, as seen from a stationary reference frame (the Earth). Experimentally, this can be verified with muons, a particle similar to the electron. High-speed muons are created in the upper atmosphere by cosmic rays. At rest in the lab they decay so quickly that they could never travel the distance to the Earth's surface. Because of time dilation, however, the particles last about five times longer, allowing them to be detected on Earth. The effect can also be verified by putting extremely precisely atomic clocks on airplanes.

GENERAL RELATIVITY After Einstein had successfully unified space and time and "refined" Newton's mechanics, he—and probably many others—noticed that there are two kinds of mass: mass that is affected by gravitation and mass that resists acceleration. And these two are always numerically identical. But are they actually the same? Einstein next attempted to integrate gravitation into his theory of relativity. This lead to his Theory of General Relativity, published in 1915. His 1905 theory was retroactively named Special Relativity, in the sense of "special case". It is special because it applies to inertial systems, systems that are either at rest or move at a constant speed; acceleration is outside its scope. His new theory generalized relativity to accelerated systems and to systems that are under the influence of gravity. He postulated that gravitation and acceleration are the same, that heavy mass and inertial mass are indeed the same concept. This is known as the "equivalence principle".

So how then does gravitation impact the notion of spacetime that was introduced in special relativity? A popular way to express this impact is to say that gravitation bends, or curves, spacetime. In this model, the gravitational pull of body A on a body B is explained as B "rolling down" into a gravitational well created by A's mass. Gravity is not something that happens "in" spacetime. The bent spacetime itself is gravity!

Another way of understanding gravity is as a field. A field is a physical quantity that has a particular value at each point in space—or more generally, spacetime. The value can be a single number (a scalar field such as the temperature distribution in a volume), a vector with direction and amount (for example, a wind field in a volume of atmosphere), or more generally a matrix or a tensor (for example, a deformation field that describes different displacements in different directions). Fields are frequently used to describe remote effects. Magnetic and electric fields are the obvious examples, but wind can be seen as a remote effect of a low-pressure area or many interacting areas.

Gravitation can also be seen as a field. In Newton's world, it is a scalar that describe the gravitational potential at a given distance from a heavy body; its gradient results in an (accelerating) force. In Einstein's world gravity is a tensor field that describes distance, volume, curvature, and angle, plus time and the separation of the future and the past.

Beyond the ubiquitous visualization above, another way of understanding the curvature of space is the following. Consider again our massive body A and its gravitational field. As we know, the force becomes less with the square of the distance. For bodies that have a non-zero extent, bodies that are not point-like, a part of that body will be in the stronger field closer to A and other parts will be in the weaker field further away. This tidal force can be interpreted as the curvature of space: bodies change their shape.

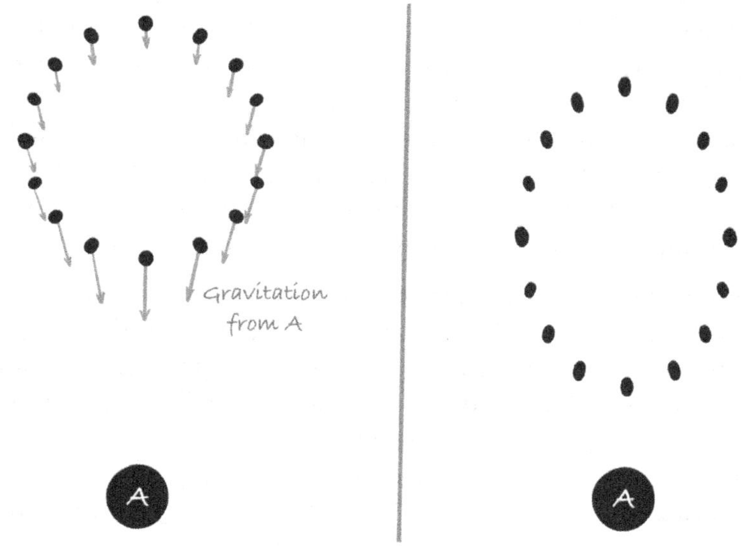

In Newton's laws, the resulting gravitational force is defined as:

$$G\frac{m_1\, m_2}{r^2}$$

where m_1 and m_2 are the masses of the two bodies, r is their distance and G is the gravitational constant, a property of our universe. For things that have no mass, such as light, the resulting force would be zero. General relativity, which interprets gravitation as curved space, predicts that light would be also be deflected by a heavy mass. And indeed it is: this has been shown experimentally when a star showed up "at the wrong time" because its light was bent around the edge of the darkened Sun during an eclipse in 1919.

As we know, the crew of the International Space Station experiences weightlessness, even though they are close enough to the Earth for gravity to be hardly any weaker than on the surface: in fact gravity is 13% weaker at the ISS's 400km high orbit. Why are they weightless? The Newtonian way of explaining this is that they circle the Earth at a speed at which centrifugal force exactly compensates for gravitational force. Taking curved space into account, one can say that the ISS just moves "on a straight line", undisturbed and unaccelerated, a straight line in the curved space created by the Earth's gravitational field. Interestingly, a system that feels no acceleration other than that from gravitation—in other words, falls freely towards the gravitational mass—can be seen as an inertial system in which special relativity applies.

WHAT ARE GRAVITATIONAL WAVES If the body that creates the gravitational field is itself accelerated, for example because it is on a circular trajectory around another body (the Earth around the Sun, for example), then its gravitational field moves with the body, of course. However, this change in the field itself has to propagate from the accelerated mass. Such propagation cannot occur instantaneously, as it too travels with the speed of light. From these observations, and certainly a lot of math to back it up, general relativity predicts gravitational waves—"ripples in spacetime"—whenever a mass is accelerated. Newton's laws did not predict such waves, because they assumed that gravitational interactions propagate at infinite speed. In this sense, the detection of gravitational would be a confirmation of general relativity.

To illustrate the effect of gravitational waves, let's assume we have a set of point-like particles arranged in a circle, all in a single plane, such as the plane of the paper (or screen) you're reading this on. Let's now say a gravitational wave passes in the direction perpendicular to that plane, into the paper. As a result, space will be distorted in a way that deforms the circle into an ellipse.

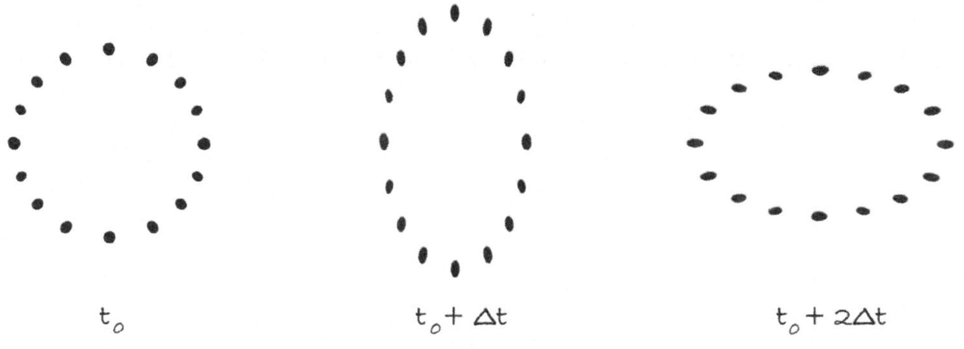

INDIRECT OBSERVATION Similarly to electromagnetic waves, gravitational waves transport energy in the form of gravitational radiation. This implies that a system that is continuously accelerated and produces gravitational waves must lose energy over time. In 1974 Russel A. Hulse and Joseph H. Taylor observed just such a system, the binary pulsar PSR

B1913+16, which earned them the 1993 Nobel Prize in physics. A pulsar is a rotating neutron star that radiates electromagnetic energy in a directed beam, in much the same way as a lighthouse. We receive this radiation only when the beam sweeps over the Earth, so we observe it as a pulse of radiation at a frequency that is proportional to the pulsar's rotational speed. This particular pulsar rotates 17 times per second, a period of 59 milliseconds. Hulse and Taylor noticed a systematic variation in the arrival time of the pulses over time. This variation is consistent with the pulsar being one half of a binary star system—two stars circling each other. The diameter of their orbits can be derived from the variation in arrival time. Herein lies the connection to gravitational waves: because the two stars are continuously accelerated around each other, they lose energy due to gravitational waves. This in turn leads to a reduction in the diameters of their orbits. Using precise measurements of the variation in arrival times, Hulse and Taylor showed that this orbital decay corresponds to the prediction of general relativity with an accuracy of 0.2%. The Hulse–Taylor pulsar was thus the first observational evidence of gravitational waves, but it was indirect evidence. It took until 2015 for gravitational waves to be detected directly.

2 | *The Detector(s)*

BASICS OF THE DETECTOR In 2017 I visited GEO600, an experiment set up to detect gravitational waves directly by measuring the distortion of space resulting from the passage of a wave. It is not the only such detector; several have been built worldwide, including LIGO in the US and VIRGO in Italy. They all rely on the same operational principle: the Michelson interferometer. A Michelson interferometer is not completely unlike my school experiment that imperiled the costly oscilloscope: you reflect a laser beam between two mirrors and measure its transit time. Based on the known and constant speed of light, you can then calculate the beam's path length. Any passing gravitational wave changes that length ever so slightly.

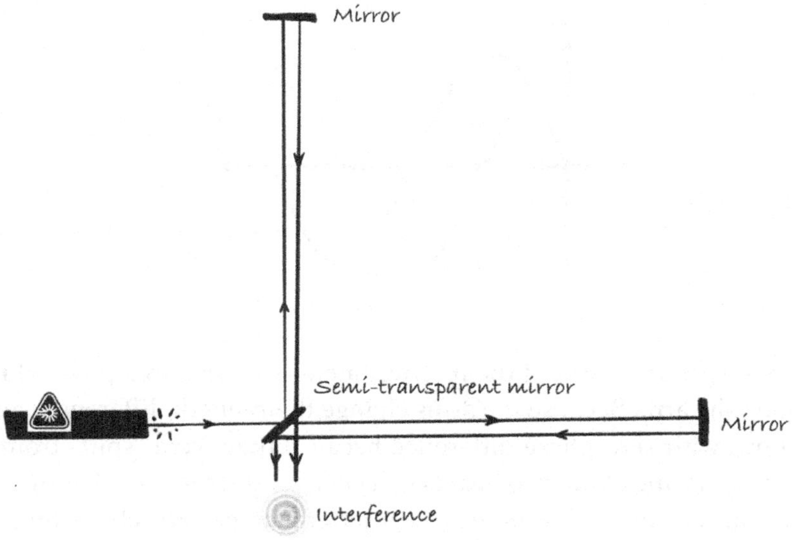

To measure this change in length precisely, you start with a single laser beam, which you then split into two beams using a beamsplitter, basically a semi-transparent mirror. You then send the two beams along the two roughly perpendicular arms: these are 600 meters long for GEO600 and 4,000 meters for LIGO. A mirror at the end of each arm reflects the light back. Because of the different orientations of the arms, a passing gravitational wave will shorten or lengthen the two arms differently, as illustrated by the next illustration that repeats the test particles from before:

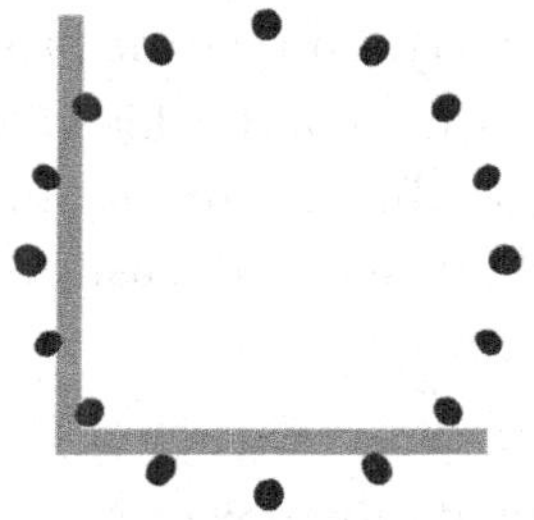

To measure this length difference you let the two beams interfere with each other. As you probably remember from school, interference happens when two waves superpose. Depending on their phase difference, they arrive at the point of measurement with both waves at their peaks or troughs (which leads to constructive interference), one at a peak and one at a trough (destructive interference), or any combination between.

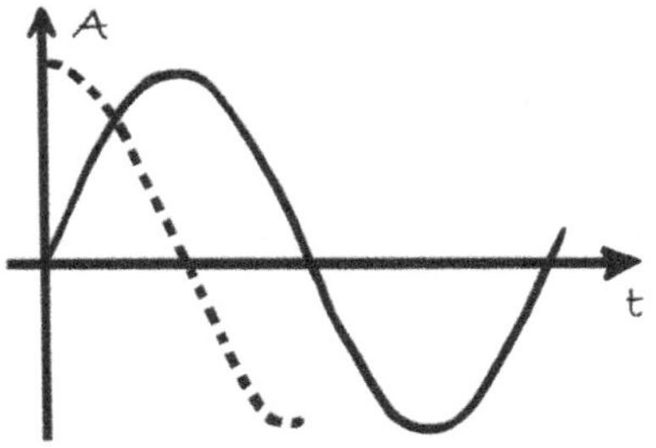

Changes in the length of an arm of the interferometer will lead to a phase shift in the beam that travels along the arm. Because the arms change their length differently, and because the beams started out with zero phase difference because they were "split" from a single laser beam, the relative phasing of the two beams will change, and the resulting interference intensity will change accordingly. The particular interference pattern characterizes the relative length change of the two arms.

An interferometer naturally indicates changes in relative phase that is in the order of magnitude of the wavelength, between 10^{-7} and 10^{-6} meters for lasers in the visible spectrum. For larger differences the system has to count the sequence of intensity changes.

So far, so good. But here's the problem: gravitational waves are extremely weak. The relative length change caused by a gravitational wave is only 10^{-21} meters per meter of measurement distance. The distance traveled by the beams in GEO600 is of the order of 10^3 m, so the absolute change in length produced by a wave is 10^{-18} m, roughly 1/1000 of the diameter of a proton. This is very much less than the 10^{-7} m wavelength of the laser, so you'd only get a very very small change in phase, and thus intensity. You see the problem. Very high precision measurement is required—especially considering that the detector should not just record the fact that a wave has passed, but should also be able to capture the wave's evolution in frequency and amplitude over time, to be able to characterize the astronomical event that created it. So let's understand how to attack this problem.

PASSIVE SEISMIC DAMPING Gravitational waves are made up of frequencies over a large range; a scientifically interesting band is between a few dozen and a few hundred Hertz. A passing wave will make space itself "vibrate" at those frequencies. To be able to detect such vibrations as phase shifts of the laser beams, you have to ensure that the mirrors that reflect the lasers do not vibrate at those frequencies for "earthly" reasons, relative to space—the detector would be merely a giant seismometer if the mirrors were not isolated from the environment. While the GEO600 detector is not near a railroad or a highway, it is also not outside of civilization: it's located next to a field midway between Hannover and Hildesheim. The unavoidable seismic background noise induced by civilization and the atmosphere has to be dealt with.

It is crucial to understand that a gravitational wave does not make the detector's mirrors vibrate in a classical way—that is, relative to the installation around them. Instead, space itself changes its shape. To detect this, any classical vibration of the mirror has to be extremely well damped in the relevant frequency range. To achieve this, the interferometer's mirrors are suspended from a pendulum.

Let's try to understand the necessary damping characteristics using seismic microtremors as an example. These are man-made vibrations above 1 Hz with an amplitude on the scale of micrometers, 10^{-6} m. We know from above that the relative length change induced by a gravitational wave is of the order of 10^{-18} m for the roughly 1,000 meter light path in GEO600. To isolate the mirrors from microtremors, therefore, we have to achieve damping of the order of 10^{12}. Let's look at the damping characteristics of pendulums:

$$\delta = \left(\frac{f}{f_r}\right)^2 \qquad f_r = \frac{1}{2\pi}\sqrt{\frac{g}{L}} \qquad L = \frac{\delta\, g}{4\,\pi^2 f^2}$$

The first formula shows a pendulum's damping, δ. This is frequency-dependent, as we can see from the f in the formula; this simple formula applies for $f \gg f_r$, where f_r is the resonant frequency of the pendulum. If we rearrange the formula, we can see in the second formula that f_r depends on the geometry of the pendulum, and in particular on its length L. Let's say that we want to damp vibration at 100 Hz, a frequency that is right in the middle of where interesting astronomical phenomena occur, and let's assume we have a pendulum with $f_r = 1$ Hz. The reduction in vibration amplitude is then $\delta = 10^4$. How long would such a pendulum have to be? Solving the second formula above for L, we get around 25 centimeters for a resonance frequency of 1 Hz—a feasible size.

However, if we try to achieve the necessary damping of 10^{12} , the pendulum would become humongous—do the calculation and see for yourself using the third variant of the above formula. Fortunately there is an alternative: stacking them. This means that the wires of pendulum two are attached to the mass of pendulum one, as the following illustration shows:

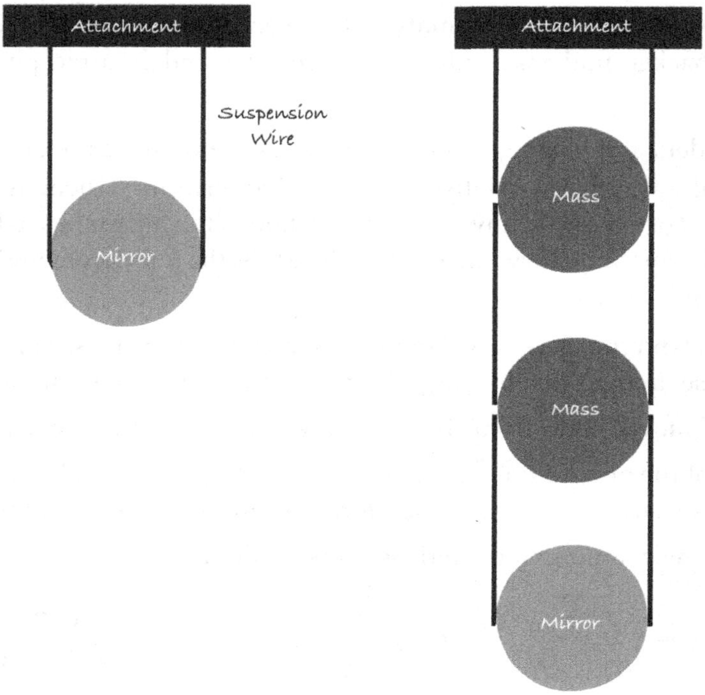

If pendulums are stacked, their attenuation is additive. So by stacking three pendulums with a damping factor of 10^4, we can achieve an attenuation of $10^{(4+4+4)} = 10^{12}$. Now we're getting somewhere.

The specific setup of the pendulums differs between the various detectors. GEO600 uses exactly this kind of triple-stacked pendulum. LIGO uses four stacked pendulums and VIRGO seven, but they all rely on the idea of stacked pendulums to achieve large damping values without requiring huge size. We'll discuss a few more details about GEO600's pendulums later.

VACUUM The arms in which the laser beams travel are evacuated. This is necessary for several reasons. First and foremost, air in the arms would transmit outside vibrations as sound waves. All the sophisticated suspension-based 10^{12} damping would not be of much use if air transmitted external disturbances.

Less obviously, the vacuum is also crucial to maintain all the optical components at the same constant temperature to prevent them expanding or shrinking. Thermal expansion varies widely for different materials, but even the best exhibit in the order of 10^{-6} meters of expansion per (cubic) meter of material per Kelvin of temperature change. This would clearly be another no-go for a detector that is supposed to resolve to 10^{-18}. Long-term temperature variations can be a problem for the mirror suspensions, causing them to sag and change the alignment of the optics. More critical are shorter-term variation that affect only part of the system; air currents could produce such variations. Finally, the movement of the air in the tunnel—thermal turbulence and sound waves—would refract the laser beams and potentially disturb their collimation, the perfectly parallel orientation of the waves. This could make the laser diverge.

The final argument in favor of a vacuum is mirror cleanliness. As we will see, very high laser power circulates in the arm cavities, which can burn any specks of dust onto the optics, destroying their surface.

Creating and maintaining a vacuum is a technical challenge in itself. The process starts by mechanically pumping out the air with turbo pumps or scroll pumps. Scroll pumps use meshed spirals to produce a continuously decreasing volume that compresses gas and then expels it through an exhaust. Turbo-molecular pumps work roughly like jet engines: rotors accelerate air molecules into successively higher-pressure stages, until the pressure is high enough for more traditional pumps to deal with it. LIGO uses ion pumps to get to the required ultra-high vacuum quality of 10^{-6} Pascal. Ion pumps ionize the gas in a vacuum vessel and then use a strong electric field to accelerate the ions towards, and then stick to, a solid electrode. Water molecules are extracted with liquid nitrogen cryo-pumps that condense water vapor onto a cold surface. The vacuum is continuously monitored by computers and actively maintained with the various pumps.

Operating all the optical equipment necessary for the detector in this vacuum is an additional engineering challenge. Operation has to be reliable, with absolutely minimal manual intervention, because every opening of the vacuum vessel requires "rebuilding" the vacuum from ambient pressure. For LIGO this takes over a month!

THERMAL NOISE AND THE Q FACTOR Every material has internal friction that transforms kinetic energy into thermal energy. For example, when you move a pendulum, the internal friction of the suspension wires damps the pendulum slightly; in exchange, the wires get slightly warmer. The quality factor Q of a material characterizes this behavior: the higher the Q-factor, the lower the dissipation, the lower the damping. However, as we have said, the pendulum does not actually move when a wave passes (it is space itself that expands and shrinks), so why is the effect relevant?

Every material exhibits internal random fluctuations of its electrons as a result of its temperature. This is called thermal or Nyquist noise. It's basically the electrical equivalent of Brownian motion in gases. The energy of these fluctuations depends on the square root of the temperature, as well as on properties of the material. These internal fluctuations can have a (very small) macroscopic effect: the mass starts to vibrate a tiny bit. At the precision gravitational wave detectors operate, these mechanical vibrations might disturb the detection, so they have to be avoided.

Enter the fluctuation–dissipation theorem. This states that the magnitude of spontaneous temperature-induced thermal fluctuations is proportional to the material's damping of an external disturbance—that is, materials with a high quality factor macroscopically vibrate less for a given temperature. This is why the components of the pendulum are made from fused silica, a type of glass; its Q-factor is several orders of magnitude higher than, for example, steel, while it still has sufficient mechanical strength to hold the heavy masses. In addition to the wires, thermal vibration can also originate from the pendulum masses themselves, so they are made from fused silica as well. Even the connection between the wires and the masses might be a source of internal friction, so the two are bonded together. Bonding is a way of chemically joining two parts such that they basically share free electrons. It results in a strong, thin, vacuum-compatible connection that is free from internal stresses. Finally, it turns out that for the pendulum as a whole, the Q-factor can also be increased by making the mass heavier. GEO600 uses 10 kg, LIGO uses 40 kg.

It is important to point out that understanding all these processes and coming up with the kind of suspensions used in today's detectors is the result of decades of basic research and experimental refinement. The Institute for Gravitational Research at the University of Glasgow specializes in building these suspensions; GEO600's and LIGO's suspensions were built by this group. They are continuously optimized. For example, LIGO was offline between 2010 and 2015 to be upgraded to the Advanced LIGO configuration which, among other enhancements, added a fourth level of pendulums. They also replaced their original steel attachments with high-Q monolithic glass. Both updates were instrumental in the first detection of gravitational waves in 2017.

ACTIVE DAMPING We have seen that the passive damping system based on three stacked pendulums achieves a factor of 10^{12} for frequencies at around 100 Hz. But a lot of the action in gravitational waves occurs at low frequencies, where the dampening is less because we get closer to the resonance frequency of the pendulums. On the other hand, these frequencies are easier to compensate with actively controlled damping, which is exactly what is done. However, it is important to ensure that the active control is not relative to a base that itself vibrates because it is connected to the "outside world". To achieve this, the overall mirror mounting system in fact consists of two pendulums, located right next to each other. Sensors mounted on the masses that ultimately hold the mirrors detect movement. Actuators mounted on the reaction masses then create stabilizing forces. Each level is damped separately.

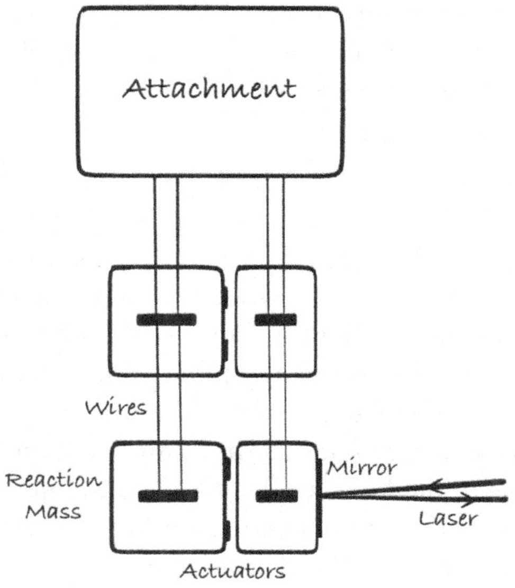

The measurement detects relative movements of the masses. This can be done with electric fields or with shadow sensors, in which a photodiode is partially shadowed from an LED on the opposing mass depending on the relative movement. However, it is not unlikely that both mass "chains" are disturbed similarly—after all, they are ultimately installed in the same environment. To detect "joint" movement, accelerometers are used; those measure absolute acceleration in inertial space, rather than relative to a reference surface. In particular, this prevents any coupling with seismic noise. The sensitivity of these sensors is limited by electronics noise and the problem that they measure acceleration, and ultimately, relative displacement, so errors are cumulative.

The actuators' forces are created with electric and magnetic fields: coils and magnets, not unlike the fine drive in SOFIA. In total, GEO600 uses more than 250 active control loops in various elements of the optical path.

All this passive and active damping leads to very stable mirrors. As Christoph Affeldt, operations manager of GEO600, says "Because the ground also just moves at sub-micrometer range, [the damping ensures that] our mirrors basically don't move at all." But there are other issues that limit the resolution of the detector, several of them related to the laser.

To allow the sensors at the output that detect any intensity variation due to interference to work effectively, it is useful if the laser has a high power, as the signal-to-noise ratio gets better with higher-power lasers. We have all heard of very high-power lasers: those used to cut metal generate a few kW, the one that was supposed to be installed on the YAL-1, a 747 modified to shoot down tactical ballistic missiles, was up to 1 MW, and those used at the National Ignition

Facility to attempt laser fusion (and do some nuclear weapons research on the side) are probably even more powerful. However, while these lasers have lots of power (and potentially only for short bursts), their quality is not necessarily suitable for gravitational wave detectors, as we will see below.

How do lasers work? The term "laser" stands for Light Amplification by Stimulated Emission of Radiation. The idea is to amplify a particular frequency of radiation to produce a beam that contains only this one particular wavelength. A laser cavity consists of two mirrors with light reflected between them, forming a standing wave. Inside the cavity is a gain medium that allows only a particular wavelength of light to form the standing wave, thus amplifying that particular wavelength. This is why lasers are monochromatic. One of the mirrors is semi-transparent, to allow some of the light to escape to form the beam. A wide variety of materials, including liquids, gases and solids, can be used as a gain medium for particular wavelengths, which then lend their name to the laser itself, as in "helium–neon laser" or "oxygen–iodine laser". The energy required for the amplification is "pumped" into the gain medium through electric discharges, light flashes or light from another laser.

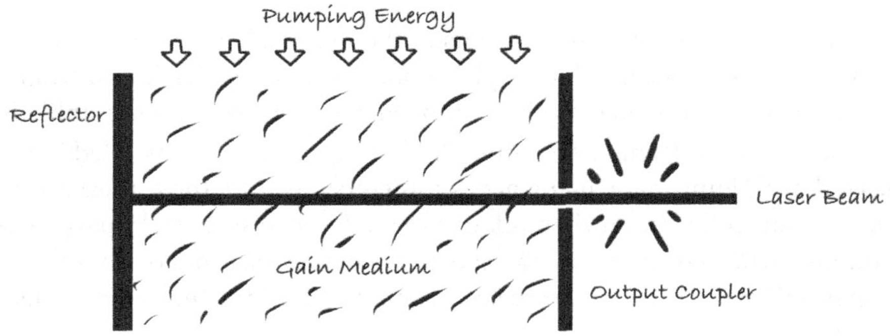

So how do you define the "quality" of a laser? A laser beam should be coherent, which technically means that it's phase, frequency and waveform is stable over time and space. Since the detector's ultimate goal is to allow the laser beam to interfere with itself to detect gravitational wave-induced phase differences, it is crucial that the undisturbed beam retains a constant frequency and phase for as long as possible. Several factors play into this. First is the mode content. A mode is a state of a system in which all components are excited jointly, in a standing wave. For an electromagnetic system such as a laser, modes refer to the field patterns of the electromagnetic waves; for example, the electric and magnetic fields may be transversally or longitudinally oriented relative to the direction of travel of the electromagnetic wave. There are also higher modes in which the patterns are multiples of the fundamental mode. Depending on the exact setup of the laser, it might produce light that contains several (higher) modes. Multi-mode laser beams tend to diverge more as they travel; ideally therefore the laser should be a mono-mode laser.

A second criterion is monochromaticity. It can be shown that a laser decoheres faster the larger the range of frequencies in the wave. Third, a high-quality beam also implies smooth wavefronts, which in turn means that the phase is strongly correlated over the profile of the beam. Generally, the higher the power of a laser, the harder it is to achieve these quality attributes. Reasons include mechanical vibrations, thermal problems, and the fact that they have to use higher-power pump sources which themselves have higher noise levels.

But even if we can create high-quality and high-power laser light, we don't want to increase power too much, because this leads to other problems. Remember that we discussed how important it is to keep the detector thermally stable? If the laser gets too powerful, the beam heats up the mirrors and other optical components, which leads to distortions and ultimately inaccuracy in the system. Furthermore, the mirrors start vibrating due to the radiation pressure the laser exerts on them. Radiation pressure is the result of the exchange of momentum between photons and an object at which they are reflected. Because of quantum effects that lead to tiny variations in the amplitude of the light wave, this pressure is not constant and can thus lead to tiny vibrations on the mirrors. The force is of course very low and not noticeable in everyday processes, but it can become a problem at the sensitivities required for a gravitational wave detector.

THE LASER IN GEO600 Let's look at the lasers used in LIGO and GEO600. They are custom-built by another group of specialists, a cooperation between the Max Planck Institute for Gravitational Physics and the LaserZentrum Hannover e.V. The components are installed on an optical table in a cleanroom, acoustically isolated to ensure they are decoupled from environment vibrations. The full system consists of several stages.

Stage one is the laser itself. It is a solid-state laser build from yttrium aluminum garnet doped with neodymium (which makes it a Nd:YAG laser). It is arranged as a non-planar ring oscillator, which means that there is a single crystal in which the light circulates. It is pumped by laser diodes at 808 nm. This first stage delivers an output power of only 2 W but very high frequency stability. As we said above, high-quality lasers don't have much power. The next stages will boost the power by factors of up to 100.

Stage two is an amplifier that raises the beam power to 35 W. The laser from stage one passes through four YVO_4 (yttrium orthovanadate) crystals, also doped with neodymium and pumped with the same 808 nm laser diodes. Note that there is no bouncing back and forth between reflectors during amplification, which is why the frequency remains stable. At GEO600 stage two is the last stage, so the 35 W is all we get.

LIGO has one more amplifier. Until 2018 they used a ring oscillator with an output power of 220 W, again Nd:YAG. It is an oscillator, which means that "new" laser light is created, with the potential quality problems attached to that. To ensure that it has the same frequency stability as the original stage one laser, the first stage light is injection-locked to this one. Injection-locking means that the stage one laser is injected through a semi-transparent mirror, and, assuming that the frequencies are close (which they are, being produced by the same materials), this forces the new laser to oscillate at exactly the injected frequency. The resulting laser's frequency stability was supposed to approach the theoretical value. However, this did not quite work as expected, as the amplifier had problems with noise. The reason was surprising, to me at least: the oscillators were water-cooled. For efficient cooling the water flow had to be turbulent, but this injected jitter into the beam pointing and size. The water feed lines on the optical table were a problem as well. In 2018, LIGO switched to a 70 W amplifier that requires less cooling, and in particular is designed such that turbulence in the water flow does not directly couple to the oscillators. It's a compromise because, while it has better quality, it has less power than its 200 W predecessor.

The last stage in all systems consists of additional optimizations necessary to ensure that the laser is stable enough for its use in the interferometer. Again three steps are involved. The pre-mode cleaner suppresses the higher-order modes mentioned earlier, improves pointing and filters radio-frequency fluctuations. The mirrors used here are also suspended from pendulums, to ensure that remaining vibrations from the laser's optical table are cancelled out. Next, a power stabilizer reduces power fluctuations in the relevant band between 10 Hz and 10 kHz. This measures fluctuations with a photodetector and uses an acousto-optic modulator, to stabilize them. Stage three is active frequency stabilization based on a frequency-stable reference cavity. I will spare you the details that I might not fully understand myself—and there's even additional magic beyond this.

Why am I telling you all this? I want to drive home the point that, even a component where you'd expect to just go to the leading professional laser vendor and pick up a high-quality laser, requires a custom design and years-long effort by specialists. I got all of this from a paper titled "Stabilised High-Power Laser System for the Gravitational Wave Detector Advanced LIGO", by P. Kwee et al. It is 18 pages long and written for laser people (by which I mean that it doesn't waste space with the basics)! This is the most stable laser ever made at its frequency range.

SHOT NOISE As we know, light is a wave. But quantum theory tells us that it comes in discrete "packets" called photons, so we can also treat light as particles. A laser beam is not continuous, but is actually a sequence of photons. There is inevitably some random variability in the exact sequencing of those photons due to the Heisenberg uncertainty principle. It states that for specific pairs of physical quantities, we cannot know both to arbitrary precision. It is most commonly applied to position and momentum:

$$\Delta x \cdot \Delta p = \frac{h}{4\pi}$$

For optical systems there is another of these pairs, where A is the amplitude and ϕ is the phase:

$$\Delta A \cdot \Delta\phi = \frac{h}{4\pi}$$

The variability in ϕ is known as "shot noise", and the effect can impact the measurement: the variability seen when the photons arrive at the detector is not just because a gravitational wave distorts space, but also because that's how physics works at the micro level. For the Michelson interferometer, shot noise is given by the following formula:

$$s = \frac{1}{L}\sqrt{h\,c\,\frac{\lambda}{4\,\pi^2\,P}}$$

h (the Planck constant) and c (the speed of light) are constants, L is the length of the arms, λ is the wavelength of the laser and P is its power. As you can see, you can reduce the noise by building longer arms, but once the detector is built, increasing the power of the laser is the last remaining leverage available. As we have seen, however, increasing laser power leads to problems with laser quality and thermal stability. In addition, P is under the root in this formula, so we get diminishing returns for the reduction in noise per unit additional power, while still getting the other problems resulting from increased power. Shot noise is a fundamental limit for the sensitivity of gravitational wave detectors, especially at higher frequencies. At lower frequencies it is the radiation pressure noise mentioned earlier; in fact, it corresponds to the ΔA term in the Heisenberg formula above.

However, we can try to play with basic physics. For the interferometer we are primarily interested in phase, not the exact value of amplitude. If we could tweak the system to increase ΔA, then we would get a lower uncertainty in phase $\Delta\phi$ (because the product has to remain constant), lowering shot noise. Is this possible? It turns out that it is. The quantum state of a wave where the uncertainty in one of the variables is increased and correspondingly decreased in the other is called "squeezed light". To really understand this state we'd need a lot of math, so we won't. But how do you squeeze a laser? You inject a squeezed vacuum field into the unused port of the interferometer. If it has the same wavelength, polarization, wavefront curvature (plus a few other properties) as the laser, the laser in the interferometer takes over the quantum noise of the squeezed beam. The squeezed vacuum field is created through pumped, nonlinear crystals. Again, explaining this in any detail is beyond the scope of this chapter.

GEO600 has been working with squeezed light routinely since 2010 and has improved their sensitivity by a factor of almost two. Christoph: "This allows us to look twice as far into the universe, and because space is three-dimensional, we can see eight times more gravitational wave events in some frequency ranges." LIGO also has been injecting squeezed light since April 2019.

One question remains: isn't the increase in ΔA a problem at lower frequencies because of the increased vibration resulting from radiation pressure? It turns out that at lower frequencies there are other noise sources that have a bigger impact, so slightly increased radiation pressure noise is not the limiting factor.

POWER RECYCLING We are still trying to increase the signal-to-noise ratio without resorting to higher-power lasers because of their quality problems. The final approach used in search of this goal is power recycling. Picture the light path in the interferometer shown in the next illustration. The light comes from the laser and hits the beam splitter, where it is split into beams A and B. Beam A passes through the splitter, travels down the arm (A1), is reflected, comes back (A2) and is then split again. Part of the light is reflected to the output port (A3) and part passes through the splitter, back to the laser (A4), and correspondingly for beam B.

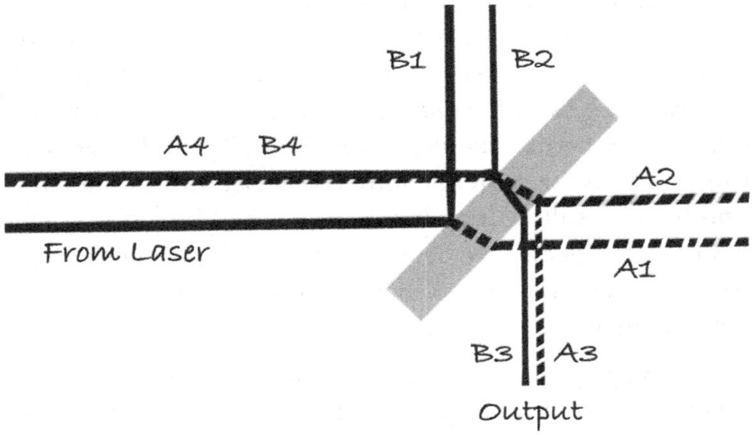

As we will see in the next section, the output of the interferometer is operated (almost) dark—that is, the two beams have almost perfect destructive interference. Active control of length offsets in the arms ensures that the phase difference between A3 and B3 at the output port is (almost) π, to ensure (almost) destructive interference.

However, conservation of energy requires that the energy that does not arrive at the (dark) output has to go somewhere, and indeed the vast majority of the energy takes the path back to the laser (A4, B4). Why is this useful? Because that light is then reflected back into the interferometer with yet another mirror. This is known as "power recycling": it increases the power of the beam inside the interferometer without requiring more input power from the laser, which would imply the aforementioned quality problems. The shot noise only increases with the square root of the power, as we have seen before, so there is a net benefit.

In GEO600 the increase in power inside the arms through power recycling is a factor of around 1,000. LIGO is designed to achieve a factor of 3,750, resulting in an effective power of 750 kW, although only around 200 kW is achieved currently. Of the 3,750-fold amplification, the first factor of 50 is contributed by power recycling, while the remaining 75 comes from an additional mechanism called laser resonance in the arms themselves.

The catch is that all of these mechanisms require lots of additional mirrors, which all have to be aligned very precisely, and this alignment has to be maintained even when mirrors vibrate slightly. Active control is necessary. In fact, when the idea of power recycling was conceived initially it was unclear how this could be achieved technically. Here is a paragraph from the 1998 paper "Power Recycling for an Interferometric Gravitational Wave Detector" by Masaki Ando:

> Though power recycling will be used in all of the interferometric detectors under construction, several problems must be solved in order to apply power recycling to these detectors. One of them is a problem in control of the interferometer. In order to behave as free masses, the mirrors of an interferometric gravitational wave detector are suspended as pendulums. Though the suspension system has another role to isolate the mirror from the seismic motion at the observation frequency band, the mirror is largely excited by the seismic motion at the resonant frequency of the pendulum. Thus it is necessary to control the interferometer in order to keep it at the operational point. However, since a power-recycled interferometer has a complex and coupled optical configuration, careful analysis and design are required to extract independent control signals from the interferometer. In addition, in a power-recycled interferometer, the lock acquisition is not a simple problem. Since the control signals are linear functions of the motion of the interferometer only around the operational point, it is not obvious whether the interferometer can be locked at the operational point from an uncontrolled state. Thus, the controllability and lock-acquisition of a power-recycled interferometer must be tested experimentally.

CONTROL Obviously the community tackled the control challenge. But this helps to explain the need for the 250+ active control loops. For example, for resonance and thus amplification to occur, the distance between the power recycling mirror and the mirrors at the end of the arms must be a precise multiple of the wavelength—which is in the nanometer or micrometer range. As we saw before, active control is also required to ensure the correct phase difference between the beams such that the output port is (almost) dark. Another aspect is their nano-degree-precision directional alignment. Any misalignment is measured using the same laser light that travels through the system. However, to have it available for measurement and control, one has to ensure that some of the light isn't reflected back, but instead leaks out to sensors. This is achieved by modulating the phase of some fraction of the input light at radio frequencies, and carefully designing the various cavities to either reflect or transmit the light. There are sensors on the output port, as well as behind the back of the end test masses.

There are other challenges. For example, the remaining slight mechanical motion of the mirror masses can scatter light into a higher optical mode in the arm. These might then be reflected back onto the test masses, creating a feedback loop which, depending on parameters, might be beyond what the control system can compensate for. The risk of such feedback increases the more laser power circulates in the arms. Advanced LIGO uses heaters on the mirrors to slightly change their shape, and thus control their optical characteristics, to avoid such resonance.

Despite profiting from prototype systems, it took GEO600 a decade get it all to work. "After the last vacuum tank was sealed we spent a long time to hunt down the expected and unexpected sources of noise", says Christoph. Even today there are only a few people who understand its intricacies and know which buttons to press if something gets out of whack.

Another important aspect of commissioning such a system is to characterize the instrument's response to a stimulus. Remember, the detector should not just indicate the presence of a gravitational wave, but also record the frequency and amplitude of the wave as it passes the detector. This is partially theoretical work, for example based on the known relationship of damping and frequency for a given pendulum, but also on lots of measurements. In particular, the remaining resonances of system components such as mirrors and pendulum wires must be characterized precisely. It is much harder to detect signals at these resonance frequencies; the control system displays the "danger zones" on a screen, and valid detection of a gravitational wave event must involve frequencies outside these critical areas.

In addition, the data acquisition system does not just record the final signal (which we explain below). It records any number of diagnostic data parameters, such as alignment, the quality of the vacuum and the temperature in the pipes, to be able track down problems if and when the system does not work correctly. Of course they also operate seismometers and magnetometers to be able to check candidate gravitational events against more mundane, earthly causes.

Today the detectors run without much manual interference. Christoph again: "A magnitude seven earthquake on the other side of the world throws us out of lock. So does a tractor on a nearby field or the postman if he doesn't respect the speed limit near the buildings that house the reflector mirrors. If the disturbance is short, the system re-locks automatically, but if it lasts longer, manual intervention on the scale of minutes to days is required." Cameras help with manual, coarse-grained realignment. To ensure availability of GEO600, some team members are on call during science runs. LIGO uses 24-hour operators.

3 | *The Groundbreaking Detection*

THE EVENT Because it changes the interference of the two beams, a passing gravitational wave produces a small variation in intensity at the dark port of the detector. This intensity change is detected with a photodiode, a semiconductor-based device that transforms light into an electrical current, which is then measured. While photodiodes are very sensitive, additional preparation is required to make it work.

To recap: our laser is running at 1,064 nm, which corresponds to a frequency of $2.8 \cdot 10^{14}$ Hz. Let's say our gravitational wave signal is at 500 Hz. It will imprint sidebands at ±500 Hz above and below the base frequency, a relative difference in frequency of 0.000000000178 %. Over the length of the arms, this results in a phase difference, and hence interference, but one that is extremely small and hard to detect. We somehow have to "recover" the 500 Hz signal. This is achieved by exploiting beating.

Beating is familiar from sound waves: if you overlay two audio waves with slightly different frequencies, you can hear periodic variations in volume. The frequency at which the beat volume changes is the frequency by which the two overlaid signals differ. Beating therefore provides a way to change variations in frequency into variations in volume or intensity.

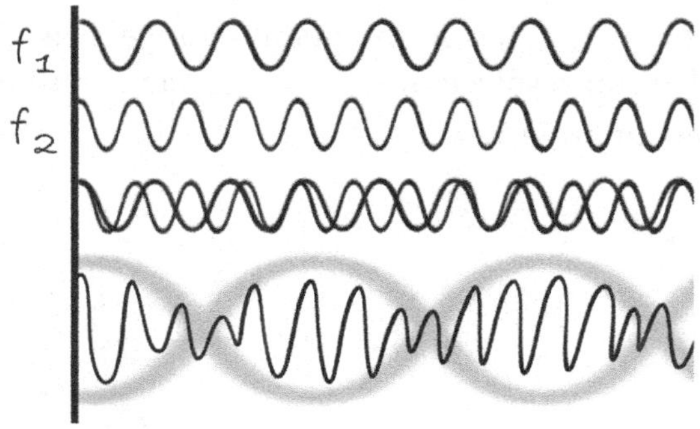

The interferometer overlays the sideband frequency (the one created through modulation from the gravitational wave) with the original, unmodulated laser. We have already mentioned that the interferometer is operated "dark"—that is, tuned to destructive interference at the output port. This is not absolutely true, however: a little bit of light is freed to serve as a local oscillator, the reference signal to create beats with the slightly differing frequency incurred by the passage of a gravitational wave. The resulting beat frequency is, of course, again the difference in the two original frequencies, 500 Hz in our example.

However, while the frequency of intensity changes is now within a useful range, and proportional to the frequency of the gravitational wave, the intensity variation is too small for a photodiode to detect reliably. Remember the power recycling we talked about before, by which the effective power of the laser in the interferometer is increased by several orders of magnitude? We can use the same approach to increase the power of just the resulting signal, the 500 Hz beat frequency. A dedicated mirror at the output reflects the signal back into the interferometer until it is amplified by a factor of 10. The system can be tuned to amplify frequencies in a particular band. This is a major reason why GEO600 has a similar sensitivity to VIRGO at higher frequencies, despite its shorter arm lengths.

WHAT WAS DETECTED Now that we have an intensity variation a photodiode can work with, let's put it to practical use. On 14 September 2015 at 05:51 local time LIGO's Hanford detector recorded a signal. The same signal was measured 7 milliseconds later at LIGO's other site in Livingston, after the wave had travelled the distance between the two sites at the speed of light. This discovery was announced in February 2016. Curiously, at the time of detection, LIGO was still in engineering mode after the upgrades to increase its sensitivity. Initially there was some confusion about whether the event could have been test data that was injected, but it quickly turned out that it was real—the increase in sensitivity had worked.

The diagram below shows the signal in terms of frequency over time, with the shading representing amplitude (darker means greater). Over about 0.2 seconds the signal increases in frequency and amplitude. Because it lies in the audible range, it is also often described as a "chirp". Notice how the detector resolved time, frequency and amplitude, not just the fact that "there was an event"—very impressive.

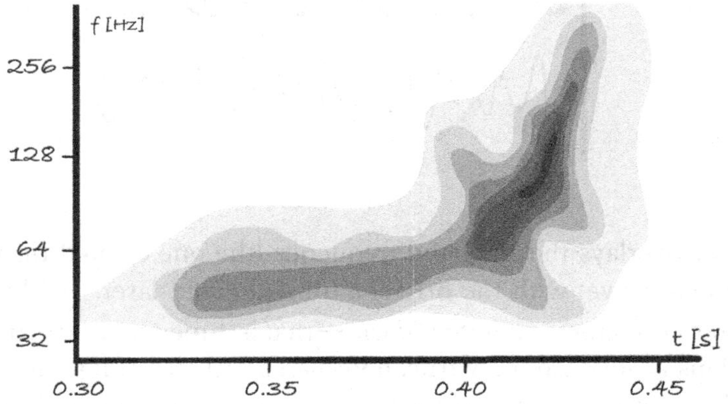

The signal was generated by the spiraling inwards and subsequent collision of two black holes. The frequency of the wave allows us to estimate the frequency with which the two objects circled each other to be 75 Hz. The amplitude and pattern of the signal suggests the masses of the objects: around 36 and 29 solar masses. From that one can derive the distance at which they circled each other, which is around 350 km. From these parameters, in turn, one can calculate the velocities: during the 0.2 seconds of the event, the velocities increased from 30% to 60% of the speed of light. Such heavy masses, at such a small spatial distance, require a density that can only be explained by black holes. The decay of the signal after the peak is also consistent with a single, merged black hole.

Picture it: 60% of the speed of light, 75 Hz, a few hundred kilometers apart, with a total mass of around 60 suns! This is the last moment before collision, an extremely violent event. During the last couple of dozen milliseconds before collision, the power of the gravitational waves generated was 50 times the power of all the light produced by all the stars in the observable universe! While this speaks for the magnitude of the event, it also demonstrates that the detectors can currently only observe the strongest gravitation waves that the universe produces.

FIVE SIGMA It took five months from the event to its publication. This time was spent ensuring that they hadn't detected something that was not actually a gravitational wave. The team worked on building confidence to the famous five sigma (actually, 5.1) factor. A confidence of 5σ means that if the "observed" result occurred by chance, they would have to run the same experiment 3.5 million times to see this particular result once. The threshold of 5σ is a convention in the high-energy physics community: a confidence of 3σ is considered good enough for "evidence", but 5σ is required for the "discovery" of something that has not been observed before.

Let's understand this in a little bit more detail. σ is the standard deviation in statistical processes. Consider a perfectly random process that produces values between 0 and 100. The average value μ would be, of course, 50. The probability for the occurrence of each particular value between 1 and 100 is the same: 1/100. Overall, however, the experiment follows the normal distribution. In a normal distribution 68.2% of the results are between $\mu-\sigma$ and $\mu+\sigma$. It is 95.2% for $\mu \pm 2\sigma$ and 99.7% for $\mu \pm 3\sigma$. This is shown in the next illustration.

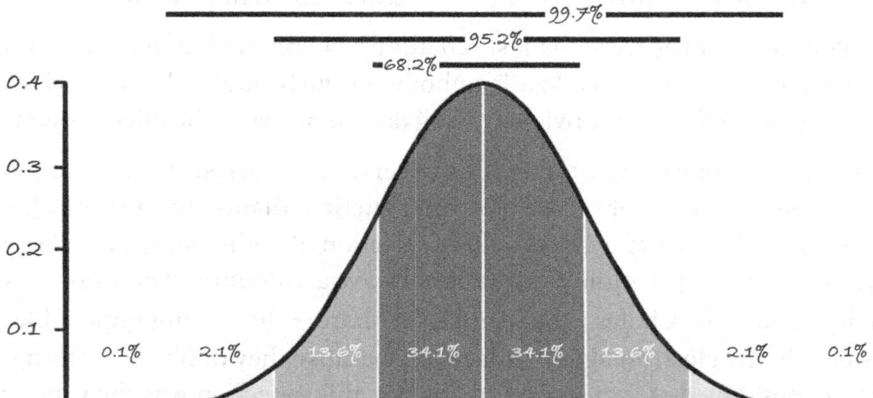

The standard deviation σ can be used for two different purposes. The first one is to characterize a distribution. If the values show less "spreading" around the mean, which could be visualized as the curve above being narrower and higher, then σ is smaller; the opposite is the case for wider, lower distribution curves. Used in this way, together with the average value μ, σ characterizes a distribution, as in "the result of the experiment is an average of 42 with a standard deviation of 7."

However, this is not the way σ is used when expressing confidence in an observation. Here we turn the situation around: we assume that the results we observe have been produced by chance, for example when the noise in the system conspires in such a way to simulate exactly what we see, such as the gravitational wave signal of two black holes colliding. Only if the probability of the noise accidentally "faking" the observed result is less than 5σ—less than 1 in 3.5 million—are we confident that the result is real: we'd have to run the same experiment 3.5 million times for it to lead to the observed result accidentally.

The problem, however, is that we don't know how well the system's noise follows a normal distribution. This is why the team performs lots of measurements that characterize how often "signals" of a particular kind are "detected" without there actually being a physical signal. This feeds into a model on which the statistical significance of a detected signal can be judged.

Note that the magical "five sigma" is not proof of the correctness of a result. This is because it only addresses statistically random processes. If you make fundamental (systematic) errors in your experimental setup, all bets are off. Remember those neutrinos that supposedly traveled faster than light from CERN to a place in Italy, where we all joked about them not adhering to the speed limit as soon as they crossed into Italy? This was a 6σ result, but it was systematically wrong: there were faults in the cables.

So why was the first observation of a gravitational wave such a big deal? For many reasons. First, it confirmed that the LIGO detector, after its upgrade to the Advanced LIGO configuration, was actually sensitive enough to detect gravitational waves. This is of course a wonderful validation for everybody involved in this project and the field in general.

Second, gravitational waves were the last remaining big prediction of general relativity that had not yet been observed directly. While nobody seriously doubted that gravitational waves do indeed distort spacetime, it is obviously a big deal to measure the effect directly.

But third, and maybe most importantly, this observation marked the start of a new kind of astronomy. In the past we've observed electromagnetic radiation in the optical, infrared, ultraviolet, gamma and microwave bands, plus a selection of high-energy particles such as neutrinos. These live within spacetime. Gravitational wave astronomy, in contrast, lets us observe in the "gravity band". Black holes are of course very hard—though not impossible, as we shall see—to observe in the electromagnetic spectrum, because they don't let anything escape, not even radiation. But they do warp spacetime. In fact, this detection was the very first observation of a merger of a binary black hole, in any band. A particular advantage of gravitational wave astronomy is that the waves interact very weakly with matter, so they can travel unabsorbed through the universe. This of course also makes them hard to detect.

We know that the event LIGO observed happened 1.4 ± 0.6 billion light years away. We cannot pinpoint the location, however, because we can't quite tell the exact direction: because there were only two detections, the direction could only be pinpointed to 600 square degrees, roughly similar to the area of the Orion constellation (with a confidence of 90%). A third measurement by VIRGO would have helped narrow down the direction, but at the time, VIRGO was offline because of an upgrade. And GEO600 is not sensitive enough in these frequencies. Now that these detectors are all working and have been upgraded in sensitivity, the community has been able to pinpoint the source of gravitational waves and then cross-reference with observations in other bands. For example, the gravitational wave GW170817 was first detected by LIGO, and two seconds later the Fermi satellite detected a gamma ray burst GRB 170817A at approximately the same location in the sky. In all, 70 observatories detected signals in various bands from the neutron star merger that created GW170817, representing a breakthrough in multi-messenger astronomy.

Generally, however, gravitational wave detectors are not particularly good at determining the precise location of the source of a gravitational wave. To compensate this shortcoming, the community is working on telescopes that are optimised for detecting the optical counterparts of gravitational wave events. BlackGEM is a project that will eventually install an array of 15 small 65 cm telescopes specifically for this purpose. When LIGO or VIRGO detect a gravitational wave, they will trigger BlackGEM to take a survey of that area of the sky that can plausibly be the source for the wave, as determined (not so precisely) by the detector. And optical signals can of course be pinpointed very precisely. When not working in service of LIGO or Virgo, BlackGEM will produce sky surveys; these are important for astronomers to learn about the statistical distribution of the various kinds of objects on the sky.

As is obvious from the description of the black hole collision, the kinds of events that can be observed with today's detectors must involve very heavy objects that travel at significant fractions of the speed of light, otherwise the resulting gravitational waves are just too weak to detect. Binary compact objects such as spiraling and merging black holes are prime candidates, but binaries composed of neutron stars (such as GW170817 above), or white dwarfs, would also work. Supernovae, a kind of "explosion" of a star, also create waves in the frequency and amplitude ranges that the current detectors can resolve. Finally, there is hope that inflation-related processes from the early universe created detectable signals.

On October 3, 2017 the Nobel prize in physics was awarded "for decisive contributions to the LIGO detector and the observation of gravitational waves". It was specifically awarded to Rainer Weiss, Barry C. Barish and Kip S. Thorne. Rainer and Kip were pioneers in the field, while Barry was the principal investigator and subsequently director of LIGO between 1994 and 2005, and was instrumental in the successful completion of the project. The original paper that announced the discovery, titled Observation of Gravitational Waves from a Binary Black

Hole Merger, has 1,012 authors from a total of 133 institution (I counted them!). This demonstrates the magnitude of the challenge—and perhaps a flaw in how the Nobel prize works.

4 | *Future Detectors*

So where do we go from here? The detectors will be continuously upgraded to increase their sensitivity. For example, an update of LIGO codenamed Voyager will install lasers of higher wavelength and cool the mirrors to 123 K to reduce thermal noise. This would halve the minimum useful frequency to around 10 Hz. New detectors are being built, such as KAGRA in Japan and a possible third LIGO site in India. KAGRA has 3 km arms installed in an underground tunnel to reduce seismic influences and will also use cryogenic technology to reduce thermal noise. Fundamentally, however, these new and upgraded detectors will be in the same frequency range as LIGO, from around 10 Hz to 1,000 Hz, with higher sensitivity at the higher frequencies.

More ambitious projects include the Einstein Telescope, a possible project proposed by European scientists with 10 km arms and multiple interferometers optimized for different frequencies. Those for low frequencies will cool the mirrors to 10 K and have a beam power of 18 kW. The Cosmic Explorer detector, envisioned by US scientists, is based on LIGO Voyager, and would go to an 40 km arm length. Even assuming the funding can be found, however, it will be decades before these could become available for science.

Many of the challenges for gravitational wave detectors I have described in this chapter—seismic noise, temperature, short arm lengths—are a direct consequence of the detectors being built on our little blue planet, Earth. How about...? Well, exactly: LISA, the Laser Interferometer Space Antenna, will get rid of this limitation and measure gravitational waves in space. Three satellites will form a triangle with 2.5 million kilometer long "arms". These extremely long arms will contribute to lowering shot noise (remember, L is in the denominator). On the other hand the laser will be much weaker, because there's not going to be much energy available on a satellite, and there won't be any mirror magic for power recycling. Overall this makes LISA less sensitive than its Earth-bound brethren, but this does not matter, because the signals themselves are stronger in LISA's detection band. The approach to detection will be different as well. In Earth-based detectors the laser frequency noise that still remains, despite high quality lasers, is cancelled out because the two arms have exactly the same length. However, on LISA, because the satellites move on slightly different orbits due to of gravitational effects, the "arms" between any pair of satellites can vary in length by up to a few percent. An approach called "time-delay interferometry" uses sophisticated algorithms to correct the errors due to different arm lengths mathematically. To ensure that there will be no effect on the test mass from solar wind or light pressure, the satellites will follow the zero-drag design: this means that the test mass will free-float inside a closed satellite shell. The distance between the mass and the shell is continuously measured; if it changes, it follows that the shell must have been "pushed" by forces other than gravitation. NASA's gravity probe B and ESA's GOCE satellite have previously used this approach. These extremely stable test masses will allow LISA to be sensitive between to 10^{-4} Hz and around 1 Hz, perfectly complementing Earth-based observatories. The 2015 LISA Pathfinder mission demonstrated achievable noise levels with 38 cm arm length. LISA is currently scheduled to launch in 2034.

6

Engineering the Big Telescopes

Astronomy can be structured into three major areas: the observation of electromagnetic waves, the detection of particles, and more recently the characterization of gravitational waves. The observation of electromagnetic waves is the oldest form and can be subdivided further according to the wavelength of the observed radiation: the range is from 10^3 to 10^{-12} meters—from the size of mountains to the scale of atomic nuclei. The overall range is split into bands (or "kinds") of light, even though these are not necessarily delineated precisely. The following illustration shows wavelength, the names of respective bands and the atmosphere's transparency for that band; lighter color indicates greater atmospheric transparency.

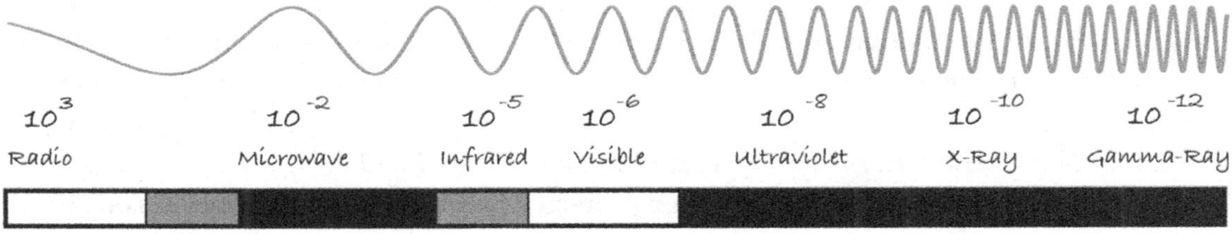

It should be obvious that very different technologies are necessary to observe at the different wavelengths. For example, the surface of the device that collects the radiation—the antenna, mirror or "dish"—must be made of materials that reflect the wavelength of interest, and its deviation from a perfect theoretical shape should be significantly smaller than that wavelength so that it does not distort the signal. The imaging resolution one can achieve at a particular wavelength is also related to the size of the antenna. Finally, as the Earth's atmosphere absorbs some wavelengths, some observations have to be done in space, or at least at high altitude: the atmosphere's limited transparency for infrared light is why SOFIA exists.

The different categories of radiation (as opposed to gravitational) astronomy are therefore aligned with the various bands of the electromagnetic spectrum. There is optical astronomy, which observes visible light and might include the near (shorter wavelength) infrared. Radio astronomy covers a broader range, encompassing the radio and microwave spectra, although telescopes are often optimized for particular sub-bands. We heard about infrared telescopes in the context of SOFIA and the Herschel satellite. Observations in the ultraviolet are performed in space by various dedicated satellites, and also by the Hubble Space Telescope. For X-ray observations the most well-known satellites are probably XMM-Newton and the Chandra X-Ray observatory. Gamma rays are also observed from space; the most recent satellite for detecting these is the Fermi Gamma Ray Space Telescope. There are also ground-based telescopes, but these observe the secondary particles or radiation produced when gamma rays impact the upper atmosphere; an example is the Cherenkov Telescope Array, which, as the name suggests, detects Cherenkov radiation. In the remainder of this section we'll focus mostly on optical and radio telescopes.

1 | *The large Binocular Telescope*

ON MOUNT GRAHAM The Large Binocular Telescope (LBT) is located 3,200 meters up on Mount Graham, a three-hour drive from Tucson, Arizona. The mountaintop site was chosen to minimize atmospheric turbulence, avoid light pollution and to get above some of the atmospheric water vapor, all of which contribute to good astronomical "seeing". I was in Tucson in October 2012 to attend the SPLASH software conference and was given permission to visit the LBT to record a podcast episode. I was planning to drive up the mountain myself, but Richard Green, the telescope's director and my guide on that day, insisted on picking me up in Tucson with his car. We started out on Interstate 10, a perfectly fine road, until we got to the base camp of the observatory at Swift Train Junction, where I had to sign a form that prohibited me from messing with the red squirrel, a protected species that apparently lives on Mount Graham. We then drove up a really scenic mountain route that became narrower and narrower for about one and a half hours, eventually turning into a gravel trail, until we finally reached the observatory at 3,200 meters. Richard's four-wheel drive was definitely much better suited for this than my small rental car!

Once at the telescope we had lunch. Guests and personnel bring their own food; there is a pleasant kitchen/living room where people can cook, eat and relax. Working at the observatory does entail some spare time, for example when the weather isn't good enough for observations. Richard and I then spent about an hour recording the basics of the observatory before we started our tour of the facility. Once in the enclosure, the dimensions of the telescope became clear: two giant 8.4-meter mirrors are mounted next to each other on a common frame. Everything is extremely rigid, to avoid deformation as the telescope moves in elevation to follow a target in the sky. When it moves you hear absolutely nothing and feel no rumbling! On first seeing the machine, a post-doc from Italy who joined our little tour of the telescope remarked "This is the Sistine Chapel of our time!"

We were planning to stay into the night to see the telescope in action, but it was increasingly unclear if the weather would be suitable for observing. Nice cumulus clouds (from a glider pilot's perspective) had been developing during the afternoon, indicating that there was moisture in the atmosphere. Moisture is bad, because condensation can damage the mirror, making little pits and streaks. The enclosure cannot be opened and no observations can take place if the atmospheric humidity is above 95%. The other weather limit is wind speed, because it might shake the telescope to the point where it cannot focus, damage the very thin secondary mirror, or deposit dust on the primary mirrors. Higher-altitude clouds such as cirrus are also a problem because they absorb some of the light, which means that it takes longer to observe to a particular signal-to-noise ratio. But at least cirrus clouds don't negatively impact the telescope for subsequent users. July and August is the rainy season around Mount Graham, so the telescope essentially shuts down for maintenance as there are few worthwhile observation days. It's October now, though, so we remain optimistic.

Let's briefly look at the constraints that govern what an optical telescope can observe using the diagram below. Its purpose is to show how many hours per night a particular astronomical target can be observed. The horizontal axis shows a couple of months'-worth of days, while the vertical axis shows hours per night. The upper line shows hours of darkness; obviously you cannot observe in the visible spectrum while it is bright outside. The length of the night of course varies over the year, which is why this line in bent. The lightest shaded area represents the geometric visibility of the target: if it does not rise above the horizon (or above the minimum elevation of the telescope), it cannot be observed. Depending on the brightness of the target and the objective of the science, the hours available for observation are further constrained if the Moon is up; its period of roughly 27 days is visible in the diagram. Finally, as we have just experienced for the LBT, the average expected weather on specific days further constrains the useful time for observations, due to excessive atmospheric turbulence or humidity.

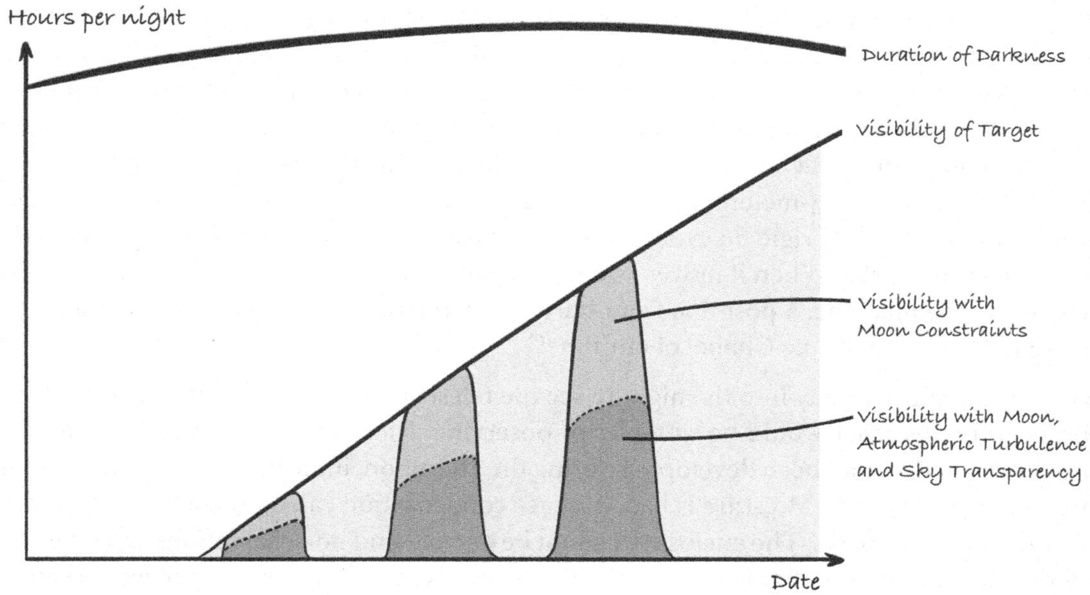

A more detailed but conceptually similar plot is available for each night. It shows evening and morning twilight—during which infrared observations are generally feasible but optical observations are not—how the target rises and/or falls in elevation, plus the target's angular distance from the Moon. The nightly plots are relevant for observation planning, because ideally a target is observed when its elevation is high. Lower elevations are less desirable because the light has to go through a larger air mass; the air mass is defined as the integral over the air density between the telescope and the object. Greater air mass means that light is more attenuated and there is more potential for turbulence. In addition, the sky closer to the horizon often exhibits greater light pollution.

Back to the LBT. Dinner came and went, and the operator announced that he planned to open the enclosure at 17:45. That time came and went as well, so Richard and I spent some time talking about the various instruments installed on the telescope and what they are used for; we'll discuss some of those later in this chapter. Just as we finished talking, Richard's radio received the operator's announcement that the enclosure would be opened in a few minutes. We hurried into the enclosure to witness the opening from the inside.

Every large machine or scientific instrument I've visited has this magical moment when you are awestruck by its sheer impressiveness. With SOFIA, the moment came when I first entered the hangar, walking up close to the beautifully polished Boeing 747. I actually had to suppress a tear or two, not least because I had been looking forward to that moment for a long time: the trip had been in planning for over six months. The opening of the LBT enclosure was another such moment.

It is completely dark in the enclosure (important for calibrating the telescope), only a few dim red lights indicating the status of various electronic devices. We use small flashlights to find our way down to the base of the telescope and forward towards the big sliding doors. It is also quite cold, almost freezing. We stand in complete darkness, and as I look up the doors move sideways with a slight rumble, revealing a clear, starry night sky the like of which I have never seen before. Behind me I can see the sky reflected in the two 8.4-meter mirrors. Amazing! The huge telescope starts moving above and behind us—I can tell because the image of the Milky Way moves across the mirrors. We can also see the dark shape of the landscape below us as we look outwards.

The whole event was deeply impressive, and clearly drove home the point that this was another masterpiece of engineering, and a "spectacular instrument for science", as Richard called it. To some extent it's "his" instrument; he's been involved for more than a decade, from design through construction, first light and—at the time of recording the podcast—validation of all of its features.

OVERVIEW The defining characteristic of an optical telescope is the size of its mirror. Their job is to collect light and focus it onto a detector, also referred to as the sensor. The bigger the mirror, the more light it can collect per unit of time and the greater the signal-to-noise ratio in the radiation collected. In addition, the larger the mirrors, the better is the telescope's resolution—its ability to distinguish small features. The 8.4-meter mirrors of the LBT are the largest single-piece mirrors of any telescope. 8.4 meters is the size of almost five average-sized European males; that's really quite huge for a single piece of highly polished glass!

The LBT has two of these mirrors, mounted next to each other on a common frame. They can be used separately, for example to observe the same target at two slightly different wavelengths, but they can also be used jointly by mounting identical instruments on each of them. The light-collecting power of the two 8.4-meter mirrors then becomes equivalent to one 11.4-meter mirror with the same area of around 100 m^2. More importantly, however, the two mirrors can also be combined interferometrically to increase the telescope's resolution to around nine times that of the Hubble Space Telescope. This is because the resolution in interferometric mode is governed by the longest baseline, the maximum separation of any two points on the two mirrors. As there is a 6-meter gap between the two mirrors, this longest baseline is 22.8 meters. Both the combined area and the combined baseline length is bigger than for any other single-mount telescope on the planet. The "Large" in its name is indeed justified.

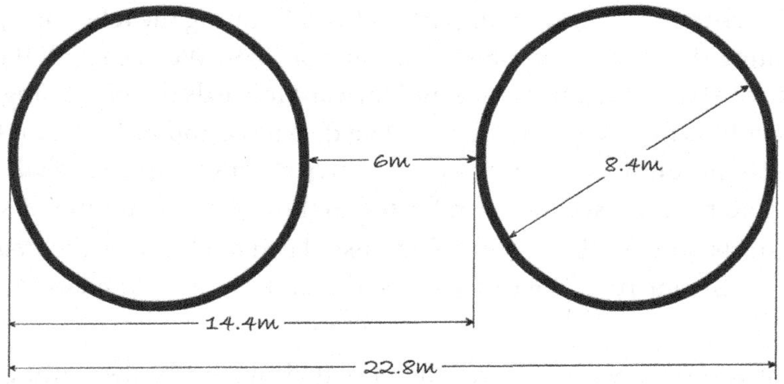

As an optical telescope the LBT and its instruments work in a wavelength range between 315 nm and 12,000 nm—that is, from near-ultraviolet, through visible light, to mid-infrared. Because of atmospheric absorption a telescope based on Earth won't see much outside this window unless it moves to the much longer wavelengths of the radio spectrum.

As is typical for large optical telescopes, the LBT is mounted on an azimuth-elevation mount, meaning that it can rotate around the compass direction as well as in elevation above the horizon. It is actively controlled to compensate for the Earth's rotation. The telescope is located in a box-shaped enclosure that rotates along with the telescope, but is mounted on a separate base to ensure that the vibrations of the enclosure do not affect the alignment of the telescope.

Let's now look at some of the engineering challenges peculiar to this telescope. I won't repeat the basic principles that I explained in the SOFIA chapter— you can always refer back to that chapter for details if you need to.

MIRRORS The mirrors in reflector telescopes have traditionally been made from massive blocks of glass. However, these suffer from the problem that they deform under their own weight; you can't make them sufficiently rigid. They also take a long time to respond to a change in temperature. "At their core such a mirror could have the temperature from a week ago, and this warps the glass", explains Richard. They are also very heavy, so they need a very beefy support structure, making the overall telescope very expensive.

Even longer ago, at the dawn of optical astronomy, telescopes were designed as refractors that used glass lenses instead of mirrors. However, this has two disadvantages. First, lenses bend light differently depending on its wavelength, an effect referred to as "chromatic aberration". This makes it very hard to build refractor telescopes that create a uniform image across a broad range of wavelengths. The other problem is more prosaic: like large and heavy mirrors, lenses bend under their own weight. Because the light passes through them, you can't simply build a support structure under one side as you can with mirrors. Reflector telescopes are also shorter overall, because the light path involves changes in direction (because of reflection), as opposed to being arranged linearly. This makes engineering easier in many respects. Refractor principles are no longer used to build large astronomical telescopes.

The mirrors in the LBT are different: they only have a one inch-thick surface and are backed by a honeycomb structure. This provides a high degree of stiffness for relatively low weight, as well as quick response to temperature changes because of the reduced volume. The glass surface is then coated with a very thin layer of aluminum; a few grams are enough for a whole mirror. The second mirror—"the better one of the two", says Richard—has a surface distortion of 19 nanometers. That's the same ratio as 1-inch waves would have relative to the surface area of the Atlantic Ocean! Bigger mirrors are of course always better. But the 8.4 meters were the maximum size "that could conveniently be made at the time", as Richard puts it: the scale of glass making, furnaces and the ability to transport the mirror to the observatory are all practical constraints. We'll see later how newer telescopes are working on mirrors that are substantially bigger, and we'll learn in more detail how the mirrors are made.

2 | *The Optics of the LBT*

OPTICAL PATH The LBT has several configurations for its light path. In the simplest case, an instrument is placed directly at the primary focus of the main mirror using a swing arm (1). An example is a 36-megapixel CCD camera that records a wide field of view, roughly the size of the full Moon in the sky. Alternatively, a secondary mirror can be installed near the focal point of the main mirror (point 2 in the illustration below). The light is then reflected back through a hole in the main mirror to an instrument that sits right behind it (point 5 on either mirror). In addition, just as in SOFIA, a tertiary mirror can be swung in to reflect the light by 90 degrees (3) to a row of five instruments mounted in what is called the "gallery" (4). In the illustration these instruments would be arranged along the axis running into/out of the page. Since this gallery is between the two mirrors, this is also the location for instruments that combine the light from both mirrors.

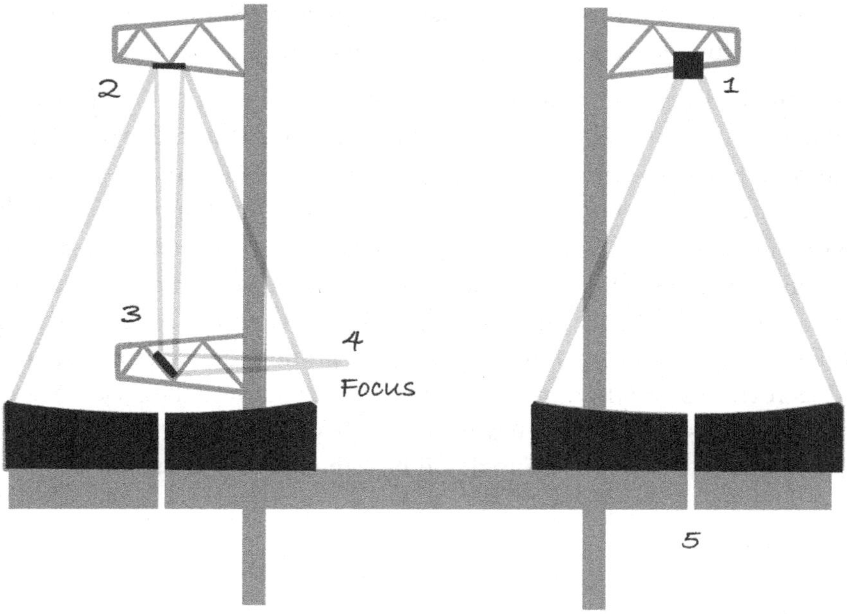

Once the LBT is completed all these positions will be permanently filled by a set of instruments, pending major updates in the future. All instruments are custom-designed for the LBT to exploit its capabilities to the fullest. Switching instruments is then either just a matter of changing the optical path or, in the case of the primary focus camera, of using the automated swing arms to replace the secondary mirror with the camera. The ability to change instruments quickly is crucial for a telescope for two reasons. The first is that the science might require that the same target be observed with different instruments during a particular observation. The other is that during the night the weather might change, constraining wavelength and sensitivity in a way that suggests switching to a different instrument and a different science target; there are always several ready for observation just in case. In both cases it is obviously good for the telescope's efficiency if such changes take as little time as possible.

ACTIVE OPTICS I have already pointed out that the LBT's mirrors are relatively thin, yet rigid, to minimize deformation caused by their weight as their orientation changes. Nevertheless, they do deform relative to the required precision, which is in the order of one wavelength. These deformations are continuously measured and then compensated—an approach called "active optics". The lightweight mirror is mounted on a rigid structure, and substantial pneumatic actuators push and pull on the mirror to maintain a perfect surface.

CLEANING AND RECOATING It is important that the surface of the mirrors remains as pristine as possible over time. We already talked about the problems of condensation, but dust also accumulates over time. "And very rarely we will get a bird inside the enclosure and they get very nervous and might soil the mirror", says Richard. In addition, Mount Graham is "invaded with moths every summer and they land on the mirror". The reflectivity of the mirror is measured regularly, and if it deteriorates they use the equivalent of a car wash wand to dust the mirrors off with carbon dioxide snow: a stream of miniscule dry ice particles removes particles of a wide range of sizes via mechanical and solvent interactions. "But we definitely need to have covers for these mirrors. That's one of the last major design elements of the telescope that need to be completed", adds Richard.

In addition, the very thin reflective layer on the mirror has to be refreshed about every other year. Most other telescopes—among them SOFIA—remove the mirror from the telescope and transport it to a separate recoating facility. For the LBT it was decided to leave the mirror in place and bring the recoating infrastructure to the mirror. After the old coating has been removed (manually) using various chemicals, a 55-ton crane lifts a vacuum bell over the mirror. It is sealed to the mirror mount's vacuum flange, forming a vacuum vessel above the mirror's front surface. To obtain a uniform new coating (that does not itself add surface roughness that distorts the image), aluminum is vaporized in the vessel; the idea is to fill the vessel with a homogeneous aluminum vapor which then settles onto the mirror as a roughly 100 nm thick layer. In practice, the devil is in the details and it took two years to develop a reliable and not-prohibitively-expensive recoating system. For example, one problem is that the walls of the vessel "steal" some of the aluminum, resulting in a thinner coating towards the edge of the mirror. To solve this problem, 28 sources of aluminum are installed in the vacuum bell at locations that were determined experimentally. These sources are cylindrical "buckets" that contain aluminum, in liquid form once heated electrically. The buckets are made from boron nitride, because, even at 2,000 K, it is not attacked by the aluminum.

ADAPTIVE OPTICS The maximum resolution the telescope can achieve is directly proportional to the size of the mirror (or the pair, for interferometry). However, in practice the resolution achievable is also limited by the atmosphere: in addition to absorbing certain wavelengths, the atmosphere is turbulent, from wind and thermal activity. This affects the sharpness of the image, blurring the light. Space telescopes do not have either of these problems, but of course their resolution is limited by smaller practical mirror sizes: Hubble's mirror is 2.4 meters.

Earth-based telescopes rely on adaptive optics to solve the problem of a "wobbly" atmosphere, actively compensating for atmospheric disturbance by deforming the secondary mirror at up to 1,000 Hz. This is not the same as the active optics described previously: the former keeps the shape of the primary mirror optimal as it moves in elevation and bends under its own weight; it operates at frequencies of one Hertz or less.

It is hard to overstate the importance of adaptive optics. I have heard two impressive illustrations. First, without the adaptive optics, the 10-meter mirror at the Keck telescope on Mauna Kea in Hawaii would have the resolution of a (perfect) 10-cm mirror. That's a factor of 100 improvement! The other comparison I was told of is that, without adaptive optics, there would be no point in building mirrors greater than 40 cm diameter to increase resolution—of course their light-collecting power would increase.

So how does adaptive optics work? Let's say the field of view of the telescope contains a perfect point source such as a very distant star, maybe the guide star that is used for tracking the sky. This reference object should lead to a perfectly flat and uniform light wave on the mirror with no distortions and no wrinkles. Atmospheric turbulence, however, distorts the wave "into something that looks like a crumpled tissue", as Richard says. How can this be detected? As usual there are different approaches, with various tradeoffs. The one used by the LBT relies on what is called a "pyramid wavefront sensor", which records four images of the primary mirror simultaneously to compute the distortions of the wavefront. However, the Shack–Hartmann approach is easier to understand. This consists of a two-dimensional array of small lenslets. A lens focuses light onto a point, and the location of the point beyond the lens depends on the angle at which the beam hits the lens. If the wavefront is completely straight (and the imagined elementary beams impact the lenses at a 90-degree angle), then all the lenses focus their light onto a point that is right in the center below the lens (the gray dashed lines on the screen in the following illustration). If the wavefront comes in at an angle, for example because it is distorted by the atmosphere, then the lens will focus the light onto a point outside the center (the solid lines with the black dots on the screen).

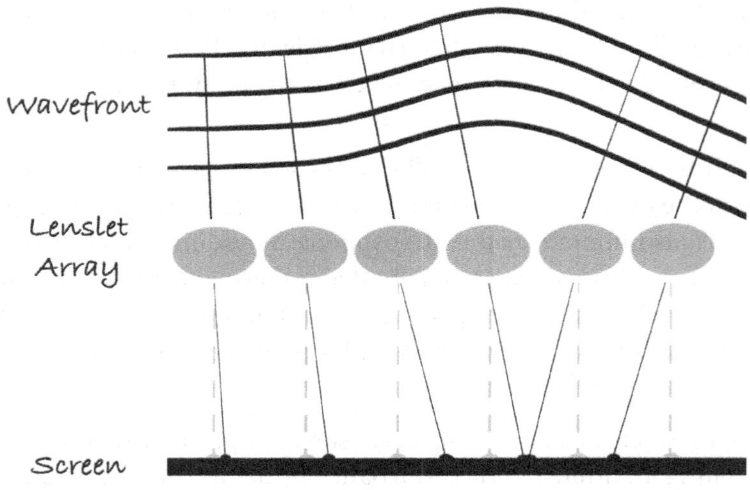

A high-speed camera captures the two-dimensional image of these dots on the screen and how they move over time, and a gigaflop of computing power calculates the difference to what would be seen by a perfect telescope with no distorting atmosphere; the next illustrations shows the perfect atmosphere on the left and an example of a distorted one on the right.

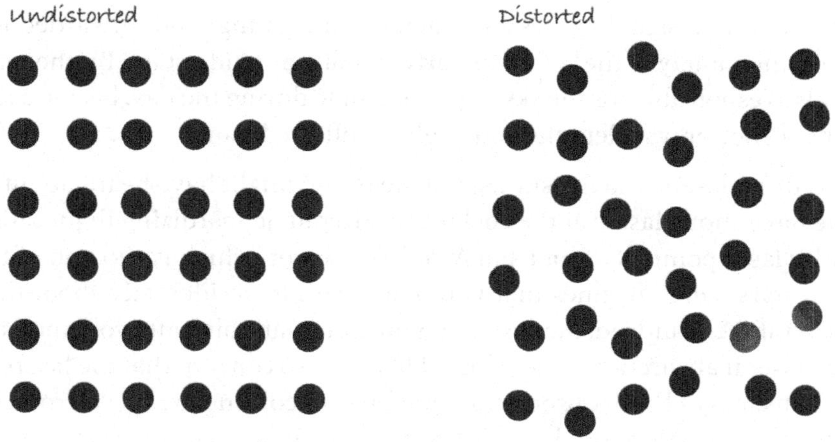

This signal is used as input to a correction mechanism on the secondary mirror, which is only 1.6 mm thick. Richard: "It's like a very large contact lens". It has 672 permanent magnets on its back. The mirror sits on top of a more rigid reference body that carries a corresponding 672 voice-coil magnets. These perform the necessary tiny corrections at a frequency of up to 1,000 Hz. The light that is reflected by this "vibrating" mirror achieves up to 90% of the sharpness that is theoretically possible based on the baseline of the telescope, depending on the wavelength: the longer, the better.

What is new at the LBT is that this correction happens directly at the 90-cm secondary mirror, and not with smaller, centimeter-sized correction elements, after a number of lenses, in a special device or instrument. This greater size leads to increased sharpening. In addition, because the secondary mirror only has the primary mirror between itself and the sky, it does not "see" thermal distortions from the warm telescope. This explains how the LBT can reach 90% of theoretical optimum sharpness; the traditional mechanism of correcting in the instrument achieves only around 50%. As Richard says, "One of the many exciting technological breakthroughs of this telescope is that it has an adaptive optics system that is uniquely high-performing among all telescopes at the moment and, everyone who competes agrees that right now this is the best system that's working on any telescope." Once fully operational the adaptive optics will give the LBT an overall resolution ten times better than Hubble.

The adaptive optics system relies on a bright point source in the field of view. But what do you do if no such source is available in the region of space you want to observe? You create your own! Several green Nd:YAG lasers with 18 W each are beamed into the sky and focused at 12 km, where they create an artificial active optics star (this is sometimes also called a "laser guide star", even though strictly speaking this is wrong, as they cannot be used for guidance because the telescope creates them). At 12 km altitude the laser light scatters on the atmospheric molecules in a process called Rayleigh scattering. Because it is a well-defined physical process, the wavelength of the reflected light is known, so it can be split from the received beam (using quite a bit of additional signal processing, such as time-gating) and forwarded to wavefront sensors, which then analyze the light similarly to natural guide stars. By the way, Rayleigh scattering is also responsible for the sky appearing blue during the day: because the process is more efficient at shorter wavelengths, blue light is reflected more.

As an aside: is firing lasers into the sky a good idea? You might have heard about incidents in which people have shone lasers at the cockpits of aircraft, jeopardizing flight safety and crew health. Regular laser pointers output 3 mW of laser power, which is 1/10,000 of that used for these artificial stars. One imagines that you don't want to accidentally shoot the guide star lasers at an aircraft! And indeed, the system has an automatic aircraft avoidance system, shutting down the laser if an aircraft comes close. There is also concern that the lasers might blind the sensitive detectors of Earth observation satellites. According to reports from 2008, the US Air Force imposes restrictions on the use of lasers. Astronomers have to get approval for their use of lasers from the Laser Clearing House at Vandenberg Air Force Base three days in advance of their intended use. Because the orbits of many of the military satellites are secret, astronomers can't really plan: they can just submit their intent and then get back a (seemingly random) set of yes/no replies. According to the report I read, the restrictions have gotten so frequent for some telescopes that significant fractions of observation time are lost. And their ability to react to short-notice events such as supernovae is limited because of the long approval times. An interesting twist to this story is that most experts suggest that the lasers wouldn't damage the sensors at all: they'd have to look directly at the laser, would be exposed for only a few milliseconds, and most sensors used at night operate at different wavelengths anyway. Which is why some suspect that the real reason for the laser prohibition times is that the lasers would briefly reflect from a passing satellite, thus helping to reconstruct the orbits of secret spy satellites. An interesting conflict between research and the military!

Back to the LBT. Despite the availability of adaptive optics, it's preferable to avoid turbulence in the first place. One source of turbulence is the mirror itself, if it is not at the same temperature as the ambient air. In particular, if it is warmer it will heat up the air directly above it, which will then start to rise, creating turbulence (aka thermals; one imagines nights with bad weather where bored astronomers fly their model sailplanes above the mirror). This is why the temperature of the front surface of the mirror is actively controlled to within a fraction of a Kelvin. Thick hoses force air into a plenum, which then has nozzles into each of the hundreds of honeycombs in the back of the mirror. The necessary cold or warm air is created by giant air conditioners in the basement of the telescope building. Richard adds, "We are working towards thermal control of the structure where we're actually going to try to get the steel into a close temperature range as well, which will help us both with focus and reducing image distortions. And we really work very hard to keep extraneous sources of heat away. For example, every bit of control electronics sits in a thermally insulated box, and chilled water is used to carry the heat out of this enclosure and down to where it can be dumped safely and not distort the images. Thermal management is very important for all modern telescopes."

POINTING Let's compare the LBT's control mechanism to SOFIA's. Just like SOFIA, some observations take a long time to achieve the required signal-to-noise ratio, up to five hours of total exposure time for the most distant object observed at the time. And just as in SOFIA, the telescope has to be locked onto its target and follow it across the night sky. This is complicated by the unavoidable bending of the telescope structure. However, this has been very well characterized, and there are lots of tables with numbers that allow for reasonably precise open-loop control. There is also an additional camera outside the observation field of view that performs optical tracking of a guide star for closed-loop control. However, in contrast to SOFIA, there is no need for accelerometers or gyros; there are no fast-paced movements to correct, such as those induced by a moving aircraft. I can also see a small, almost amateurish-looking telescope next to the big one. Richard: "It is indeed a small commercial telescope. In the early days, the control system was not nearly as sophisticated and perfect as it is now and so we actually occasionally would get genuinely lost and had to fish in a bright star and reinitialize the coordinates."

The moving part of the telescope weighs 700 tons. It is built from very rigid, heavily braced steel to minimize bending or wind-induced vibrations. If technicians have to repair something in the structure, "... we require people to get into harness and attach themselves to a tie point so that a fall will not be an injury."

In terms of elevation, the mirrors are mounted on what are known as C-rings. Like SOFIA, the telescope is balanced and slides on oil films, so that the movement in elevation can be performed with only 3-hp electrical motors that connect to gear teeth. "You can move it with your hands", says Richard. The maximum elevation rate is 1.5 degrees per second; it takes one minute from horizon to zenith. The telescope also slides on oil bearings in azimuth. In fact, the whole platform on which we stand rotates. The enclosure, all 1,700 tons of it, which rolls on rail tracks, sits on a separate foundation to keep its vibration from interfering with the telescope. Richard: "During construction, they actually needed to work at night, because the thermal changes of the rail track were too big during the day; they shot it with a laser surveying instrument to level it to a fraction of a millimeter." To make sure that the enclosure openings are aligned with where the telescope looks, it moves in sync with the telescope "... and it has four sets of really large shock absorbers just in case it doesn't."

I've mentioned before that changing the light path is automated: the operator can swing out the prime focus hub that hosts the camera, swing in the secondary and tertiary mirrors and route the light to one of the five focal stations with different instruments mounted between the two mirrors. This configuration change requires the telescope to be rebalanced; instead of moving stacks of lead weights and steel plates manually, there are tanks of liquid at the top, bottom front and bottom rear of the telescope. Liquid can be pumped from top to bottom and front to back to keep the whole telescope balanced. All this happens while the arms holding the optics are in motion. "In 20 minutes or less we're ready to go with a whole new optical path and new instrument."

3 | *LBT Instruments*

OVERVIEW On our trip from Los Angeles airport to SOFIA's Palmdale base we visited the 100-inch Hooker telescope on Mount Wilson. It was the largest optical telescope for its time—first light was 1917. Edwin Hubble used it to make the observations with which he showed that the universe is expanding. It is a Newtonian reflector with observing stations at the Cassegrain focus behind the main mirror/lens and at right angles to the primary focus. It actually has eyepieces at these two locations, with a neat little wooden chair on a raised platform for the Newtonian focus (I wonder how many people fell off the chair and down from the platform while they were observing in the dark of night). This is a historic telescope, and of course I was aware that modern telescopes don't use eyepieces: imaging and spectrographic instruments are used instead. A few days later I would see one of those: the 600-kg FIFI-LS mounted on the SOFIA telescope, during my flights. However, when I visited the LBT, the size and complexity of the instruments was something else.

LBT hosts a range of instruments, most of which are permanently installed; they receive their input by redirection of the light path, as I have explained. The two Large Binocular Cameras are installed in the prime focus of each telescope in place of the secondary mirror. LUCI 1 and 2 are infrared instruments that support imaging and spectroscopy, installed on the gallery between the two mirrors. PEPSI is another spectrograph that can also be used to determine the polarization of the received light; this is in turn an indicator of the presence and orientation of magnetic fields around the target. PEPSI is installed away from the telescope in a pressure- and temperature-stabilized room, receiving its light via 45-meter-long glass fibers. In 2019 two new instruments called SHARK will be installed. These will be especially suitable for detecting exoplanets—one of the main missions of the LBT—because of their use of coronagraphy, which blocks out the light from the central star and greatly improves the contrast in the region around it. MODS is a multi-object optical spectrograph, and LINC NIRVANA is a near-infrared interferometer. We'll look at these two in more detail.

MODS MODS is a spectrograph; it analyzes the wavelength content of the received light, as we saw in SOFIA. For this task, the resolution of the telescope is not that important; what counts is the light-gathering power, because this reduces integration times. This is why, when used for MODS, the two mirrors are operated in non-interferometric mode. LBT has two separate MODS instruments, mounted on the Gregorian focus behind each primary mirror (position 5 in the illustration a few pages back). MODS is several meters long and wide, and is roughly the size and shape of the Mercury spacecraft, but at 3,000 kg is twice as heavy, to provide the rigidity necessary for the precise optics inside. The large size is to ensure a light path that is as straight as possible, thus requiring fewer optical elements. This, in turn, is important for violet and ultraviolet radiation, which is absorbed by optical elements. MODS' mount is derotated, which means that the orientation of the image (or spectrum) remains stable as the telescope rotates to follow the night sky. As you would expect, the detectors inside the instrument are cooled to keep thermal noise low, in the case of MODS to "only" around 170 Kelvin, which can be achieved with liquid nitrogen. The coolant is brought up to the telescope once a week with a pickup truck. MODS is another example of we-can't-built-it-precisely-enough-so-let's-not-even-try: the alignment of the instrument with the telescope will change slightly as the telescope moves (and changes the orientation of the instrument). This is why MODS has a little infrared laser beam and a separate detector that actively controls the alignment of the optics.

MODS is an abbreviation for Multi-Object Double Spectrograph. The "spectrograph" is obvious, so let's understand the other words. "Double" refers to the fact that inside each instrument there are two separate detectors, one for blue light (300 to 570 nm) and one for red light (570 to 1000 nm). In each channel the optical elements such as mirrors, dispersers, cameras and CCD detectors are optimized for the respective blue or red wavelength region.

"Multiple Object" takes a little more explanation. First, since MODS is an instrument on an optical telescope, it observes both in the visible range and the adjacent near infrared and ultraviolet bands. For these bands you don't need fancy pressurized crystals as they do for the (further) infrared light in FIFI-LS; the spectrum can be recorded by a CCD chip which, as we know from digital cameras, can be packed very densely. MODS has an 8192 × 3088 chip, about 25 megapixels. The reason why it is wide instead of square (even though the beam received from the telescope is circular) is because the spectra are spread laterally across the chip. Let's call this the x axis, and the vertical, shorter side we'll call the y axis.

We know from SOFIA that one way of producing a spectrum is to let the light fall onto an optical grating, which then reflects each of the component wavelengths at a characteristic angle. This wavelength-separated light falls onto the respective x-location of the CCD, and the brightness recorded there then corresponds to the energy content at the respective wavelength; the brighter, the more energy. Importantly, if one let the whole image fall onto a grating every point would produce a spectrum, and these would overlap, resulting in a useless mess. Therefore, to create useful spectra, the area onto which the spectrum is projected must be kept dark (in the sense that no image light strikes it, only the light dispersed by the grating). This is why MODS can insert masks into the light path: they make sure that only a thin slit of light reaches the grating and the rest of the CCD is kept dark, allowing the spectrum created by the grating to be captured by the CCD.

This particular kind of grating-created spectrum expects the slits to be in the middle of the x axis of the mask; the spectrum is then spread out over approximately 6,000 pixels along the x axis. The approach can be used to capture the spectrum of a single target object or, by arranging several slits vertically, an object's extent along the y axis can be resolved into several spectra arranged beneath each other on the CCD. MODS has a set of predefined masks with five slits arranged vertically at x center. Different slit widths are available to take care of different intensities of the target object.

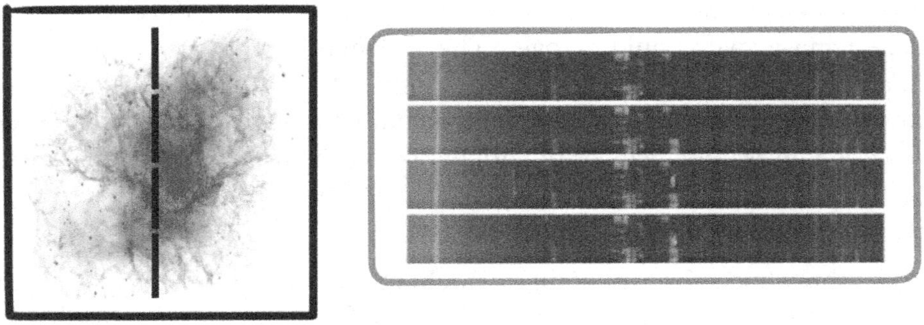

However, as the "MO" in MODS suggests, the instrument can also capture the spectra of multiple objects in the field of view, ideally even when they are not aligned in a straight line down the middle of the image. For this use case, scientists can use custom masks with mini slits all over the square region of the image. Here, the spectrum is not created with a grating, but with a prism; this produces a spectrum that has lower resolution, resulting in a spread in x direction of only around 600 pixels. This allows dozens of mini slits to be placed on the surface of the mask as long as they are either at different locations on the y axis, or far enough apart in x direction for the 600-pixel-wide spectra not to overlap. The masks also contain a couple of square holes that help with alignment of the observation.

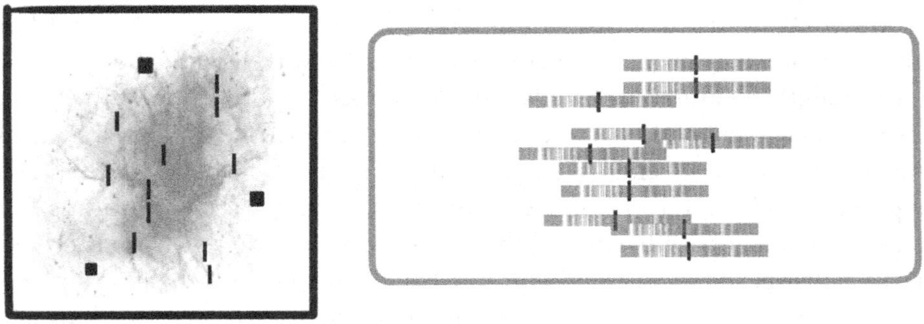

A software tool helps with the design of the masks; users can put an image of the target region as the background and then "draw" the slits onto interesting regions of the image. The tool also helps to ensure that the resulting spectra do not overlap. Masks are made of thin black sheet metal cut by a precision laser cutter, done in a clean lab at the University Research Instrument Center on the University of Arizona campus in Tucson. Once at the telescope the masks are inserted into a jukebox mounted on the side of the instrument; this allows the observers to exchange masks automatically during an observation. Switching from the grating to the prism happens by rotating a roughly one-meter diameter selector turret, which also contains a flat mirror that can be used for a (non-spectrum) imaging mode that is particularly useful during alignment of the telescope.

The diagram below shows the overall optical path of MODS, illustrated for the blue band; the red band has a similar structure. The two bands are separated at the entry of the instrument using a dichroic mirror, which reflects light of a particular wavelength (range) and lets other wavelengths pass through. In each of the two channels the light then strikes a collimating mirror that turns the diverging light rays from the telescope's focal plane into a parallel beam about 420 mm wide. This then hits the disperser (the prism or the grating) that spreads the light.

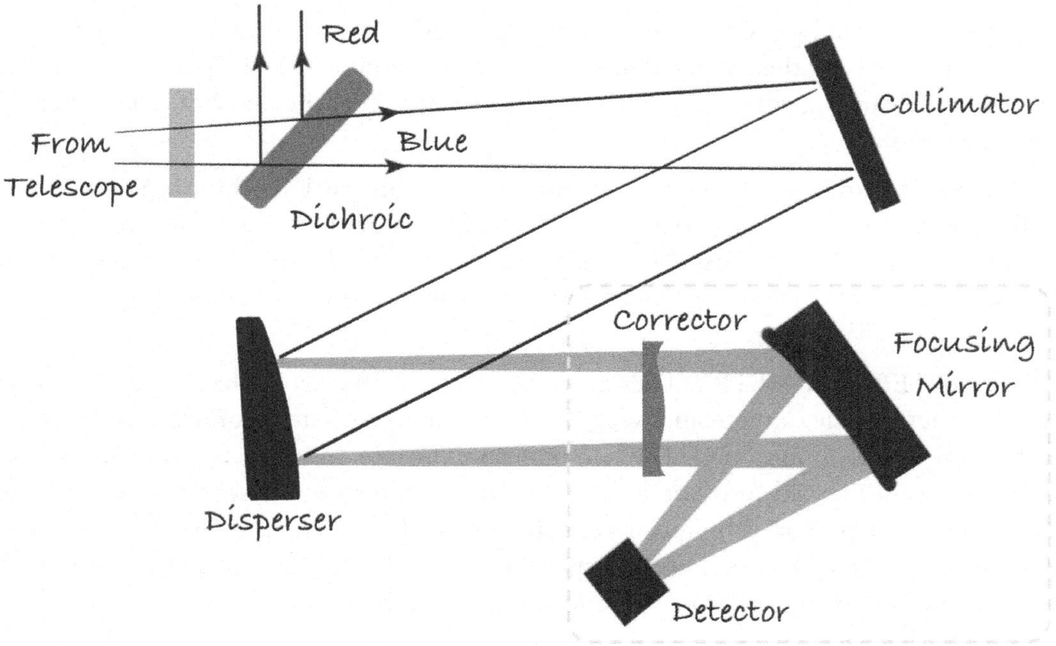

The remainder of the optics (enclosed in the dotted box above) is a Maksutov–Schmidt camera, a design similar to the Schmidt–Cassegrain telescope that is popular with amateur astronomers. It has a concave, spherical main mirror that focuses the image onto the detector. Such mirrors have various aberrations, the most important being the spherical aberration: a spherical mirror does not focus all incoming light onto the same point. The further out from the optical axis, the closer the focus is. This means that if you insert a sensor at the focus, only a part of the overall field of view will be "sharp".

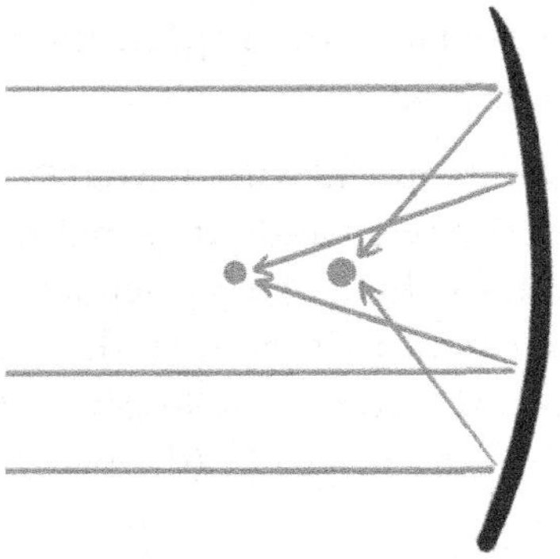

The corrector lens in the MODS camera, and the Schmidt plate in Schmidt–Cassegrain telescopes, compensate for this aberration. The final component is the CCD detector that sits at the focus of the camera, which is located off axis to ensure that the detector does not create a shadow on the mirror.

A mirror with a parabolic shape does not suffer from spherical aberration. You might ask, therefore, why they didn't use a parabolic mirror. This is because it exhibits another distortion, called "coma". We'll discuss this later in this chapter, but for now let's just say that coma is not necessarily easier to correct than the spherical aberration of spherical mirrors.

OPTICAL INTERFEROMETRY Let's move on to LINC NIRVANA. The abbreviation stands for Lbt INterferometric Camera and Near-IR/Visible Adaptive iNterferometer for Astronomy; we'll abbreviate this to just "LN". The two-part name results from the fact that two separate instruments were merged into one because of strong synergies between the two proposed technologies (and probably financial considerations): LINC, an interferometer, and NIRVANA, an instrument that exploits an especially sophisticated form of adaptive optics. We will discuss both in turn. Let's start by understanding interferometry.

As we now know, the maximum resolution achievable by a telescope is governed by the baseline, the largest distance of any two points on the telescope. This is not an engineering limit, it's something that physics tells us, called the "diffraction limit". Adaptive optics helps us to get the practical resolution of the telescope very close to this theoretical limit, but there's no replacement for a larger baseline. The idea behind interferometry is to use "points" from multiple mirrors to create a baseline that is larger than the individual mirrors involved. In the case of the LBT, this longest baseline is 22.8 meters. Interferometry with two mirrors is not quite a replacement for a single bigger mirror: if you picture the two, most of the light is coming from points that are between 8.4 meters (one mirror) and 14.4 meters apart (the inner edge of one mirror to the outer edge of the other mirror, or center to center). Only some of the light has the highest resolution produced by the 22.8 meter baseline—the telescope's light collecting power at the high resolution is limited. We'll get back to this below.

Interferometers are instruments that derive information from the interference of (separate beams of) light, usually by analyzing their phase difference. Interference can be exploited in many different ways. One is to detect the precise difference in the distance travelled by two beams of light. We've seen this in the Michelson interferometer used in gravitational wave detectors. Optical stellar interferometers work differently, and there are many different variations, all with different tradeoffs regarding the engineering of the telescope and instruments, as well as the data they produce. A major distinction is how exactly they combine the light. The interferometer in the LN is a Fizeau interferometer, which means that the beams are combined in the image plane, basically on the detector. To understand this in more detail, we have to go back to basic physics, and in particular, the Young double slit experiment.

In this experiment, light falls onto a mask with two parallel slits, effectively forming two separate beams, or waves, emanating from the two slits. These two waves then hit a screen where interference produces the characteristic fringe pattern we all know from school. As evident from the illustration below, the location on the screen at angle ϕ is further away from the top slit than from the bottom slit, which is why the two beams (which start out with the same phase, because they result from the same wavefront) arrive with a ϕ-dependent phase difference. Depending on that phase difference they interfere constructively (bright), destructively (dark) or any value in between.

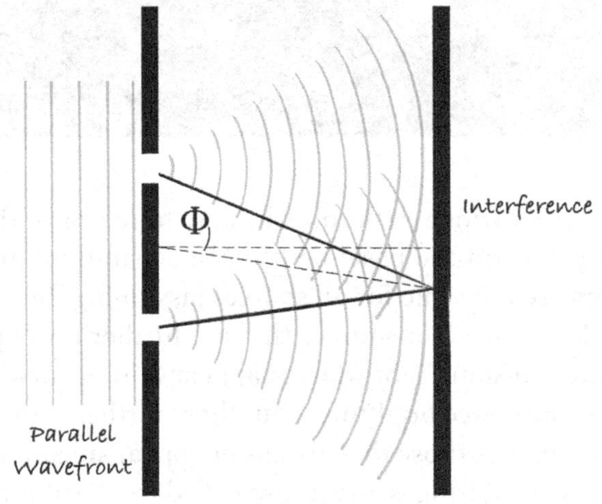

This experiment is confirmation of the wave nature of light. By performing the experiment with single photons you can also demonstrate the particle nature of light (because every photon can be shown to go through one of the two slits), as well as some of the weirdness of quantum physics, because once you detect that a specific photon goes through one of the slits, the interference goes away—you have "decided" that you want to treat it as a particle, and as a single particle cannot go through both slits, the quantum wave function collapses and no interference takes place.

In the context of LBT's interferometer, however, we are interested in the optical patterns themselves. The particular layout of the fringes—their distance, width, intensity and contrast—depends on various parameters of the setup, such as the baseline (the separation between the two slits), the distance of the screen from the mask, the distance of the light source from the mask, and of course the wavelength.

In the book "An Introduction to Optical Stellar Interferometry" by Lipson and Nisenson I found a very nice experiment that you can do yourself. Prepare a mask with two slits separated by about 1 mm. Now wait until it's dark and go somewhere where you can see street lamps or other urban point light sources from different distances. Hold the mask closely in front of one of your eyes and close the other one. You should see interference patterns for the lights as shown below.

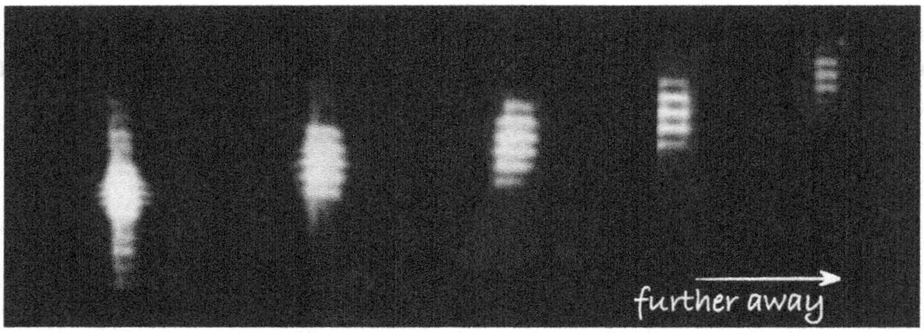

Each light source produces a fringe pattern. All the patterns have the same fringe spacing, because the spacing depends on the baseline (the slit separation) and the wavelength, and both of these parameters are the same for all sources (assuming they are all the same kind of street lamp). You will also notice that sources that are further away produce fainter fringes. That's not surprising; more distant light sources appear fainter, so a camera has to integrate for longer to record the same image brightness. But those farther sources also produce clearer, more contrasted stripes, whereas those that are closer appear smeared, "blurry". This contrast is an important characteristic of fringes and is also called the "visibility".

What is the reason for the observed difference in visibility? Does it actually relate to distance? Only indirectly: it really depends on the angular size of the source relative to the fringe spacing (and thus, the slit separation). The angular size is the apparent size of an object as seen by a telescope:

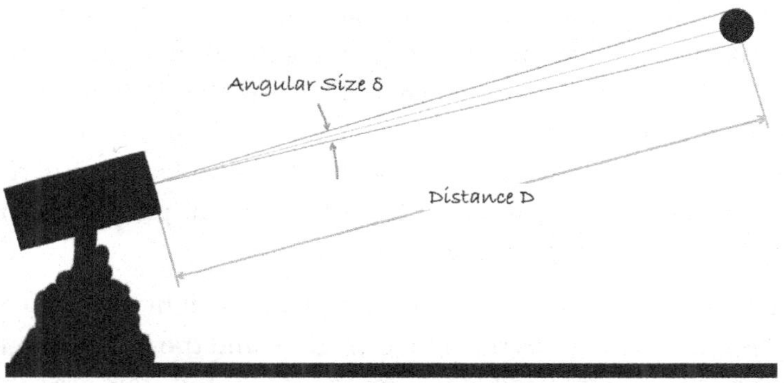

If the angular size is small you get clearly separated fringes; the pattern becomes smeared for angular sizes that approach the slit separation, and for sources that are larger (in angle) than the slit spacing, you get no fringes at all. This is the key to how optical stellar Fizeau interferometry works: the two apertures represent the two slits. An "interferometric image" of the sky shows fringes for each star, just like the lamps in the picture above. By quantitatively analyzing the visibility of the fringes one can determine angular size. Determining the angular size of celestial bodies or features on those bodies is precisely what "resolution" means.

Let's quantify this a bit. Assume a wavelength λ= 660 nm, red light. As in the double slit experiment, the angle ϕ_m of fringe m (as counted from the center) is given by

$$m \cdot \lambda = s \cdot \sin \Phi_m$$

where s is the slit spacing (the baseline). As long as we assume small angles (where $\sin \phi \approx \phi$), the spacing of the fringes can be calculated by λ / s. In our experiment, s is 1 mm, so the angular spacing of fringes becomes $6.6 \cdot 10^{-4}$. The angular size of an object of geometric size d at distance D can be computed as $\tan \delta = d / D$, as should be apparent from the illustration above. The approximation we used for the sine function also works for tangents, so with $\tan \delta \approx \delta$ we get $\delta = d / D$. For a street lamp with s = 30 cm located at distance D, its angular size is 0.3 / D. Fringes appear if 0.3 / D is much smaller than $6.6 \cdot 10^{-4}$. Solving for D, we get 450 meters as the minimum distance for seeing nicely contrasted fringe visibilities.

What about the astronomical use case? From our vantage point on the Earth, the largest angular size of any star is around $2.5 \cdot 10^{-7}$. The math then turns out to require a minimum baseline of around two meters for visible light: the LBT's 14 to 22.8 meters well exceeds this. The relationship discussed here also illustrates why bigger baselines are better; they allow resolution of smaller angular sizes.

Let's briefly consider how astronomers determine the actual size d of an object once they have measured its angular size δ. Of course they rely on the same $\tan\delta = d / D$, but how do they figure out the distance of the object D? Multiple approaches are used. One relies on parallax, the apparent change in the position of the star when viewed from two different locations, such as from the Earth, six months apart on opposite sides of the Sun; it's the same effect that makes your thumb jump when you view it one eye at a time with an outstretched arm. A second mechanism relies on Cepheid variables, stars that vary periodically in their luminosity. Discovered by Henrietta Swan Leavitt in 1908, a Cepheid's period is directly related to its actual luminosity. Knowing the actual luminosity from the period, plus the apparent luminosity from observation, we can estimate distance.

The two remaining mechanisms rely on the spectrum of the light we receive from that particular star. The first approach analyses the light to understand the relative abundance of particular elements (we've seen spectra used for this purpose before). Stars are categorized into several spectral types: O, B, A, F, G, K and M. Those in turn are characteristic of particular physical processes, with corresponding surface temperatures (at 50,000 K the O category is hottest, while M is coolest at 3,000 K). Stefan's Law then associates the surface temperature with its absolute luminosity, which we can then use, as before, to estimate the distance by comparing with the star's apparent luminosity.

The final way to estimate D relies on Hubble's Law. This expresses a proportionality between the speed at which a body recedes from us and its distance: $v = H_0 \cdot D$, where H_0 is the Hubble constant, which is around 70 km/s per megaparsec of distance; a parsec represents a distance of around 3.3 light years. The reason for this relationship between distance and recession speed is that the universe expands and so every point moves away from every other. How do we know the recession speed v? We can compute that from the Doppler shift: light sources that speed away from an observer are shifted to longer wavelengths, towards the red. And how do we know the red shift? Spectra to the rescue! We can look at known spectral patterns (the relative distance of the lines in a spectrum) and measure by how far they have moved towards red in absolute wavelength.

The latter two approaches for determining the distance D (which then allows us to calculate the actual size from the angular size) show how the interferometer's capability to resolve small angular sizes can be combined with a spectrometric measurement to determine absolute size. It is also an example of where it might be useful to observe the same source both with an instrument like LN and with a spectrometer like MODS, and why it is therefore beneficial that the LBT can switch instruments merely by reconfiguring the light path by inserting and removing its tertiary mirror.

So far I have provided an intuition of how a Fizeau interferometer achieves high resolution by interferometrically combining the beams from multiple apertures and analyzing the fringe patterns. However, it is important to understand that this is a tool for measuring angular size; what you see on your detector is not a regular image but just with higher resolution; it has fringes all over. We call the mapping from the point source to what the detector sees the "point-spread function". So let's think a bit about the relationship between interference fringes and images. To do this, we return to the regular, circular aperture that we find in any camera or telescope. When a lens or a mirror creates an image of a well-defined point-like object, we don't get a perfect rendering of the point in the image. Rather, the point-spread function produces an Airy disk:

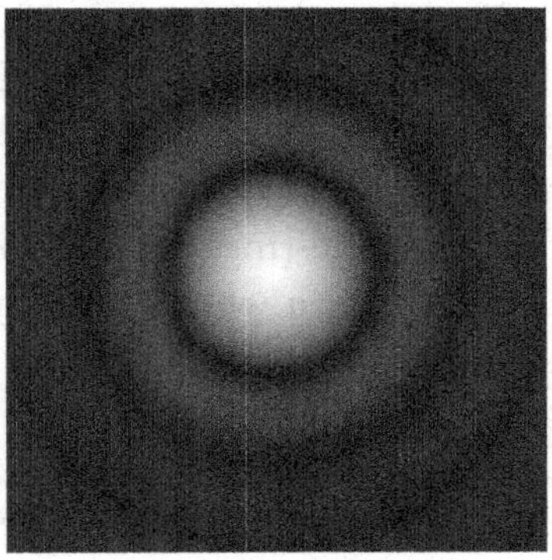

An Airy disk has a central maximum and progressively fainter rings around it. The central peak is what us laypeople expect the image of our point-like object to be. But where do the rings come from? By now you've probably noticed that they look a lot like the fringes from the double-slit experiment. Indeed, a full circular aperture can be seen as an infinite number of double-slit baselines: every point on the aperture interferes with every other one. Importantly, however, each baseline has a different length and orientation. A few are shown below.

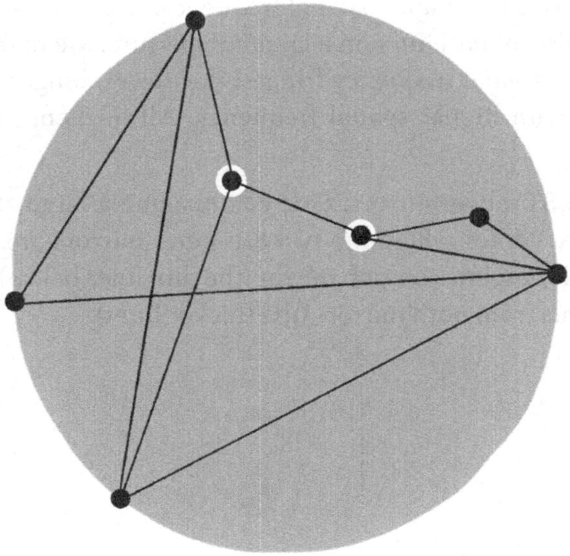

So we learn that one way of looking at an image created by a full, circular aperture is as the sum of the fringe patterns from all possible baseline lengths and orientations (between all points) on the aperture. But this still does not explain the central maximum and the fainter rings in the Airy disk of the full image. To understand those, let's look at the statistical distribution of the baselines in terms of length and orientation. There are comparatively few of the longest edge-to-edge baselines, because only the points on the edge are available to form these long baselines. Assuming that the world is discretized (which it isn't), we could say there are 180 of them, one for every degree of "compass direction". How often do shorter ones occur? For example, the one between the two white-ringed points? We can still get 180 of these baselines through rotation, but we can also get a whole bunch more by moving the two points around while keeping their relative position. It is also easy to see that the shorter the baseline is, the more copies we can obtain from moving it around: there's more "space" available to place them. Thinking in terms of the extremes helps to see this: for the maximum length there's only one per degree of orientation, but for a distance of zero we can get infinitely many.

Let's go back to our street lamp experiment. We said that the "fringiness" depends on the relationship of the angular size of the object and the separation of the slits, which corresponds to the length of the baseline in our case here: the smaller the angular size (compared to the baseline), the clearer the fringes. We have also seen that the "spread" of the fringes, their spatial frequency, depends on the slit spacing via

$$\Phi_m = \arcsin\left(\frac{m \cdot \lambda}{s}\right)$$

So larger slit spacing gives us a higher spatial frequency. Let's combine this observation with the statistical distribution of baselines on a circular aperture: the many short baselines result in relatively bright, low spatial frequency fringes; the (fewer) longer ones give us the clearer (but dimmer) fringes with higher spatial frequency. All of them are overlaid optically, of course.

Let's get back to the LBT's two mirrors. One can imagine a large 22.8-meter mirror that is masked everywhere except for where the two 8.4-meter mirrors are. A detector will receive data from the "internal" baselines of each mirror (the thin lines below), but also from the baselines that connect points from both mirrors (the thicker lines).

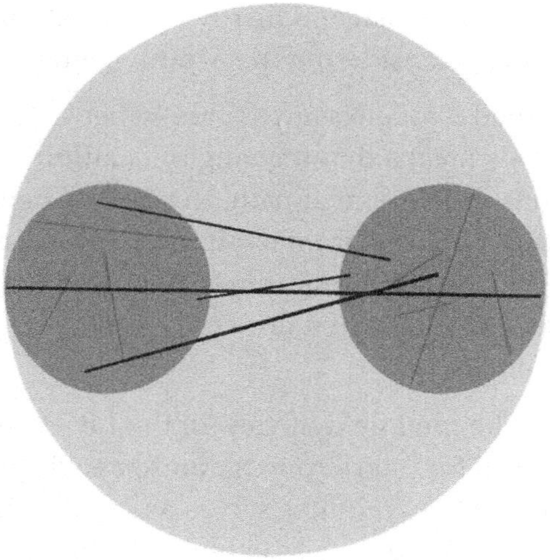

The longer inter-mirror baselines will lead to fringes with high spatial frequency, but there are only (comparatively) few of them, so these fringes will be relatively dim. This explains the statement from the introduction where we said that only some of the light has this highest resolution produced by the 22.8-meter baseline.

With all this in mind, it is now obvious that the LBT setup does not get us a perfect (Airy-disk) 22.8-meter image of our object, because there is no data available for many of the baselines of that imagined 22.8-meter mirror. What can we do? One thing is that we can rotate the whole system. In fact, such a rotation occurs naturally, because the sky rotates around the Earth. Because interesting astronomical targets are often faint, an observation often integrates for long enough to make this effect practically relevant. Astronomers talk about "filling the u-v plane": we'll go into more detail later, but for now, let's just say that the each point in the u-v plane represents a particular baseline length and orientation. Once "filled", we have captured all available data. Generally, interferometric data is analyzed by measuring the contrast ("fringiness") per baseline length and orientation: we basically analyze the energy content per spatial frequency. Does this sound familiar? Remember multi-beam sonars? Exactly—the interferometric data is the (spatial) Fourier transform of the image. So we can use the inverse Fourier transform to get from interferometric data to the image. To explain this more thoroughly we'd actually need the math, so we'll stop here and move on to the engineering of the LN interferometer. We'll come back to (inverse) Fourier transforms in the context of radio interferometry later.

LINC-NIRVANA LN is a really big instrument: its largest dimension is over six meters. It weighs around 10 tons. It was under development for 15 years by a team from the universities of Heidelberg, Cologne, Bonn, Padua, Firenze, Bologna, Rome and Genoa. In 2015 it was transported to the LBT in a huge wooden box, first by ship along the Neckar river and over the

Atlantic, then through the Panama canal and finally by flatbed truck up Mount Graham. The latter was a non-trivial endeavor, considering the winding and progressively narrower road.

The instrument is now installed on the Nasmyth gallery between the two mirrors (position 4 in the earlier illustration). This means that it changes its attitude as the telescope moves. To avoid unnecessary distortions in the optical path, the optical bench, the 6-meter-long backbone of the instrument, is constructed extremely rigidly, yet with relatively light weight, from three carbon-fiber panels separated by aluminum honeycomb spacers. It is optimized to have a thermal coefficient of zero, meaning that it does not shrink or expand with the temperature changes experienced in the enclosure.

Let's understand the optics. The total field of view of the LBT is six arcminutes; but only the central two arcminutes are used for interferometry and passed along the light path described below. The "ring" from two to six arcminutes serves as input to the ground-layer wavefront sensor which, after analyzing the wavefront, as explained earlier, controls the 672 actuators on the LBT's adaptive secondary mirror.

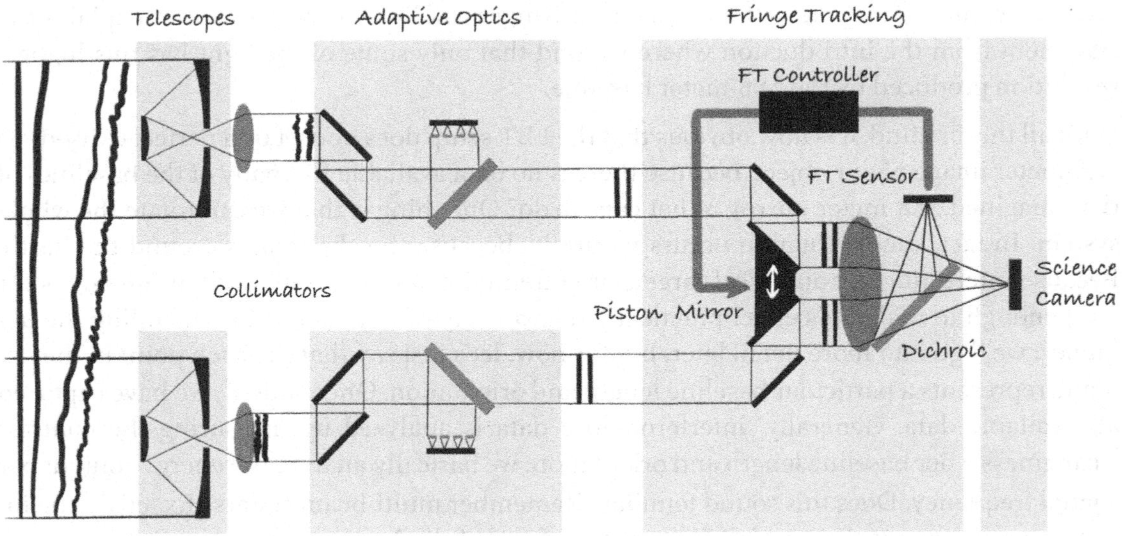

adapted from The LINC-NIRVANA Fringe and Flexure Tracking System by T. Betram et al.

On with the central two arcminutes of the beams from each main mirror. Similar to what we saw in MODS, the first step of "light processing" is collimation: forming a parallel, non-diverging beam. In LN the collimator is made from several lenses (and is thus much more complex than shown), and mounted on a movable stage to compensate for slowly changing temperate variations in the telescope enclosure.

LN is an interferometer for near-infrared light. You might ask: why not for optical? I asked that question of Tom Herbst, the principal investigator of LN. He replied that it simplifies the engineering. At shorter wavelengths (and correspondingly higher frequencies), adaptive optics becomes more challenging: the correction on the adaptive mirrors must be finer-grained, which requires more actuators on the deformable mirrors. They must also be considerably faster, which in turn requires more computing power. Adaptive optics is just much easier for wavelengths above one micron. But concentrating on near-infrared light is not necessarily a limitation from a science perspective, because lots of interesting phenomena can be observed there. For example, infrared radiation can pass through dust without being scattered; the center of the Milky Way and the heart of star-forming regions can only be observed at longer wavelengths. Targets that are far away also radiate in the infrared band because of the redshift caused by Hubble expansion.

Therefore, after collimation the light is separated by a dichroic mirror into a visible band and a near-infrared band. The near-infrared light is reflected by 90 degrees into a cryostat, where it is interferometrically combined and then analyzed. That is the drawback of working with infrared light: the detectors have to be cooled.

However, before this separation, one has to compensate for the differential length of the optical paths from the two mirrors; the two beams have to be overlaid precisely in phase. As with every control loop, this step consists of a sensor and an actuator; the overall process is called "fringe tracking". The actuator is known as a "piston mirror". It has two reflecting surfaces, one in each of the two optical paths. It sits on a piezoelectric nano-positioning stage, and when it moves it shortens one optical path and lengthens the other. It moves by 75 micrometers (achieving a phase delta of twice that). It weighs 3 kg, which limits the frequency to around 140 Hz. The piston mirror sits outside the cryostat, but the sensor that controls the mirror sits behind the beam combiner, inside the cryostat. The sensor gets some of the combined light for analysis—the rest goes to the science detector. It looks at the fringe pattern at a sampling rate between 50 and 200 Hz. A difference in the path lengths manifests itself in particular asymmetry patterns in the intensity distribution of the fringes (imagine much more math here). The control loop works to keep these asymmetries to a minimum. The goal is a residual difference of $\lambda / 10$; the wavefronts from the two mirrors are now aligned.

The final component of LN is the actual science detector. It is called HAWAII-2; as you can probably guess from the name, it was (co-)developed by the University of Hawaii. However, the name is officially an abbreviation for the HgCdTe (mercury cadmium telluride) Astronomical Wide Area Infrared Imager. Those guys should get a special award for extra-clever naming!

The detector only looks at the central 10 × 10 arcseconds of the available 2 × 2 arcminutes field of view of LN. The reasons for this are interesting. When designing the detector, you start out with the goal of a high enough pixel density to not degrade the precious high resolution that is created by all the magic described so far. You want a densely packed detector. However, you cannot make the single detecting elements arbitrarily small in relation to the optics of the camera. To understand this, we have to look at a physical quantity you've probably never heard of, called "étendue": it characterizes how spread out a beam of light is. Take a look at this illustration:

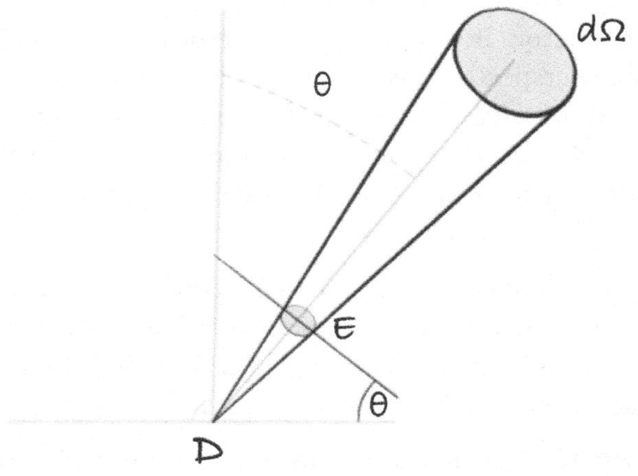

Imagine a detector at location D. It receives light from a beam that has a solid angle (opening, spread) of dΩ. The light beam is tilted by an angle θ off the vertical from the sensor. The light is received via an entrance pupil (telescope opening area) of a size A at point E. Assuming θ is zero (which it approximately is for a telescope), then we can compute the étendue basically as the product of the area and the solid angle:

$$\iint A \, d\Omega$$

As we can see, étendue increases if we either increase the opening angle or collect the light through a larger pupil. Now, the key here is that an optical system cannot reduce a beam's étendue, only increase it or keep it constant. Reducing étendue would correspond to a reduction in entropy, which, as we know from the second law of thermodynamics, isn't that easy (a bit of hand-waving going on here). Basically you cannot built a "light funnel". So the requirement to not reduce étendue guides the dimensions of a telescope's optical path, and, in particular, the minimum size of a detector: étendue conservation doesn't just apply to the whole image (and thus, the whole detector surface), but also to every infinitesimal sub-area, such as the pixels of a detector. This is why in the HAWAII-2 detector one "pixel" is 20 microns. Etendue conservation implies that you cannot "funnel" the beam into a smaller detector pixel area, considering the (fixed) dimensions of the optics in front of the detector, including the telescope. For comparison, the CCD pixels in a typical cellphone camera are around 1–2 microns, because a phone's optical system, and in particular, the lens, is much smaller.

So if your detector size and pixel size cannot be made arbitrarily small, but you still want to capture the high resolution produced by the telescope, you necessarily have to increase the overall size of the detector. But this leads to another problem. Unlike the detecting elements in a CCD, those in an infrared detector are not made from silicon, but from an HgCdTe semiconductor. However, they are still mounted on a silicon chip. Because HgCdTe and silicon have different temperature coefficients, having too many of them glued to a single chip results in mechanical problems when the temperature changes. At the time of construction of LN only 2,048 × 2,048 pixels were feasible. Of course one could use several such chips together. However, this quickly becomes very expensive: the equivalent of Moore's Law for infrared detecting elements hit the buffers about ten years ago. Ultimately, this led to LN only "seeing" the previously mentioned central 10 × 10 arcseconds.

Let's revisit the adaptive optics. We said above that LN analyzes the wavefront and interfaces with the LBT's adaptive secondary mirror to correct for the wobbly atmosphere. But if you look closely at the illustration of LN's light path above, you will notice that not all the wavefront jitter seems to have been compensated by the flexible secondary mirror, otherwise there wouldn't be any jitter at the collimators. So what's going on? LN performs additional corrections using an approach called Multi-Conjugate Adaptive Optics, or MCAO. The atmosphere typically has several distinct layers of turbulence; a ground layer, a medium layer between 4 and 8 km, plus a high layer between 8 and 15 km. Around 70% of total turbulence is in the ground layer, which is why the adaptive secondary mirror "looks" at around 100 meters above the telescope for correction. The idea of MCAO is to compensate separately for the medium and high layers as well. This requires two ingredients. First, the turbulence in the various layers must be separately detected. Second, we'll need several deforming mirrors in order to compensate for each layer separately.

Let's talk about the latter first. I mentioned above that the collimator consists of several lenses and two folding mirrors. Subsequent upgrades of the instrument will replace these with deformable mirrors that are available off the shelf: Xinetics DM349PMNS4, with 349 piezo-actuators, around 15 cm in diameter. The first is already installed and compensates for high-layer turbulence, the second one, when installed, will address mid-layer turbulence.

How does layer-specific detection of turbulence work? To understand this, we have to go back to basic optics. A lens, or its reflective equivalent, a curved mirror, projects a point from the object space into the image space. Where the lens projects the image depends on the optical power of the lens, characterized by its focal length f. An object at distance o will be projected at a distance i according to Gauss' thin lens law; for mirrors with the same f it works the same way, except that i is in front of the mirror instead of behind the lens.

$$\frac{1}{f} = \frac{1}{o} + \frac{1}{i} \qquad\qquad i = \frac{1}{\frac{1}{f} - \frac{1}{o}}$$

If we solve for i and look at large values of o (such as stars, which are effectively at infinity) we can see that 1/o approaches zero, so we get i = 1 / (1/f), or just i = f. This is what we expect: a picture that is "far away" will appear at a distance f behind the lens. This is what focal length means. However, for smaller values of o, the corresponding distance i becomes larger. Let's say we use o = 100 m (the height of the ground layer turbulence) and f = 8 m (which is roughly the LBT's focal length), we get i = 8.7 m. So if we analyze the light 70 cm behind the focus of the lens, we can "see" the ground layer turbulence. The plane 70 cm behind the focus is called the "conjugate plane" of the 100 m layer. You can start to see where the term "multi-conjugate adaptive optics" comes from.

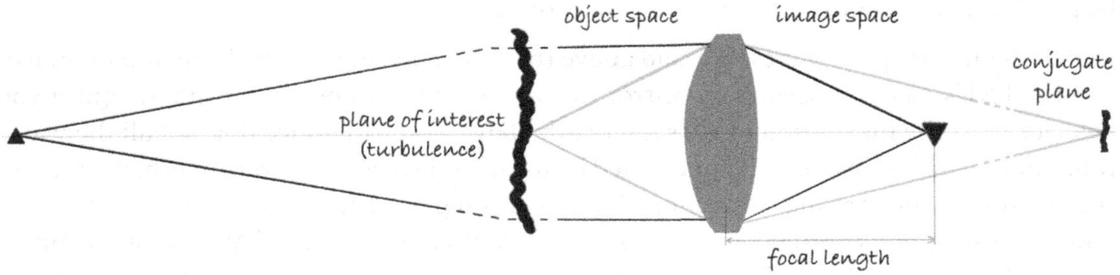

So if we want to "see" the jitter of a point source at various heights in the atmosphere, the wavefront sensor has to sample the light at the corresponding conjugate distance from the mirror. Technically this means that in the optical path light is separated at the respective distance and its wavefront is analyzed.

A perfect system would be able to adjust that distance, because the distance o from the mirror is not the same as altitude, depending on the elevation of the science target. Such a system was planned for LN. However, the team was unable to find mechanical components with the required precision. This means that for lower elevations of the telescope the sampling altitude of the turbulence changes and the performance of the MCAO system degrades slightly.

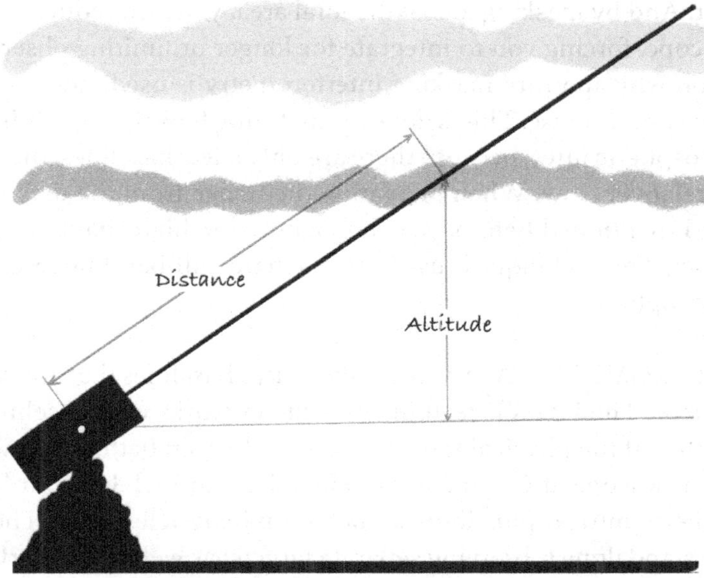

It was surprising to me to learn about the many steps required to actually exploit the big mirrors of LBT, to make sure the practically achievable resolution comes close to the mirror's diffraction limit. The resulting complexity of the instruments is staggering: LN has over 130 motors to control the optical path. Control of one deformable mirror alone requires a 4-processor Xeon PC with fast IO hardware. No wonder LN was under development for 15 years and is still in the process of commissioning. As Tom Herbst told me, interferometry still hasn't been performed on the sky, only in the lab.

LN is considered among the first of a new generation of instruments built for the coming generation of telescopes. We'll take a look at this new generation after we briefly look at other forms of optical interferometry.

4 | *Other Optical Interferometers*

APERTURE MASKING Earlier we understood the LBT's two mirrors as two circular "holes" in an otherwise opaque mask over a 22.8-meter diameter mirror. Aperture masking generalizes this approach by putting masks with holes in front of actual large mirrors, such as the 10-meter segmented mirror of the Keck telescope on Hawaii (or, more pragmatically, in front of their smaller secondaries). Instruments then analyze the interference patterns created by light that is collected by these smaller, virtual apertures. Why would one do this? You don't increase the resolution, because the largest baseline cannot be increased beyond the physical size of the original mirror. And by masking lots of the total area, you also reduce the light collecting power of the telescope, forcing you to integrate for longer or limiting observations to bright sources. The reason why aperture masking interferometry is used, however, is that it helps with reducing atmospheric noise. This is done by analyzing how the signals from the baselines change due to atmospheric jitter; because there are only a few baselines, the required computational power is relatively low. When recreating the image from the data using the inverse Fourier transform I mentioned before, you then ignore the high spatial frequencies created by atmospheric jitter. This technique is useful to reach the full-baseline resolution even if you don't have adaptive optics.

THE VLT INTERFEROMETER Aperture masking uses baselines that are smaller than those of the physical mirror. The LBT's Fizeau interferometer works with baselines that are effectively three times that of the physical mirrors. Can we do even better? Yes, we can. Enter the ESO's Very Large Telescope at Cerro Paranal in Chile. The VLT has four "unit" telescopes, each with an 8.2-meter mirror, plus four 1.8-meter auxiliary telescopes. The unit telescopes can be used either standalone (observing separate targets) as well as in interferometric mode; this is referred to as the "VLT Interferometer", or VLTI. The unit telescopes are arranged such that, when combined, all baselines have different lengths and orientations, the longest being around 130 meters. The auxiliary telescopes can be moved on rails to add "custom baselines" depending on their position.

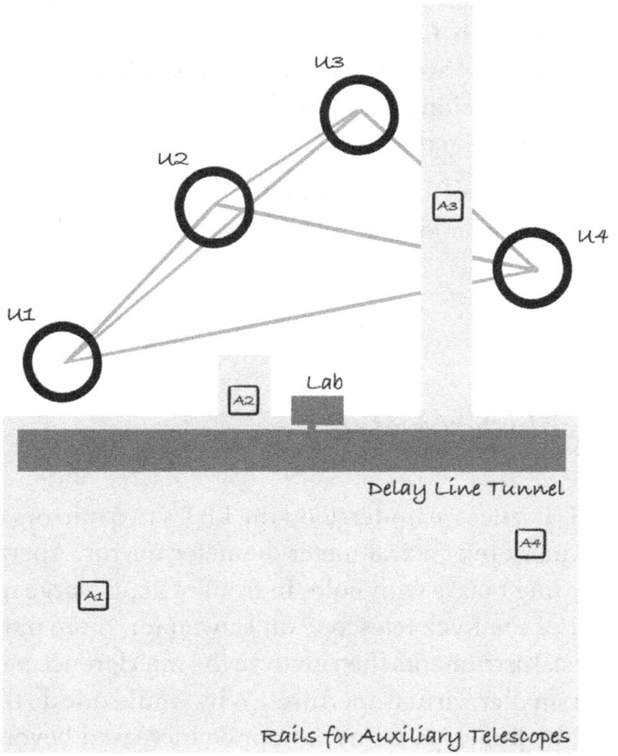

A quick remark on astronomical competition: if you suspect that the LBT's additional 20 cm in mirror size has been added specifically to be able to claim the single-biggest-one-piece-mirror record, you are probably correct. In fact there's even more data: the Gemini telescopes on Hawaii and in Chile have 8.1 m, the VLT has 8.2 m, the Japanese Subaru telescope, also on Hawaii, has 8.3 m, and finally the LBT has 8.4 m—and their construction timeline indeed aligns with a mine-must-be-bigger-than-yours theory.

Back to interferometry. Remember that for interference to work meaningfully, the wavefronts from the involved apertures have to be precisely aligned to effectively zero phase difference. In LN this is achieved with the piston mirror. Its very small range of motion is enough to compensate for the very small differences in the lengths of the two light paths because the two mirrors are mounted on a common frame. At VLT, however, the mirrors are independent telescopes, so the variability in the path lengths will be much greater. More importantly, as a target object moves across the sky (between t_1 and t_2 in the figure below), the geometry changes: the distance to one telescope will become larger, while that to the other will get smaller.

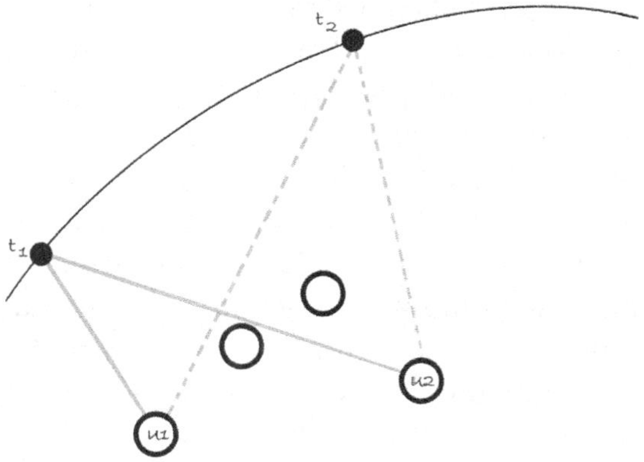

To keep the phase differences resulting from the changing geometry at zero, it would seem good enough to adjust by at most $2\pi\lambda$—after all, waves are periodic. However, the phase difference, computed as $\phi = d / \lambda$, depends on the wavelength: every spectral bandwidth in the beams would arrive with a different phase difference. So to observe interferometrically with any meaningful spectral bandwidth, the only practical way to achieve zero phase difference for all spectral bandwidths is to make the paths exactly the same length, $d = 0$. This is achieved with delay lines in which the light travels the respective additional distance before it reaches the beam combiner. It is easy to see from the above illustration that the maximum required delay distance is the length of the longest baseline, around 130 meters. Which is why at VLT, there is a 60 meter long tunnel in which the light travels back and forth; they can compensate up to 100 meters. As the target moves, the length of the delay lines for each beam is adjusted continuously to keep the two beams at the same exact length, and the phase difference at zero.

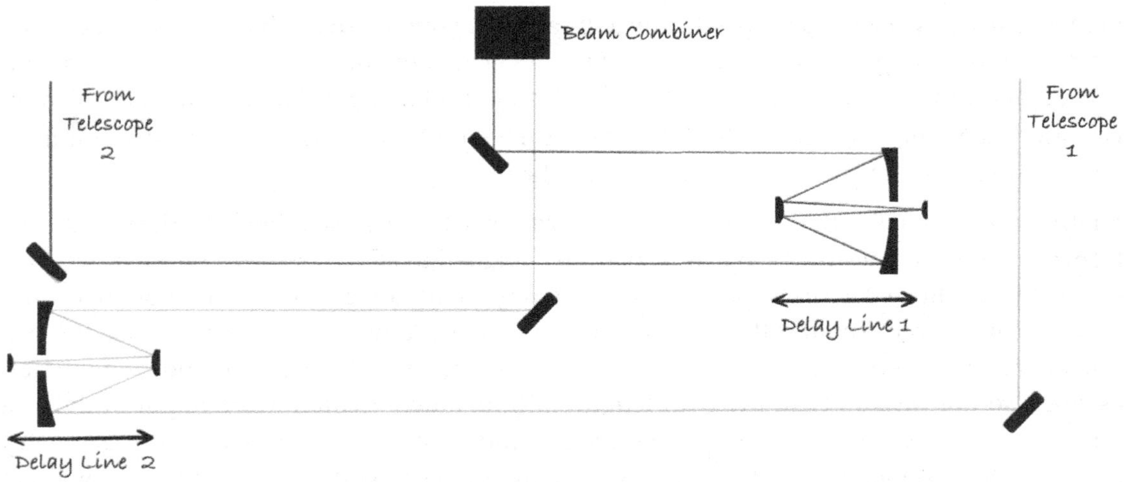

Adjusting the additional length traveled by the light is done by means of mirrors mounted on moving carriages. These carriages, nicknamed the "Paranal Express", are moved via a stepper motor controlled by an open loop algorithm. The reflectors are mounted on voice coils for fine adjustment, which are faster and more precise; the exact alignment is measured using laser interferometers as part of a closed loop controller. The tunnels that contain the delay lines are temperature-controlled to avoid "atmospheric jitter" in the tunnel. The overall optical path is really complex: the lab that contains the instruments that analyze the combined light is of course stationary, over ten mirrors are involved for each of the telescopes and all of them must be controlled precisely. No wonder the VLT's tag line is "The world's most advanced visible-light astronomical observatory". Currently, one instrument, GRAVITY, combines up to four beams interferometrically; a maximum of six are potentially possible, because there are currently six delay lines.

There are other restrictions regarding the observations with such an interferometer. In particular, the field of view is very limited. There are two reasons for this. The first one relates to the geometry of the optics. Because of the very long optical paths, the sensor is effectively looking through a very long tube, and the opening angle in that tube is by definition small. In fact, the interferometer gets the light from the two telescopes through very thin single-mode optical fibers! Conservation of étendue plays into this again: in LINC-NIRVANA, the size of the sensor could not be made arbitrarily small because you cannot "funnel light". Here, the conclusion is in the other direction: because of the thin, long tube (and its very low étendue) the size of the telescope pupil and angle from which you collect light—in other words, its field of view—must be limited.

This is an interesting difference to the LBT's Fizeau interferometer. It maps the pupil to the imaging plane of the interferometer, basically by ensuring that the path lengths of the off-axis light rays are exactly similar to those of the on-axis ray. This is why the LBTI has a much wider field of view than the VLTI—with much shorter baselines of course. There VLTI also attempts to remap the pupil from the frontend to the backend using variable curvature mirrors in the delay lines. The original plan was to build a Fizeau interferometer for the VLT. There is a several meters deep pit in the beam combiner room that was supposed to hold an imaging beam combiner; it is named "Beckers Pit" after Jacques Beckers who was the head of ESO's interferometry group at the time the VLTI was conceived. But it is just much more challenging (read: more expensive) because of the larger dimensions. Today the pit is filled with sand.

The second limiting factor for the VLTI's field of view is that, because of their geographical distance, they see different atmospheric turbulence. They each uses their own adaptive mirror to correct it, but this correction only has the required precision for interferometry very close to the (guide) star that is used for measuring the turbulence in the first place; this is only a small area of an individual telescope's field of view. Even though the VLTI currently does not use the laser guide stars for interferometry, they could potentially help, because you can point them closer to the target you want to observe. Working with multiple guide stars, or multi-conjugate adaptive optics also improves the situation because both provide more data for modeling the turbulence above the telescope.

There is also a limit regarding the brightness of the observed target. The atmosphere also wobbles in piston (away from and towards the telescope), and this cannot be detected by the wavefront sensors as part of adaptive optics. For regular imaging, the scale of these wobbles does not matter. But when combining two beams interferometrically, those wobbles lead to a noticeable phase difference, jittering the fringes. So the integration time of an observation has to be shorter than the "wobbling period", which is a problem for low brightness targets. To solve this problem, the differential path lengths must be measured and corrected at nanometer scale with millisecond precision. The PRIMA instrument did exactly this: the idea was to look at a bright source close to the to-be-observed darker target and use this for realtime fringe tracking. Lots of additional components were required: a laser metrology system that measures very precisely the optical path difference between the target and the bright source; additional high-speed differential delay lines; and a fringe sensor unit.

So, to sum up: these optical interferometers are really still at the leading edge of technology: The VLTI requires multiple active control loops, and often the instruments and beam combiners are custom-built for particular kinds of observations; PRIMA is an example. As we will see later in this chapter, interferometry is much easier for radio astronomy where the wavelengths are longer—but it also much more necessary.

NULLING INTERFEROMETRY The final flavor of interferometry I want to introduce is "nulling interferometry". This is relevant for detecting exoplanets. The challenge here is to detect a small and relatively dim planet next to much brighter star; it has been compared to observing a firefly next to a searchlight. Various approaches exist and several of them rely on

interferometry. An approach called "single-pixel nulling interferometry" allows detection of the presence of a planet around a star even if the aperture cannot resolve the star, let alone the planet. Here is how it works: the path lengths from the two apertures are set up to have a phase difference of π, and have the same amplitude. When the two beams are combined, this leads to complete destructive interference: no light is received from the star in the single-pixel detector. Let's assume the baseline is long enough to resolve a planet (it is some distance away from the star, so the required resolution is lower). Because it is at some angular distance from the central star, its light will not interfere completely destructively. So if we receive any light in our single-pixel detector, it must come from a planet. The approach basically gives a yes/no answer: planet versus no planet. An alternative approach is based on imaging Fizeau interferometers such as those available on the LBT. If we introduce a π phase difference between the two beams, then a dark fringe will cross the image right at the center. Assuming we point the telescope exactly at the star, then this dark fringe will hide the light from the star. If the planet lies in the bright fringe, it will become visible.
This concludes our overview of optical interferometry, and we move on to the next generation of optical telescopes. Who needs 22.8-meter baselines from combining two smaller mirrors if we can instead just build a monster mirror of 39 meters?

5 | *The Next Generation: the ELT*

INTRODUCTION The European Extremely Large Telescope, or ELT, is currently being built at Cerro Armazones in Chile, not far from Cerro Paranal, where ESO's Very Large Telescope telescope has been in operation since 1998. The ELT will have a 39-meter mirror, making it the first of a new generation of much bigger telescopes. An ESO video states that the ELT's light collecting power is more than that of all current astronomical research telescopes combined! It is projected to cost about 1.1 billion EUR. This is roughly ten times the cost of the LBT; however, it also has 11 times the collecting area, so the cost per area remains approximately constant.

Two other telescopes of comparable size are being designed, both schedule to see first light in the late 2020, just like the ELT. The Thirty Meter Telescope, driven by a consortium around Caltech, is planned for Mauna Kea on Hawaii; as I write this in early 2020, the construction is still blocked over conflicts with Native Hawaiians who consider the mountain sacred. The other one is the Giant Magellan Telescope. It will have a 25 meter mirror, composed of seven mirrors of 8.4 meters each; in fact, these mirrors will be very similar in design and manufacturing to those of the LBT. Its construction has started at Cerro Las Campanas in the Atacama desert, a mountain that already hosts the Magellan Telescopes, two separate telescopes of 6.5 meters.

Do you see a pattern emerging in these telescopes' names? Large Binocular Telescope, Very Large Telescope, Extremely Large Telescope, Giant Magellan Telescope, and there was once a plan for the Overwhelmingly Large Telescope (cutely abbreviated to OWL, a bird that can see clearly at night). The optical astronomy community is on a path towards a critical decision: do we stop building larger telescopes, or are we willing to adopt politically incorrect names such as the F***ingly Huge Telescope and the Holy-S**t-Is-This-Big-Telescope. Or do we follow the example of the TMT, the Thirty Meter Telescope, and make it boring. By the way, putting the size into the name is also a bad idea, because of the inevitable funding cutbacks that will reduce your mirror size. The ELT had been planned to be 42 meters (which, it seems to me, was probably motivated by Douglas Adams and not by science). You'd have to change your name or live with the embarrassment, like Germany's "Fighter 90" aircraft that became operational in 2006 because of delay after delay. Anyway. For now we'll stay with the ELT.

THE SEGMENTED PRIMARY MIRROR A 39-meter mirror is way too large to be manufactured from a single piece of glass: it would be impossible to transport, very hard to keep stable under its own weight, and also way too big for existing manufacturing processes and facilities. So once again, instead of building something large and precise, let's "virtualize" it and use active control to achieve the necessary precision. The ELT mirror is built from hexagonal segments. The illustration shows how hexagonal segments can approximate a circular shape; however, instead of the seven segments shown in this picture, the ELT uses 798.

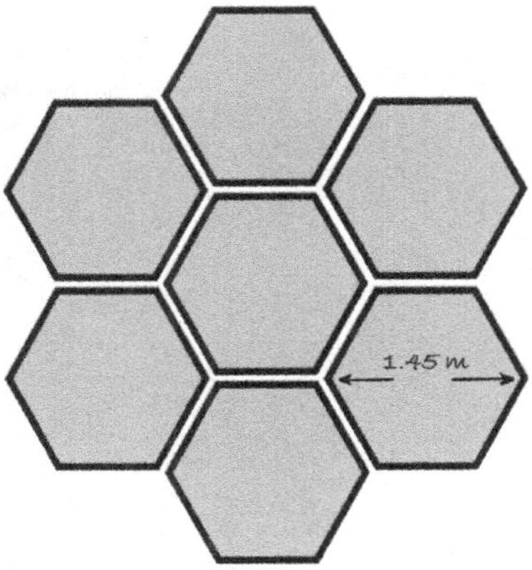

Each segment is 1.45 meters in size and weighs around 350 kg. They are spaced 4 mm apart. Because they are all located at different places in the "big parabola" of the main mirror, they are grouped into six groups of 133 different shapes each. One more of each particular shape will be built—931 in total—as spares for the average of two segments per day that undergo recoating, guaranteeing parallel continuous operation and ongoing maintenance.

Obviously the big 39-meter segmented mirror requires the equivalent of active optics: its shape has to be controlled continuously to compensate for gravitational bending of telescope structure's at different elevations. To achieve this, each of the 798 segments is mounted on a "segment assembly". Each of these has three actuators for adjusting the position of the mirror in piston (up/down) as well as in tilt/tip. The actuators have an electromechanical regime for coarse-grained, persistent changes, plus a high-speed, finer-grained piezoelectric mechanism. Each of the assemblies also has a warping harness that helps to correct curvature, astigmatism and trefoil, aberrations induced by thermal and gravitational effects during operation, as well as errors from manufacture and test. They rely on 12 actuators for the shape and a further three for piston, tip and tilt. You can imagine that controlling 798 mirrors with thousands of actuators isn't a trivial task. We will look at this in some detail later.

OPTICAL PATH The overall architecture of the telescope is not unlike that of the LBT: the primary mirror sits on a rocking-chair assembly for elevation control, which in turn resides on a "tank-turret" for control in azimuth. However, while this structure at the LBT weighs 700 tons, that for the ELT is estimated to be 3,000 tons. On either side of the mirror there are two tennis court-size Nasmyth platforms that hold the instruments. Unlike at the LBT, there are no other instrument mounts at the ELT except for a Coudé focus. A Coudé focus is stationary and does not move with the telescope; it is thus suitable for extremely heavy instruments. In addition to M1, four more mirrors are involved, as shown below. Note how the light path from M2 to M3 has its focus, the beam's thinnest point, right at the location of M4—clever.

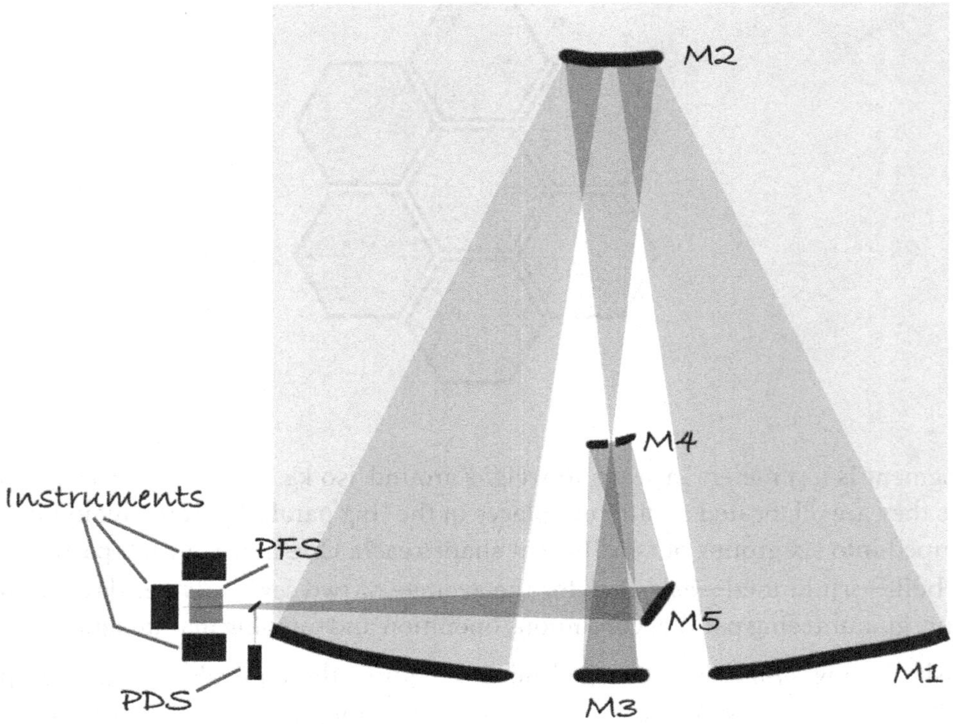

Most large telescopes, including the LBT, use three mirrors to bring the light to a focus for the instrument. ESO has decided to use five, four of them optically powered, because this allows the telescope to compensate for more of the aberrations that are inherently introduced by the primary mirror. In a three-mirror design some of these aberrations have to be corrected in the instruments, potentially in several.

In particular, M2 and M3 help with the creation of a sharp image from the light collected by M1; jointly they form what is called a "three-mirror anastigmat". Through their particular shapes, these mirrors minimize spherical aberration, coma and astigmatism. Spherical aberration means that the light that hits the surface of a mirror at different radii from the optical axis is not focused at exactly the same distance: the picture is in focus at only one particular radius from the center, not over the whole image plane. Coma means that light that is not perfectly aligned with the optical axis blurs slightly—point sources are imaged with slight "tails". Astigmatism refers to light from different planes being focused differently; for example, only the vertical or horizontal line of a cross would be in focus, but not both.

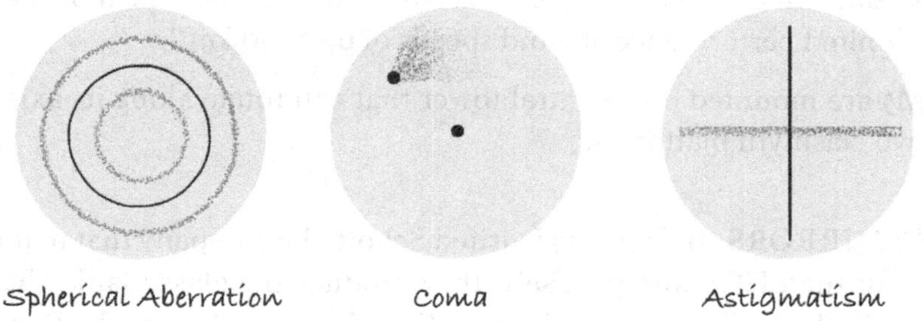

Let's look at the mirrors other than M1. M2 is convex and 4.1 meters in size. It is a thin meniscus that corrects the primary mirror's spherical aberration. Thirty-two pneumatic actuators act as lateral supports. These are necessary to keep the mirror precisely in the required position as the telescope moves in elevation; the gravity vector pulls on M2 in a different direction depending on the elevation angle. Active compensation is required because the structures would become too heavy if they were built sufficiently rigid to not deform under gravity. The heavier a structure becomes, the more power you need to move it, and the harder it becomes to build supports and bearings that do not introduce vibrations or "stickiness" into the system when it moves. In any case you need active control for fine adjustments, so why not build it less rigid and actively compensate more? In addition to its position, the shape of M2 is also actively controlled, by 84 axial actuators. However, they are rather slow, compensating for environmental conditions and external and structural loads.

Like M2, the tertiary mirror M3 is also an actively controlled meniscus, 10 cm thick (which is thin for an almost 4-meter mirror). It weighs around five tons, and must be able to move up and down by around 25 centimeters to a precision of 0.01 mm to achieve a variable focal length for the telescope. The vertical movement is achieved by a hexapod built from roller-screw actuators with an external linear encoder; just like the hexapods used in the motion platforms of simulators, these are used because they allow movement in all six degrees of freedom.

M4 is flat and 2.7 meters in diameter—the same size as the main mirror in SOFIA! It is the location for adaptive optics, with over 5,000 voice-coil actuators adjusting the shape of the 1.95 mm-thin glass surface, corrected at a maximum of 1 kHz. Each actuator projects to around 50 cm^2 on the primary mirror. The control strategy has been validated on the LBT's adaptive secondary mirror. M4 is conjugated to a few meters above the telescope; it is expected that the huge telescope and its equipment will produce considerable thermal turbulence. Higher-conjugate-adaptive optics will have to be performed with additional adaptive mirrors in the instruments. M4's position can also be adjusted to help with collimation to direct light to the various Nasmyth stations.

Finally, M5 is 2.6 × 2.1 meters in size and elliptical, because it is tilted at roughly 45 degrees. M5 is a relatively "boring" design: a high-quality flat mirror. However, it can move in tip and tilt at up to a few Hertz using piezo actuators the thickness of your arm. This is to compensate for the wind shake that affects the telescope, especially M2. The telescope must be able to deliver its diffraction-limited performance at wind speeds of up to 40 km/h.

M3, M4 and M5 are mounted on a central tower that can rotate along its axis to point M5 at either of the two Nasmyth platforms.

MAKING THE MIRRORS In late 2019 I visited Schott, the company that manufactures most of the mirrors for the ELT. More precisely, they produce the glass blanks that are then polished to the required high-precision surface by REOSC, a subsidiary of the French Safran conglomerate. Once the surface is sufficiently precise, a thin aluminum layer is added to provide the necessary reflectivity, just as with the LBT.

I talked to Thomas Westerhoff, who is responsible for the department at Schott that deals with Zerodur, the brand name of the special glass–ceramic compound used for large astronomical mirrors. The crucial characteristic of this material is that it has a very low thermal expansion coefficient of the order of 10^{-9}/K. This means that the mirror retains its very precise optical shape as it transitions through different temperatures during an observation or over the course of several. It is also a very homogeneous material that can be polished to the extremely smooth surfaces necessary for optical mirrors—with roughness below 1 nm RMS. Zerodur is also chemically stable and does not degrade over the decades-long lifetimes expected of telescopes.

Schott developed Zerodur in the 1960s specifically for astronomy. The Max Planck Institute for Astronomy asked Schott for a 3.6-meter mirror that would be installed at the Königsstuhl observatory near Heidelberg to catch up with the US astronomy community, which at the time had just "upgraded" to the first generation of 4-meter mirrors. This first Zerodur mirror is still in use, although it has since been transferred to the Calar Alto observatory in the Sierra de los Filabres in southern Spain. Since then, Schott's Zerodur glass–ceramic has become the material of choice for astronomical mirror substrates, even though the majority of Zerodur business comes from other domains. For example, it is used as the optical "backbone" of laser gyros in inertial measurement units in airplanes, and is also used for the picometer-scale position measurement that enables chip-manufacturing wafer scanners to position wafers and projection lenses with nanometer accuracy. "These other domains provide the reliable revenue necessary to keep the knowhow alive for the much rarer astronomical mirrors", says Thomas Westerhoff.

So how does the material achieve the feat of retaining its shape and volume as the temperature changes? Most glasses have a relatively low thermal expansion coefficient, of the order of 10^{-6}/K. To get to 10^{-9}/K, additional engineering is required. Zerodur is made from over ten different ingredients, the dominant ones being lithium, aluminum and silicon. Like all glasses, Zerodur is amorphous, which means that—in contrast to crystals—its constituent atoms are not arranged in regular systematic structures. However, when solid, 70% of Zerodur's volume consists of nano-crystals of 50 to 70 nm in diameter. As the material gets warmer, the expansion of the amorphous glass is compensated by the shrinking of these crystals. The precise ratio of the various ingredients is crucial to ensure that the two processes compensate for each other exactly in the relevant temperature ranges.

The ingredients are mixed as solids with specialized mixers that have minimal abrasion—you don't want material from the mixer in your glass. The amounts are measured to gram precision even for typical batches of several hundreds of kilograms. The grain sizes of each ingredient are also crucial to ensure that they melt uniformly, as they experience a common temperature profile during the melting process. Next, the raw materials mixture is heated to above 1300°C, where it becomes a liquid, which is then poured into a mold of the right shape and size. The following steps are key for ensuring that the resulting solid has the exact ratio and size of embedded crystals that lead to the required very low thermal expansion coefficient.

The glass is cooled through a slow and controlled process to ensure that its complete volume cools down uniformly, "Otherwise it will be cold and solid on the outside and hot and liquid on the inside; once the inside cools, the resulting stresses will transform the mirror into a 3D puzzle", as Westerhoff put it with smile. The cool-down rate is much slower than the natural process; the glass is placed under a dome and heaters moderate the process. This so-called "coarse annealing" takes several weeks for the 1.52-meter ELT M1 segments; for the larger 4-meter mirrors it takes several months. Visual inspection for streaks and air bubbles follows—you don't want any of those, especially in thin mirrors such as the adaptive M4, which is less than 2 mm thick at 2.4 meters in diameter; the typical diameter of a bubble would be greater than the thickness of the glass, so a bubble qualifies as a hole!

If everything is ok the glass is heated again, very slowly, to a temperature at which the crystals grow; the exact temperature, profile and time at the high temperature govern the exact ratio and size of the crystals. This second heating–cooling cycle is called ceramization and again takes from several weeks up to months, but this cooling stage is quicker than the first, because the material is now much more robust against temperature change due to the low thermal expansion achieved by its precisely adjusted crystal content. Once the final Zerodur structure has formed, samples are taken from the outer regions of the glass volume to verify its thermal expansion properties in the lab—for absolute values, but also for homogeneity over a single blank and the whole set of M1 segments. Samples are also taken while the glass is still liquid: if the precise composition of the mixture is understood, small deviations from the ideal mixture can be compensated for in later phases of the production process.

This whole process is empirical, based on years and years of experience. Schott has worked with this material for decades, so they know how to achieve the necessary precision. What is unique for the ELT is the large number of similar segments for M1: 798 are used by the ELT's main mirror at any given time, but Schott is tasked with building 949, as 133 are needed as spares to support the continuous two-segment-per-day re-coating process. Another 18 were built as a pre-series run to qualify the manufacturing process. It is the first time that Schott has engaged in series production of astronomical mirrors to such an extent—until now the largest set of similar mirrors for a single telescope was the 36 segments of two meters each for the Gran Tecan and the two Keck telescopes.

Defining and then executing a repeatable process to deliver the same quality over almost 1,000 mirrors is the central challenge for Schott; Schott will spend the next four years producing on average one segment per working day! Considering that the pour/cool/ceramization-process takes around a month, and taking into account that one mold is filled to yield about five mirrors, many molds will be being coarse-annealed and ceramized in their own enclosures at any given time, while dozens of mirror disks will be in various stages of production every day. The team spent quite a bit of effort on production logistics. Westerhoff pointed out this example: "How do you store the segments in a way you don't trip over one as you handle another, but at the same time retain the ability to access each of the current batches?" You want to handle each disk as few times as possible; after all, it's still glass, so it's brittle and has a tendency to break. Schott also developed handling tools, as the glass blocks weigh 270 kg each. Workers can't just carry them around, but on the other hand, around 1,000 is still too few to justify robotics-based automated production.

Let's walk through the production steps of an M1 mirror segment. The process starts with a cylindrical mold with a diameter of 1.52 meters, around 30 centimeters high. Each of the mirror segments will be around 5 cm thick, but Schott molds five at a time; the cylinder is cut into 5 cm disks when it's solid. They use a wire saw for this, just like masons. The blanks are cylindrical in shape; the required 1.40-meter hexagonal shape is cut out of the circular disk after polishing at REOSC. Schott also produce a 30 mm-diameter 25 mm-deep blind hole on the back of the mirror: six actuators that move the segment radially will be glued to the edge of this hole. The grinding is performed using CNC grinding machines similar to those used in metalworking. However, the tool used by the machine does not chip away material, as is typical for metal—instead it grinds away material using small diamonds embedded in a matrix, more like a jackhammer.

The susceptibility of glass to develop cracks and eventually splinter is, of course, a feature of the material, but it also depends on the quality of the surface. Small microscopic scratches can develop into cracks that compromise the integrity of a mirror if driven by external disturbances such as earthquakes. These are possible in Chile, and considering ELT's projected 50-year lifetime it's not unlikely that one will hit the ELT. Not surprisingly, the jackhammer-style drilling, as well as the wire saw, create such scratches. The front surface of the mirror is of course polished to very high precision for optical reasons, so there won't be any of those dangerous scratches left. The back, however, must also be free from scratches. To achieve this, the back is also ground with increasingly finer tools, then treated with hydrofluoric acid. Hydrofluoric acid so corrosive that it even attacks glass. It is a hazardous substance and has to be handled with extreme care—respirators, gloves, full suits and face shields. Dealing with such substances is not Schott's core expertise, so it is performed by a subcontractor.

Next comes the shaping of the front: every radial ring of segments of the M1 mirror requires a slightly different shape to create the overall spherical shape of the 39-meter mirror; after polishing there will be 133 different segment types, all aspherical but with slightly different specifications. However, Schott only delivers three different "rough" spherical shapes, with at most 2 mm deviation from a flat surface. This shape is created via the CNC-based grinding machine as well; the challenge is to perform this grinding without damaging the back, which has already been de-scratchified by the hydrofluoric acid. Schott also glue 27 connectors to the back of each mirror that will receive the axial actuators that form the warping harness. The 5-cm thick glass disk is bent by nanometers to make sure the overall mirror remains "in shape" as it rotates in elevation.

The last step is packaging into a crate—hopefully without inflicting any new scratches. Six of these crates are put into a 20-foot container for road transport to REOSC in Poitiers in western France. REOSC creates the final high-precision shape of the mirror's optical surface using successively finer milling and polishing machines. The final step down to 15-nanometer accuracy is done with high-precision ion beam figuring. The surface quality is continuously verified optically, with measurement equipment that, ironically, relies on Zerodur in its optical path.

Schott has also manufactured the other ELT mirrors, with the exception of M5. This is because M5 is a planar mirror: it has no optical power; thermal expansion will not change its shape, only its volume. However, it has to be very light and stiff because it will move at several Hertz, so it was easier to build it from silicon carbide ceramic.

The production process for the 4-meter M2 and M3 mirrors was fundamentally the same as for the M1 segments. However, because they have a much higher curvature, the raw volume of glass out of which the curved shape would be ground was larger and the cooling times correspondingly longer. The ground volume was also much larger, because both mirrors are 10-cm thick meniscuses (to reduce the weight of the suspended M2, and because they both are active), so both the front and the back had to be ground.

The adaptive M4 was yet another story. With its 2.4-meter diameter and a thickness of just 1.95 mm it is really thin, which is necessary because 5,000 actuators continuously adapt its shape to correct for atmospheric turbulence. The actuators have a stroke of 400 micrometers, run at 1,000 Hz, and each pair of actuators is 27 mm apart, so in the worst case the glass body must be deformed by several tenths of a millimeter over a distance of around 3 cm. It must therefore be thin. The mirror is made from six piece-of-cake-shaped petals. The reason for this segmentation is that segments are easier to manufacture than one large 2.4-meter disk, and also cheaper to replace if one of them gets damaged; Schott has actually produced a seventh. The sector-shaped petal is cut from a circular disk of 1.4 meters diameter and 35 mm thickness. It is then milled down to its 1.95 mm final thickness. Because of the mechanical stresses from the adaptive optics actuators, and its thinness, Schott were given an extremely tight specification for this mirror regarding streaks, air bubbles and imperfections in the crystal structure. "Producing this mirror was a high effort", says Westerhoff, implying a lot of material inspection time was required to produce relatively little "mirror volume".

Before we move on with the ELT, let's briefly look at another way of creating mirror blanks: centrifugal casting. This is especially useful if a thin but very large surface must be created, as for example in the case of the four VLT mirrors. These are the largest mirrors Schott has ever cast: 8.2 meters in diameter and 18 cm thick, thin enough to support active shaping, just like the mirrors of the LBT. Schott spent two years working on this mirror, reducing 46 tons of raw glass to 22 tons of beautiful mirror. With centrifugal casting, a mold that approximates the ultimate parabolic shape is rotated. The molten glass is poured slowly into this mold and, because of the centrifugal force, spreads out over its surface. This avoids casting a huge block and then milling away lots of the volume; as glass millings cannot be remelted and reused, this makes production a lot cheaper. In addition, because of the lower total volume the cooling periods become shorter, speeding up production. A curious detail is that the rotating turntable used for the VLT mirrors was built by a company that normally builds railway turntables—although that for the VLT was made to rotate faster.

The LBT's 8.4-meter mirrors, manufactured by the Richard F. Caris Mirror Lab at the University of Arizona, were also built using centrifugal casting, with three differences to the ELT mirrors. First, each has a one-inch-thick front surface that is backed by a honeycomb structure. This provides high stiffness with relatively low weight. Second, the material is borosilicate glass rather than Zerodur. Its thermal expansion coefficient is 1,000 times higher than that of Zerodur, but this doesn't matter in this case because the mirror's shape is actively controlled. The third difference is that the glass was not poured directly from a melting oven after mixing. Instead, it was cast into shards by a manufacturer in Japan, and those shards were poured into the mold in solid form. The rotating mold was then heated, liquifying the glass and allowing it to move out from centrifugal force, filling the honeycomb ribs and the front surface. The mold's rotation was tuned to 7.42 rpm so that the surface came out as a paraboloid close to the final desired shape. Of course, additional grinding and computer-controlled polishing with laser-based measurements were still required. To make it an actual mirror, the final surface was then coated with a very thin layer of aluminum; a few grams are enough for a whole mirror. The second mirror, "The better one of the two", said Richard Green, has a surface distortion of 19 nanometers; that's the same ratio as 1-inch-high waves would have relative to the surface area of the Atlantic Ocean!

PREFOCAL STATIONS The illustration of the optical path of the ELT above shows several items in addition to the mirrors. The prefocal station (PFS) is central to the operation of the telescope. It has three major roles.

First, it mounts the instruments—three in total. One is mounted behind the PFS, two at an angle of 90 degrees. The straight-through port gives access to the full ten arcminutes field of view; the ports on either side are supplied by the 1.5 meter M6N mirror that is lowered into the light path from the top in the prefocal stations; this reflects the central five arcminutes. The PFS steel structure has instrument derotators that ensure mechanically that the orientation of the instrument's field of view remains stable as the sky rotates.

The second task of the PFS relates to wavefront detection in the context of the active optics of the M1 mirror; it helps to guide the telescope so as to ensure pointing stability for the target of interest. At the rear of the PFS, in the path to the straight-through instrument, there's an "adapter" that holds probes that pick off the light from guide stars and feed it into Shack–Hartmann wavefront sensors. One reason why the left and right PFS ports only receive the central five arcseconds of the whole field of view is to ensure that these probes can continue to pick up light from the outer rim of the field of view even when the M6N mirrors are lowered. The probes are mounted on circular rails that allow them to move around the field of view; they can also move in and out. In this way the four probes can effectively reach any point in the ten arcminutes field of view to "look at" natural guide stars and analyze their wavefront. Like everything else on the ELT, these things are biggish: the probes are 1.9 meters long, have a 23 cm mirror and weigh 130 kg each.

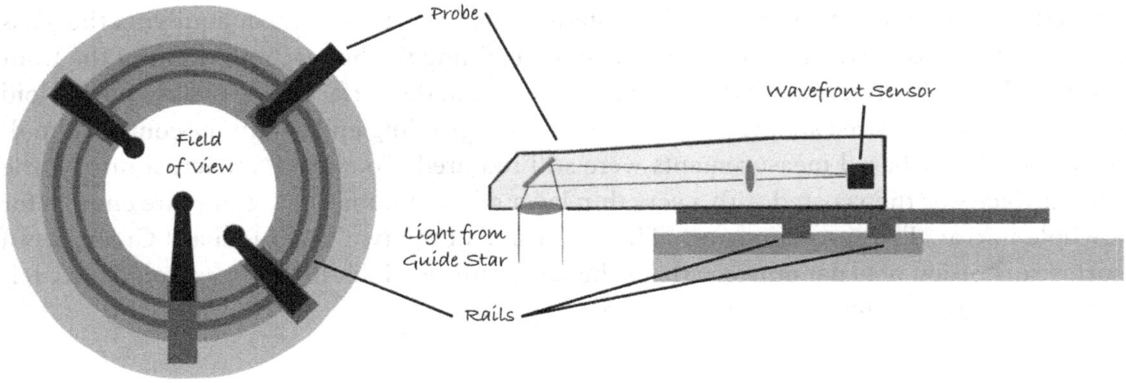

The rear of this same adapter has four more similar sensors whose purpose is to track and analyze laser AO stars. Because their location on the image is fixed (as a consequence of the lasers' installation at a particular location) they cannot move around the field of view, only in and out.

The ELT plans to use a different approach to producing an artificial AO star than the LBT. The laser is tuned to 589.2 nm (orange) to activate sodium atoms in the mesosphere at around 90 km altitude. The approach is more technically complicated (and costly) than the Rayleigh-scattering-based approach used at the LBT, but because it produces a "star" at 90 km instead of 10 to 20 km it "captures" more atmospheric turbulence, potentially allowing for better correction. This approach is already used successfully at one of ESO's VLT unit telescopes, where four lasers are installed around the main mirror; each has around 20 W of power and is 30 cm in diameter. The lowest 10 km are visible as orange beams—I'm sure that's a pretty cool sight.

The third task for the PFS is optionally to insert yet another mirror, M6C, that picks up the inner 20 arcseconds of the field of view. This either routes the light onto a train of mirrors towards the Coudé focus. Alternatively, M6C can also route the light to the phasing and diagnostics station (PDS), which is used for setting up the optical path of the telescope. We'll come back to the PDS later.

INSTRUMENTS As we have seen with LBT already, a telescope is nothing without its instruments: a diverse set of instruments and the ability to switch between them quickly gives it versatility and longevity. The ELT can host eight instruments at a time: each of the two PFSs can host up to three instruments, while two additional ones can be placed at the Coudé focus. The scientific justification for a new telescope can only be made if the design includes a convincing set of instruments. In parallel to the design and construction of the ELT telescope, therefore, there were various phases of proposals and reviews for instruments that exploit its huge mirror. Eight instruments were proposed by different research consortia during Phase A.

When categorizing instruments, the primary criterion is always the wavelength: is the instrument intended for visible light or for infrared? The proposals ranged—not surprisingly—between 380 nm and 10,000 nm. And related: is an instrument specialized for a narrow range of wavelengths, or can it process light in a wider range? Then field of view; ten arcminutes is the absolute maximum the telescope can provide. The instrument proposals ranged from 0.5 arcseconds to about 7 arcminutes. The third criterion is the size of the pixels (also known as "spaxels" for imaging spectrographs) relative to the image. These range from around one milliarcsecond to one arcsecond. The final criterion is the spectral resolution, given as $R = \lambda / \Delta\lambda$. The range here is from 100 to 100,000. For 500 nm (cyan) this corresponds to between 0.05 and 5 nm. I will highlight a few interesting aspects of some of these instrument proposals below.

The first category of instruments aren't really instruments: they are post-focal units supporting multi-conjugate adaptive optics at 4 and 12 km altitude. In contrast to the LBT, these are not part of an instrument; they are separate devices with additional deformable mirrors dedicated to AO; different instruments can be installed on instrument adapters to exploit the "perfect" resolution.

CODEX was a study for a high-resolution optical spectrograph whose goal was to achieve a Doppler precision of 2 cm/s. This would allow the instrument to measure the increasing red shift due to the accelerating expansion of the universe directly—although it would have to observe for 4,000 hours over a 20-year period, while relying on extremely stable wavelength calibration based on laser frequency combs. CODEX would also allow the detection of Earth-sized planets in the habitable zone around Sun-like stars. Both of these are in line with the science goals specified for the ELT. Its high spectral resolution also supports nucleochronometry, in which the age of stars is determined by detecting spectral lines that belong to physical and chemical processes that are in turn related to the age of the star—the equivalent of carbon dating on Earth. A final stated science goal is to test whether the laws of physics have been stable over the lifetime of our universe, by measuring the wavelength of particular atomic transitions of stars back to 90% of the age of the universe. To perform this measurement, the instrument must be temperature stabilized, which is why it will be installed at the Coudé focus.

EPICS is an instrument optimized for imaging exoplanets directly. It also comes with its own "extreme adaptive optics" system. This demonstrates that adaptive optics really are the key to the ELT's success, and that this is still an active area of research, for example in the area of wavefront sensing.

EAGLE is an imaging spectrometer for near infrared that also comes with its own multi-object adaptive optics system. It is optimized for studying galaxies at very high redshift. They are far away (because they have had a long time to "expand" away from us), and the light we receive from them was therefore emitted a long time ago. This should allow EAGLE to observe galaxies when they were forming during the epoch of reionization of the universe.

METIS will be a multi-purpose instrument for the thermal infrared band. It will offer long-slit spectroscopy (not unlike the vertically aligned slits in the LBT's MODS), imaging spectroscopy, coronagraphy—in which the light from a star is blocked, so you can spectrally analyze its corona, or see anything in its vicinity such as planets or circum-planetary dust—and polarimetry, in which the radiation's polarization is measured, which can reveal the content and structure of the atmosphere of exoplanets without resolving the planet optically.

METIS will be one of the first-light instruments, those that will become available as soon as the ELT itself starts observing. Since there's not that much time for development, and because they have to be useful for a wide range of observations, first-light instruments are naturally not the most ambitious and risky ones. Two more first-light instruments have been approved by ESO. One is MICADO, a diffraction-limited near-infrared camera. Its sensitivity will be comparable to the James Webb Space Telescope, but with six times the resolution. Again the scientific focus will be high-redshift galaxies, and—in coronographic mode—the detection and characterization of exoplanets. To help with the latter, MICADO will also have a single-slit spectrograph and will be fed by a multi-conjugate adaptive-optics system called MAORY. The second first-light instrument is HARMONI, a spectrometer suitable for a broad range of wavelengths. This will be able to obtain spectra of typical Jupiter-mass exoplanets that are detectable with ten-meter class telescopes. It's a large instrument with its own set of three-story platforms to provide maintenance access. Generally, many of these instruments are several meters in each dimension and weigh several tons—some up to 30! Their design to fit together around the prefocal stations was part of the proposal.

6 | *Controlling the ELT*

INTRODUCTION Controlling 10,000 tons of steel and glass spread over a thousand mirrors, moved and positioned by 20,000 actuators, plus a total of 60,000 input/output data points, to micrometer precision at (in some places) up to a kilohertz is a seriously non-trivial task. Let's try to get a grasp of how to tackle this challenge. We will look at how to form a working telescope from all the optical components, and also at the bigger picture—the systems engineering of the control system.

The first order of business is to break the overall system down into subsystems that have some degree of autonomy in terms of how they are controlled: modularization is not just best practice in software, but also in systems engineering. The most complex subsystem is the optical path itself: there are several separate but closely synchronized control loops, with rates from 0.01 Hz to several kHz. Some of these loops have significant computational and bandwidth requirements. For example, the adaptive optics requires 700 Gflops of computing power and realtime data rates of up to 17 GB/s. Overall, this is a control task with many thousands of degrees of freedom.

THE MELT AND THE PDS Let's look at the control of the optics in more detail. There are two main aspects: the alignment of the 798 segments of M1 to produce one 39-meter virtual mirror, and the control of the other four mirrors to form a precise optical path. A team of five engineers is currently developing the control strategy for both—which they have to do without an actual telescope. As a subsitute they have developed the Mini-ELT (aka MELT), an optical bench setup that has all the important components of the future ELT.

Central to the MELT is a 15-cm mirror built out of 61 hexagonal segments. It has been developed as part of ESO's Active Phasing Experiment, which was on sky as an "instrument" on the VLT. It also had various kinds of wavefront sensors, to evaluate their capability to detect the aberrations produced by a segmented mirror.

The MELT's optical path then has a hexapod-positionable lens that emulates M2, a deformable mirror for adaptive optics that corresponds to M4, as well as a fast tip/tilt mirror that simulates M5. M3 is omitted, because it does not play an important role in the control strategy. A movable light source that emits radiation in the relevant wavelength range through a turbulence generator emulates the ELT's main axes and the atmosphere. The MELT also has ways of introducing some of the aberrations that the ELT will exhibit naturally, such as beam instability due to pupil movement as the telescope moves across the sky.

The MELT serves three purposes. One is to understand the aberrations of the ELT's optical path and develop optical components to diagnose them. For example, the MELT has various kinds of wavefront sensors that work in different wavelength ranges and detect different kinds of aberrations, plus several cameras. Once understood, this will help with the design of the Phasing and Diagnostics Station (PDS), a simplified "instrument" that is inserted into the optical path to help calibrate the ELT. The PDS is small enough to be calibrated on the bench using lasers, sighting telescopes and other lab-size equipment; it then acts as a pre-calibrated reference during calibration of the full telescope. The second goal of the MELT is to define an approach to commissioning the telescope. What steps must be taken to get from a telescope that is initially aligned only to mechanical tolerances to one that has the necessarily precise optical path? What algorithms are necessary, and how will they exploit the PDS? The third goal is to validate the control strategies, a set of software algorithms that maintain the telescope's focus and pointing during an observation. The MELT has the same software interfaces as the future real telescope, which means that the control programs can be adapted with very little effort after they have proved their efficacy on the MELT.

PHASING M1 Let's understand how to make sense of the main mirror. Initially, its 798 segments are installed to mechanical tolerances for piston, orientation and shape. Consequently, each segment will project its own separate image of a point source instead of one integrated, focused point (the purpose of a large mirror is of course to collect lots of light and project it all onto one bright image). And each projection will probably be a donut instead of a point, because the telescope will initially be out of focus. Based on calculations, the team expects that all these donuts will be distributed over a five arcminute area in the focus. Initially it will be unclear which mirror segment creates which donut, because of the random orientation of each segment. As the telescope follows the sky, each donut will move and warp in undefined ways, because there will not yet be a pointing model, no understanding of how gravity impacts the alignment and shape of each segment.

To "assemble" an integrated M1 from this disjointed collection of 798 separate 1.42-meter mirrors, the first step is moving M2 up and down to get the telescope into focus as much as possible—to get the donuts to better approximate points.

Step two is called "segment capture". The goal is to identify which segment produces which donut. As Thomas Pfrommer said in our interview, "We have to develop an efficient algorithm that wiggles each mirror and observe the changes in the image plane. Initially we cannot even use the PDS, because it can only see the central arcminute. We plan to put a 2 × 2-meter screen at the focus of the telescope and then use basically a better webcam to capture the image."

Once captured, the segments have to be stacked: the goal is to ensure that all 798 segments project their image exactly on top of each other. Assuming they all project to the central arcminute, stacking can be done with the PDS using its wavefront sensors and the algorithms developed in the MELT.

To understand the position and attitude of each segment relative to its neighborhood, each segment shares 12 contactless inductive edge sensors with its six neighbors. These sense differences in piston, gap and along-edge with picometer accuracy. They are glued to the segments, and one challenge is to make sure that the glue does not creep too much over the years. Statistical approaches are used to detect edge sensors that report unreliable data and cut them out of the control loop until they are repaired.

The next step is coherencing, which means that the segments are positioned so as to align their wavefronts to within one wavelength, around 700 nm. This mostly means moving the segments in piston and using the edge sensors to detect offset. For seeing-limited observations, where the atmosphere determines the resolution of the image, a coherent M1 is good enough. However, for diffraction-limited observations, where the telescope has to perform at the limit of its optical performance, the segments have to be aligned even more precisely such that even the phase of all segments is aligned. To achieve this level of precision, it is no longer enough for the segments to be aligned perfectly in piston and position; the shape of each segment has to fit with the ideal shape of the virtual 39-meter mirror. An error called "scalloping" describes the situation in which the edge sensors are all within limits but there are still low-order deformations over the whole surface.

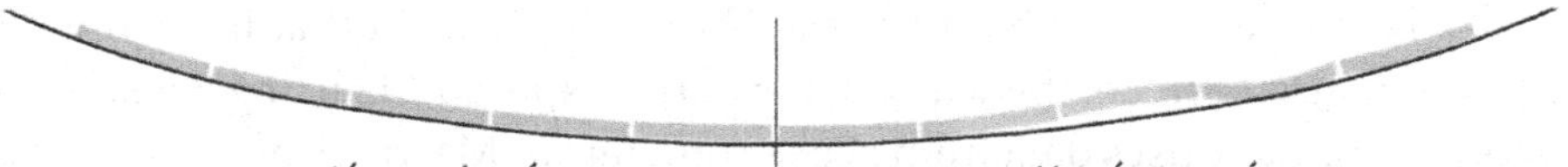

Such deformations lead to errors in the wavefronts not unlike those produced by the atmosphere: the angle at which the light hits the sensor changes. Consequently, errors from deformed segments can also be detected in the same way, using wavefront sensors. The PDS has lots of them to help "phase" M1. The warping harness underneath each segment mirror is then used to adjust the shape of the glass of each M1 segment until the wavefront sensors are happy. You might wonder why the (probably not yet corrected) atmospheric distortions don't disrupt the phasing of M1. The reason is that the wavefront errors that result from the atmosphere are at different granularities than those produced by a scalloping M1. In addition, the atmosphere wobbles: the patterns change quickly over time. Errors from M1, however, change only very slowly as the telescope rotates.

Once the PDS has been used to diagnose the mirror, and the positions and shapes of all segements have been adjusted, the respective settings are stored in huge lookup tables. During observations an open-loop control system based on these tables keeps the mirror roughly "in shape"; a second-level finer-grained closed control loop relies on the wavefront sensors in the adapter probes mentioned earlier. The calibration has have to be performed for the different elevation angles, because of the varying direction of the gravity vector. Because of drifts in sensors and other accumulating errors, the team expects that it will be necessary to "rephase" the mirror every two weeks once the telescope is operational.

CONTROLLING M2 THROUGH M5 The approach for controlling the position of M2 also relies on the wavefront sensors: if the mirror is out of position it produces characteristic wavefront errors that are different than those produced by, for example, a non-phased M1. The PDS can detect these and instruct M2's position actuators to correct them. Once again the parameters are put into elevation-dependent lookup tables, and used by a relatively slow open-loop controller that keeps the position of M2 stable during an observation—because of its position roughly 40 meters above M1 it is especially sensitive to elevation-induced bending.

M3 is not expected to be actively controlled during regular operation. Because it sits near the bottom of the telescope, its mount is much stiffer, and no need for elevation-dependent correction is expected. It does have actuators through, because during commissioning, of course, its shape and position has to be determined and adjusted.

Just like the segments of M1, the six petals of the adaptive M4 have to be phased: M4 has its own control system to achieve this phasing, but the PDS can detect the resulting characteristic wavefront errors as well, for example, using pyramid wavefront sensors in the infrared band.

M5's main task is to reflect the light out to the Nasmyth platform that hosts the PFS and the instruments, but it also corrects the wind shake of the exposed M2 using a fast tip/tilt mechanism. The control strategy uses data from accelerometers in M2, but it also receives input from the wavefront sensors in the instrument or the PDS; specific patterns are characteristic of M2 wind shake. M5 is also involved in M4's adaptive optics. The voice-coil actuators in M4 have limited range; they are optimized for speed, not travel. So when the M4 actuators get close to their range limits, the tip/tilt mechanism in M5 is used to offset M4, effectively giving back travel to the voice coils.

As we have seen, errors in the alignment of each mirror in the optical path lead to characteristic patterns in the wavefront, which are detectable by the PDS. If it turns out that the differences are not distinctive enough, another means of disentangling the aberrations produced by the various mirrors is available. If you look at the optical path diagram shown previously, you can see that there is an intermediate focus between M2 and M3, more or less at the location of M4. There are provisions for placing an artificial light source at this location. This would allow the impact from M1 and M2 to be separated from the M3-M4-M5 train.

SYSTEM HIERARCHY Let's zoom out from the control of the optical path to look at other subsystems that need to be controlled and at the systems and software that does that. This section is based largely on a 2018 SPIE paper by Chiozzi et al. titled "The ELT Control System". There are twelve subsystems in total, amongst which are the opening and closing of the dome, various power supplies, air conditioning and cooling systems, as well as an interlock and safety system that guarantees system integrity independent of what the control system instructs the telescope to do.

Each of the telescope subsystems comes with its own local control system to monitor and command its own hardware, and each is contracted out to a different supplier. Likewise for the instruments: each is developed by a separate consortium, and comes with its own instrument control system.

A central control system, developed in-house by ESO, makes use of the local control systems to coordinate overall system behavior. An important function of the central control system is system-wide mode management, and in particular coordinated startup, shutdown or instrument procedures. To play well with the central control system, every subsystem (or more precisely, every function implemented by a subsystem) has to implement a state machine with predefined states that represent these modes, and react to predefined events.

The control of the adaptive optics is especially challenging, because several subsystems are involved: primarily the wavefront sensing in the instrument and the adaptive mirror M4, but also M5 for offloading. Worse, the components are physically distributed, and the timing requirements are especially tight. This is why the realtime computing that performs adaptive optics is part of the central control system. This is an example of intentionally compromising modularity in order to achieve an overriding design goal, performance in this case.

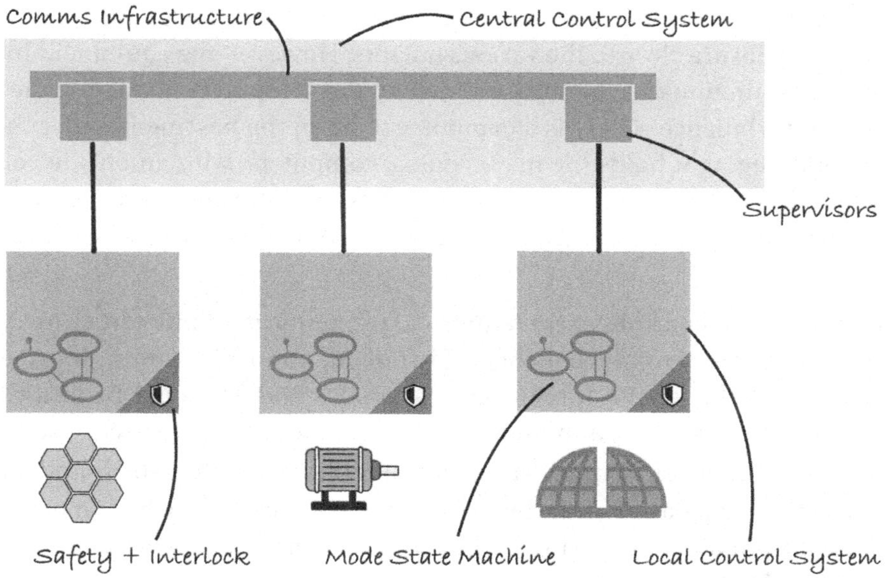

To interface with the local control systems, the central control system runs a supervisor for each of them. This is an example of a widely used software engineering principle called "separation of concerns". The idea is to ensure that the parts of the system that have different rates of change, different engineering drivers or are relevant to different communities are kept separate in the software. In the ELT, a local control system changes when its hardware changes (probably quite rarely), must be reliable and fast, and is relevant to its suppliers. The supervisor, in contrast, changes if the capabilities of the subsystem changes, addresses science use cases, and is relevant to those who integrate the local control systems into a "telescope workflow". For example, the main structure local control system is able to follow a trajectory specified as positions and velocities in elevation/azimuth coordinates, controlling the actual actuators on the mirrors. But the tracking of targets in the sky, and the necessary conversion from scientific to technical coordinates, is done by the main structure supervisor.

Another example of separation of concerns can be found within each local control system: safety and interlock is kept separate from routine control. Its job is to ensures that, whatever the control system instructs the subsystem to do, no humans are hurt and no equipment is damaged. It has much simpler—and therefore less error-prone—logic than the full control system, is implemented mostly in hardware and PLCs, and has minimal interfaces to the control system to avoid common-cause failures. Here the separation is an example of dissimilar redundancy (with degradation), not unlike the safety architecture in flight control systems. We'll further elaborate on this hardware/software work share in the last chapter of this book, where we look at the Large Hadron Collider.

COMMUNICATIONS The control system is physically distributed. Some parts are located in the telescope enclosure close to the various mirrors. However, the equipment in the enclosure is kept to a minimum, to avoid heat created by the computers impacting the sensors or creating thermal turbulence. There is a computer room in the basement under the dome, in the auxiliary building, that holds the main control computers, with an on-site control room next to it. This will be used mainly for commissioning, maintenance and diagnostics; the actual science operations will be run from the control room at the VLT, 30 km away on Cerro Paranal.

A physically distributed control system requires a communication infrastructure to exchange data. This has to address several challenges. First of all, it must be robust. Messages sent by one party should reliably arrive at their destination. Second, it should be fast. Closed-loop controls have to run at a specific minimum rate for them to work as expected, so the communication infrastructure must transport the control messages with a small and—just as important—deterministic delay. Third, the system must be scalable, which means that these properties should remain stable with an increasing number of messages. Finally, all this should be achieved with manageable software complexity, or it will become too buggy, too expensive, or both.

The communication infrastructure consists of five layers. At the bottom are the physical networks, in the form of cables and active components such as switches, routers or network cards. One level up are the protocol stacks that provide a relatively low-level software interface for sending and receiving data over the particular physical network. Examples are Ethernet or TCP/IP, the protocol stack used on the internet. On top of that, distributed systems usually run some kind of middleware that provides higher-level abstractions to the client programs in the layers above them. Such abstractions might include different levels of reliability, realtime guarantees or optimizing use of the lower layers to achieve scalability. Another layer that is often found in large systems is an application framework that handles concerns that are common between the various applications that participate in the system; monitoring is a typical example.

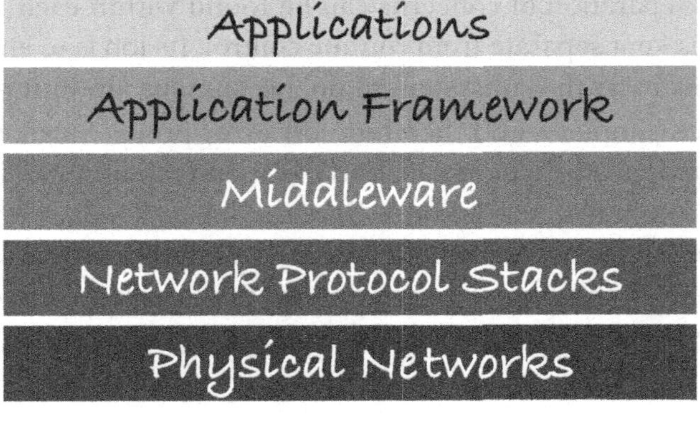

A major reason for using such a layered approach is that the client programs—the supervisors and control systems in the case of the ELT—do not have to be concerned about the details of the lower layers. On the other hand, the layers are not completely independent: a middleware layer can only give timing guarantees if the underlying network protocol and hardware is fundamentally fast and deterministic enough to fulfill them.

It was initially expected that different physical networks (plus their protocol stacks) would be used: standard Ethernet for the non-realtime communications, plus PROFINET IO and EtherCAT for those high-speed control loops that require realtime guarantees. Currently the plan is to build most of the infrastructure on top of very fast Ethernet connections; the trick is to run high-speed data over dedicated subnets with low utilization to ensure that no congestion occurs, such that even Ethernet is adequately deterministic. Some local control systems use other network technologies internally, but do not expose those beyond their subsystem boundary.

For the middleware, conversely, the initial idea was to rely mostly on a single industry standard called Data Distribution Service (DDS). This relies on a publisher/subscriber architecture in which publishers of data submit messages to a channel and subscribers subscribe to channels that transport data in which they are interested. Whenever a publisher posts a message, a copy is delivered to each subscriber. A particular supervisor can of course be a publisher and a subscriber at the same time, although usually however for different channels.

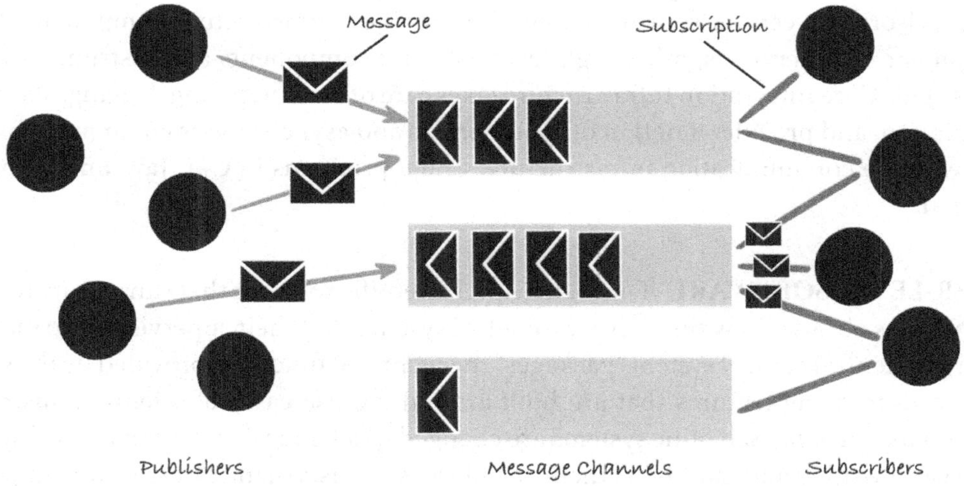

Relying on a single middleware obviously helps reduce overall complexity: only one technology must be mastered. A publisher/subscriber architecture is also useful, because publishers do not have to know who wants to receive a message—everybody just has to agree on the message types and the channels. This reduces global knowledge during the development of the system and is also more flexible as the system evolves. In addition, different channels can make different quality guarantees regarding reliability or timeliness. DDS claims to be able to support a wide variety of these; however, it did not quite work out this way in practice. DDS still plays a role in ELT, but ZeroMQ and OPC/UA are most widely used. An in-house-developed UDP-based infrastructure called MUDPI plays a role as well.

A separate infrastructure for distribution of a highly accurate time signal is also required. The internet's Network Time Protocol can provide an accuracy in the tens of milliseconds range. However, some systems require accuracy of up to 0.5 microseconds. The Precision Time Protocol will be used there. Finally, to avoid common-cause failures, the interlock and safety systems will run on separate networks—or at least on separate logical networks on top of the same physical layer.

INTEGRATION The development of such a large hierarchical control system requires clearly defined interfaces, otherwise there is no hope of the overall system working once the subsystems are "plugged" together. The interfaces specify data types and formats, logical addresses, data rates and timing requirements. The ESO team that is in charge of the central control system coordinates the process of agreeing on interfaces among the various subsystems (or more specifically, their supervisors). They also specify the formalisms to define these interfaces; an XML schema is used here. If the "languages" were different for each subsystem one couldn't even find out whether any two behaviors would be compatible.

To further ensure compatibility between local control systems, the central control system and the supervisors, the central control system also provides infrastructure components for the development of supervisors, other high-level software components and instrument control systems. This Core Integration Infrastructure takes care of error handling, logging, alarms and configuration, and provides a notion of synchronous and asynchronous commands for interaction with the communication infrastructure. Client programs in C++, Java and Python are supported.

HIGHER-LEVEL SOFTWARE The higher-level scientific software that runs the instruments and schedules observations sits above the control systems and their supervisors. To make this simpler, the central control system "packages" the technical functions provided by the subsystems into user-facing features that are built around the use cases of scientists, instrument teams and technicians. Scientific systems are changed much more often than control systems, so the ease of writing and modifying them, ideally by scientists themselves, is more important, especially in the early years of a telescope. This is one reason why the astronomical community has a rich tradition of using a wide range of programming and scripting languages to write their observation software. Python is a favorite, but you can also see Java and Motif GUIs on the computers used by observers and scientists. For many existing telescopes it is often a diverse mix of tools, connected by a variety of standardized and custom data formats. At the ELT the software for core science workflows is built with C++ and Python, relying on the Qt library.

This software takes care of scientific proposal submission and refereeing, long-term scheduling, observation preparation, observation execution, night reporting, quality control, archiving, data reduction and archive exploration. Some parts are also available over the web; for example, observation preparation can be done programmatically using RESTful APIs.

SUMMARY The current and—even more so—coming generation of optical telescopes are true marvels of engineering. Like all such large-scale science experiments, they are also marvels of project management, especially considering how many different organizations, cultures and countries are involved. I experienced some of these challenges when I was involved with ALMA, which we look at later in this chapter. It's a little bit of a miracle that these systems do what they are imagined to do during the proposal phase. It should also not be surprising if they run a little bit over budget.

7 | *Radio Astronomy*

IN THE DESERT After about 90 minutes of driving through the New Mexico desert we are getting close to Magdalena, a little village with around 900 inhabitants. Those who have been here before told us that we really have to stop at the burger joint on Main Street, as they have the world's best chilli-burger. I suspected that this might be a case of strange-circumstances-make-boring-food-great, like when you eat a dry bun and a pepper on a mountaintop after a three-hour climb. But that thing was really good! After around 45 minutes we continued our drive west on US-60 through more scrubland. Fifteen minutes later, as the road makes a slight right turn and the scrub and rolling hills stop, we enter a large plain. After a few more minutes of driving under a big blue sky we can see the first radio antennas. A few minutes later we turn left onto NM 52S, and after another two and a half miles, finally, there's a small sign that reads "NRAO VLA". We have arrived at the Very Large Array.

The Very Large Array, since renamed the Karl T. Jansky Very Large Array, is a radio interferometer with 27 antennas, each 25 meters in size. You've probably seen it it before, with Jodie Foster kneeling in front of one of its antennas: it's where the movie "Contact" was filmed in 1997. Now it's 2002 and I'm here with a couple of colleagues from ESO and NRAO to visit a prototype antenna for the ALMA telescope array. We are spending the week at the Array Operations Center in Socorro, the nearest city, for a software architecture workshop. This is part of my three-year on-and-off consulting work for ESO in the context of the ALMA project. The Atacama Large Millimeter Array, to give it its full name, is a 66-antenna radio array located at 5,000 meters in the Chilean Atacama desert. Unfortunately I've not been able to visit it so far, but it's certainly on my list for future travel.

RADIO ASTRONOMY As we saw on our wavelength chart at the beginning of this chapter, radio astronomy encompasses a wide range of wavelengths, from kilometers to below millimeters. Some ranges are blocked by the atmosphere, which is why most Earth-based radio telescopes operate in the range between dozens of meters and tenths of millimeters. Which part of the spectrum they observe exactly is a function of dish size, geographic altitude and scientific interest, usually materialized through receivers and instruments. The VLA operates from 6 mm to 4 meters, or 50 GHz to 73 MHz; ALMA operates between 0.3 mm and 9.6 mm, or 31 to 950 GHz. The 100-meter telescope in Effelsberg operates from 3 mm to 70 cm, or 0.4 to 95 GHz.

Some of these frequencies are within the range of transmissions we create on Earth: you probably know that FM radio is broadcast between 87.5 and 108 MHz, VOR navigation beacons are directly above that, and civilian aviation radio communication is between 118 and 136 MHz. Amateur radio is on various bands in the 100s of MHz, wifi occupies bands between 2.5 and 5 GHz and radars use bands between 1 and 200 GHz. It is thus useful to build radio telescopes in areas where there are few such transmissions; the middle of New Mexico, on a plain surrounded by hills, seems like a good start. A desert plateau at 5,000 meters in the Chilean desert is even better, especially for wavelengths below a centimeter or so, because a (wet) atmosphere has limited transparency at these wavelengths (they overlap SOFIA's range). If you absolutely want to build a radio telescope in a densely-populated country such as Germany, the people who built Effelsberg did quite well: they chose a rural area in the Eifel, in a valley that uses the surrounding hills to give the telescope some degree of shielding.

Being in the range from tenths of millimeters to almost meters, the wavelengths in radio astronomy are much longer than the (hundreds of) nanometers in optical astronomy. As a consequence, the surface quality of the dish does not have to be quite as precise as with optical telescopes; the rule of thumb is that the maximum surface roughness, or divergence from the ideal shape, must be below a tenth of the observed wavelength. For Effelsberg this is around 0.5 mm. For ALMA, because of its shorter wavelengths, the mirrors are accurate to within 25 micrometers. These dishes are often manufactured from separate panels with visible boundaries between them. This in turn also means that one can build much larger antennas without resorting to actively controlled segmented mirrors, avoiding the resulting complexity. The 100-meter dish in Effelsberg is testament to this.

There are even larger antennas, such as the 305-meter Arecibo Telescope in Puerto Rico and the 500-meter FAST telescope in China. But those have a more or less fixed orientation; the dishes don't move. FAST is interesting. Like Arecibo, it is built into a natural hole in the ground. The surface is built from 4,450 panels, and 2,250 winches adjust the shape to form a parabolic antenna that can point to different positions in the sky in a relatively limited cone vertically above the telescope. A cable robot moves the receiver cabin to the corresponding focus above the parabola. This makes it a hybrid between a fixed and a movable antenna, with an effective size of 300 meters.

Why are radio telescopes so much bigger than optical telescopes? One reason is simply that it is possible to build them bigger. We've already mentioned that the precision requirements for the reflective surface are much lower than for optical telescopes because of the longer wavelengths. However, it is also necessary to build bigger antennas to get to the same resolution. The smallest angular resolution achievable with an aperture of diameter D is proportional to λ / D: the longer the wavelength, the bigger the smallest measurable angle. Making the dish bigger can compensate for this, but only to a degree. For example, for the LBT's eight-meter mirror the ratio is around 10^{-7} for visible light; for Effelsberg, at 10 mm wavelength, the ratio is 10^{-4}, three orders of magnitude worse, despite a 100 m dish! We know what to do about this: interferometry allows us to obtain resolutions that are governed by the distance between the involved antennas instead of their sizes: most modern radio telescopes are interferometers. We'll look at two of them, ALMA and the EHT, after we've first learned a bit about the 100-meter dish in Effelsberg.

WHY OBSERVE IN RADIO But first: why would you observe in radio? What is there to see? Lots of interesting things! I'll mention a few of them briefly, before we go back to the engineering of the telescopes. For example, certain classes of stellar objects are only detectable in the radio band. The pulsars we mentioned earlier in the context of gravitational waves are one example. Quasars are another; these are basically the accretion disks around black holes in the center of galaxies. More generally, radio galaxies, as the name implies, are visible in the radio band. They emit synchrotron radiation because they accelerate particles perpendicular to their velocity vector. Radio telescopes can also detect the remnants of supernovae, which are basically explosions of stars resulting either from a heavy star running out of "fuel" so that it collapses under its own gravity, or from a relatively light star that becomes a white dwarf and is then ripped apart due to the carbon fusion process that results.
Radio astronomy can also detect clouds of ionized gas roaming through the universe. It can see neutral atomic hydrogen, which in turn can help uncover the structure of galaxies. A picture of the night sky in the optical band is basically black, with discrete sources of light, mainly stars. In contrast, the radio sky looks more like crumpled aluminum foil, because one can see much of the "stuff" between the stars and galaxies.
The cosmic microwave background was also initially detected with Earth-based telescopes in the radio spectrum, and has since been mapped with NASA's COBE and ESA's Planck satellites. The cosmic microwave background is weak radiation created during the early development of the universe. Because of the universe's expansion it has been shifted to longer and longer wavelengths, and is detectable today mostly between one and five millimeters. Because of its ancient history, it has helped our understanding about the origins of the universe. Its discovery by Arno Penzias and Robert Wilson in 1964 was accidental: they first thought it was noise in their receiver, stray radiation from New York, or caused by bat and pigeon droppings in their antenna. After they eliminated all these problems and still detected the radiation they

gave up and talked to more senior astronomers, who suggested it might be the predicted background radiation. The rest is history—which includes a Nobel Prize for Penzias and Wilson.

8 | *The 100 m Telescope in Effelsberg*

I'VE BEEN THERE BEFORE I first visited the radio telescope in Effelsberg when I was maybe ten years old. My family was visiting my mother's relatives in the Rheinland region of Germany, and I guess my father decided that we might as well do something geeky while we were in the area. We went to see the monster excavators for brown coal and also checked out the telescope in Effelsberg. I don't remember much from the visit except the view of the telescope and the control building from the visitor platform. In 2019 I decided I'd redo this visit for omega tau.

The map below (from opentopomap.org, CC-BY-SA) shows the vicinity of the telescope. You leave your car at the parking lot (A) and walk to the telescope, slightly down into the valley that shield it from interference such as radars. Modern cars, with all their electronics, are a problem as well, which is why the car park is 800 meters away and higher than the valley. They also ask you to turn off your phones as you approach the telescope. As soon as you turn onto the path towards the telescope you can see its top peeking above the surrounding hills. There's a visitor center (B) and a viewpoint (C). Turns out I remembered correctly from 35 years ago: you see the telescope (E) right in front of you and the control building (D) to the left.

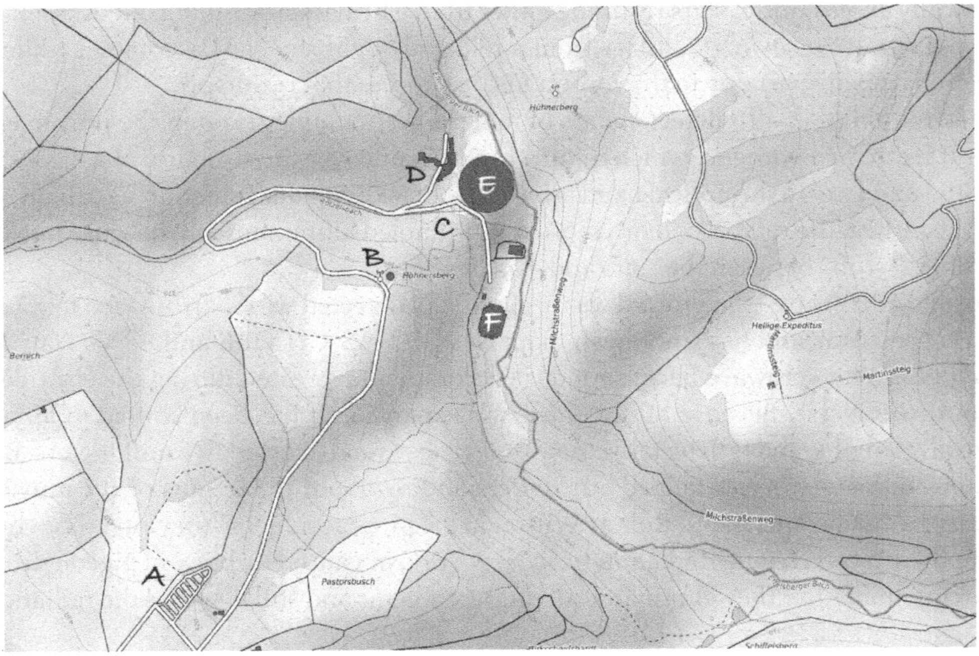

There were other reasons why this particular site was selected other than the shielding properties of the surrounding hills. This particular valley is open to the south, so the telescope can observe the center of the Milky Way at low southerly elevations—at least for around two and a half hours per day. Another reason was the composition of the ground: it is able to bear the 3,200 tons of the telescope.

MECHANICAL STRUCTURE The Effelsberg telescope was built between 1968 and 1971. At the time it was the largest movable-dish radio telescope in the world. The diameter of the antenna is 100 meters, and this is also the diameter of the entrance aperture (unlike at FAST, where the physical size is much larger than the usable aperture, as we saw before). Since 2000 the new Green Bank Telescope in West Virginia has held the world record: it has a 100 × 110-meter elliptic offset antenna, in which the receiver is outside the aperture so that it does not throw a shadow on the primary reflector. This instrument replaced the 90-meter telescope built in the early 1960s after it collapsed in 1988 due to failure of a structural member. You can maybe see a pattern here: US 90 meters, Germany 100 meters, US 100 × 110 meters. Apparently the radio guys are no different than their colleagues in the visible.

The dish is mounted on a rather beefy azimuth/elevation mount. The rotation in azimuth is facilitated by a central pin and four bogeys that travel on train tracks. Sixteen electrical motors of 23 hp each rotate the 3,200-ton telescope around the track. The elevation axis is supported on two sides and driven by a gear ring and four similar electrical motors. Each group of motors forms a backlash-free propulsion system (the motors work against each other) to ensure that the telescope can be moved to between two and three arcseconds of precision. The dish has a counterweight, otherwise 92 hp would not be nearly enough to move it in elevation.

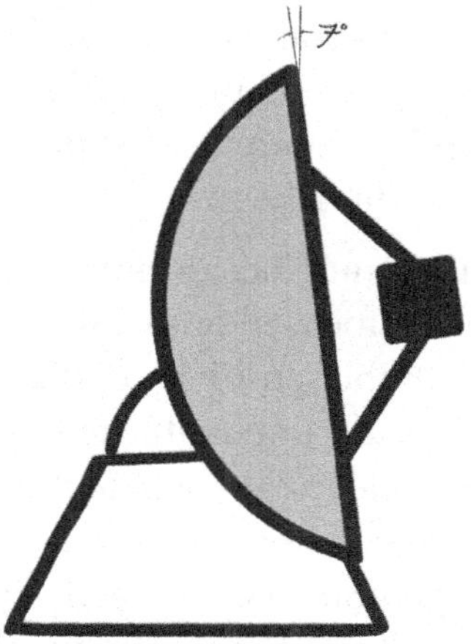

Standing directly under the telescope on the elevation platform is awe-inspiring. When the telescope is in its vertical position (seven degrees elevation is the minimum) it is more than

100 meters high; that's more than half as high as the typical altitude for glider's traffic pattern! A complex mesh of structural members of decreasing thickness holds the dish in place. When the telescope moves, which it does very slowly, it looks like some kind of industrial dinosaur.

When I visited in August 2019 the telescope was offline because of painting work. The observation schedule has slots reserved for painting and maintenance every summer, and the dish is moved to its horizontal position so that painters can crawl around inside it. Huge lifts are used to let them access the underside of the antenna. While it's probably not very varied occupation, working between steel beams 80 meters above the ground is certainly exciting! The same guy has been doing it for years, and it takes 18 years for every part of the telescope to be repainted. Repainting the structural members is crucial to prevent corrosion and ensure stability. Repainting the mirror panels, however, is important to maintain the precise shape of the mirror: darker surfaces absorb more heat and thus expand more.

THE HOMOLOGOUS MIRROR Effelsberg operates in a wavelength range down to 3 mm. This is because shorter wavelengths cannot be received at a location in central Europe at around 300 m above sea level—there's too much water vapor in the atmosphere above the telescope. This limit means that the required precision of the mirror is roughly 0.5 mm. Its surface is made up of 2,360 separate aluminum panels mounted on a steel structure. The lower limit of the frequency range, 300 MHz, or 100 cm, is determined by the fact that below this frequency radiation can be observed using grids of dipoles, which are much less expensive to install. The LOFAR array is an example of such a system, with 30 stations throughout Europe that are combined interferometrically. One such station is located right next to the 100-meter telescope in Effelsberg.

In our discussion of optical telescopes we saw that the size of the mirror is limited not only by manufacturing constraints, but also by the need to keep the shape stable as the telescope moves in elevation and gravity pulls in varying directions relative to the orientation of the mirror. The optical community is moving to segmented mirrors that actively control the required shape for this reason.

The radio community has a similar problem: if the shape diverges from a perfect parabola, radio energy isn't focused onto a well-defined focal point and thus escapes from the sensor. This causes both the sensitivity of the telescope and the sharpness of the image to suffer. As you can imagine, it is impossible to build a 100-meter dish rigid enough not to deform under gravity without making the whole thing impossibly heavy. In fact the Effelsberg mirror deforms by several centimeters as it moves from the vertical to the horizontal. To avoid these deformations, you can of course actively control the shape, just as optical telescopes do; this is how the Green Bank Telescope solves the problem. However, such active control technology was unavailable in 1970 when Effelsberg was built. They opted instead for a passive system that relies on a technique called homologous deformation.

Assume that the mirror has its nominal parabolic shape when it looks straight up. In this horizontal position the panels are adjusted with screws to achieve the nominal shape of the mirror, with at most 0.5 mm deviation from a perfect parabola, as shown to the right of the illustration below. This nominal parabolic mirror has a particular focal length, 30 meters in the case of the Effelsberg telescope. The receiver is held by arms at exactly this focal point to obtain a sharp image.

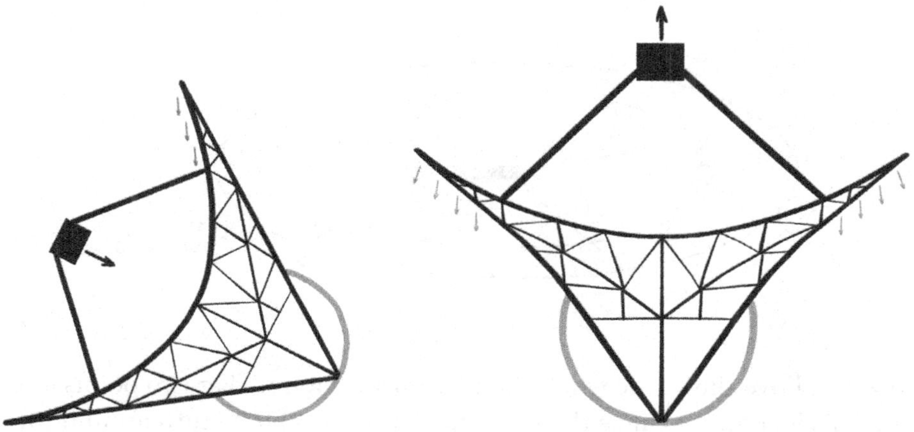

If you now swing the telescope down to lower elevations, as shown in the left part of the illustration, gravity deforms the mirror. Here's the cool thing: the homologous steel structure that backs the mirror has been designed, using the finite element method, to redistribute the forces in a way that ensures a parabolic shape of the mirror. It's a different parabola, however, so the focal length changes and the receiver has to be moved in or out. It's also not quite at the 0.5 mm deviation from the perfect parabola, but it still produces a reasonably sharp image.

Using calibration data obtained by observing a point source, the correct position for the receiver on the telescope's axis is determined for each elevation angle. No special equipment is required; they just move the receiver in and out and detect where the beam is sharpest. Once the receiver position is known for all elevations, the controller, when moving the telescope, will automatically adjust the receiver position during observations based on a lookup table. This works because the distortions due to gravity are repeatable.

Let's get an idea of how you design an homologous mirror. We'll start by observing how an idealized beam bends depending on where it is supported (we're talking about a mechanical beam here, not a beam of light). If you support it at each end, it will sag in the middle. If you support it in the middle, the ends will sag. No surprise! It turns out that supporting the beam at 63% along its length minimizes the average distortion from the ideal beam.

While a telescope's mirror is not a beam but instead a spherical sector, the same observation is basically true. So instead of supporting the dish at its outer edge, Effelsberg's is supported at around 63% of its diameter.

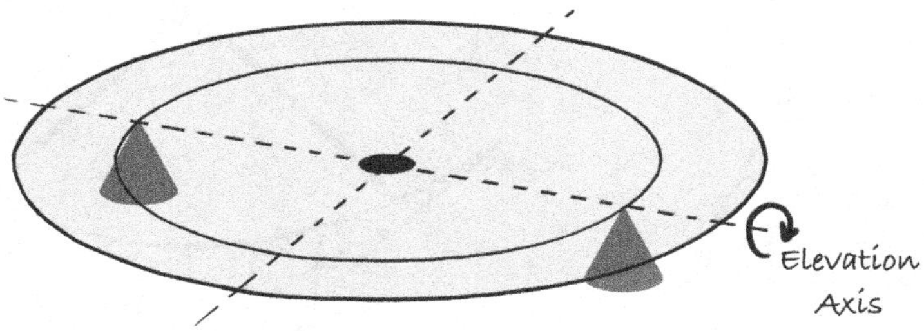

The illustration above shows the whole circular mirror mounted at two points along the elevation axis. This is intuitive, since the telescope has to be able to turn around this axis, and early radio telescope dishes were supported this way. However, the problem is that the dish will now sag along the axis that is perpendicular to the elevation axis. The solution to avoiding this deformation is conceptually simple—just add more supports around the circle:

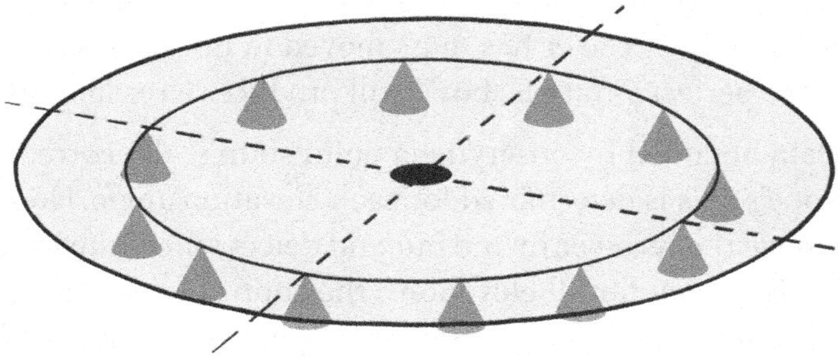

To achieve homology the forces from all 12 supports have to end up at a single point. How can you achieve this? One way is to build a pyramid of beams, almost like an inverted umbrella:

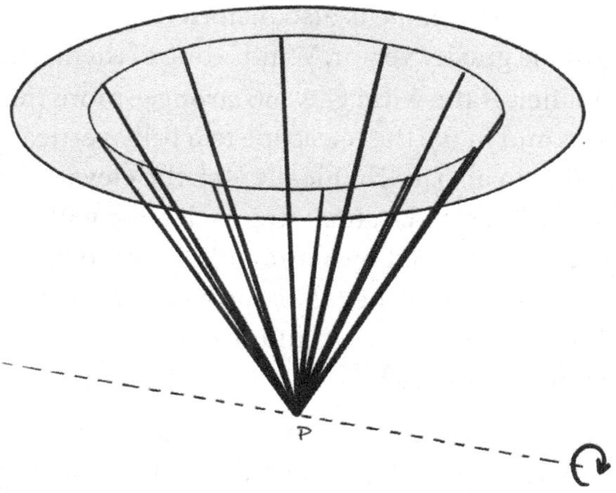

So far so good. But how do you mount the whole thing so that it can rotate in elevation? A naive approach would be to mount the whole telescope at the force concentration point, P. This would result in huge asymmetry on the elevation axis, however, and it would be very hard to control the telescope precisely in elevation (although telescopes have indeed been built this way in the past). So the final step in this design is to move the elevation axis closer towards the dish and mount a counterweight at P. To make this possible, you hang the whole construction from two massive beams, labelled H in the illustration below. The umbrella beams and the hanging beams do not connect anywhere except at P, so the umbrella "stands" on a platform held by the two H beams. The elevation axis itself is supported by two bearings (B) that are mounted on triangular stands (S). At Effelsberg, which pioneered this clever approach, those are in turn mounted on the platform that rotates in azimuth.

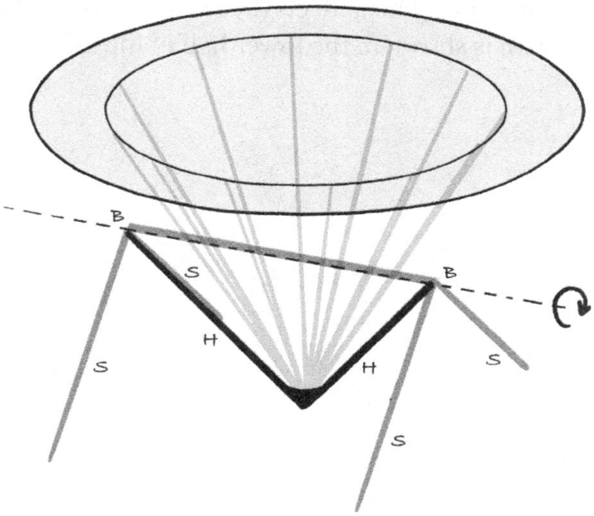

Let's zoom out again. The mirror's shape is also distorted by other factors that are not as repeatable as the influence of the gravity vector. Wind is one of them. However, these distortions or vibrations are much smaller. If the wind gets too strong—more than around 50 km/h—the operators stop observations and move the telescope to a fully vertical position with its narrow side towards the wind. They can also mechanically lock the elevation axis and ensure that the electrical motors don't bear all the load. A final step in dealing with wind loads is that the four outermost rings of panels are perforated by 6 mm holes to reduce drag, making them appear semi-transparent. This also means that the telescope has a correspondingly smaller aperture for short wavelengths. However, because resolution is proportional to λ / D, a large value for D is not quite so necessary for a smaller λ.

Weather is more generally an issue: thunderstorms are bad, because they might be accompanied by hail that could damage the panels in the mirror. Moving the telescope to the vertical with the steel structure towards the wind solves that problem; the mirror panels are pointed towards the lee side. Lightning might also strike the telescope and damage the receivers; again, a vertical position makes sure that, if lightning hits what is effectively a giant lightning rod, it hits the steel structure rather than the receiver cabin at the primary focus. Snowfall also requires a move to the vertical, otherwise the dish could collect snow and become too heavy. Clouds and normal rain are not necessarily a problem: water in the atmosphere does not prevent observations for low frequencies, such as the 21-cm hydrogen line. For higher frequencies the atmosphere has to be drier. The observation schedule needs some flexibility to be able to react to the changing weather in central Europe.

THE SIGNAL PATH How does the received radiation get to the instruments? After it hits the primary mirror, it is reflected to the primary focus. A cabin installed there (A in the illustration below) contains a receiver. Alternatively, the receiver can be moved out (it is movable anyway because of the adjustment to compensate for the changing parabola) and a secondary mirror can be inserted into the signal path. This reflects the radiation to another receiver cabin at the secondary focus (B); that path is shown in the lower half of the illustration.

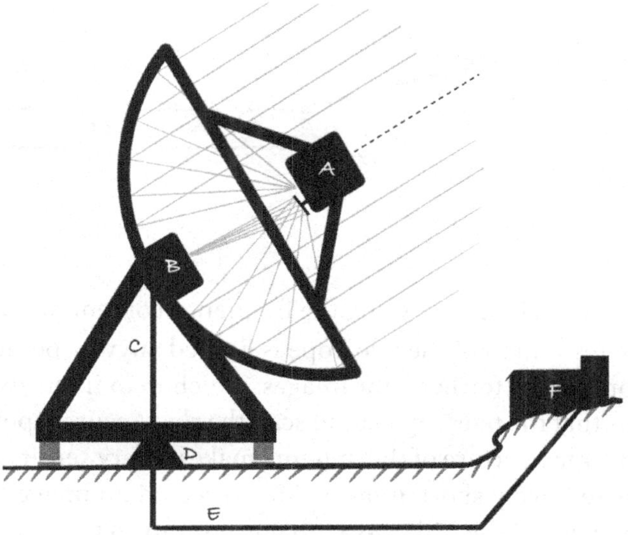

The receivers transform the radiation into an electrical signal. The electrical signal is then transported through a cable (C) to ground level and into an underground room through the telescope's central pin (D). From there the cables lead through a tunnel (E) to the control building (F), where the signal is analyzed, often by computers after digitization.

The single receiver installed in the primary focus cabin is typically that for the highest frequencies. This is because the secondary mirror has traditionally been too imprecise to be able to focus high frequencies well (another example of the dependency between wavelength and surface quality). In addition, manual intervention is required to exchange this receiver, which makes a change cumbersome. In contrast, the cabin at the secondary focus holds several receivers that can be switched from the control room. Moving the primary focus receivers out to enable the secondary mirror to be moved into the path (and thus forward light to the secondary focus) is also automated.

Most of the receivers are single-pixel: radiation is collected with a horn antenna that focuses it into a waveguide. A horn antenna is a gradually narrowing conductive structure that can collect radiation over a wide range of wavelengths. This is because it does not require resonant—and thus wavelength-dependent—elements: it gradually adapts the impedance of free space to the impedance of the waveguide. Waveguides allow a wave to propagate in one direction with minimal loss; the principle is similar to that of a glass fiber for light. Waveguides are used to transport the radiation to the receiver, which is typically installed close to the horn to allow the waveguides to be relatively short. We will later see how the actual receivers work in the context of ALMA.

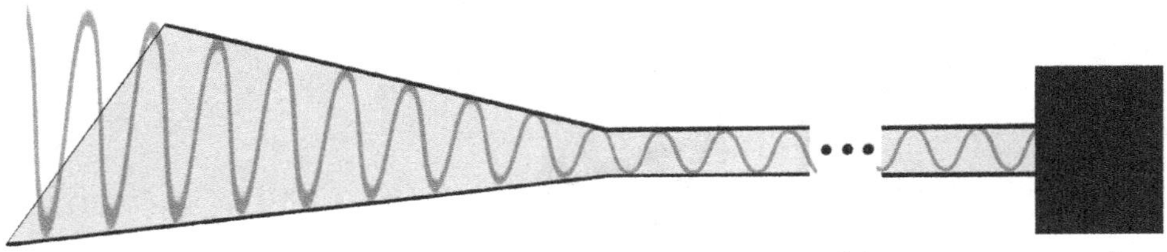

For spectroscopy, which is about analyzing the frequency content of radiation, a single pixel is perfectly fine—the resolution of the telescope is limited anyway because of the long wavelengths at which it operates. However, for images, which map intensity to position, a single pixel "camera" means that the telescope has to scan the sky. Accurate pointing and tracking is necessary. While the large aperture of the antenna makes it very sensitive and the integration time for each pixel is relatively short, it takes time to record an image this way. An extreme example is the radio image of the Milky Way that is proudly displayed in the Effelsberg control building: it took a total of 3,000 hours of integration time, spread over years.

In 2006 the Max Planck Institute for Radio Astronomy, which owns and operates the telescope, installed a new secondary mirror. This has an active surface driven by 96 actuators. It is used to correct imperfections in the parabolic shape of the main mirror as it moves under gravity, as the homologous deformation does not retain a perfect 0.5 mm-surface-error parabola. This has two advantages. First, the telescope becomes sharper overall, and it retains good sharpness at a wider range of elevation angles. This is because the combined optics of the primary and secondary mirror approximate to a much more perfect parabola than the primary mirror alone. The second advantage is that the receivers for the shortest wavelengths can now also be installed at the secondary focus. This allows them be switched automatically as well. These improvements make observations faster and the telescope more efficient as a whole.

OUTLOOK When I chatted with Norbert Junkes at the telescope he was optimistic that large single-dish telescopes like Effelsberg will continue to remain relevant, even though the radio telescopes developed over the last 20 years are essentially all arrays that rely on interferometry; ALMA is the prime example. He cited two reasons for his optimism. First, through incremental upgrades receivers become more sensitive and of lower noise, so the same dish can observe fainter objects. Relatively cheap upgrades (compared to building a new telescope) such as the active secondary mirror help too. Higher sensitivity also means that large dishes have a higher temporal resolution: because each snapshot requires relatively little integration time, phenomena that change over time can be "filmed" relatively well.

The second reason for his optimism is that the trend towards interferometry doesn't mean that large, single dishes become obsolescent. This is because it is possible to form interferometers from telescopes that were originally designed as separate; we will cover this astronomical technique in the last part of this chapter.

He was also quite confident that the structure of the Effelsberg telescope will not meet the same fate as the original Green Bank Telescope and will retain its structural integrity for years or decades.

9 | *The ALMA Array*

INTERFEROMETRY REVISITED Let's go back to the problem that radio telescopes require huge apertures to achieve reasonable resolution because resolution is proportional to λ / D, and λ is rather big. We now know how to solve this problem: interferometry allows us to obtain resolutions that are governed by the distance between the antennas involved instead of by their sizes. Interferometry in the radio band is in some sense easier, because the longer wavelengths and lower frequencies are easier to process. It also works differently. In optical astronomy interferometers "overlay" the actual light beams from the apertures onto a sensor and then analyze the interferometric "image". In radio interferometry, in contrast, the signals from the various antennas are interfered computationally, after they have been recorded and digitized. This is a big advantage with respect to the potential separation of the antennas involved: there is no longer a need for complex and size-limited delay lines like those in the VLTI: instead you can record and digitize the data locally at each antenna, then transport it over computer networks to a central computer that performs the "overlaying". Such computer is known as the "correlator". The maximum distance between two antennas at the VLA is 36 kilometers; at ALMA it's 16 kilometers.

Before we look at the engineering, let's recap how interferometry works. We use a slightly more mathematical approach to help us understand the correlator. This section is based on a series of YouTube videos made by Aaron Parsons, called "Introduction to Radio Interferometry".

The basics are of course the same as in optical interferometry: we observe the sky from two (or more) locations separated by a distance B; these are our two "slits". We know that the target we observe is far away, so we can assume planar wavefronts. This leads to the following situation:

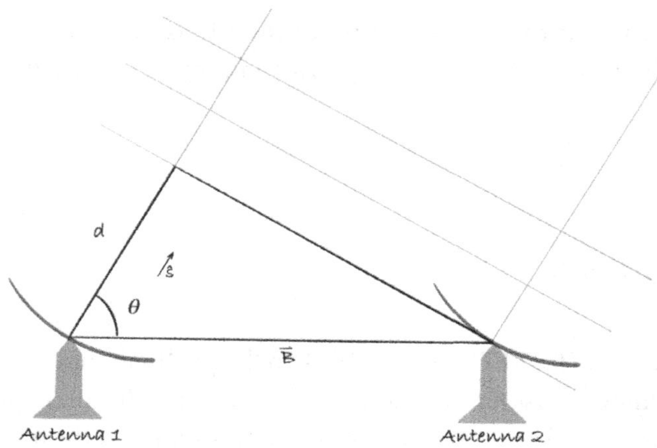

Because Antenna 1 is farther from the source than Antenna 2, the radiation has to travel the additional distance d to reach Antenna 1. We can express this distance as the product of the delay time τ and the speed of light c:

$$d = c\,\tau$$

The distance d also depends on the angle Θ at which we observe; the next diagram illustrates that d is longer for smaller angles.

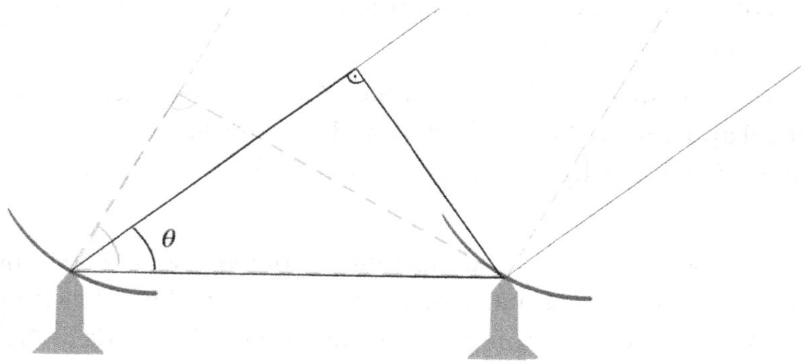

The dependency of d on Θ can be captured by

$$d = B\cos(\theta)$$

This projection is mathematically the dot product between two vectors; the cosine function is "implicit" in the dot product. The vectors involved here are the vector of the baseline B and the unit vector that represents the direction of observation, s. Using vector algebra and rearranging the equations from above, we can express the delay τ as

$$\tau = \frac{\vec{B} \cdot \hat{s}}{c}$$

Alternatively, we can express it in terms of the number of wavelengths. This is useful, because ultimately we are interested in the phase difference between the two beams, which depends on multiples of the wavelength:

$$N = \frac{\vec{B} \cdot \hat{s}}{\lambda}$$

The intensity of the received interferometric signal V for two antennas j and k at a frequency f is given by the actual source intensity of the sky S multiplied by the phase difference ϕ:

$$V_{jk}(f) = S(f)\, \phi$$

For a harmonic oscillator, the phase difference ϕ is as follows (you can either just take this as a given or Google for complex exponential signals and Euler's formula).

$$\phi = e^{-2\pi i\, N} = e^{-2\pi i \frac{\vec{B} \cdot \hat{s}}{\lambda}}$$

Note the i in the exponent; this term is a complex number, but we'll not dwell on that here. Let's take a closer look at the baseline vector B: like any vector, it can be broken down into components along the axes of a coordinate system, say east/west, north/south and up/down on the sky. We can also express the baseline length as multiples of the wavelength. Again this is useful, because interference depends on the phase difference. So we get

$$\frac{\vec{B}}{\lambda} = (u, v, w)$$

where u is the east/west component, v is north/south and w is up/down, each expressed as multiples of λ. The source direction, the angle at which we observe, can also be expressed as a vector with three components: l, m and n. Because we use it only as a direction, we know this is a unit vector with length one, and we can use the Pythagorean theorem to express n in terms of l and m:

$$(l, m, \sqrt{1 - l^2 - m^2})$$

The number of wavelengths of the phase difference can now be expressed in terms of these componentized vectors, as the dot product produces a scalar value from vectors by summing the products of their components.

$$\frac{\vec{B} \cdot \hat{s}}{\lambda} = (ul \ + vm + w\sqrt{1 - l^2 - m^2})$$

This allows us to rewrite our previous equations for phase

$$\phi \ = \ e^{-2\pi i \frac{\vec{B} \cdot \hat{s}}{\lambda}}$$

in terms of these components as well:

$$\phi = e^{-2\pi i \left(ul + \ vm + w\sqrt{1 - l^2 - m^2}\right)}$$

We now have all the ingredients to write a formula that gives the response of an interferometer V_f (which depends on the frequency) for each point in the sky, given by its coordinates l, m and n, for a particular baseline defined by the components u, v and w. We assume that the two antennas are at the same altitude, so the up/down direction can be ignored; this allows us to drop the w and n terms:

$$V_f(u, v, w) = \iint A(l, m) \ \ I_f \ (l, m) \ \ e^{-2\pi i (ul \, + \, vm \,)} \ dl \ dm$$

Despite this simplification it's still a big formula. Let's understand intuitively what it represents. First we integrate over the whole sky by "infinitesimally iterating" over l and m, the componentized direction of observation; this is what integrating over dl and dm means. Inside the integral we multiply three quantities. I is the actual intensity of the source in the sky at position l, m. A is the main beam receiving power, a characteristic of the antenna that also depends on the angle of the target. The product of A and I is basically the brightness of the sky measured with a given antenna. The third quantity is the complex phase term from above. We need this because the angle-dependent phase difference between the antennas separated by the baselines vector (u,v,w) "modulates" the brightness of what we receive—remember the Young slit experiment! The phase difference is also frequency dependent. We don't see that in the formula, but u, v, and w were defined in terms of the wavelength at which we want to observe.

Why go through all this mathematical detail? If you are familiar with calculus you might recognize the shape of the equation for the visibility V above as a Fourier transform. The formula lets us see mathematically that an interferometer takes the true intensity I, "mixes" it with the characteristics of the antenna A, and then transforms the result into a representation of the sky in the spatial frequency (Fourier) domain. It transforms the intensity at coordinates l and m into their Fourier counterparts u and v which, if you remember, represent the direction and length of the baseline. You may remember the Fourier transform as transforming a signal from the time domain (a classical sine wave that represents intensity over time) to the frequency domain (a spectrum that captures relative intensity versus frequency). However, it can also be used in the space domain, where it detects the spatial frequency contents of "waves" in space.

Let's understand the phase term. This takes the baseline and the observing direction as parameters. Assuming a particular fixed baseline, the phase term depends on the direction: the relative phase difference between the two antennas that make up the baseline depends on the direction, just as we expect. Another way of expressing this is to think of each baseline projecting a "sampling sine wave" onto the sky at a frequency and orientation that depends on the baseline. These sampling sine waves are effectively used to detect the spatial frequency content of structures in the sky. The longer the baseline, the higher the spatial frequencies that can be resolved, because the sampling sine wave becomes of higher frequency; and according to the Nyquist sampling theorem, a higher sampling frequency lets us sample higher signal frequencies.

So what are we going to do with the Fourier transformed sky? When applying the Fourier transformation to a signal in time (or space), there is no loss of information: the representation of a signal in the frequency domain is equivalent. This is why the transformation can be reversed. So: our interferometer "measures" the sky in the frequency domain, and we can, in principle, use an inverse Fourier transform to recover the image (its representation in the space domain).

However, there is a catch: an inverse Fourier transformation will only reproduce the original signal in the space domain (aka the image) if we have sampled data for all spatial frequencies in all directions. But practically, with a limited set of baselines, we can never fully sample the frequency domain. The abstraction that describes which baselines we did sample is called the u,v-plane. Consider the following array of four antennas:

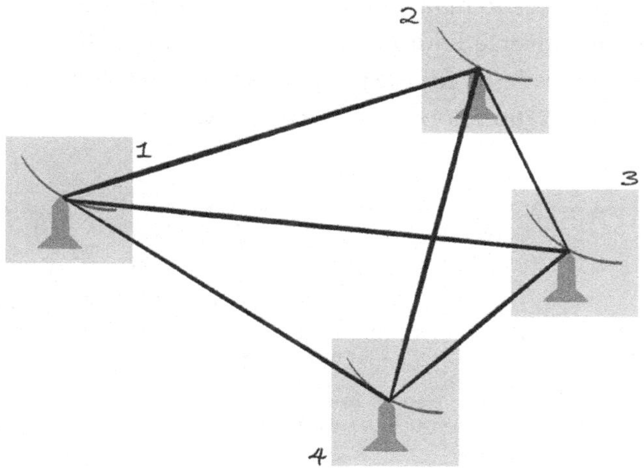

Four antennas give us six baselines: B12, B13, B14, B23, B24 and B34. We also get the reverse ones, but they don't contain additional information so we'll ignore them for convenience. Let's draw a diagram that plots the lengths and orientations of each of the baselines vectors, using u as the horizontal and v as the vertical axis. We basically move each of the vectors so that they start at a common origin, the u,v-center.

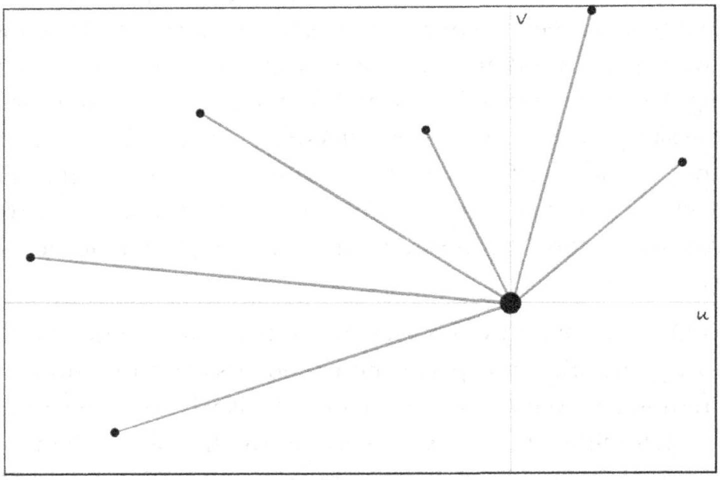

Each point in the u,v-diagram represents data captured for this particular baseline length and orientation. Note that this is not some representation of the image—it's a visualization of the completeness of our data. To be able to reconstruct the actual image we'd have to completely fill this plane. We can either achieve this through more baselines (that is, more antennas), or we can just wait and exploit the rotation of the Earth. However, this only changes the orientations of the baselines, not their lengths, leading to characteristic arcs in the u,v-diagram.

In practice we will always have to make do with a partially filled u,v-plane, so we cannot just perform a straightforward inverse Fourier transformation to get to an image. So what do we do? We could assume zero visibility for the u,v-points where we don't have data, then perform an inverse Fourier transformation nonetheless. Here is what we get:

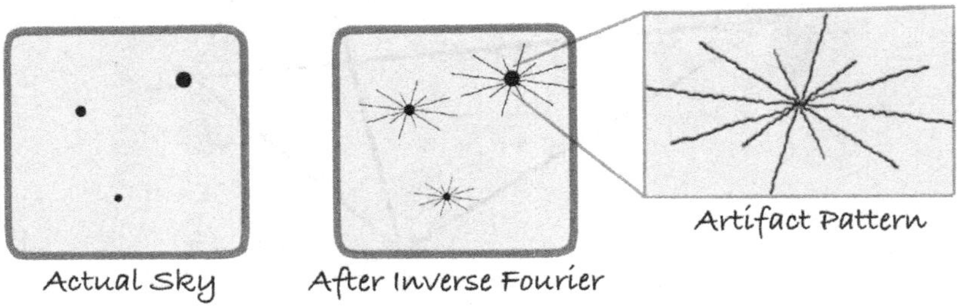

Assuming the sky has the intensity distribution shown in the left image (three point sources of different size), the inverse Fourier from the incomplete u,v-plane will give us what's known as the "dirty" image, in which each point source has small artifacts around it known as sidelobes. If you look at the sidelobes closely you will notice that their shape looks similar to our baselines. Indeed this is the case: another way of understanding the Fourier transformation of the sky performed by an interferometer is that it convolutes the intensity distribution of the sky with the baseline pattern. The convolution theorem tells us that if we take the Fourier transform of the intensity of the sky I multiplied by our sampling pattern P, this is equivalent to the convolution of the intensity with the Fourier transform of the sampling pattern.

$$\mathcal{F}[I \cdot P] = I * \mathcal{F}[P]$$

$$= I *$$ 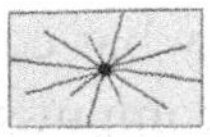

Without going into the details of what convolution is, the process basically "imprints" the structure of the sampling pattern (also known as the "dirty" beam) onto each point of the image, producing the "dirty" image. So the artifacts visible in the dirty image are not random: they relate to the lengths and orientations of the baselines of the interferometer. As we will see, we exploit this when we reconstruct the image. We might also have prior information about what we expect to see; in the example above, for example, we might know that we expect point sources. Or when we image neutral hydrogen in a galaxy, we might expect to see more or less continuous areas of gas. We can now take these three ingredients—the dirty image, the known dirty beam and the assumed prior knowledge—and reconstruct the actual sky:

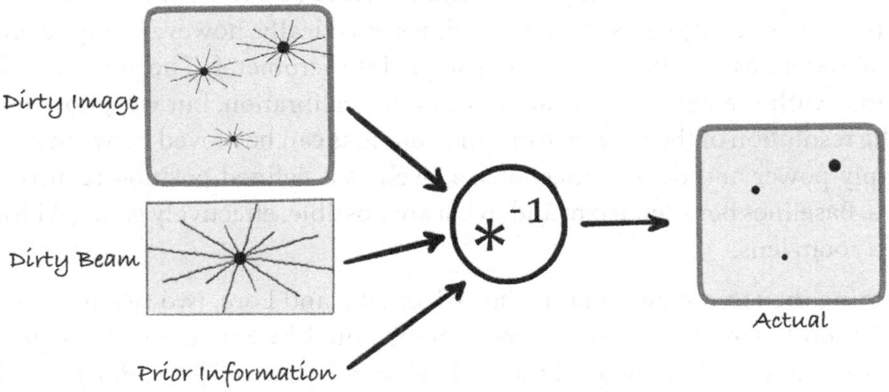

This process is known as "deconvolution". Deconvolution is an example of an inverse problem, one for which there is not a closed analytical solution. Instead you are basically "searching": you're trying to find input data (the real sky) which, after it has gone through the interferometer's sampling process, would look like the dirty image. The challenge is that there might be several such inputs, including the dirty image itself! You need the prior information to be able to decide which of the many possible inputs is the correct one.

Going with our assumption that the observed sky consists of point sources and is otherwise dark, we can use an algorithm called CLEAN to remove "dirty" artifacts from the image. This is an iterative process. First you guess where the point sources are, based on the dirty image. You will probably start with the brightest locations; finding these is a trivial image processing problem. Next you compute the Fourier transform of the image to get the true (full, complete) u,v-plane. You then mathematically sample this with your antenna pattern to acquire a new dirty image. Because this more closely resembles the real sky, it will have fewer sidelobes. Then you iterate until your image is sufficiently CLEAN. Other algorithms for other (assumed) image structures are available: for example, if you know that you're observing nebulas, gas clouds or other continuous areas of emission, you could use an algorithm that tries to fit the smoothest sky that is consistent with the observed data. Since the invention of CLEAN in 1974 the radio astronomy community has collected a whole catalog of such algorithms and their (efficient) implementations on computers.

THE ARRAY Now on to ALMA. The Atacama Large Millimeter Array is the world's largest and most sensitive telescope (array) at millimeter and sub-millimeter wavelengths. It has a total of 66 antennas, 54 of 12 meters diameter and 12 of seven meters. Larger is generally better, of course, because it offers more collecting power for a given integration time. The reason for the different sizes is that smaller antennas can be packed more closely to help image sources that are wider in angular extent. If I remember correctly from my involvement back then, the smaller antennas were not planned from the beginning: they came into play later when the Japanese joined the project.

The 54 antennas of the main array can form 1,431 baselines. If the 12 smaller antennas are included, this theoretically sums to 2,079 baselines. Practically, however, only 64 antennas, with their 2,016 baselines, can be joined into a single interferometer. The antennas also create 64 correlations with themselves; they are relevant for calibration, but we ignore those here. To adjust the resolution of the overall array, the antennas can be moved between a set of 192 pads that supply power and data connections, as well as a defined position relative to the other antennas. Baselines between 150 m and 16 km are possible, effectively giving ALMA the equivalent of a zoom lens.

Transporting the 120-ton antennas is the task of Otto and Lore, two 130-ton 20 × 10 × 6 meter 28-wheel 1,700 horsepower transporters, manufactured by Scheuerle. As is characteristic of Scheuerle's heavy load carriers, all their wheels can be steered and adjusted in height separately, leading to a precise and smooth ride for the delicate cargo—albeit not a very fast one: their maximum speed with an antenna is 12 km/h. The transporters grab the antennas in a way not unlike modern aircraft tugs "embracing" an aircraft's nose gear, and position them on their pads with millimeter precision. At this stage the transporter is controlled externally, allowing the operator to visually check alignment with mechanical guides on the pad that force the antenna into a precisely known position. The antenna is then attached to the base with six M70 bolts, locking it into a position that is precise to around a millimeter. This is good enough for interferometry, considering that coordination in time is much more precise and actively controlled, as we will see later.

The VLA has mobile antennas as well, even though they are twice as large and heavy as ALMA's. VLA's transporters travel on rail tracks that run for 40 miles. They are laid out as double tracks so that the transporter runs on two parallel, twin-tie tracks at any time for additional stability, resulting in around 80 miles of "regular" rails for the VLA. This is a non-trivial railroad network, so they have their own tamping machine and other maintenance equipment. For the rail geeks, there's an additional quirk to "VLA Rail": it has 90-degree turnoffs. For example, each antenna stand is accessed by a short stretch of rail, 90 degrees off the main line along the VLA's Y-shaped axes. This creates two challenges. First, how can you have crossing rails without them mutually blocking the transporter's wheels as they drive on the other rail? The main line is a normal, continuous track, but the 90-degree antenna tracks are segmented so that they do not block the main line. They are also a little higher. Whenever the antenna track is used by the transporter, the gap is closed by a bridge; because the antenna track is slightly higher, the bridge goes over the main line tracks.

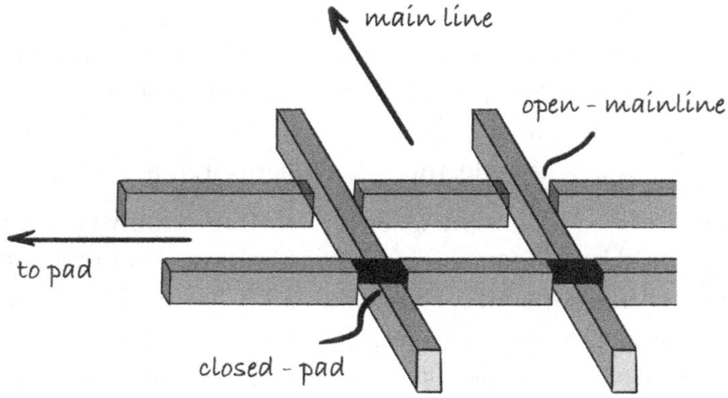

How does the transporter do the 90-degree turn? It drives over the crossing, lowers hydraulic stanchions, lifts each of its four bogies, rotates them through 90 degrees, then lowers them onto the 90-degree offset track. Clever!

Back to ALMA. The facilities are spread across two sites. One is the Array Operations Site on the Chajnantor plateau at 5,000 meters, which hosts the antennas and the ALMA Technical Building. The Operations Support Facility that hosts the control rooms, the "hotel" and antenna maintenance shops is located at 2,800 meters. It's also Lore and Otto's job to transport antennas along a 28-km mountain road between these sites that has a maximum gradient of 10%. Especially robust brakes had to be installed for downhill transport, and the 28 separate tires provide redundancy should one fail. You really don't want to lose the precious cargo because the transporter has a problem.

The reason for the lower site is that staying at 5,000 meters altitude for extended periods of time is tough for untrained people. In fact Lore and Otto's driver's seats are designed to allow a driver to carry an oxygen tank on their back! Some of the horsepower installed in Lore and Otto vanishes at the higher site as well.

While rain and other forms of humidity are rare—the main reason for the high desert location—the weather at the higher site can still interfere with observations, as evidenced by this report from the European ALMA regional center in 2017:

The ALMA Observatory has experienced back to back severe winter storms, making it difficult to recover the 12-meter array for PI observations. This has had an even more detrimental impact on the relocation to the long baselines. Many roads are blocked with two meters of snow, high winds often return snow to cleared locations, and snow is compacting into ice. We have returned to routine observations utilizing the 7-meter and Total Power Arrays, and we are working hard to clear the remaining antennas to return to 12-meter array PI science and reconfigure to the longest baselines. We will issue an update regarding the revised configuration schedule in the coming weeks.

ANTENNAS Like all large science projects, ALMA is an international collaboration, with contributions from different countries and their respective astronomy organizations. These contributions are often financial, but also in kind. You can probably guess who contributed the two transporters from their names: Otto and Lore. They are also a good illustration of why countries contribute in kind: Scheuerle, their manufacturer, is a German company.

The main components of ALMA are the 66 antennas; if one country and manufacturer had provided them all it would have been one huge contribution. That was not feasible; as a consequence there are four different kinds of antennas. ESO designed and built 25 of the 12-meter antennas with a consortium formed between Alcatel Alenia Space France, Alcatel Alenia Space Italy and MT Aerospace. Another 25 of the big antennas were contributed by the United States' NRAO and their manufacturing partner, VertexRSI. The ALMA compact array, with its four 12-meter and 12 seven-meter antennas, is managed by the Japanese NAOJ and was built by Mitsubishi Electric.

To ensure that all antennas are to the same standard, the ALMA consortium specified detailed requirements. Their surfaces must be accurate to 25 micrometers, to support operations to up to 1 THz, the pointing accuracy must be 2 arcseconds, tracking must be accurate to 0.6 arcseconds, they must be able to cope with the temperatures, wind and the occasional snow at the Chajnantor plateau without an enclosure, they must be compatible with the mounting pads, and they must conform to standardized interfaces to integrate the receivers and calibration equipment, which are the same for all antennas. While these requirements led to relatively similar designs, the various kinds of antennas do look different, even those of the same size; for example, the way the front-end cabinet is mounted over the dish is different between the

European and American antennas. The need for transportability also led to specific requirements; for example, the antennas must have standardized transportation mounts and weigh not more than 100 tons. Antenna mobility also mandated an especially beefy design: it is crucial that the optical path is not deformed as a result of repositioning. The requirements were quite detailed. For example, one specified that:

> ... the antenna should have a non-repeatable/repeatable path length variation of at most 15 / 20 µm, respectively, within the time between expected consecutive calibrations of the interferometer array (15–30 minutes).

A variation in the length of the optical path will of course impact the phase of the signal received by an antenna as seen by the correlator, and therefore must be carefully controlled. Calibration can compensate, of course, but you can't recalibrate all the time. So another driver for beefiness of structure came from this and similar requirements: wind pressure, thermal expansion and the changing gravity vector when the antenna moves in elevation must not have too much of an effect on the optics. Several observation modes require regular switching between the target and the calibration source, and so quick moves with relatively high accelerations must not "bend" the antenna.

The prototype antenna we visited in the New Mexico desert in 2002 was one of two, one for each of the 12-meter antennas. These were used to validate both the antenna's observation performance and its mechanical properties. Laser detectors were installed to measure how components changed their position and angle, and to measure length changes precisely. The dish was then moved in elevation and azimuth to discover whether accelerations would change its geometry. Around 100 sensors were installed to measure temperature distribution in the system and to correlate temperature change with changes in the optical path. The antennas fulfilled the requirements for relative stability over typical integration times. While wind speeds below the specified limit do not materially change the path length, temperature does in the longer term. For example, the path length could change up to 200 micrometers during a day. However, the correlation between path length change and a small number of temperature measurements was strong enough to be used as a basis for forecasting models.

Building the antennas must have been quite a challenge. I remember that the software I was involved in back then was late, and we joked about the fact that nobody would notice because the antennas were even further behind schedule.

RECEIVERS Aside from the visible part of an antenna—the dish and the frame—the real magic lies in the receivers. These are located in a cabinet located above the dish at the antenna's primary focus. The cabinet is roughly one cubic meter and cylindrical; most of it is evacuated and cryogenically cooled to 4 Kelvin to minimize thermal noise. ALMA can observe between 31 and 950 GHz, with a few gaps where either the atmosphere is opaque or the receivers are lacking. Because the receiver technologies are slightly different for different frequency ranges, the overall spectrum is subdivided into ten bands. The cryostat contains a modular receiver cartridge for each of these bands, each of a different size and with different feed horns and/or mirrors. This modularity allows for stepwise installation during commissioning, and will allow independent upgrades of each band in the future. Only one band observes at a time, but up to three receivers can be powered on; if more were switched on the cooling system would be overwhelmed. Alternating between receivers that are already powered on is achieved in a matter of seconds; powering up an additional receiver takes about a minute, plus five minutes for thermal stabilization. The subsystem that schedules observations optimizes the sequence to ensure a minimum number of receiver swaps and associated cooling delays.

The next illustration shows a schematic of how the receiver cartridges are installed in the plane of the receiver; the numbers represent different bands. In this illustration radiation, as reflected from the dish, is moving into the plane of the paper. The antenna has to be offset a predetermined amount for each receiver to ensure that the focus lies exactly at the entrance window of the particular cartridge.

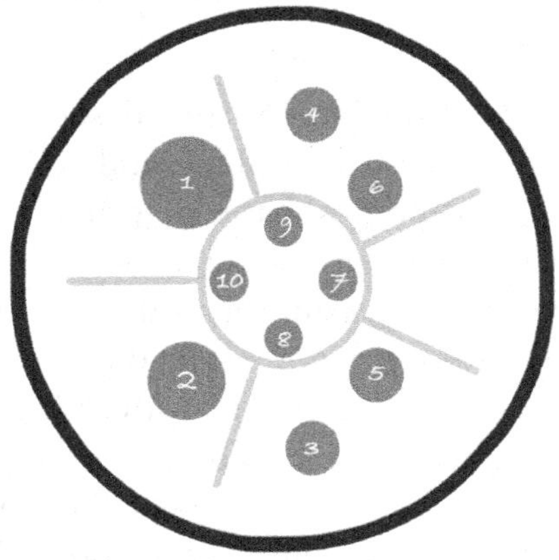

It is important to understand that a single ALMA antenna is not an imaging sensor; it does not pick up structure within the radiation it receives. Instead, its single pixel signal is used for interference with its partner antenna at the other end of each baseline, together forming a single point in the u,v-plane. Structure is recovered from the interferometric combination of all baselines by deconvolving the u,v-plane. In that sense, every antenna's signal contributes to all pixels in the final image.

The frequencies of the received radiation—hundreds of gigahertz—are hard to work with directly, as all clock rates and processors would also have to operate at these high frequencies. To avoid this problem, the receivers apply the heterodyne principle, in which the received signal is mixed with a reference to reduce it to a frequency that is easier to deal with downstream. In the case of ALMA this frequency is the range between 2 and 4 gigahertz—the same bandwidth but at a lower absolute frequency band. The reference frequency that is mixed with the raw signal must be tuned specifically for the specific frequency at which the scientists want to observe to map the particular received frequency (band) to that same 2 to 4 GHz. This process is not unlike an AM radio: turning the dial adjusts the reference frequency of its heterodyne receiver to select the frequency of the station to which you want to listen.

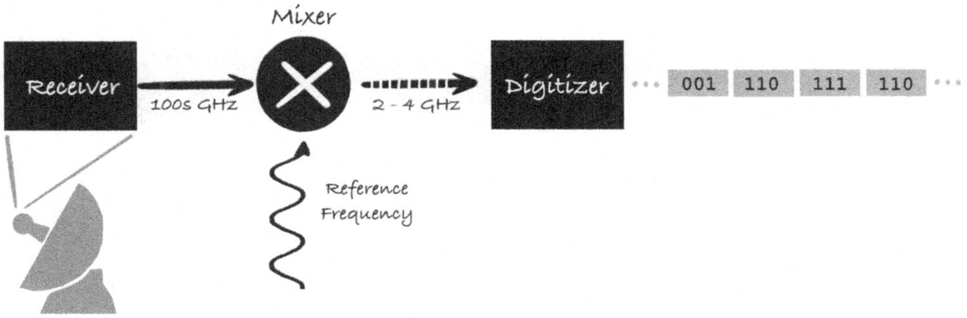

Once downshifted, the 2 to 4 GHz signal is amplified and then digitally sampled using three bits, resulting in eight levels of intensity, much like a gray-scale image with eight levels of grey. This relatively low number is a compromise. More bits would be better, because it would remove quantization noise, more accurately reflecting the actual intensity on sky. But more bits would also lead to more data, and the data rate into the ALMA correlator is already at the maximum of what was technically feasible when it was built, almost two decades ago. Future upgrades are looking at adding more bits as technology improves. Ultimately the received signal is very weak and its dynamic range low, so a limited number of bits is not a significant limitation.

It is also important to understand that the three bits refer only to each individual sample; in the correlator these three-bit signals are integrated over time into a 32-bit number. You might ask how this makes a difference if we assume the sky to be constant during any particular observation? The integrated value will only be a multiple of eight, so surely we don't gain anything? That's not the case, however, because every measurement is noisy for all kinds of reasons, the atmosphere being the primary contributor. So there is a lot of variability in the three-bit signal; the distribution of random noise during an integration is expected to be Gaussian. The "value" of the integration is represented by the peak in the distribution, and the width of the peak tells us something about the statistical significance of a particular integration.

CORRELATOR ARCHITECTURE The heart of ALMA is the correlator, a special-purpose supercomputer that performs multiplication and integration for all baselines and for the whole sky. It basically runs discretized calculations of the big formula we had seen before:

$$V_f(u,v,w) = \iint A(l,m)\ \ I_f\,(l,m)\ \ e^{-2\pi i(ul\,+\,vm\,)}\ dl\ dm$$

The diagram below shows the context in which the correlator does its work:

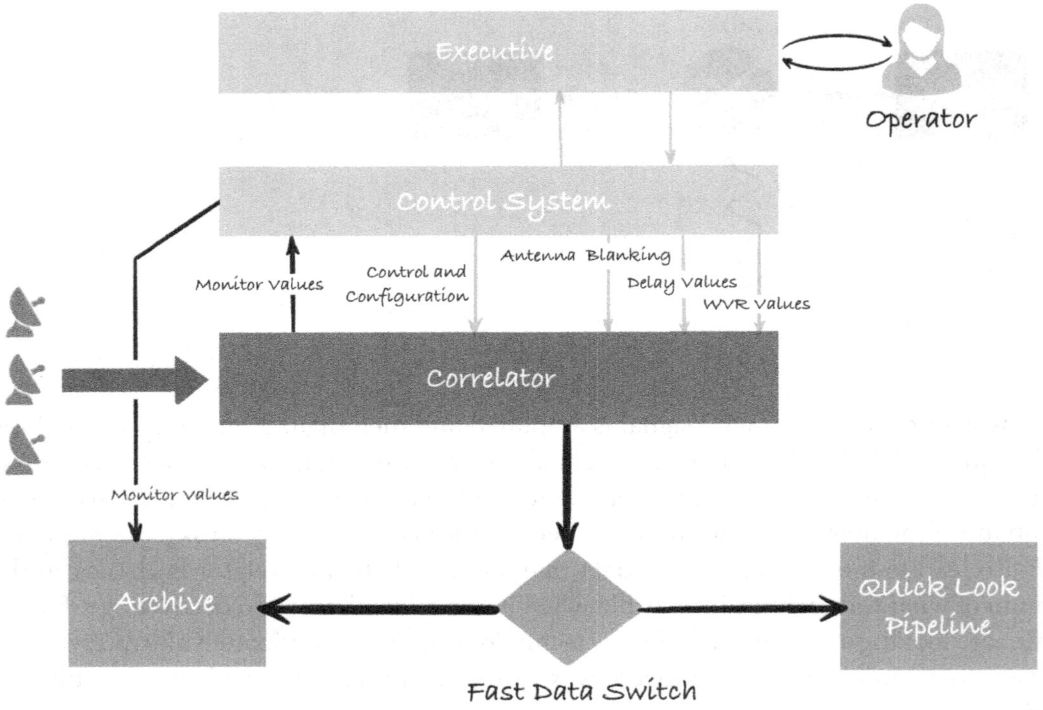

The correlator receives the digitally-sampled data from the antennas, performs its calculation in real time, then outputs the results to the archive for later scientific processing. It also pushes the data to the quick look pipeline, which provides a near real-time view of a lightly processed data stream for plausibility checking and quality control.

The control system is in charge of running observations. It controls the antennas and their receivers, and also sets up the correlator for each particular observation. ALMA can be divided into subarrays, and the correlator has to be told about this segmentation. In addition to sending their received data, antennas also report their technical status to the control system. If the antenna or the control system detect an error with an antenna, the control system reports this to the correlator, which marks the data for the baselines that involve this antenna as invalid. A related problem occurs at low elevation, especially if antennas are arranged in a compact configuration: one antenna might shadow another, rendering received intensities bogus.

The control system is also responsible for monitoring the correlator computer itself and storing the monitor data in the archive, as it might be relevant for the science pipeline that conceptually follows downstream from the archive. The final data is stored in the form of data cubes: two dimensions for the position in the sky, one dimension for the frequency, and the value at these three coordinates to represent the intensity. Various metadata, including the time of the observation and the setup of the observatory, complete the data files. Like every other astronomical archive, the ALMA science archive is accessible to the public, at https://almascience.eso.org/asax/

As the Earth rotates the antennas move in azimuth and elevation to keep track of their target. Keeping the antennas locked onto a target is one of the tasks of the control system. However, the rotation also means that the relative geometry of the antennas with respect to the target—and with it the delay values between the two antennas involved in each baseline—change as well. The control system computes the geometric delay for each baseline and reports it to the correlator continuously to enable it to calculate the correlation correctly; we'll discuss below why this is crucial. The antennas also move physically from pad to pad during a configuration change; at the time of writing this typically happened roughly every three weeks. There's a configuration database in the control system that tells the software which of the fiberoptic inputs from the pads actually contain antennas, and thus have to be used in the correlation. The geometric delays are also sent to the digitizer, which injects a delay per antenna to compensate for the time it takes the data to travel along the fiberoptic cable to the correlator, which is different for each antenna pad.

The delay seen by an antenna also depends on the atmosphere, and in particular on the water vapor content along the antenna's line of sight. As antennas can be up to 16 km apart, their local water vapor conditions can be quite different. Radiometry receivers installed on each antenna perform the measurement. Together with separately measured pressure and temperature, a model is built that predicts atmosphere-dependent phase changes. These have to be taken into account by the correlator; the control system supplies these values once per second. The correlator also corrects various errors that are a function of the signal's digitization process, bandpass filtering and the structure of the correlator hardware itself.

Finally, the executive is the high-level program that runs an observation. This contains the scheduler, as well as subsystems used by the telescope operators to control an observation.

CORRELATOR COMPUTATIONS ALMA's antenna design and receiver system allow it to observe between 31 and 950 GHz. Based on the science requirements for any particular observation, the receivers are tuned to four base frequencies at which to actually record data. Only one receiver cartridge is active at a time at each antenna, so for a given subarray this has to be the same receiver in all antennas. Say we select the receiver for band 7, from 275 to 373 GHz, these four frequencies could be 280, 300, 330 and 370 GHz. ALMA can collect data in a 2 GHz band around each of these base frequencies—for example 298 to 302 GHz or 368 to 372 GHz.

It does this for two different polarizations, resulting in eight signals from each antenna—four base frequencies with two polarizations each—at a bandwidth "around" the base frequency of 2 GHz. So we get a total data bandwidth of 16 GHz per antenna. Each signal is digitally sampled with three bits at twice its frequency, 4 gigasamples per second. This is a consequence of the Nyquist–Shannon sampling theorem that says that, to sample an analog signal adequately, the sampling rate has to be at least twice the highest sampled frequency. Ultimately therefore the total data rate per antenna becomes 96 gigabits/s. That's a lot!

During an observation the antennas produce data at this rate continuously and the correlator has to process it in real time—it has no capability to store raw data from the antennas. The fundamental integration time is 1 ms. So every 1 ms the correlator computes the data for one point in the u,v-plane. It then takes many such integrations to fill the u,v-plane sufficiently to enable meaningful image reconstruction. Depending on the details of the observation and the geometry of the array configuration, it might take several hours before the Earth has rotated enough to provide sufficient u,v-plane coverage.

To keep up with the stream of data from the antennas, the correlator has to perform on the order of 10^{16} million integer operations (MIPS) per second. For comparison, modern desktop computers can manage between 10^5 and 10^6 MIPS. It's obvious then that the system was not going to be built from off-the-shelf general-purpose processors (especially 20 years ago): you would need too many, which would make the system expensive. It is also not a good idea conceptually: the correlator's job is very specific and is not expected to change much in the future: it's basically a giant multiplication machine. It was therefore much cheaper to use special-purpose hardware.

A key observation that guided the correlator design is that, in principle, the correlation for each baseline and for each frequency band can be done separately, without dependencies on the other baselines. This means that the overall computation can be parallelized; the ALMA correlator exploits this on several levels.

At the top level the correlator is broken down into four quadrants, each processing the 2 GHz of data from one base frequency for both polarizations and for all antennas. Each quadrant is hosted in a several dozen-meter-long 19-inch rack with hundreds of cards. The quadrants are managed by one central correlator control computer.

One hierarchy level down, each quadrant then has 32 correlator "planes" that use custom ASICs to perform the actual correlation function. These run at 125 MHz, 32 times slower than the system's sample rate of 4 GHz; consequently there are 32 planes to parallelize the whole workload. The planes contain 32,768 correlator chips, each with 4,096 multipliers. A total of 16,384 shielded twisted-pair cables establish the necessary connections.

The correlator can be operated in two modes, each using the 32 planes differently: time-division multiplexing and frequency-division multiplexing. Let's look at frequency division first. Here, the 2 GHz of total bandwidth is subdivided into 32 sub-bands of 62.5 MHz each and each of them is correlated separately. We get results in 32 distinct frequency bands (within the recorded 2 GHz range), which the correlator software, in a stage after correlation, joins into a contiguous spectrum. The extraction of the 32 sub-bands is performed by tunable filter bank cards that were custom-developed for ALMA. Each card has sixteen FPGAs, and there are 64 × 2 cards in each correlator quadrant.

In contrast, time-division mode does not use the sub-bands. Instead, the stream of samples is allocated sequentially to each plane: sample 1 to plane 1, sample 2 to plane 2 and so on, until it starts over after plane 32. The correlation outputs of the 32 planes are then averaged to produce the result sent to the correlator software. In effect this sums 32 consecutive samples into one u,v-pixel, but does not distinguish the particular frequencies within the 2 GHz range. This mode therefore has high sensitivity (you'd need 32 times more integration time in frequency division mode to get to the same intensity), but a very low spectral resolution. The decision of which mode to use is driven by the science; for example, one might want to observe specific molecular emission lines of an astronomical source, so frequency-division mode is recommended and one has to pay the price of the longer integration times. Time-division mode is also used for a lot of calibration work.

There's another step between the tunable filter banks and the correlator planes: the station cards. These are partly responsible for removing the geometric delay, together with the digital samplers at the antenna (which remove the finest delay) and the filter banks. To understand why this is necessary, let's review this illustration:

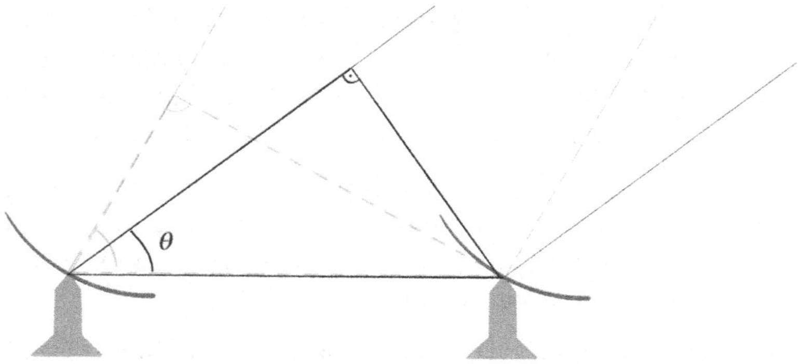

The geometric delay is the part of the total delay between the two antennas that results from the baseline's geometry; it is constant, even as Θ changes. From a computational perspective, one can separate dealing with the part of the delay that is geometry-dependent and the part that depends on the spatial frequencies observed on sky, expressed by Θ.

When all antennas are correlated the input data rate to the correlator is around 715 GB per second. After the hardware correlator has done its job of multiplication and integration, the data rate is reduced to at most 1 GB/s. This stream flows into the Correlator Data Processor, a set of PCs that performs additional processing based on observation parameters; for example, it creates the contiguous spectrum in frequency-division mode. Its data output rate is 70 MB/s or less; the networking infrastructure into the science archive cannot deal with higher rates. This is a limitation that is currently being addressed; higher rates will be supported in the future.

The correlator is located at the high site and its altitude, coupled with the lower air density, has led to problems for the computers. For example, it turned out that hard disks cannot operate there reliably, because the air pressure is too low to form the air cushion necessary for the heads to "fly" over the metal disks. This is why all the data storage happens at the support facility at 2,800 meters. The reduced air density also leads to issues with cooling; additional fans had to be installed. All in all, the ALMA correlator is the largest supercomputer used in astronomy, the fastest single-purpose computer generally, and the only supercomputer that operates at this altitude. Around 1,000 km of fiber-optical cables were installed for ALMA. These are used to transport the data from the antennas to the correlator, from the correlator to the archive (at the low site), as well as to distribute laser light for timing purposes. We will look at this next.

TIME CYNCHRONISATION Interferometry is ultimately based on removing the delay between signals received by two antennas on a baseline due to its orientation and length, such that the two signals are coherent at the correlator input and are useful for interference. For this to work the antennas have to agree on a common time, referred to as "array time" in ALMA. Clock synchronization is a hard problem in computer science generally, because it takes time to transport the information about what time it is. Worse, you cannot easily measure that delay, because to do that you'd already have to have a common time reference. Generally, the finer the required time synchronization, the harder the problem.

For ALMA, time precision on the order of one picosecond is required at the highest-frequency observation modes, so it should be no surprise that a lot of engineering has gone into the time subsystem. ALMA clock synchronization is a two-step process that relies on (computer-network) broadcast messages and on a local oscillator at each antenna.

The broadcasts originate from a central computer that uses GPS time as its reference. During initialization of an observation it sends messages to all antennas specifying that "at the next time event, the time will be such-and-such". Time events then arrive at the antenna computers every 48 milliseconds via an interrupt, with the computers counting these events to move their local view of array time forward. This enables all antennas to be realigned to a common absolute time every 48 milliseconds. Missing a time event is a major fault, and the respective antenna immediately goes into recovery mode and is "blanked" to ensure the data is marked as faulty by the correlator. The absolute, GPS-based time message is transmitted through computer networks: in the technical building and inside the antenna cabin it is transported via an

RS-485 bus and is forwarded to the antennas through the fiberoptic data connections. The 48 ms time event arrives via a laser—we'll see exactly how this works below.

The second ingredient in ALMA's time subsystem is an antenna-local oscillator that has a high and stable frequency, which acts as a very fine-grained "clock tick" between time events. This clock must also be precisely aligned between all the antennas in terms of drift and phase. The distribution of a stable time reference to all antennas is based on a dedicated fiberoptic connection that transmits a laser at a precisely maintained frequency. The frequency of this laser acts as the fine-grained clock tick at each antenna.

The laser can be distributed to the antennas relatively easily using a glass fiber, allowing all antennas to share the same fine-grained clock. But transporting the signal over several kilometers comes with its own challenges.

First, we need a high degree of frequency stability and sufficient coherence over the kilometer-long range required in ALMA. But we don't need too much power (in contrast to the gravitational wave detectors) so this is not too much of a challenge. The other challenge is of course that the wave takes time to travel through the fiber to the antenna, and each antenna is at a different distance from the central laser source. While there is no need for absolute alignment (in contrast to the delay lines in optical telescopes), the phase has to align perfectly. This is achieved with a line length correction module in the path to each antenna that runs as part of a closed control loop—effectively a mini-interferometer—that actively controls the phase of the lasers at each antenna. The line length correction is achieved with a 5 mm-long piezoelectric fiber stretcher, a ceramic body whose volume can be changed through a voltage, around which the optical fiber is wound.

How is the local oscillator actually used? Remember our discussion of the heterodyne receivers that require a reference frequency to mix down the high-frequency signal on sky to a band between 2 and 4 GHz? The laser supplies this reference frequency. This clever approach ensures that the receivers of all antennas are tuned to precisely the same frequency and are also phase-aligned.

Let's look at this in a little more detail. In the ALMA technical building, a master frequency standard produces a stable 10 MHz signal based on the 2-photon transition of rubidium 85. Through a couple of oscillators in the central reference module this is converted to reference frequencies at 5 MHz, 25 MHz, 125 MHz, 2GHz and 48 ms—the time event—and encoded onto a 1,532 nm laser. The 48 ms time event and the 125 MHz signal also go to the correlator; the latter is the clock rate for the correlator's 32 planes.

The master reference also frequency-locks a microwave synthesizer, which in turn drives a laser synthesizer. This contains four semiconductor slave lasers that allow continuous tuning over different frequency ranges. It uses four slave lasers because, as we have seen, ALMA can receive at four independent base frequencies, and therefore we need four reference frequencies to drive the heterodyne receivers. Finally, the 1,532 nm laser plus the four tuned reference frequency lasers are encoded onto a single fiber via a wavelength division multiplexer and—after going through the line length correction modules for each antenna—travels to the antennas. There it acts both as an ultra-precise clock and as a precisely tuned reference frequency for the heterodyne receiver. Based on the paper "The ALMA Photonic Local Oscillator System", the whole assembly seems comparable in complexity to gravitational wave detectors.

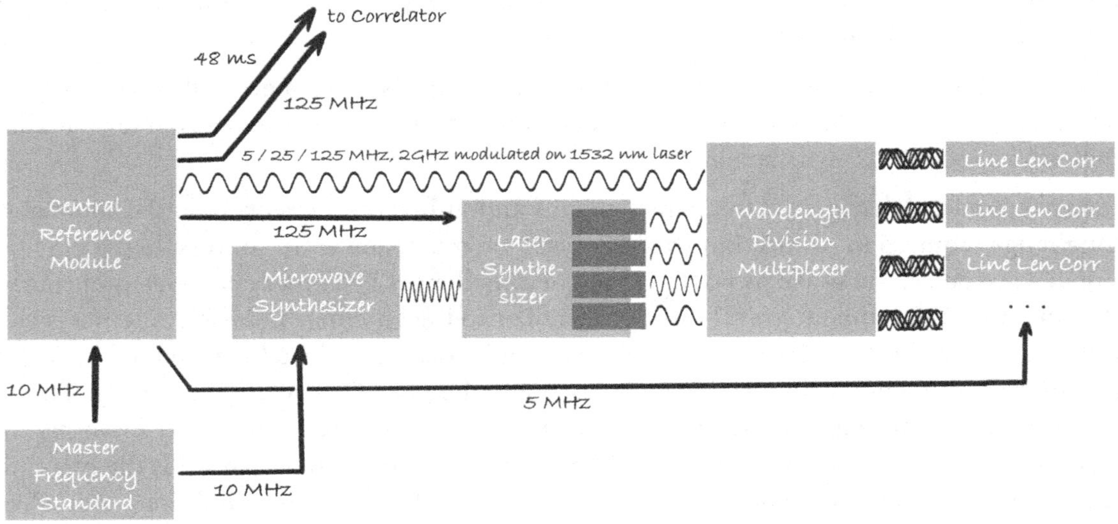

The system is even more flexible than what I've explained, because ALMA supports multiple subarrays that can receive at different frequencies. There are in fact several of these laser synthesizers, each of which can be set to different frequencies, and whose output can be routed to the various antennas via a single thermally-stable rack.

SCIENCE To conclude our discussion of ALMA, we should at least briefly take a look at the science that can be observed with all this amazing technology? One example is the detection of very faint red-shifted radiation. In 2017 ALMA observed a galaxy from a time when our universe was only four percent of its current age. Interestingly, resolving the object was only possible because of gravitational lensing provided by a galaxy cluster, which basically acted as a cosmic telescope and magnified the size of the "old" galaxy to a size where ALMA could resolve it. We will come across this effect again in the next part of this chapter. A collaboration with the VLT and its X-Shooter experiment was used to confirm the age of the object.

Another interesting observation used ALMA's spectral capabilities to identify the complex organic molecule methyl isocyanate in a distant star system. This is relevant because it is considered a prebiotic, meaning that it is a precursor to organic molecules that are necessary for life to form (although it itself is quite toxic). In a nice example of the scientific process at work, the team also performed lab experiments to show how the molecule can form in the cold conditions of interstellar space, thereby validating their observation.

When massive stars explode in a supernova, a twisted knot of new molecules forms at the core of the explosion. Researchers used ALMA to render a 3D image of the structure of this core. ALMA's resolution is crucial, but so was its ability to observe in the millimeter range, which allows it to see through the gas and dust that fills space between the Earth and the supernova. In particular, silicon monoxide and carbon monoxide, which radiate brightly at ALMA's wavelengths, were detected.

In 2016 ALMA was used to measure the mass of a black hole at the center of a nearby galaxy, by measuring the speed of carbon monoxide gas in orbit around the black hole. Its characteristic spectral line will be red- or blue-shifted as it recedes or approaches an observer on Earth, allowing for an estimate of the rotational speed. The result was around 500 km/s. Using numerical simulations, they derived the mass of the black hole to be 660 million solar masses, ± 10%. This is two orders of magnitude bigger than the black hole at the center of our own galaxy, but still much smaller than the heaviest black holes known. Let's stay with black holes for a bit.

10 | *The Event Horizon Telescope*

BLACK HOLES To understand black holes we have to return to general relativity. As outlined in the previous chapter, this explains gravitation as the bending of spacetime. Gravity therefore affects everything, even things that have no (rest) mass—in particular, it bends the path of electromagnetic radiation. According to the field equations of general relativity, the strength of this bending depends on the mass of the body that creates the gravitational field. The observation of this effect during a solar eclipse in 1919—in which stars showed up at the "wrong" place on the edge of the darkened sun, because their light was very slightly bent around the sun by its gravitational field—was a major experimental confirmation of general relativity.

From a more classical perspective, we know that if an object such as the International Space Station is in an orbit around a planet such as the Earth, its speed is such that the centrifugal force is equal to the gravitational pull created by the Earth's mass. In the following equation, G is the gravitational constant, a property of our universe; G is different from g, which represents the acceleration created by the Earth's mass.

$$F_C = F_G$$

$$\frac{m_{ISS} \cdot v^2}{r} = \frac{G \cdot m_{ISS} \cdot m_{Earth}}{r^2}$$

If you solve this for the orbital speed v, you get:

$$v = \sqrt{\frac{G \cdot m_{Earth}}{r}}$$

So flying a higher orbit (larger r) requires less speed. How does this fit with the fact that a spacecraft like the ISS has to accelerate to boost itself into a higher orbit at regular intervals? The reason is that when a spacecraft accelerates, what it essentially does is increase its energy. That energy is used to "climb up" in the gravitational field; it increases its potential energy. There's much more to orbital mechanics, including the fact that a single acceleration on a circular orbit makes that orbit elliptical, and only a second acceleration at the highest point of that ellipse makes it circular. But the important point to take away here is this: to get further away from a massive body, we have to increase our energy. Let's say we want to take this energy from a body's kinetic energy. Then we can say that, for a given gravitational mass M, a particular minimum speed is required to leave the gravitational influence of this body; the kinetic energy that corresponds to this speed is completely converted into potential energy in the gravitational field of M. This speed is known as the "escape velocity" and can be calculated as

$$v_e = \sqrt{\frac{2\,G\,M}{r}}$$

where r is the radius at which the object currently orbits (or just passes) the massive body. This formula is easy to derive from the conservation of energy. We start with the object in its low, orbiting state (index o in the following formulas); assuming no other forces, we know it must have the same energy in its escaped state (index e), where the speed is then zero (we have just barely escaped, using up all the kinetic energy) and the gravitational pull of the large body M is also zero, because the distance is infinite:

$$W_o = W_e$$

$$K_o + G_o = K_e + G_e$$

$$K_o + G_o = 0 + 0$$

$$\frac{1}{2}\, m\, v^2 + \frac{-\,G\,M\,m}{r} = 0$$

$$\frac{1}{2}\, m\, v^2 = \frac{G\,M\,m}{r}$$

$$\frac{1}{2}\, v^2 = \frac{G\,M}{r}$$

$$v_e = \sqrt{\frac{2\,G\,M}{r}}$$

We can see that as M gets bigger and r gets smaller, the faster the object must travel be to be able to escape. This makes intuitive sense: the closer we are and the stronger the pull, the more speed/energy we need. However, we know that an object's speed can never exceed the speed of light. So let's solve for r and plug in c for v_e:

$$r_s = \frac{2\,G\,M}{c^2}$$

This radius, known as the Schwarzschild radius, is the radius for a body of mass M at which anything that travels at or below the speed of light will be pulled towards that body and cannot ever escape. This is independent of the mass of the moving object—its mass is not in the equation (only the mass M of the body that creates the gravitational field). Even "things" with zero (rest) mass, such as photons, are affected: even radiation that comes closer than r_s is trapped inside the Schwarzschild radius. So if we were to observe any body whose actual radius is smaller than the Schwarzschild radius for its mass, it would be completely dark—no radiation would be emitted from it. This is where black holes, the bodies we are talking about here, get their name.

The radius r_s is also known as the "event horizon" radius, because anything—any event—that happens inside it can never have any impact outside of r_s. This is because the information that the event occurred can never be transmitted—after all, such a transmission would require some kind of electromagnetic signal traveling at c.

How big are Schwarzschild radii? Of course this depends on the mass of the gravitational body M, as we can see from the formula above. For the mass of the Sun, it's 3 km; and for the Earth

r_s is 9 mm: all of the Earth's 10^{24} kg of mass inside a 9 mm radius! That is an insane density for an Earth black hole:

$$\rho_{EBH} = \frac{m}{V} = \frac{m}{{}^4/_3\ \pi\, r^3} = \frac{10^{30}\ kg}{{}^4/_3\, \pi\ (9\ mm)^3} = 10^{36} \frac{kg}{m^3}$$

For comparison, the most dense materials on Earth, such as uranium, gold or osmium, are around 21 kg/m^3, which is 35 orders of magnitudes lower.

This back-of-the-envelope calculation assumes that the mass would fill its Schwarzschild volume uniformly. Is this the case? What actually happens inside a black hole? Unfortunately, we cannot really know: any probe we might send in could never communicate its findings back out (and, as we will see shortly, would not survive the mission). But we can still imagine what happens inside: if we were to cross the event horizon, we'd inevitably be pulled further towards the center by gravitation. On our way in, gravitation becomes stronger as we get closer to the center and the pull increases. So a black hole is "self-compressing": matter is effectively compressed into a point of infinite density. Because of the ever-increasing gravitation gradient "on the way in", any object will be spaghettified (technical term!) by the differential gravitation between its near and far ends. Ultimately we expect everything, from space probes to atoms, protons and quarks, to be torn apart. No known force can resist the increasing gravitation. General relativity, let alone classical mechanics, cannot really explain what happens then. Quite probably quantum effects will come into play to avoid the problem of infinite density. But we don't know how general relativity and quantum theory relate—a quantum theory of gravitation has eluded us for decades. We will discuss this a little bit in the last chapter of the book.

Other strange phenomena occur: because of the strong gravitational force, time slows to a standstill from the perspective of an outside observer. Gravitation will also stretch the wavelength of light further and further into the red, a phenomenon known as gravitational redshift. From an external perspective, therefore, matter never actually arrives at the center of the black hole; it gets stuck on the event horizon, fading away in redshift. From the perspective of the matter itself, time continues normally and the matter simply falls through the horizon into the black hole.

Mathematicians call black holes a "singularity": the known laws of physics don't apply, and we cannot describe this state with our existing theories. However, the field equations of general relativity predict that such singularities should exist; basically, we run into a division-by-zero as we approach the parameters that characterize black holes. However, a singularity in a theory doesn't necessarily have to represent a physical singularity in the real world—they might be an artifact of the math. So for a long time it was unclear whether black holes actually exist in our universe, or some as-of-yet unknown physical process would prevent their formation.

So how do black holes form? The basic process is gravitational collapse. Stars are stable because their own (inward) gravitational pull is compensated by the (outward) pressure of the fusion reaction that burns inside. As a corollary, the gravitational force increases pressure at the core enough for fusion to occur. When a star runs out of fuel and its temperature reduces, the radiation pressure that resists the gravitational pull decreases and the star will inevitably collapse. For stars with an original mass of around 20 solar masses the resulting body will be a black hole; lighter stars collapse into a dead compact star. Black holes that form in this way are called stellar black holes and can have masses of up to several dozen suns. There are also supermassive black holes, the heaviest of which are estimated to be 10^{10} solar masses. Their formation is controversial; theories suggest they form by smaller black holes accreting material over time, when the black holes at the center of galaxies merge during galaxy mergers, or by the collapse of gas clouds in the early universe.

Throughout this section I have assumed that we know that black holes actually exist. But how do we know, except by watching Hollywood movies? Initially the evidence was indirect, as it often is for new theories: the motion of some of the stars near the center of our own galaxy, the Milky Way, can only be explained if there is a very heavy—and at that time, invisible—gravitational mass at its center. In 1998 it was inferred that a body of 2.6 million solar masses must be contained in a volume with a radius of 0.02 light years. Over time the estimate was refined to 4.2 million and 0.002 light years. A supermassive black hole is the only plausible way of putting so much mass into such a small volume. More direct evidence came in 2015 when—you already know this—LIGO detected a gravitational wave signal that could only come from the collision of two black holes. Finally, in 2019, the astronomy community announced the direct observation of the event horizon of a black hole. In terms of a direct observation this is as good as it gets, because the black hole itself cannot be seen: no radiation emanates from it.

So what exactly did the scientists mean when they said that they have "seen" (the event horizon of) a black hole? To answer this question we have to explain two aspects: what kind of radiation we see and how is it created, and the size and shape of the visual effect. Let's start with the radiation. A black hole accretes matter and gas because of its strong gravitational pull. Matter that's on course into the event horizon has the honor of contributing to the monster's growth. The trajectory of matter that is a bit further away will be bent towards the black hole, and, depending on geometry and energy, will start orbiting it, over time forming an accretion disk. Because of the strong gravitational field, the high speeds and the resulting friction forces, the matter in the disk will form a plasma. And plasmas radiate at many frequencies (black body radiation), some of which can be observed in the radio band. A secondary effect is synchrotron radiation: as black holes accrete matter, they expel some of it as powerful plasma jets; the charged particles in the plasma are accelerated by the magnetic fields that form in the vicinity of the black hole. How exactly this works is still a subject of research.

Let's address the second aspect, the shape of the radio source we can observe. We know that there is an event horizon from which no radiation can ever escape. This is the size of one Schwarzschild radius r_s around the core of the black hole. Then there is the radiating plasma that orbits the black hole; the innermost stable orbit turns out to be 3 r_s (this is a consequence of general relativity that we won't discuss further). So do we see a plasma glow beyond 3 r_s and blackness inside of it? Almost. Consider the following illustration, which shows light rays coming from the right-hand side of the page and traveling towards a black hole.

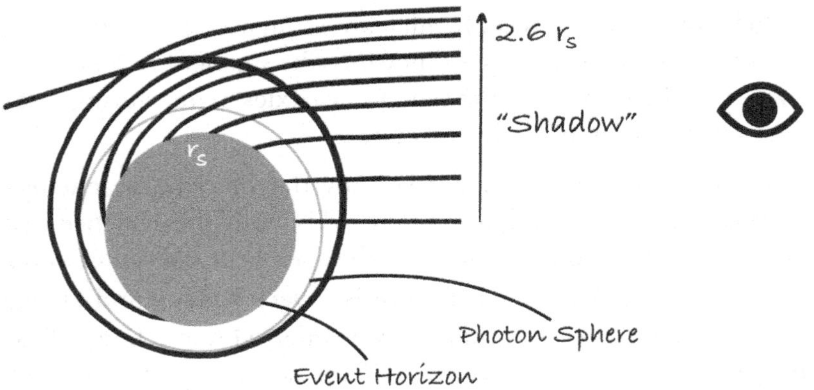

Those rays that are inside r_s will disappear into the event horizon: we see blackness. However, those that are further out will reach the event horizon as well, because the strong gravitation bends space, and with it the trajectory of light. There's a critical radius at 1.5 r_s at which light, traveling at c, can orbit the black hole. However, this is an unstable orbit; if a photon crosses it even a little bit, it will inevitably spiral into the event horizon. This radius is known as the "photon sphere". Physicists have calculated that at up to 2.6 r_s light rays are bent so strongly that they will cross the photon sphere, and thus eventually "hit" the event horizon, potentially at the rear of the event horizon sphere—which is dark as well. So what we see when we observe the black hole with sufficient resolution is this: between $0 < r < r_s$ we see the (black) side of the event sphere that points towards the observer, then from $r_s < r < 2.6\ r_s$ we see the (also black) rear of the event sphere, and for $r > 2.6\ r_s$ we can see all kinds of radiation produced by the orbiting plasma. We therefore observe a "radio donut" with a black region from the center out to 2.6 r_s. The fact that we see a black area that is larger than the cross-section of the event horizon is called "gravitational (self-)lensing".

There are many more details about the shape of the donut, mostly as a consequence of the very strong bending of space close to a huge gravitational mass and its impact on light, as well as the fact that we don't necessarily look at the accretion disk perpendicularly. In addition many black holes have a spin; things are even more complicated for such rotating black holes. I suggest you watch "How to Understand the Black Hole Image" by Veritasium on YouTube, who explains it much better than I can.

Historically the donut could not be resolved, and only an undefined "radio blob" could be detected—for example the Sagittarius A* radio source that corresponds to the location of the black hole at the center of the Milky Way. In 2019, however, the appropriately-named Event Horizon Telescope, EHT for short, was able to resolve the donut when they observed M87*, a black hole at the center of the radio galaxy M87. Let's understand how.

VLBI The EHT is an interferometer, so let's briefly review the basics of interferometry. In an optical interferometer the light beams from the two antennas are superimposed optically—two in the case of the LBT and four in the case of the VLTI. The interferometry happens optically: we make the photons interact, and the detector then picks up the resulting (fringy) image for analysis. Piston mirrors or delay lines remove the geometric delay. For this to work the beams have to be brought together physically, as light beams. This is different in radio astronomy, as we saw with ALMA: the signals are down-converted and digitally sampled at the antenna, then the "digital wave" is transported through data networks to the correlator, which performs the interference computationally. While the optical approach has inherent limitations for the length of baselines—you can transport and delay a light beam only so far—this limitation does not exist for radio interferometers. In fact, you can record the wave onto a hard disk and ship it halfway around the globe to interfere it with a wave recorded there.

Enter very long baseline interferometry, or VLBI. VLBI has been possible since the 1960s, but this was at wavelengths in the centimeter range. Shorter wavelengths are desirable, because, as we know, the smallest angular resolution achievable with an aperture of diameter D is proportional to λ / D. For a given baseline, therefore, shorter wavelengths result in better resolution. However, shorter wavelengths are more challenging because of the need for higher precision in timing and the tuning of receivers, because of increased noise in the receivers, because the atmosphere is more opaque (requiring higher sensitivity or longer integration times) and because observations are more susceptible to atmospheric fluctuations. The first experiments at 1.3 mm were done in the late 1990s and early 2000s, initially between IRAM's Pico Veleta and Plateau de Bure telescopes. As well as the need for a precise timebase, another problem encountered was capturing sufficient data quickly enough. Improvements in networks and disk recorders between the 2000s and the advent of the EHT increased the observation bandwidth, and thus detection sensitivity.

EHT OVERVIEW In the last few years, telescopes from all over the globe have joined forces to form the EHT: ALMA in Chile, the APEX prototype right next to it, the IRAM 30-meter telescope at Pico Veleta in Spain, the Large Millimeter Telescope on top of the (inactive) Sierra Negra volcano in Mexico, the Submillimeter Telescope in Arizona located next to the LBT on Mount Graham, the James Clerk Maxwell telescope on Mauna Kea on Hawaii, the Submillimeter Array located on the same mountain, and the South Pole Telescope that is part of the

Amundsen–Scott station located—as you might have guessed—a few hundred meters from the geographical South Pole.

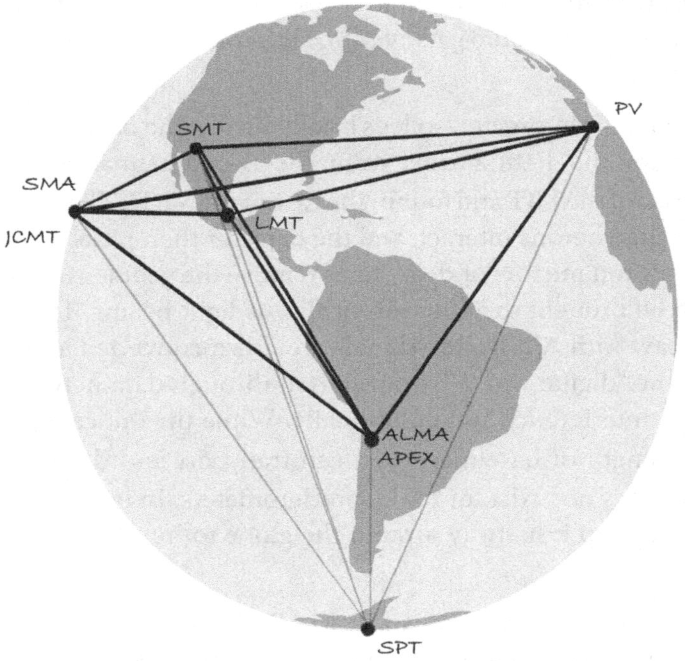

The telescopes that make up the EHT were chosen based on several criteria. Individually, each must be able to observe at the 1.3 mm wavelength that is used in the EHT, which means that their location had to be sufficiently high and dry; this ruled out Effelsberg, for example. Jointly, they had to be able to form a broad variety of baselines. The current set of participating telescopes can form fifteen baselines (the lines in the illustration above), from 160 m to 10,700 km and with various orientations. The longest baselines give the EHT a theoretical diffraction-limited resolution of about 25 microarcseconds, which hypothetically allows the EHT to read a newspaper in Paris from New York! More importantly, taken together with computational methods, an 120 × 120 μas black hole will be imaged with around 36 × 36 pixels, which is enough to resolve the donut shape. Finally, all participating telescopes must be able to see a sufficiently large set of interesting science targets at the same time. For the observation of M87* the South Pole Telescope and its baselines were not used, as it cannot see this particular target; M87* is in the northern sky. However, it was used for calibration during these observations.

TIME SYNCHRONIZATION As we know from our discussion of ALMA, time synchronization is crucial for an interferometer—a picosecond time base is required for millimeter interferometry with samples captured every 0.25 nanoseconds. However, the EHT cannot use

ALMA's approach of distributing a single reference frequency to all the antennas via fiberoptic cables: no such cables are available across the globe. Even if they were, the various active components necessary in such a network would make timing and phasing practically infeasible. So the local oscillator at each antenna is really local, and not an actively phase-corrected replica of a central laser.

To ensure that time synchronization is precise nonetheless, each site has its own atomic clock based on a hydrogen maser. It has two important features. First, it is stable against drift for the duration of a typical EHT integration (10 to maybe 100 seconds); if the drift on this scale was extrapolated, it would drift on the order of seconds in every 10 million years. The second important feature is that it has very low phase jitter. During downsampling from 230 GHz to a reference frequency of 10 MHz, any phase instability would be multiplied by 230,000. But for a meaningful observation, the signal must still be phase-stable after downsampling. For an interferometer like ALMA that distributes a single reference signal to all antennas, phase stability is less important, because all antennas would see the same instability.

Because the stability of the maser is crucial for EHT observations, its stability is cross-checked regularly. The ideal approach would be to run three masers in parallel, so you can reliably detect if one of them drifts. However, three masers (at each site!) would be too expensive. A combination of two other approaches is used instead. First, the maser is compared to a high-precision quartz oscillator that has the same frequency stability, albeit only at 1 second timescales—extrapolation then characterizes the maser's stability over 10 seconds. The second approach uses GPS, which also relies on precise timing. Each GPS satellite runs its own atomic clocks, and the overall system time is actively managed by the GPS ground segment. While its stability on the scale of seconds is not so great due to the variability of the ionosphere (through which the GPS signal has to pass), it turns out that on timescales of the order of 10,000 seconds its stability is better than that of the EHT masers, which is why GPS is used for verifying long-term maser stability. The GPS signal is analyzed over several days to determine—and then correct for—the influence of the ionosphere. GPS is also used as the absolute time reference, just as in ALMA.

However, despite the precision of the maser and the corrections, there is still some residual drift in the clocks that disturbs the precise determination of the delay between telescopes. There are also about 20 additional factors that affect the geometric distance of the telescopes that form a baseline, including the telescopes' movement from continental drift (Hawaii, for example, moves 6 cm per year), tidal forces, wobbles in the Earth's axis, as well as more obscure effects such as ocean loading, the gravitational delay and imprecisions in the antenna itself. The largest contribution to distortions of the measured phase difference, and the one that changes on the shortest timescales, is the state of the atmosphere, and its water vapor content in particular.

Instead of trying to characterize all of these effects precisely, their correction is achieved via the observation data itself. The correlator starts out with an initial assumption about the current baseline length, the "correlator model". If this model is correct, then fringes can be found immediately. This is especially likely if the target is bright and the atmosphere is dry. This is because a brighter source allows for shorter integration during which water vapor varies less. If fringes cannot be found, the correlator tries the process with different delays that are offset by fractions of microseconds; once fringes are found, the model is updated. All of this is carried out while looking at a calibrator, a point source that is close to the science target and whose position is well known. Once the calibrator fringes are understood, the updated correlator model can be used to analyze the science target. During an observation, the telescopes alternate every few minutes between the science target and the calibrator.

There's an additional technique to dealing with the varying atmosphere: its effects can be canceled out by global fringe fitting and the notion of a closure phase. To get an idea of how this works, consider three antennas that form a triangle, plus the three baselines represented by the sides of that triangle.

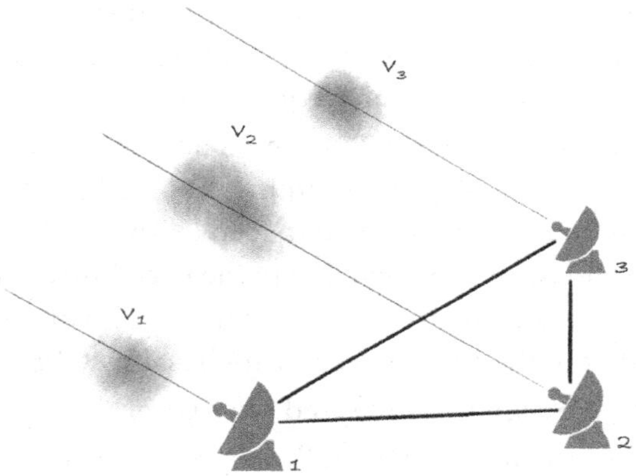

The phase difference Δ between any two antennas is given by subtracting their respective phase angles ϕ:

$$\Delta_{12} = \phi_2 - \phi_1$$
$$\Delta_{23} = \phi_3 - \phi_2$$
$$\Delta_{31} = \phi_1 - \phi_3$$

Now let's assume that the phase of each antenna is affected by some amount of water vapor in the line of sight of the antenna, represented by V. The resulting total, distorted phase difference Θ can be written as follows by adding the water vapor effect to all the phases:

$$\Theta_{12} = (\phi_2 + V_2) - (\phi_1 + V_1)$$
$$\Theta_{23} = (\phi_3 + V_3) - (\phi_2 + V_2)$$
$$\Theta_{31} = (\phi_1 + V_1) - (\phi_3 + V_3)$$

We now sum these three distorted phases into the closure phase:

$$\begin{aligned}\Theta_C &= \Theta_{12} + \Theta_{23} + \Theta_{31} \\ &= (\phi_2 + V_2) - (\phi_1 + V_1) + \\ &\quad (\phi_3 + V_3) - (\phi_2 + V_2) + \\ &\quad (\phi_1 + V_1) - (\phi_3 + V_3)\end{aligned}$$

Now a bit of math: we remove the parentheses and rejoin the ϕ-differences into the original phase difference terms Δ, which gives us the following:

$$\begin{aligned}\Theta_C = &\Delta_{12} + V_2 - V_1 + \\ &\Delta_{23} + V_3 - V_2 + \\ &\Delta_{31} + V_1 - V_3\end{aligned}$$

Can you see what happens? For each of the water vapor contributions we have a positive and a negative term, so they cancel out. The reason why this happens is that water vapor above an antenna "shortens" one baseline and "lengthens" the other:

$$\Theta_C = \Delta_{12} + \Delta_{23} + \Delta_{31}$$

By cleverly combining baselines, therefore, the effect of water vapor on the observation can be removed.

Before we move on to backends: continental drift, the wobbling of Earth's axis, deformation of the Earth from tidal effects ... somebody's bug can be somebody else's feature. Geodesy is the science of measuring the Earth and understanding its shape. They use a wide variety of techniques, several them space-based. For example, they reflect laser pulses from mirrors on satellites to determine the position of the ground station based on the travel time of the pulse and a precisely known satellite orbit, they analyse the phase of GPS signals to determine positions down to millimeters precision, and they also rely on VLBI. Several of the world's astronomical radio telescopes spend some of their time with geodetic VLBI measurements. But there are also dedicated geodetic observatories such as the one in Wettzell, Germany. They haveradio telescopes for VLBI, optical telescopes for laser ranging, precision GPS receivers as well as ring gyros to inertially measure the rotation of the Earth. There's a very nice (German language) episode on this topic on omega tau, check it out!

CUSTOM BACKENDS The EHT makes use of the participating telescopes' receivers: relying on the tuned reference frequency, they map the received signal to a telescope-specific baseband for analysis. However, to be able to correlate the data from multiple telescopes, all signals must be mixed down to a common baseband. This is why EHT requires a custom backend. Backends are comparable to instruments on optical telescopes: they get the signal from the telescope and handle all the processing. In particular, the EHT backend contains an additional mixer that downsamples the signal to the common baseband agreed upon for EHT, the digitizer and the recording system. The timing system—the maser and its correction mechanisms—can also be seen as part of the backend, in the sense that all the other components are slaved to the maser clock.

The recorder at the end of the pipeline was a particular challenge, because of the high data rate required. In contrast to ALMA, we don't just have to record the relatively low-volume data after correlation; instead, the data stream that comes from the antenna has to be recorded directly so that it can be correlated with the data from the other telescopes at a central location. Suitable recorders are developed and sold by the Haystack observatory, together with a commercial company; the current generation is called the "Mark 6". This receives data through Gigabit Ethernet cards and then time-slices the data stream across 32 hard disks using a round-robin algorithm, because a single hard disk could not support the data rate. For the overall 64 Gbps data rate at each EHT site, four Mark 6 recorders are used in parallel, recording on 128 hard disks total, each with 1 to 2 TB in capacity. The resulting capacity of more than 1 Petabyte supports VLBI observation campaigns of between three and six days, depending on the particularities of the telescope.

PHASED ALMA ALMA and the SMA are interferometers themselves—how does they fit into a VLBI array? You could just use one antenna, but this would of course waste a lot of sensitivity. This would be especially painful in the case of ALMA, which has very good receivers and is located a very good (high and dry) site. A much better approach is to use join all antennas into one large virtual antenna when ALMA is used as part of the EHT. How exactly is this different from their normal use as an interferometer? When used interferometrically, the phase difference between the various antennas (as a function of observation direction) is exploited to fill the u,v-plane; the corresponding multiplication and Fourier transformation is performed by the correlator. In contrast, for their use as a single station in a VLBI array, the signals from the various antennas have to be summed, not correlated.

However, there's a problem: as we know from the interferometric use case, the phases between the various antennas are more or less random; so if we were to sum the signals they would produce more or less one antenna; in the unlucky case the signals would average to more or less zero. Not very efficient, but this is how ALMA was used during VLBI tests in 2015. Ultimately we have to make sure that the complete phase differences between the antennas are removed before summation, so they can be added coherently. We have to remove not just the geometric difference, but also the angle-dependent difference that drives interferometry. An antenna array where the phases of all antennas align is called a "phased array".

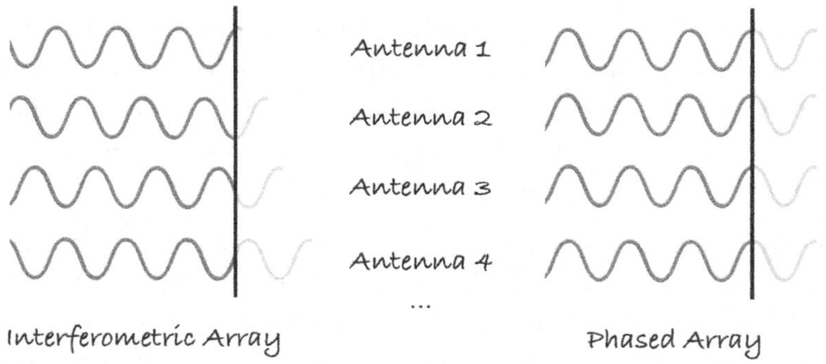

The science case for ALMA has always included VLBI. One reason is that its location in the far southern hemisphere makes it a desirable partner for large single-dish telescopes in the northern hemisphere, such as Effelsberg, in terms of baseline length and orientation. However, the capability to phase ALMA was de-scoped during the initial development for budgetary reasons. Driven by the EHT requirements, this capability was added to ALMA between 2011 and 2017 as part of the ALMA Phasing Project. If you read up on phased ALMA, you will also find the term "beamformed" ALMA. And in fact, the process, is very similar to what we have encountered in the discussion of Enterprise's multi-beam sonar.

So what needed to be done to phase at ALMA? As with all the other EHT sites, a very precise maser-based timing system had to be installed, as the rubidium-based system is neither precise nor stable enough. In ALMA they subsequently decided to use the hydrogen maser generally as the new timing standard, not just when used as part of a VLBI experiment. The Rb 85 system is now used as a backup in case the maser fails. The high-speed recording system based on Mark 6 recorders also had to be installed, because the existing science archive cannot accept the raw uncorrelated data rate. The recorders were placed at the low site, connected to the high site with a fiber link. When ALMA was built, a whole bunch of different fiberoptic cables were put into the ground to make sure that sufficient capacity with suitable specifications would be available for future projects; one of them has been designated for supplying the Mark 6 recorders.

The (now misnamed) correlator has to do four things: measure the phase difference between the involved antennas, correct it to zero ("phased"), sum up the signals from the antennas, and then transfer the sum to the Mark 6 recorders. The detection of the phase difference was already part of the correlator, because this is also needed for interferometry; in fact, the regular correlation logic runs during phased operation, and the resulting interferometric data is inspected to determine the phase difference. The phase correction for each antenna is computed by a software module in the telescope calibration subsystem, and the correction is performed by the tunable filter bank cards—they are also involved in the correction of the geometric delay in interferometric mode. This basically forms a control loop that keeps the phase difference effectively at zero. The only major modification was to install additional cards, called phasing interface cards, that perform a bunch of data conversions, encode the sum signal in the standard format for VLBI observations and forward it to the network interface of the fiber link to the low site, where the recorders persist the data. Since ALMA anticipated the need for phasing when it was built—remember the original science case—there was space available in the correlator to house these new phasing interface cards.

In theory a perfectly phased array has a sensitivity comparable to a single dish with the same area. For ALMA's 54 antennas with 12 m diameter each, this would equate to a single antenna of 90 meters; this is still less than Effelsberg, which drives home just how huge that telescope is. 90 meters is also bigger than any other antenna that is currently part of the EHT, or is every likely to join. The M87* observations in 2017 were among the first uses of the phased ALMA, and only between 33 and 41 antennas were available during the nights of the observations, 37 on average. This corresponds to a virtual 75-meter dish. However, as usual, practice isn't quite as beautiful as theory and there are additional losses in efficiency. For example, the phasing itself is imperfect due to tropospheric fluctuations, remaining delay errors and latency in the phasing corrections due to limits in data transfer rates in the correlator; these amount to roughly 10% of the loss in efficiency on average, depending on weather and elevation. In addition there's a 12% quantization error from converting ALMA's 3-bit representation of the visibilities to the 2-bit representation used in EHT. There are additional losses, the reasons for which are unknown and continue to be investigated. Nevertheless, ALMA is currently the most sensitive VLBI node in the millimeter wavelengths; no other station has a similar combination of collecting area and atmospheric dryness.

In the context of the EHT, ALMA acts as the reference for the backend: all EHT participants use the same data rates and mixing frequencies. In other words, once phased, ALMA needed the smallest amount of changes; for example, it does not need an additional, EHT-specific mixer. However, there was a problem with ALMA's polarization. All other EHT telescopes record left and right circular polarizations; ALMA instead records dual linear polarization. Correlating signals with different polarizations does not work directly, so ALMA's polarization had to be converted to circular. One way to achieve this is to use hardware, so-called "wave plates". Wave plates are made of transparent birefringent materials, which means that their refractive index is different for differently polarized radiation. By sending light through an appropriately configured wave plate, its polarization can be changed. However, this hardware solution would have had to have been installed on all ALMA antennas. A hardware solution would also have meant that the received data could no longer be used for regular ALMA correlation in addition to its use in the EHT. To avoid this effort, the EHT team used software that achieves the same goal after EHT correlation.

CORRELATORS Apropos correlation: in the same way that the EHT "reuses" existing telescopes, it also made use of existing correlators: the data is processed by a dedicated VLBI correlator at the MIT Haystack Observatory, as well as by the Max Planck Institut für Radioastronomie. Interestingly, these correlators are implemented in software that runs on general-purpose Linux clusters, connected by 40 and 100 Gigabit data links. Each correlator has more than 1,000 compute cores. The correlators mirror the telescope array: it uses the same Mark 6 data recorders to play back the data as if it came directly from the backends at each of the telescopes.

IMAGE PROCESSING On to image reconstruction. Let's recap briefly. The data collected by the interferometer in the u,v-plane is fundamentally equivalent to the spatial data we call an image; an inverse Fourier transform goes from the frequency domain to the spatial domain. However, the frequency data is only equivalent if the u,v-plane is completely filled by observations—that is, the telescopes have collected data for all possible combinations of length and orientation of baselines. This is impossible in practice. Consequently, image reconstruction is always an under-constrained inverse problem—many different images are potentially consistent with the observed data.

ALMA has over 2,000 baselines when all 54 of the 12-meter antennas are used, while the EHT only has 10. So even though the EHT makes heavy use of Earth rotation synthesis, the u,v-plane for the EHT is much more sparsely populated compared to a single-site interferometer like ALMA. Another challenge is that the data from the various telescopes lacks absolute phase and amplitude calibration. While the above mentioned closure-phase approach can be used to get rid of the dependency on an absolute phase, this is not quite as simple for the amplitude; significant uncertainties remain in the amplitude of the fringe visibilities. Because of all these factors, EHT's image reconstruction is harder, the set of images that match the data is even larger larger than, for example, for ALMA. So how did the EHT team approach the imaging challenge?

One cornerstone of the approach was to use three different algorithms. One was the standard CLEAN algorithm. As we saw earlier, this relies on deconvolving the u,v-plane. The second is called RML, short for "regularized maximum likelihood". This class of algorithms searches for images that are consistent with the data but that also have particular properties such as smoothness or compactness. For the M87* observation, the EHT used two different RML algorithms. One reason for using different algorithms for this first major EHT observation was to ensure that some weird combination of (sparse) data, array geometry and image reconstruction algorithm would not accidentally produce a bogus result.

Both classes of algorithms must be parameterized to find good solutions and do so in reasonable time, as is typical for inverse problems. To avoid bias, the first step in the development of the imaging pipeline relied on four different teams trying to image early engineering data. The teams worked independently and then cross-checked their results. The results were similar and plausible—they all produced donut-shaped images with a diameter of around 40 μas.

In the second phase the EHT collaboration set up three different imaging pipelines that used the three algorithms mentioned before, implemented by three different software packages. Each pipeline then scanned a range of imaging parameters, producing between 1,000 and 10,000 images. These were images of the observed data, but also of synthetic data that represented other geometric shapes in addition to the donut, such as crescents or filled disks. This is important to ensure that "donut-ness" isn't "programmed" into the parameters. This approach allowed the teams to select a set of parameters that were best suited, taking into account the observed data. Again, in all three pipelines, this respective best set of parameters transformed the observational data to donut-shape images with a diameter of around 40 μas.

It is important to understand that the selection of parameters is not a way to cheat, to create arbitrary images, because the image has to fit the observed data convincingly. This means that a good selection of parameters will minimize the errors for the observed data while taking into account physics constraints. The process is fundamentally similar to curve fitting. Consider the following illustration.

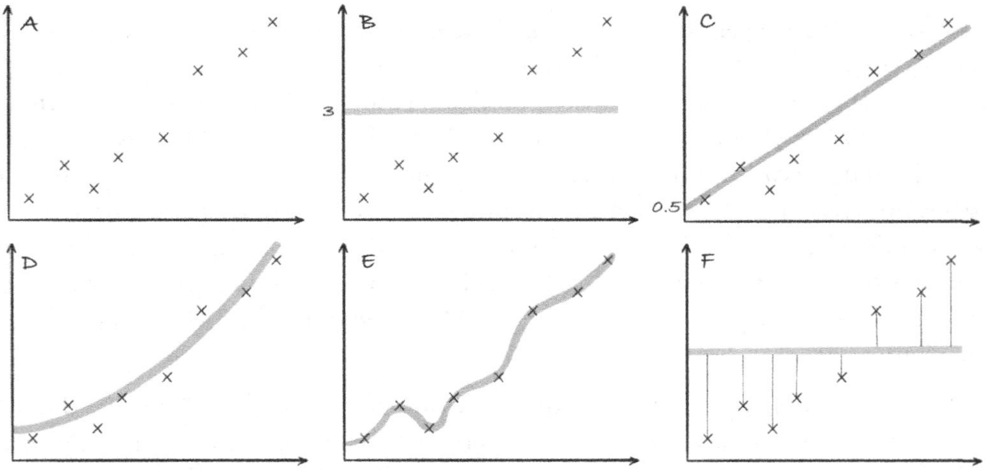

Diagram A shows a set of measured values for which we want to find a curve that fits. We expect that the solution is some kind of polynomial function of the shape

$$y = a_n x^n + a_{n-1} x^{n-1} + \ldots + a_1 x + a_0$$

because polynomials can in principle fit any data: they can describe very complex shapes. The parameters we have to decide on in this example are n (the degree of the polynomial) and all the a_i for $0 <= i <= n$ (if we decide $n = 2$. we have to define three more values a_2, a_1 and a_0). Let's say we decide $n = 0$. Then we have to set only one additional parameter, a_0; let's say we chose

3. This gives us a straight line with y = 3, as shown in diagram B. Obviously this is not a good fit. With n = 1 and $a_1 = 0.75$ and $a_0 = 0.5$, we get

$$y = 0.75\,x + 0.5$$

which is the straight line with a gradient of 0.75 and offset by 0.5 shown in diagram C. This is better. We can try a quadratic equation, such as

$$y = 0.5\,x^2 + 1$$

shown in diagram D, or an even higher polynomial that "catches" all measurements exactly, shown in diagram E.

How do we know which is correct? Generally, we want to minimize the total error. The total error is the sum of all the differences between each measured value and the one approximated by the fitted curve. In diagram F the total error is illustrated as the total length of all the thin black lines. If we applied only this criterion, then the higher-order polynomial in diagram E would win—in fact, for n+1 data points, an n-th order polynomial will always exactly fit the data. However, usually we know, to some degree, what the values represent—we know the underlying physics. In the example, we might know that the underlying physics leads to a quadratic equation. Thus we know that n = 2, and only the values for a_i are free to select in order to minimize the total error. A "better" fit with a higher order polynomial would have a smaller error, but is useless because it lacks physical meaning. We have a similar situation in the case of a black hole, where we know, for example, that the resulting image must be smooth to some degree. We also know some of the physics: a (dark) object at the center of the image, plus an accretion disk and jets around it. Various existing models predict how these should look. In actual practice, the approach is slightly more sophisticated because we often know an error bar for each of the measured values fit. This error bar plays a role in the calculation o fthe fitting error as well.

How can we get an understanding of the underlying physics, and the way it might impact the observed data (and thus, the image we expect to see)? The team ran general-relativity magnetohydrodynamics to simulate a black hole's effect on surrounding radiation. When this model was (synthetically) sampled with the EHT and then imaged with the three pipelines, it also produced the donut shape. In fact, the team learned how some of the particular features of the image can be explained: gravitational lensing, relativistic beaming and the black hole's spin all play a role.

To further increase the trust in the observation and the reconstructed image, they processed four different data sets from four different days. A final means of validation is that the team calculated the mass of M87* based on the observed radius of the photon sphere, and this mass was close to prior estimates obtained via different methods. So, to sum up, image reconstruction for the EHT used established approaches, but added processes and mechanisms to increase trust in the reconstruction, to compensate for the especially sparse data.

WHAT WAS OBSERVED So what did the EHT actually look at? The well-known first observation of a black hole's event horizon targeted M87*, the supermassive black hole at the center of galaxy number 87 in the Messier catalog. The Messier catalog is a list of 110 stationary objects, mostly star clusters, nebula and galaxies, assembled by French astronomer Charles Messier and first published in 1771. The M87 galaxy is at the center of the Virgo galaxy cluster and is Virgo's brightest radio source; it is approximately 55 million light years from Earth. The mass of the black hole is between three and seven billion solar masses, depending on how it is measured, and its Schwarzschild radius is between 20 and 40 μas as seen from the Earth.

Why this particular black hole? Most importantly, the angular size of the candidate's Schwarzschild radius had to be big enough to be resolved by the EHT at the 1.3 mm wavelength that is currently the state of the art. The candidate's position in the sky also had to be such that enough existing (and participating) telescopes could see it—if too far north or south this would have become a problem. The problem is illustrated by the fact that the South Pole Telescope is unable to see M87* and was used only for calibration.

Four separate observations were performed on 5, 6, 10 and 11 April 2017. Data was integrated for around two hours per day, on average, spread over eight hours. For calibration purposes they used the quasar 3C279. The resulting picture has made it all round the world; it is included in the photo collections of both the Museum of Modern Art in New York as well as in the Rijksmuseum in Amsterdam. It's probably the most famous donut ever.

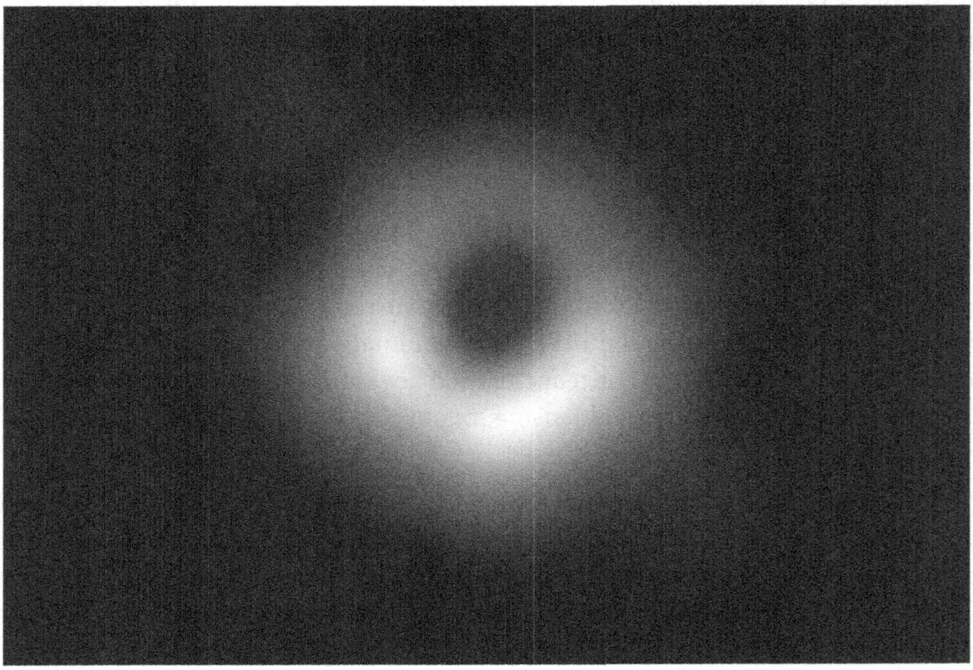

PRACTICAL PROBLEMS There were lots of practical challenges for the EHT project. For example, the custom backend needs space, and you have to find room for the cables from the receivers to the backend. Both are rare commodities at telescopes.

Because the EHT observed jointly only once per year, it was always a bit of a challenge to restart the equipment: things had to be reconnected, cables were switched and in one case the polarization was wrong because of a setup gaffe. To minimize the risk of errors each site would run through a detailed checklist. This includes verifying the operation of the mixers, the digitizers and the recorders, as well as the frequency and (non-)drift of the masers. An end-to-end test injects a test tone of a known frequency into the receiver, verifying that the downstream pipeline works correctly. The tone is also mixed down with a separate local oscillator and compared to a similar tone derived from the maser frequency reference. If the two are phase-locked when observed on an oscilloscope it demonstrates the stability of the whole system.

Some of the telescopes are not so easy to reach: Hawaii, Atacama Desert, Antarctica. Getting a total of 25 EHT team members to the respective telescopes to perform the observations wasn't necessarily easy. The amount of data captured by an observation was also huge; the maximum data rate per day was around 350 TB per telescope! This amount of data cannot easily be transported by data links, especially not from remote locations where fast connections are unavailable. The data was therefore transferred by physically shipping disk modules. The Mark 6 recorders are accordingly designed with removable disks. It took between days and months (South Pole) to bring all the data together. In one case the manual transfer of the hard disks constituted the fastest data transmission rate in history. Because of the manual data transfer, however, it was not possible to correlate in (near) real time, preventing end-to-end verification of the array during the observation periods while everybody was at the telescopes to fix potential problems. This exarcerbates the need for local verification and quality control, such as the checklists mentioned above.

There was also a scheduling challenge, because the weather—especially the water vapor in the atmosphere—had to be acceptable at all telescopes at the same time. The team reserved ten days; if the weather was good everywhere they performed EHT observations, otherwise the local staff ran their own science program that did not require coordination with the other EHT telescopes. The team built a monitoring tool for (not yet quite) all the telescopes that reported weather, pointing and receiver status, as well as the state of the atomic clock.

ORGANIZATIONAL CHALLENGES Previous work had established that, with Earth-sized baselines, M87*'s photon sphere should be visible with 1.3 mm VLBI from Earth. So the overall motivation for the EHT was clear. However, actually getting the project off the ground was not a trivial task nonetheless. It was not clear what exactly they would see, how much of the internal structure would be resolvable. Lots of theory, modeling and simulation work was done to shed light on these questions; workshops were held, papers were written. The candidate approaches were refined and narrowed down to a single one. Technology was developed and pathfinder experiments were conducted, for example regarding receiver and clock technology. In 2004 the development of the necessary high-speed broadband digital equipment was started at the Haystack lab.

Now the real work began. People and their organizations had to be convinced that it was worthwhile to spend some of "their" valuable telescope time on the EHT. The idea had to be sold as a groundbreaking physics experiment, comparable to discovering the Higgs Boson. A catchy name was needed: the Event Horizon Telescope. The project had to become part of science cases for new telescopes and instruments; it had to get into vision documents such as the European Astronet Roadmap or the US Decadal Survey. Around 2011 a letter of intent was signed between many of the future participants of the project. Grant proposals were written, submitted and awarded, including one called BlackHoleCam, funded by the European Research Council, that focused on the development of the backend, as well as a corresponding grant financed by the American National Science Foundation that focused on EHT infrastructure.

THE FUTURE OF EHT So what does the future hold for the EHT? Three additional telescopes are in the process of joining: the Greenland Telescope at Thule Air Force Base, the Kitt Peak 12-meter Telescope in Arizona, as well as NOEMA, an expansion of the Plateau de Bure interferometer from 6 to 12 antennas. While this doesn't necessary lead to longer baselines (the Earth's diameter is fixed), it leads to more baselines (with reduced reliance on the rotation of the Earth), which means that the u,v-plane will be filled more and image reconstruction will be of higher quality and faster. The team is also working on a remote control capability to run the observations without sending people to every telescope—a really important feature to keep cost under control as more telescopes join. The team is also working on additional observations, such as Sagittarius A*, the black hole at the center of the Milky Way.

As I write this in September 2019, the US National Science Foundation has announced a $12.7 million award to design and prototype the next-generation EHT (ngEHT). This includes the integration of even more telescopes, especially smaller ones at high and dry sites. ngEHT is also looking at improving the data recording rate by a factor of ten, ideally by using off-the-shelf hardware to keep cost down. Image reconstruction algorithms will also be optimized. For example, the team succeeded in reconstructing the EHT image with an algorithm called CHIRP that uses (among other things) Bayesian reasoning and machine learning; it requires fewer assumptions and parameters than the established algorithms to reconstruct the donut. ngEHT includes funds for her group to improve this technique. The grant also funds experiments that record "videos" of the event horizon.

More generally, the VLBI community is moving towards higher frequencies, because this will lead to higher resolution with the same baselines. This makes the aforementioned challenges with timing precision, receiver tuning and receiver noise worse—the shorter the wavelength, the higher the precision required to "detect" the phase. Luckily the maser is good enough down to 0.87 mm, so there is some potential left before new hardware is required. In addition, higher frequencies require really low water vapor in the atmosphere, so it will be a challenge to get enough baselines at any given time.

Space-based interferometry is an obvious idea to get to longer baselines. Even one antenna in space that forms an extremely long baseline with an antenna on Earth will be useful. Such experiments have been done in the past: between 2011 and 2019 the Russian RadioAstron/Spektr-R satellite had its 10-meter antenna combined with various Earth-based telescopes, including Effelsberg, to form baselines of up to 350,000 km. A big problem is the transfer of the huge data volumes from space. Maybe they will have to reactivate a technology used by early spy satellites that dropped cassettes with exposed film from orbit, only this time they'd drop hard disks. Nonetheless, more such missions are being considered, envisioned or fantasized.

The observations of M87* by the EHT were a big sensation all over the world, and not just in the science community. The 2020 Breakthrough Prize—sometimes referred to as the "Oscar of Science"—was awarded "for the first image of a supermassive black hole, taken by means of an Earth-sized alliance of telescopes". It was awarded to all 347 scientists whose names appeared on any of the six papers that documented the various aspects of the observation. Each of them will receive USD 8,645.53, one 347th of the three million prize money.

THE SKA We finished our discussion of optical telescopes earlier in this chapter by looking at the 39-meter ELT as one of three members of a new generation of huge future telescopes. The radio astronomy community is also working on the next generation, but they seem to have joined forces in a single project—the Square Kilometer Array, or SKA. The name refers to the total collecting area of 1 million square meters, spread over thousands of antennas. For comparison, ALMA totals 6,000 square meters, Effelsberg 7,800 and the FAST telescope's effective area is 70,000 square meters. So just like the ELT, the SKA's ability to collect radiation will be in another league compared to the current generation of arrays. The SKA is also a jack-of-all-trades, in the sense that it will have dedicated antennas for different wavelengths, and comprising baselines from dozens of meters to thousands of kilometers.

Let's discuss antennas. For the lower and mid frequencies, from 50 to 350 MHz, the SKA will have low-frequency aperture antennas. These are relatively simple and cheap dipoles similar to those used by LOFAR today, but improved in resolution and sensitivity. Over 100,00 antennas will installed at 500 stations, in Western Australia, up to 65 km apart. These arrays will use beam-forming to observe a range of directions at the same time. Remember how the delay between the signal received from a point in the sky depends on the angle of that point? In interferometry, the variation of this delay across the sky is what allows the interferometer to detect the spatial frequencies that are the basis for the inverse Fourier transform that in turn results in an image.

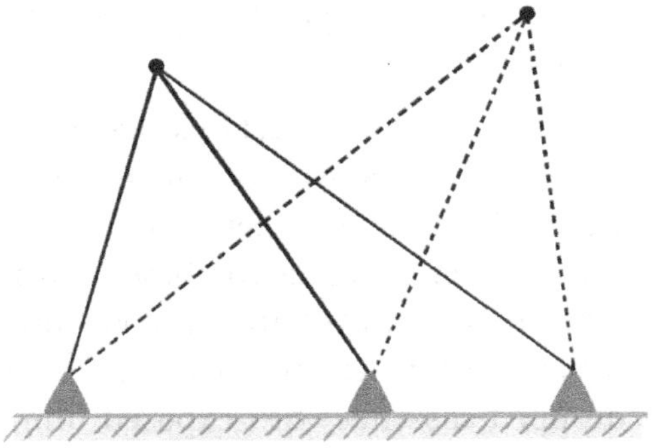

As we saw in the context of phased ALMA, when you phase an array of receivers, you remove this particular angle-dependent delay: the waves from that angle interfere constructively, whereas the waves from other angles interfere randomly. You therefore effectively amplify the signal from a particular direction—you focus the telescope. What is really neat about this approach is that it can be done for many angles during post-processing in software, using the same raw data: you can effectively observe many places at once. This approach will significantly increase the SKA's speed for surveys.

For the higher frequencies above 350 MHz, the SKA will have an array of 15-meter dish antennas. A site in South Africa will have 200 of them (2,000 in a later phase), with a maximum distance of 150 km. Unlike previous dish antennas, they will be made from carbon fiber, to provide a very high surface smoothness and make them suitable for frequencies up to 350 GHz or 0.85 mm, the wavelength down to which maser-based time synchronization is currently feasible. Efficient and consistent mass production is a particular challenge, because the SKA will need so many of these antennas. Cost, both of manufacturing and operation, is also a major concern at these quantities.

The dish antennas will be steerable, although a research program is under way to use beamforming for the receivers. When we discussed Effelsberg and ALMA we said that a radio telescope typically only records a single "pixel", capturing the data from the beam formed by the physical orientation of the antenna. The SKA antennas will eventually feature a phased-array feed in which the receiver acts as its own little array of antennas, using beam-forming to "look at" different locations of the dish. Mechanical steering will take care of coarse-grained positioning, whereas the phased-array feed will allow fine adjustment of direction or scanning within the region pointed to by the dish. Again, the same benefits: (slightly) different directions can be focused during beamforming, improving the survey speed significantly: it will be 60 times more efficient that the current fastest one, the VLA.

With all the interferometry and beamforming, the SKA will rely heavily on computing and the ability to transfer data in realtime to central correlators: together, the two arrays produce around 8 TB/s, which is $8 \cdot 60 \cdot 60 \cdot 24 \cdot 31 = 21{,}427$ PB per month—according to Wikipedia/Cisco, this is roughly the total monthly internet traffic from 2010! But of course data transfer and computation gets cheaper every year, and this tradeoff is a trend in most areas of science.

Although construction of the SKA has not yet begun, pathfinder experiments for its various new technologies have been going on for a while; for example, 64 of the 200 dish antennas are already built (with 13.5 meters) as part of the MeerKAT telescope, and prototypes of the phased-array-based receiver are being developed.

As with the TMT, there is opposition to the use of land in Africa, and a budget has not yet been committed. Let's hope these road blocks can be resolved and this fascinating project has a chance to be realized. Watch out for a future episode of omega tau.

7 Models in Science and Engineering

Before we continue with more big physics and its engineering, I think it is appropriate to discuss what a model is and why models are so central to both science and engineering.

Let's start with model railways. This might be the first time in a child's life where the word "model" is used. In this context, a child probably understands the term to mean "small": it's a smaller version of a real railway. Being smaller is often a characterizing feature of a model, because smaller usually means cheaper and faster to build. Examples include both model airplanes such as those used for play by children and those used by researchers in wind tunnels, as are models of buildings and other civil engineering works. Of course there are exceptions: models of atoms or other microscopic phenomena are very much bigger than the real thing.

For a model to be a good one, however, it has to be "like" the real thing: it has to be realistic. As a child I was very annoyed if the proportions of an H0 locomotive such as the Deutsche Bahn BR 103 didn't quite resemble those of the real thing, or if the livery was off. Or when the landing gear on a model glider didn't extend as far down as on the real thing. Apart from serving as a toy for children, a model that looks like the real thing can be used to illustrate something to people who have never seen the original. That's useful, but not very. We build models because they help us understand something about how the real thing works, beyond just the visuals. What else can we do with a model train? We can drive it into a railroad station and stop, wait a bit, continue on to the next station, stop again, and so on. So we can replicate what the real thing is used for—but only to a degree. For example, the proverbial aliens would probably ask themselves why the train stops periodically. After all, the model railway doesn't include model passengers that enter and leave the train when it stops at a station, so the purpose of a real train, and why it stops, is not apparent from the model. Unless you know the real thing, as most children do, because then you can use your imagination to fill in the missing details such as passengers. The whole thing then makes sense.

What else can we learn about the real world from the model? Assuming we have a similarly down-scaled version of the Stuttgart station, can we use the model to understand how many trains can fit into the station and what their maximum length can be to still fit onto the platform? Yes, we can do this, because the only property we require to be realistic is the relative sizes. Can we measure how many railcars a locomotive is able to pull and then extrapolate to the real thing? Probably not, because the power-to-weight ratio is not the same as in the original. Can we use it to find out, experimentally, how fast a locomotive can safely drive across a switch? Very clearly not, as the material properties of the model and real-world tracks are different, and so is the dynamic behavior of the locomotive and the railcar.

Can we use our model of the Stuttgart station to come up with a scheme for positioning the switches in a way that guarantees that trains never collide? This is actually a non-trivial problem. Train tracks are divided into block sections, and only one train can be in each block at any one time. Signals enforce this rule by requesting trains to stop before a block that is already occupied by another train. Any such red signal must be preceded by an advance signal that tells the driver to expect a red signal, so he can slow down. If the driver ignores a red signal, magnets signal this to sensors on the locomotive and it stops automatically. Sensors on the tracks count axles as they enter and leave blocks to detect railcars that are separated from a train and stay back in a block. Flank protection ensures that no switch from a parallel track leads into a protected block from the side, independent of whether a train is allowed onto that parallel track. An interlocking scheme prevents combinations of switch configurations and signal indications that would be dangerous. It's a really complex system, and using models to figure out configurations that are provably safe and at the same time maximize utilization of a given set of tracks is a good idea. So could we do this with our model? Assuming we have a faithful replica (same tracks, switches and signals) and we are only interested in the algorithmic aspect of the switching scheme, then yes, the model could be used.

What is there beyond the algorithm? What are the limits of our model? The switches in our model probably switch faster than real-world ones, and we likely don't simulate the impact of ice on switches, which might prevent them from changing position, or other technical malfunctions. So if we were to use the algorithm in the real world, we might want to install a separate system that verifies that a switch has actually moved to the commanded position after a timeout, and that signals actually show the color or state that the algorithm requests. This is done in real railway infrastructures.

Do we need a physical model to do this? Not really. We can perform the same task with a pen-and-paper model. Or more realistically, with one expressed in software. Software models are great, because they allow us to tune their fidelity: we can model a switch as a binary variable with two states and an immediate change of position. Alternatively, we can model the position of a switch as a real number between zero and one, and when we command a change we can make it change its position slowly from zero to one, simulating the fact that a real switch takes time to change position. We can add a random element that prevents a position change when commanded to simulate faults. We can simulate system faults such as icing by relating simulated faults of multiple switches that are near each other, maybe even based on simulated weather. It gets complicated, but it's possible in principle.

So whenever we use a model to understand or predict aspects of the real world, we have to be sure that our model considers that particular aspect, that it is able to make a meaningful statement about what we are interested in. This is the most important caveat when using models. In the rest of this section we'll explore models and their use further.

1 | *Mathematical Models*

The best-known kind of models are mathematical models. One of the simpler ones is the model for the simple pendulum. Here it is:

$$\frac{d^2\theta}{dt^2} = -\frac{g}{\ell}\sin\theta$$

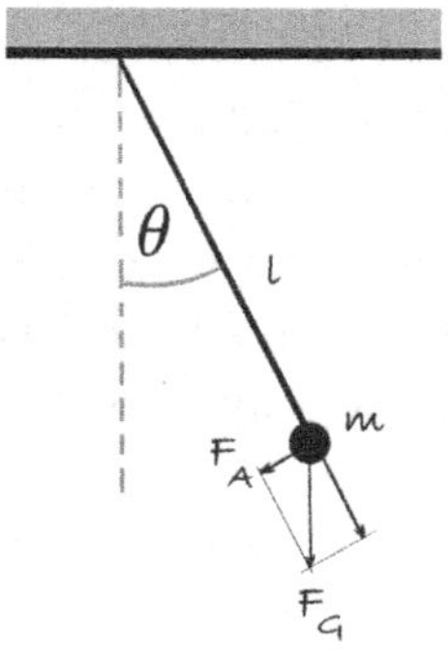

Here, l is the length of the pendulum, g is the acceleration due to gravity and θ is the pendulum's angle at time t. The mass m of the pendulum does not appear in the model because it not relevant. Relevant for what? What are we asking the model to explain? We want to be able to describe the motion of the pendulum. We know that it goes back and forth, and so one important question is: what is the frequency of that back and forth motion?

The formula above is a differential equation. A differential equation relates a quantity to a derivative of the quantity. A derivative represents the quantity's rate of change. The d^2/dt^2 means that this is the second derivative. The second derivative of an angle can be understood as the "angular acceleration", the rate at which the rate of change of that angle changes. It is intuitive to understand that for bigger θ we get a greater acceleration: F_A gets larger with $\sin(\theta)$. If this is not obvious, it helps to consider the extreme values: for $\theta = 0$ we see that $F_A = 0$ and for $\theta = \pi / 2$ (corresponding to 90 degrees) F_A becomes maximal. The minus sign indicates that the acceleration works "against" the angle, which is also intuitive, otherwise there would be no back and forth motion—it would just be a continuous acceleration.

Can we do something with the model above? For example, can we compute the frequency or period of the pendulum? It turns out that this is not possible in the general case, because this differential equation is non-linear. There are a bunch of mathematical tricks, including elliptical integrals and series, to help compute the period. Just for the fun of it, I'm including this screenshot from Wikipedia (CC A-SA) that computes the period depending on the initial deflection after applying some of these tricks:

$$T(\varphi_{\max}) = T_0 \sum_{n=0}^{\infty} \left(\frac{(2n)!}{(2^n n!)^2} \right)^2 \sin^{2n}(\varphi_{\max}/2)$$
$$= T_0 \cdot \left(1 + \left(\frac{1}{2} \right)^2 \cdot \sin^2\left(\frac{\varphi_{\max}}{2} \right) + \left(\frac{1 \cdot 3}{2 \cdot 4} \right)^2 \cdot \sin^4\left(\frac{\varphi_{\max}}{2} \right) + \dots \right)$$

It is not uncommon for it to be easy to write down a mathematical model but then hard to solve it. For example, I chatted with Ralf Kleiber, a theoretical physicist at the Wendelstein 7X fusion experiment. I asked him how complex the math is that represents the model of the plasma in the reactor. He replied that this depends on how abstractly you formulate it. The most abstract form is $df/dt = 0$, where f is the distribution function of the plasma. I was slightly perplexed, and asked him what they do all day. "Solve it", he replied. There are hundreds of effects that mess with the plasma. Each has its own theory, and some of them have their own math behind them. Taken together it would be too complex to solve them. We will return to this conundrum later.

2 | *Applicability*

Can we simplify the math for the pendulum? We can if we limit ourselves to small angles of θ. If you look at the sine function, it turns out that around zero $\sin\theta = \theta$. With a particular starting angle θ_0 and assuming that the pendulum is not moving at $t = 0$, one can compute the period T as:

$$T = 2\pi\sqrt{\frac{l}{g}}$$

We have seen before in the context of the pendulums in gravitational wave detectors. But it is important to stress that this is only valid for small angles; certainly true for seismic vibrations in those detectors, but probably not for traditional Black Forest cuckoo clocks.

All formulas for the period of pendulums have ignored aerodynamic drag. Can we justify this? It depends. If you have a thin threadlike pendulum with a lead ball on the end, it will take many swings to practically measure the impact of drag. If instead you replace the ball with a feather, the pendulum will probably never swing beyond the vertical even once. Why? Because the accelerating force depends on the mass, but the drag force depends on the area and the shape (represented by the drag coefficient c_D). These couldn't differ more between a lead ball and a feather. Make sure you can justify a simplification if you make one!

The idea of restricting the range of applicability of a model and then making simplifications that are valid in that range is a common approach for dealing with complex models. Simplifications, in turn, are often necessary to make an analytical solution feasible in the first place, or because it will speed up computations (especially if you have to do lots of them). Assuming a planar wavefront in our discussion about multi-beam sonars is another example of such a simplification; it is valid if the receiver array is far enough from the reflecting surface that the radius of the sphere is "approximately enough" planar.

In many cases it is the simplification that has first been discovered historically, and people were not aware that it is actually a simplification of a more complex model. Consider Newton's $F = m \cdot a$. We can rearrange this to $a = F / m$, which says that if we apply a force F to a mass m, an acceleration of a will result. From $v = a \cdot t$ we can derive $v = F / m \cdot t$. If we assume a constant force (for example, an ion engine on a space probe) and a constant mass (which is another simplification because a spacecraft becomes lighter as it burns fuel), then over time its speed will increase forever. This is in line with our experience on Earth (assuming another simplification: no drag or other forms of resistance), which was the realm of Newton's experiments. It is even valid for "normal" spacecraft. But we know from our discussion of relativity in the chapter on gravitational waves that mass increases with the γ-factor:

$$m = m_0 \frac{1}{\sqrt{1 - \frac{v^2}{c^2}}}$$

So, assuming constant engine force, acceleration will decrease as v approaches the speed of light c, and ultimately stop altogether because m approaches infinity. Newton wasn't wrong, it's only that his laws are limited to $v << c$.

3 | *Numerical Models*

Let's go back to our pendulum. Is our differential equation from a few paragraphs back a good model? Does it faithfully represent the real world? Probably not. We all know that over time the amplitude of a pendulum's motion will become less, and at some point the pendulum will stop its back and forth motion. We don't see this effect in the formula above. In fact, this formula represents the mathematical pendulum, not the physical one, which is a serious simplification. It assumes that the rod (or wire) is massless, the mass at the end of the rod/wire is point-like, the motion is purely along one dimension, and, importantly, that there is neither aerodynamic drag nor other kinds of resistance within the pendulum's material. We know that both cannot be true; we've learned about aerodynamic drag in the chapter on aviation, and internal friction (and the creation of heat) in the context of the damping pendulums of gravitational wave detectors.

I have no idea of the mathematical model of the internal frictions inside the rod/wire material. But we can see how we might take care of aerodynamic drag. We know its formula is $F_D = 1/2 \cdot \rho \cdot v^2 \cdot A \cdot c_D$. The speed v can be derived approximately from the angular speed (the first derivative of θ). Let's see how this turns out:

$$mg\sin(\theta) - \frac{1}{2}\rho\,\dot{\theta}^2\,c_D\,A = m\,\ddot{\theta}$$

F_A F_D $m*a$

This equation is fundamentally a variant of $F = m \cdot a$. Here, a is expressed as the second derivative of the angle; instead of writing d^2/dt^2 we use the two dots above θ, which is a common shorthand for a derivative relative to time. The overall force is then composed from the accelerating force F_A (downhill-slope force created by the angle) minus the aerodynamic drag F_D; we represent speed as the first derivate (rate of change) of the angle. Unfortunately this differential equation cannot be solved analytically; we have to resort to a numerical approach.

Solving the equation numerically means that we step through time increment by increment, and compute the quantities we are interested in (θ in our case). How do we deal with derivatives? We discretize. As mentioned, the derivative represents the change of a quantity over time for an infinitesimally small time delta. When we work with a discretized version of a differential equation, we use a small but finite delta. To make this concrete: the first derivative of θ, the angular speed, can be computed numerically as:

$$\frac{\theta_t - \theta_{t+\Delta t}}{\Delta t}$$

This formula calculates the change in angle between two points in time separated by Δt, divided by that timespan Δt. Assuming that we specify starting conditions, we now simply iterate through time and, for example, plot the result. Let's use a steel sphere of 2 cm diameter that weighs around 33 g. With no drag a pendulum will swing forever; the following plot shows amplitude over time with no damping for the first 50 seconds. Since I am a physicist and not a mathematician I'll take the liberty of extrapolating from 50 seconds to forever.

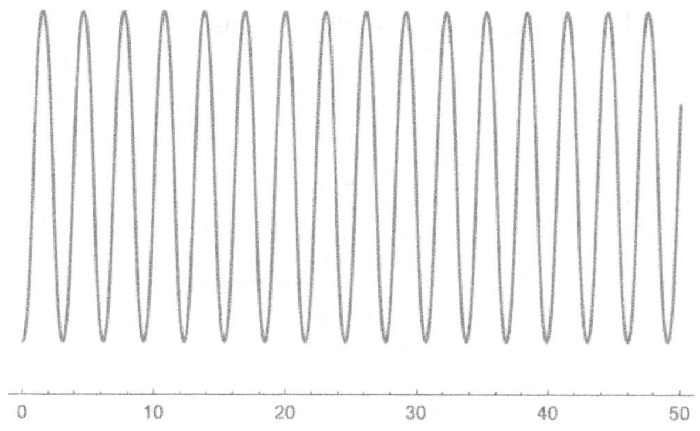

Once we consider drag (with the standard ρ=1.225 kg/m^3 and c_D=0.15 for spheres) we get this damped oscillation:

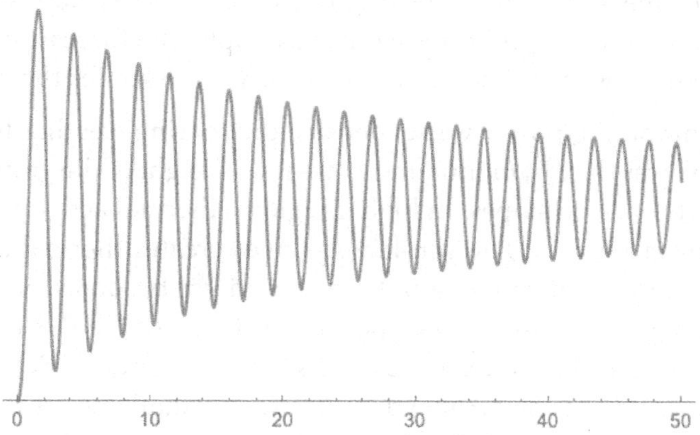

Numerical methods are used for the vast majority of interesting mathematical models, including the models of plasma behavior inside the fusion reactor mentioned above, for the kinds of finite element models that were used to calculate Effelberg's homologous mirror, as well as those used for weather forecasting and climate simulation. Weather modeling that is of particular interest to me as a glider pilot are the forecasts that tell us when we can expect strong thermals. Those models don't just discretize time, they also discretize space: the forecasting area is broken down into segments that are treated as one unit in terms of weather parameters (altitude above sea level, sun angle or cloud cover). The vertical atmosphere is also discretized, using single values for parameters like humidity or temperature. An important aspect of these simulations is to compute the "spatial derivatives", to understand how the values influence neighboring atmospheric volumes through exchange of temperature, humidity or other physical parameters relevant for the development of thermals.

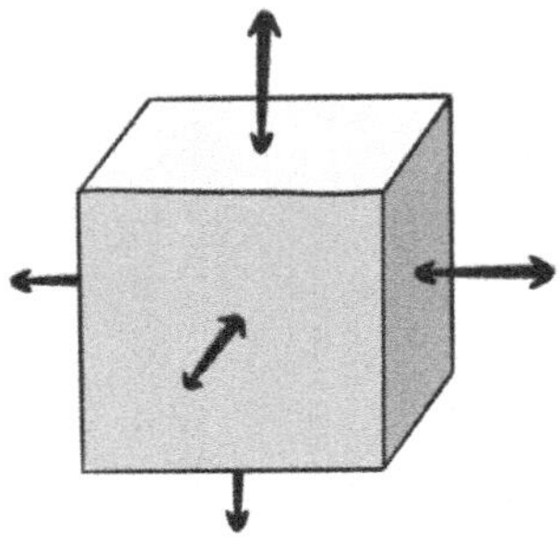

The smaller those discrete volumes are, the more accurate the forecast will be; there's less averaging going on. On the other hand, the required computing power also increases as the volume size decreases, because there are more volumes for which the parameters have to be computed for each time slice. An interesting question to ask is how small these volumes should be to produce a meaningful forecast. For example, the forecast created by the European Center for Medium-Range Weather Forecasting uses 16 km for their standard forecast.

In the days when computing power was expensive the soaring weather forecast for the Alps was really poor. Why was this? In mountainous areas the height of the ground changes quickly because there are lots of steep slopes. In higher regions, thermal activity begins earlier in the day compared to lower regions. The sun-facing slopes creates thermal activity that is more intense than for flat surfaces. If your computational grid size becomes too coarse-grained, you effectively average over all those slopes: the model does not "see" the mountains. Consequently the forecast was as if the Alps were just as "hilly" as, say, the Schwäbische Alb, a low rolling mountain range. That was wrong: thermals were usually better than the forecast and they started earlier. Nowadays specialized gliding forecasts use smaller grid sizes and take the data produced by the European Center as boundary conditions.

So, in addition to coming up with useful discretization, numerical modelers must also find out what the minimum discretization step is to make sure the model "sees" the effect that it is intended to predict or analyze.

4 | *Precision*

The tradeoff between accuracy and computational effort is a really important one for mathematical models, particularly those that must be solved numerically. This is because the code that represents the model is often executed many times: weather services run the same code for each of the volume boxes and for a large number of time slices. They execute the code even more often to "run the model forward", to integrated new measurements. Computing power gets cheaper over time, of course, but it is not free, so making the code more efficient is crucial.

One way to do this is parametric abstraction. Let's look at cloud formation in the context of thermals, which we discussed earlier. The sun heats the air, warm air is lighter, so the air rises, and as it rises it expands adiabatically. This reduces pressure, which in turn lowers the temperature, and that ultimately stops its rise. Depending on the relative humidity of the air bubble, the amount of cooling and a few other parameters, it will reach 100% relative humidity at a some altitude, at which point a cloud forms. So how do you determine the amount of cloud you will get on a particular day? One option is to numerically "run" a whole bunch of air bubbles and follow their way up through the atmosphere to "form" every cloud separately. Obviously this is very expensive computationally. An alternative approach is to parametrize: you define a very simple function that takes a few inputs (such as temperature, dew point, humidity on the ground, date and time) and then computes a number between 0 and 1 that represents the percentage of sky that is covered by cumulus clouds resulting from thermal activity. If you cannot capture the relationship as a formula you can always create a lookup table; while they consume more memory, they are even cheaper than functions computationally.

The approaches can be combined, as illustrated by the flight mechanics model of the Heatblur F-14 Tomcat plugin to the DCS flight simulator. Let's say you want to compute how the Tomcat rolls into a turn when the user deflects the stick. Stick deflection leads to the flight control system using the hydraulics to extent the spoilers and move the differential elevators with a particular rate of change (the F-14 does not have ailerons). That deflection, together with the current wing sweep and flap setting leads to a particular differential lift coefficient. With the current speed you can compute the resulting aerodynamic force. That force, working against the roll inertia that depends again on wing sweep and on the fuel in the wings then produces a roll rate.

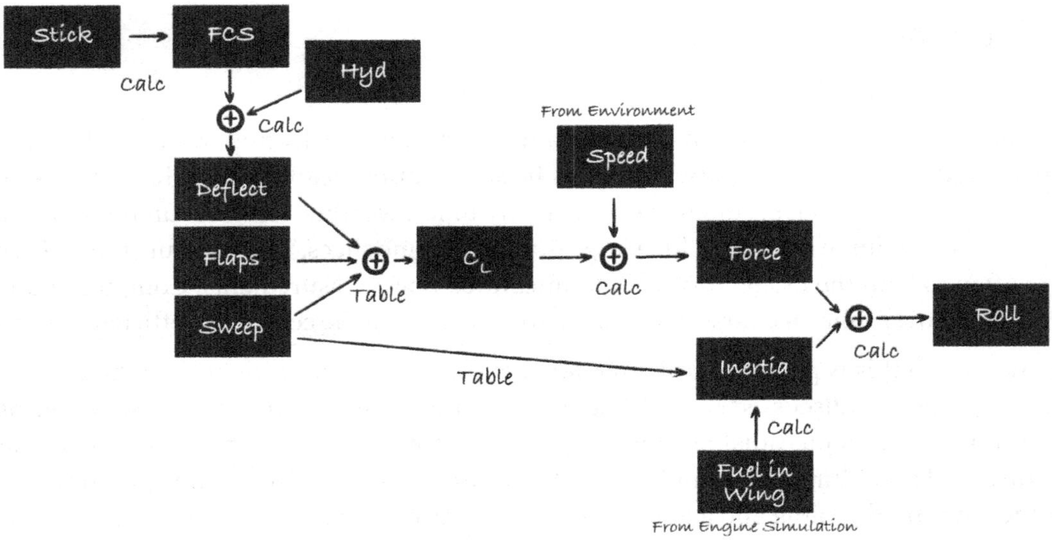

So how are these different computations implemented? The flight control system and how it translates stick deflection to a command can be simulated easily with a few formulas and if-statements. These are parametrized formulas, because the original system was an analog computer, and you don't want to actually simulate the circuits. Taken together with the hydraulics, you can calculate how far the aerodynamic surfaces will deflect, and how fast they get there; another parametrized calculation. At this point you know the configuration of the wing and you need to find out the lift coefficient for this configuration. To calculate that, you'd have to run a computational fluid dynamics simulation. That is too expensive to run in realtime on a PC flight simulator. Instead, the simulator uses lookup tables based on NASA wind tunnel data to find out c_L. Once we have that, we can calculate the rotational force, or moment, that the aero surfaces produce. In order to roll the aircraft, they have to work against roll inertia. This depends on the wing sweep via factors taken from another lookup table as well as the current amount of fuel in the wings; this can be known from based on the initial fuel load and the consumption in the flight so far as integrated by the engine simulation. Interestingly, the last two three calculations (fuel to inertia, coefficients to force and force + inertia to roll rate) use the actual Newtonian formulas, there's no approximation based on parameters. A nice property of this approach is that you can easily simulate faults realistically: let's say your hydraulic system has a leak and your pressure is only 50% of the nominal pressure. Then the flight characteristics will change "automatically" in the sense that the surfaces move more slowly and the whole downstream chain of quantities adapts automatically; you will roll more slowly as well.

If you parametrize your calculation, how do you come up with suitable parameter values? You can either run experiments (that's what NASA did in the wind tunnel for the F-14), or systematically observe the real world (a meteorologist taking notes over the years, or more likely, performing a statistical analysis of lots of observational data). Or you can run the higher-fidelity, more computationally expensive models once for all combinations of inputs and capture the resulting correlations as parameters.

It is crucial that your abstracted model captures the relevant inputs. If you have too many, your model will be unnecessarily complex. More importantly, if you forget inputs that are relevant, your model's predictions will be wrong. In the example above I mentioned temperature, dew point, humidity on the ground, as well as date and time as inputs that determine cumulus cloud development. One parameter that is definitely missing from this list is the density of higher clouds: a layer of altostratus or cirrus covering an area will severely reduce the development of thermals because the energy imparted by the sun is lower.

Dealing with clouds is a crucial issue. Early climate models got this wrong, which is why some of the early global warming predictions were inaccurate. A similar problem occurred in the context of arctic ice cover and its impact on global temperature. We know that ice has a high albedo—a high reflectivity: lots of solar energy is reflected back out to space and not absorbed by the ground. However, as ice melts, water puddles form. These are much darker than the ice—they have lower albedo—and therefore absorb more energy, accelerating the melting process in a positive feedback loop. Getting the parameters that represent the puddle-forming activity right was an important step forward in climate models.

This is a good place to cite two different kinds of "unknowns", a distinction commonly attributed to Donald Rumsfeld: known unknowns and unknown unknowns. The former are things for which we know we have to do more research to understand correctly: we are aware of the shortcoming. To better understand their effect on a model we can perform a sensitivity analysis (also known as ensemble predictions in weather forecasting): just run the model with a range of values for that known parameter. If the model's outputs vary widely, we know that it is important to better understand the parameter. We can then perform experiments or run more detailed models to find out which value is accurate. And we know that the model's current outputs are not very trustworthy. A related technique involves error calculations. For each input value you also specify an uncertainty range; the model then takes this into account to tell you the level of trust you should put in the result.

Unknown unknowns are harder, because we are not even aware that they are relevant: we don't know that we must investigate their impact on the model. For example, early models of thermal activity considered only the impact of temperature on thermal strength, the speed with which an air bubble rises. They did not take into account differences in humidity. Unknown unknowns are the primary source of (serious) errors in models, and it is hard to give a general recommendation for avoiding them other than checking the model against the real world.

Rumsfeld also talked about the "known knowns" (things we consciously understand) and the "unknown knowns". These are effectively biases, things we "know" to be true without consciously considering or questioning them.

This continuum of model precision, from high-fidelity "physical" models to parametrized ones, was nicely illustrated by Wendelstein's Ralf Kleiber. I asked him which model they use to represent the plasma in the toroid. He replied that "There's not a single model; we use a wide range of models". The most obvious model treats the plasma as a mixture of two gases, each at different temperatures. The physics is based on the kinetic gas theory, in which the constituent particles exchange energy when they collide due to temperature. An even more precise model takes into account the fact that each particle moves along tiny circular tracks around the magnetic field lines. A simplification treats them as little disks, which increases their collision probability compared to "normal" electron or proton gases. A less detailed model (called magnetohydrodynamic) considers the plasma as a liquid; the particles are not treated separately.

This latter is used in optimizations. We talked about sensitivity analyses before, in which inputs are systematically varied to understand their impact on the model's behavior. In an optimization we effectively vary all of them. We do this because we want to reach a particular goal, usually expressed as a maximum or minimum in one or more of the output parameters. For example, we might be interested in maximizing the gap between the plasma and the walls of the toroid. Do we have to increase the magnetic field, or reduce it? Inject more gas to increase density? Inject more neutral particles? Make the plasma hotter using microwave beams? To find out, we have to try all combinations of input parameters—dozens or hundreds for complex phenomena like plasmas. This results in a huge space of combinations and a correspondingly large number of optimization runs. Let's say we have 100 parameters with 255 values each, then we have a data space of 255^{100} combinations to try.

To make such an optimization computationally feasible, you must make sure that the model is as efficient as possible, to reduce the cost of each iteration. For such optimizations the Wendelstein team typically uses the model that treats the plasma as a liquid. But you also have to be clever about trying. You can't really try out all 255^{100} combinations—you can't afford the necessary computing power. A wide range of mathematical algorithms is available that try to be more deliberate in how parameters are modified (hill climbing, simulated annealing or generic algorithms, to mention a few).

Just as importantly, however, you have to respect physics. There is no point in trying a low value for pressure and a high value for temperature, for example, as physics tells us that they are causally connected. You can even argue that a model should not include both as independent parameters; you should only parametrize one and let the model compute the other. Another common way of checking that a model respects physics is for values: for example, you cannot have pressure in a tank that is twice as high as the tank can withstand, and, importantly, values typically cannot jump for successive points in time, or between neighboring volumes (in a space-discretized model). More generally, where any physical quantity has a range of values that is physically meaningful, this is also true for its derivatives.

5 | *Prediction vs. Explanation*

Let's go back to our tabular parametrization of cloud development depending on temperature, dew point and humidity on the ground, as well as date and time. Does this model actually explain how clouds form? Can you derive a physical mechanism from the parameter values in the model? Of course you cannot. The model allows you to predict an outcome based on a set of inputs that are found to be relevant, but there is no formula that causally relates physical quantities according to physical processes. Generally, models that contain parametrized abstractions have no explanatory power, even though they can predict an outcome.

Here's another example. As we know, the universe has been expanding since the Big Bang. Based on the mass distribution in the universe and the resulting gravitational effects, calculations show that the rate of expansion should slow down. However, observation shows that the rate of expansion is increasing. So physicists became creative and invented "dark energy". Dark energy is basically a parameter introduced into the calculations that "magically" propels the expansion. Does dark energy explain a mechanism? There are a few hypotheses based on the notion of vacuum energy, but nothing has been confirmed through experiments or observations. It's really a mathematical trick: it makes an existing model of the universe that does have explanatory power (relativity, gravitation, expansion) fit with the observations.

The ultimate example of models that can predict but not explain are neural networks. As the name suggests, neural networks are loosely modeled on how our brain works (or was thought to work when they were named): neurons and connections. Neurons are connected, and electrical signals propagate along the connections. A downstream neuron fires depending on the signals that come in through its upstream connections. Neurons are organized into layers, and connections are always between layers; a neuron on layer n is (usually) connected to all the neurons on layer n–1. How many layers there are and how many neurons are in each layer depends on the problem, and is a parameter that can be adjusted by users.

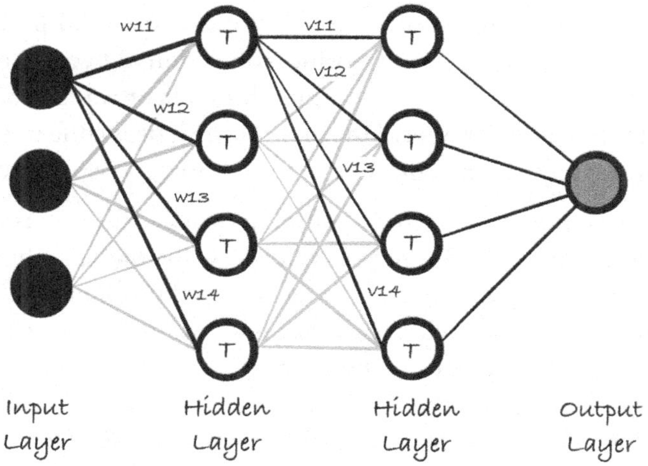

The decision whether a particular neuron fires is made based on a nonlinear transfer function, T in the illustration. Each connection has a weight. When making the decision about whether to fire, a neuron typically sums its inputs and uses the transfer function to compute its output signal. The "knowledge" of the model is encoded in the weights of the connections. After (manually) encoding a real-world problem (speech, image, data) as a pattern on the input neurons, the network "predicts" an output. It might recognize a spoken word, detect a cat in an image, or identify your buying patterns and recommend other stuff for you to buy.

Before the model is able to do this, the weights have to be set "correctly". This is done through learning. The default approach to learning is to use an iterative process based on examples. The set of examples is separated into a training set and a test set. The network is fed with each training example, one by one, together with the expected output: the to-be-recognized word, cat/no cat or a liking for blue shoes. From a starting point of random weights, the network then adjusts the weights in a way that minimizes the error of its prediction. Validation then happens by measuring the average error for the test examples. This process is called supervised learning because a human supervises the process (through the preparation of the learning set and the validation set).

An alternative approach is to use a more generalized cost function. Imagine a neural network that controls a car. Whenever the (still learning) network crashes the car, the cost function rises; the weights are adjusted so as to minimize the cost. This approach resembles general optimization strategies and is also known as "unsupervised learning".

The crucial point is the adjustment of the weights as a consequence of an error or rising cost. For this to work, the nonlinear transfer function is differentiated; the derivative then guides the adjustment of the weights (lots of mathematical details here).

Once the weights are set, the network is able to predict outputs based on inputs, as one would expect from a model. However, the weights have zero explanatory power for how it does this. Tests show that it works, but you don't know why; you cannot learn about how the world works by "inspecting" the neural network. You don't even necessarily understand which layer does what, even though they often evolve into a pseudo-sequential process, for example in image processing: edge detection, relative position detection, detection of areas, and finally, recognition of "semantic" categories such as tables, humans or walls in a room. Contrast this with our analytical mathematical model of the pendulum: it uses concepts from the real world (angle, mass, area, drag, shape of the body) to "explain" how the world works.

As a consequence you cannot argue about a neural network's validity. You can only test, hoping that you test all relevant cases. You have no idea how the model behaves for all cases that you do not test. Ensuring that you test everything is a hard problem: the Tesla autopilot that drove into a truck whose trailer looked like the sky is an example. In the analytical model you can look at the formula and say "Ah, well, there is no term that represents the internal friction of the pendulum. Shouldn't we add that?" A numerical model that "runs" the real physics has the same property. However, as you parametrically abstract some of the real-world processes, prediction might still be good enough, but explanatory power diminishes. In some sense, a neural network is the ultimate parametrized numerical model: a completely generic engine that encodes all knowledge in opaque, semantically meaningless parameters.

6 | *Past and Future*

There is another interesting aspect to the distinction between explanation and prediction: the former looks at the past, while the latter concerns the future. If you build a model based on data from the past (as you have to), then any prediction of the future is only valid to the degree to which the future will (statistically) be similar to the past. The following anecdote, which I first heard from Nassim Nicholas Taleb, illustrates this nicely:

Imagine you're a turkey. You live happily on a farm, and every day the farmer comes to your cage and brings food. Every day you use your accumulating experience from past days to build a model of the world in which the farmer is a good person, because he brings you food. The degree of goodness, based on this Bayesian reasoning, becomes stronger every day. And then, one day, probably a few days before Thanksgiving, the farmer kills you.

Your model would never have predicted this event, because there is no experience in the past that lets you derive any particular probability for it. Taleb calls these improbable but consequential events "Black Swans". He emphasizes that they cannot be predicted, because, as a consequence of their rarity, models built on past experience cannot infer a probability.

More generally, if a model is based on detailed information of the past, it might be very good at reproducing—and thus explaining—the past, but it might not be as good at predicting the future because it was "overfitted". I explained the notion of (over-)fitting in the context of image reconstruction for the EHT: in some sense the model is too precise. It is built on too much of what turns out to be irrelevant data.

To avoid this it's often better to use only very little but highly relevant data, organized as a simple decision tree, instead of, for example, a neural network that has been trained with lots of historical data: finding out what data is "highly relevant" is not necessarily easy though. I first heard about such "fast-and-frugal" trees from Gerd Gigerenzer, who described a case in which one was used to make treatment decisions in an emergency care unit. Patients would come through the door with suspected heart disease and doctors had to decide whether to put them into the intensive care unit or onto a regular ward. A fast-and-frugal decision tree that has proved highly predictive is the following, which is described in a 1997 paper by Green and Mehr:

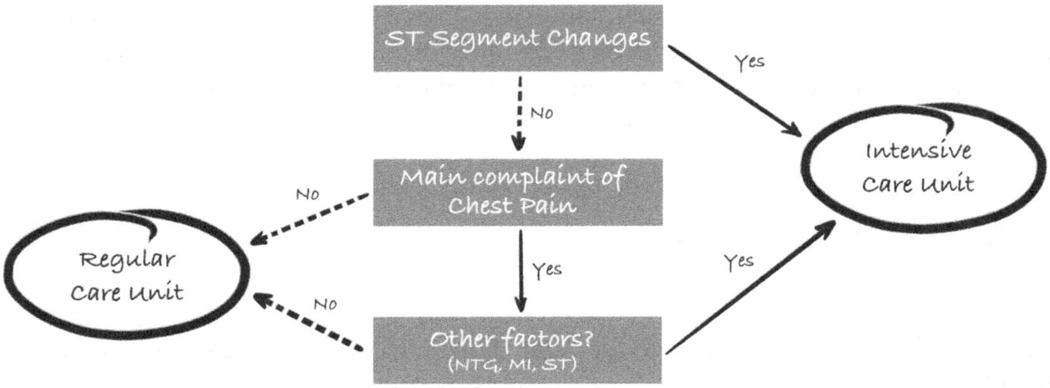

The first criterion, "ST segment changes", refers to a particular pattern in the ECG diagram; if it is present, put the patient into the intensive care unit, otherwise check whether the patient complains about chest pain. If yes, check whether any of three other factors is present, and if so also move the patient into the ICU; in all other cases, the patient can go to a regular care unit. This tree is so simple that no overly complicated diagnostics are necessary, and it can be executed by emergency medical staff. The performance of this and other similar trees is comparable and in many cases better than predictions based on a more extensive data set. This is probably because such trees avoid overfitting to ancillary data that "polluted" historical data sets. It's also a bit ironic, because doctors have used such decision trees forever, long before Big Data became available, and now we learn that at least some of them perform better than analyses based on more extensive data.

7 | *Statistical Models*

Parametrization approximates some of the underlying processes in a way that reduces computational cost. Depending on the quality of the approximation, the predictions made by the model will be more or less precise: a pendulum will "stop" after 10 minutes instead of 11, there will be clouds at 10:15 instead of 10:00, and thermals will start at 09:00 above a small mountain. When defining a parametric abstraction you have to decide how much error you can accept. The models are usually deterministic, however: they give the same result for the same inputs every time.

There are other models, though, models that only make statistical predictions. For example, neural networks will occasionally get it wrong. They may misidentify a dog for a cat, maybe because they have learned to associate cats with a feature in the image that is not actually exclusive to cats. One particularly painful example was when one such network identified black people as monkeys because the training set only used white-skinned people as examples of humans. Another one, when trained to recognise sheep, actually learned to use the rolling green hills (on which the sheep grazed in all the pictures the used for learning) as the pattern for "sheep". So neural networks (and other learning models) are correct to some statistical percentage, which is often well above 95%—as long as the task is within the bounds of the training material, which is sometimes hard to know.

Quantum processes are similar: here the actual physics is probabilistic. For example, a particle can cross an energy barrier even if that particle does not have enough energy to climb over it: it can tunnel through the barrier. The probability of this happening is very low but not zero. In the real world we don't see these effects because, for example, all the atoms of your cat would have to "decide" to tunnel onto that four-meter-high shelf at exactly the same time.

Working with statistical models requires that you run the model lots of times and analyze the results statistically to determine if the model is "good". It also requires you to be able to live with the occasional wrong result. Not like the average imprecision you might get from parametrization, but a completely wrong decision in some cases.

Parametrization, statistical models and the physical limitation of detecting signals in high-noise environments, which often leads to the use of parametrized and statistical models as part of the measurement, are a major reason for the five sigma requirement for physical experiments we discussed in the context of gravitational waves.

8 | *Agents and Emergence*

You've probably heard about the butterfly that takes flight somewhere in Brazil, slightly disturbing the energy balance in the air directly above the ground, making an air bubble rise, which then becomes the deciding factor in the overdevelopment of a cumulus cloud into cumulonimbus, which creates a thunderstorm, which then, ultimately—somehow—affects the weather in Europe. That particular chain of events is probably apocryphal, but it is true that there are systems in which small changes can lead to large effects on the outputs. If you plot the system's evolution for a long time, such small changes can make the difference between very different trajectories in terms of the system's evolution. This effect is a major reason why weather forecasts beyond six or seven days ahead are approaching chance and are thus of limited use.

Weather is a complex system. It has lots of "moving parts" that all (transitively) affect each other. Because of the way these parts are causally connected, the prediction of the system's evolution based on a model is hard. At first glance, this is similar to the kinetic gas theory model of a plasma: there are also lots of moving parts—the gas molecules—that interact with each other. This is why we can only predict on average how gas will behave in terms of its pressure and temperature. For weather, however, we are interested in more local predictions. These are harder the further into the future we try to forecast. The analogy goes further: we can reasonably predict weather averages, which we then call climate forecasts. This is the reason why the global warming skeptics' argument that "If we can't predict weather for a few days, how can we hope to predict the long-term climate" is nonsense.

Systems that are extremely sensitive to initial conditions are called "chaotic". They usually contain a variety of interactions and positive feedback loops between system components. A simple and well-known system is the double pendulum, appropriately dubbed the "chaos pendulum". It has only two degrees of freedom, but its motion is chaotic nonetheless.

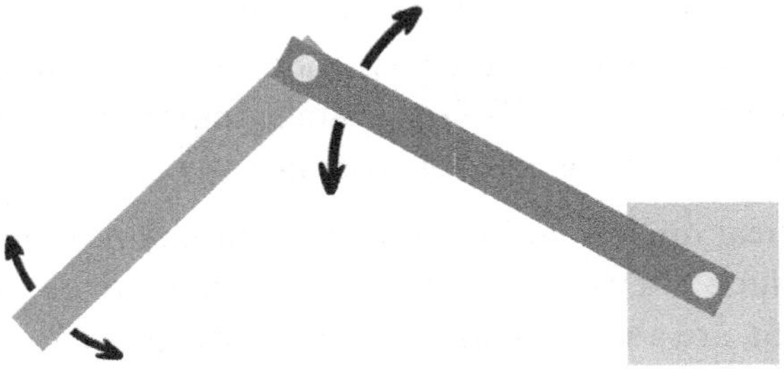

Chaotic systems are fundamentally deterministic, meaning that they follow rules that are well understood; there is no randomness involved. If we could measure all initial conditions to infinite precision, and if we could model every interaction precisely, then we could predict the system's evolution. In practice of course we cannot measure to arbitrary precision, and simulations are often limited by numerical precision issues and rounding. Consequently such systems are extremely hard to model. The only way to find out their behavior is to run them, step by step, and see how they develop. This is called path-dependence: you have to run, you can't predict.

Here's another story you might have heard: an ant colony is able to find food and transport it back to the anthill in what looks like a manner that is orchestrated by one big master ant. In reality, however, each ant has a very simple "program" that has to do with how it follows pheromone tracks on the ground, and when and how it creates such tracks by excreting pheromones. This phenomenon is called emergence: seemingly complex coordinated behavior emerges from simple rules. There are many other examples. Stephen Wolfram's New Kind of Science explores this idea in detail; he demonstrates how cellular automata, a particular kind of mathematics, can produce complex patterns from simple rules. He extrapolates this idea to all kinds of other real-world systems (which makes the book quite controversial). A primary example of emergence, however, is our society. Each human and each organization behaves according to relatively simple rules that (mostly) take into account only the local context. The evolution of the overall society—in terms of joint decision making—is emergent. We can witness this effect all the time when policies do not have their intended effect. Assuming we want to "control" society to evolve it in a particular direction, for example towards lower energy use, we have to deal with this problem: how do we design the rules for each citizen and organization so that we stand a chance of achieving society-wide goals?

Agent models can be used to try to understand these effects. An agent is a single entity that behaves (interacts with others) according to a set of rules. The rules can be algorithmic or probabilistic, and take into account the local context, such as connected actors and the state of shared data. Different categories of actors might follow different rule sets.

Agent models are usually executed as a simulation: time proceeds in increments; in each increment, each actor makes a behavioral decision based on its rules. By running these models for a long time, one can see the emergent overall behavior for the given rules. This helps to understand the consequences of rules, for example in the context of policy-making. The system can also be used the other way round, by changing the rules until the outcome is as expected.

A challenge of this approach is of course validation of the rules: how do you know that real-world actors behave according to the rules? We all know that people and organizations don't always stick to rules, rational or otherwise. We're back to Rumsfeld: for the known unknowns we can do a sensitivity analysis in which we vary the rules to understand their impact on the evolution of the model. For the unknown unknowns we can't do much. You also can't easily design experiments—you can't just clone a society and try out different futures. Social scientists try to extrapolate from "accidental experiments" run in other societies, but of course it's almost impossible to control for all the (unknown unknown) factors that might impact the results. Social science, including economics, suffers greatly from this fundamental limitation to modeling.

9 | *Model and Experiment*

Models are a way of predicting how the real world will behave. Depending on the explanatory power of the modeling formalism, a model can also help to explain how reality works. In science, however, models are often used together with experiments. How do models and experiments relate?

But first: what is an experiment and how is it different from the "real world" in general? The real world is complex. Lots of things change all the time. For example, even though two days might have the same temperature, sun angle, atmospheric humidity and wind, on one day the ground might be wet because it had been raining the previous day. Such days often start with low and dense clouds, which makes them less suitable for gliding. This leads to a typical exchange between an eager young pilot and an older guy:

Young pilot: The weather forecast is really good for today, let's hurry!

Old guy: Aaah, slow down. It rained all day yesterday. The weather will start at 14:00 earliest.

This complexity makes it very hard to predict cause and effect accurately. At the very least it makes it computationally expensive: in the concrete example, the weather forecast for a particular day would have to take into account historical data.

In contrast, a well-designed experiment is controlled: this means that one or very few parameters are systematically varied to understand their impact while all others are kept constant, or at least are closely monitored if they are out of the experimenter's control. This helps us define which physical quantities are relevant for a particular model and which parametric abstractions might be applicable.

This is the most obvious connection between experiments and models: experiments guide the development of models. If you use the model to make predictions about how the real world works, experiments are used to verify the model, to "ground" it in reality. A typical approach is to run the model for those regions of the parameter space for which experiments have been or can be done to verify the model's validity, then exploit the predictive power of the model by running it for parameter ranges that are outside the realm of experiments. Once a model is good enough you can get rid of experiments completely, or at least use them only for "final validation". A great example is from the world of aerospace: reliance on wind tunnels has been reduced significantly since the introduction of computational fluid dynamics modeling. They are mostly used to calibrate the simulations.

But the relationship can also be reversed: models influence experiments in many ways. As we have seen, a model usually has limited applicability: at the boundaries of its applicability its predictions are unreliable. This might prompt experiments to better understand what's going on there. More generally, you can design experiments specifically to fill in "holes" in your model. Models can also help explain "unexplained" signals seen in an experiment.

Finally, models help shape the experimental setup. For example, models predicted how much energy would be required to produce the Higgs boson. This helped scientists and engineers to design machines likes the Large Hadron Collider and its experiments like ATLAS or ALICE in a way that allowed detection of the Higgs. Similarly, models are crucial in shaping fusion reactors and their magnetic fields, which then let the community run experiments to further the understanding of fusion and its potential for future use as an energy source.

10 | *Meta Models and Languages*

I've been a space and aviation geek forever. As a young child I built a big Space Shuttle Orbiter out of Lego. I used the generic Lego blocks: the 4x2, 2x2 and 8x1 blocks, the 4x8 plates, and a few of the joints. We are all familiar with those; this is "classical" Lego. However, you would have needed a degree of imagination to recognize this thing as the Shuttle. Years later, I was gifted a Lego set for a spacecraft (the Galaxy Commander, Lego 6980). This looked much more like a spacecraft. Recently, Lego released a kit for the Saturn V moon rocket that looks amazingly realistic—even some of the internal structure resembles the original.

Why do these things look more and more like the real thing? A major reason (apart from the young Markus' limited talent) is that the kits contain more and more specialized bricks. The Galaxy Commander set came with transparent canopies, small jet-engine shaped pods, antenna dishes and triangular plates to build wings. The Saturn V comes with rounded bricks of the right radius for the cylindric body, accurate replicas of the engine bells and fins, as well as specialized bricks for the capsule and the crew-escape system.

What we can learn from this analogy, which I first saw in a piece written by Dirk Leopold, is that we can build more accurate, precise and detailed models if we use the right building blocks—if we use the right language. The standard Lego blocks with which I built my original Shuttle were very generic. You can build more or less anything with them, but whatever you build will look very "Lego-y". And you need lots of those generic parts, combined in creative ways. In contrast, the custom bricks provided by the Saturn V kit let you build a very accurate model with many fewer parts. On the other hand, those parts are not very useful to build anything other than the Saturn V rocket. You can't for example build a Lego farm with them. You could with the generic blocks.

That same analogy is true for models in general: for a model to be expressive and precise, it is crucial to choose the right language to express it. But what exactly is a language? We all know intuitively what a language is: we use one (or more) every day to communicate, and I am using one right now to write this. However, what I mean by language in this context is a formal language, one we can use to build models whose meaning we understand precisely. Math is one such language, which is why it is used a lot for models.

Formal languages have three ingredients: structure, notation and semantics. We'll look at each of these in turn, starting with structure. The structure are the building blocks we use to create models. For example, in the pendulum formula we used earlier:

$$mg\sin(\theta) \; - \; \frac{1}{2}\rho\,\dot{\theta}^2\,c_D\,A \; = \; m\,\ddot{\theta}$$

we can see concepts such as equality, mathematical operators such as multiplication, division and subtraction, the use of named variables to refer to physical quantities (m, g, c_D, θ, A, ρ), as well as the notion of derivatives. The structure captures the foundational abstractions required to express the model, they form the vocabulary. Try to express the pendulum model without relying on the abstraction of first and second derivatives and you will have a really hard time. Similarly, try to do it without multiplication: it's basically impossible. We never think about this in practice because these notions are so ubiquitous in math that we take them for granted. We will revisit this ubiquity later.

Let's look at the second ingredient: notation. Notation is about how we represent the concepts in a language, how we write them down. For example, writing two variables next to each other signifies multiplication, the fraction bar represents division and the dots above a variable mean differentiation with respect to time. Here is the same formula written with a different notation:

$$m \cdot g \cdot \sin(\theta) - 1/2 \cdot \rho \cdot \left(\frac{d\theta}{dt}\right)^2 \cdot c_D \cdot A = m \cdot \left(\frac{d^2\theta}{dt^2}\right)$$

I have made the multiplication explicit using the dot, I have replaced the fraction bar with the division operator, and I have used the more general variant of expressing differentiation over time. Note that this formula uses the exact same concepts—it is not a mathematical transformation of the previous one, no computation or rearrangement of terms has been going on between these two variants. It really is just a different notation. We see that the second variant is more verbose—it takes more space. On the other hand, it might be easier for some people to understand because it relies on fewer specific conventions (such as the dots for differentiation with regards to time). Here is another version, one that I might write as the input to the computer algebra system Mathematica:

```
equate[
   m * g * sin[phi] – 0.5 * rho *
     diff[phi, t] * c_D * A,
   m * diff[phi, t, 2]
]
```

Again these are exactly the same concepts, but expressed as 7-bit ASCII. No special symbols, no fancy layout such as superscript or fraction bars, no Greek letters. It is clearly harder to read, but it can be written more quickly, especially with a regular keyboard: you can just type things linearly, character by character. If you don't believe this, try entering the mathematical notations above using Word's Equation editor or, if you prefer, LaTeX. You will suffer! So there is a tradeoff between writing and reading: this is why Mathematica uses the ASCII-version as input and then renders a nice mathematical notation as the output. It is also much easier to build the editing tools: the 7-bit ASCII version is linear, it can be parsed easily, whereas the math symbology requires editors with more powerful rendering capabilities and more advanced user interactions.

The third ingredient in a language is semantics, the meaning of the language or its constituent concepts. What does it mean to multiply two numbers $x \cdot y$? One way of defining this expression is to say that we perform y additions of x to itself. So $5 \cdot 3$ can be expressed as $5 + 5 + 5$. We have defined the meaning of $\cdot$ in terms of +. Defining a more powerful language concept in terms of simpler ones is a common approach. How could you formalize this definition of multiplication? One example is by using recursion:

$$x * y \vdash \begin{bmatrix} y = 1 & \rightarrow & x \\ y > 1 & \rightarrow & x + x * (y - 1) \end{bmatrix}$$

This definition relies on case distinction: if we multiply x by 1, the result is x itself. If y is greater than 1, we add x to the result of applying the same definition again, but with $y - 1$. We will eventually reach the case where $y = 1$, in which case we return x itself and the recursion terminates. Here is an example execution of this algorithm:

```
5 * 3 = 5 + 5*(3-1)
        5 + 5*2
        5 + 5 + 5*(2-1)
        5 + 5 + 5*1
        5 + 5 + 5
      = 15
```

How could we define differentiation? Intuitively, a derivative is the amount of change of one quantity (here θ) per change of another quantity (here t). By making Δt smaller and smaller, we get to the derivative. Mathematically, we use the notion of a limit:

$$\lim_{\Delta t \to 0} \frac{\theta_t - \theta_{t+\Delta t}}{\Delta t}$$

Let's visualize. The following shows a parabola (the thick black line) that is described by $y = x^2$. The derivative $\Delta y/\Delta x$ is overlaid as the stepped line.

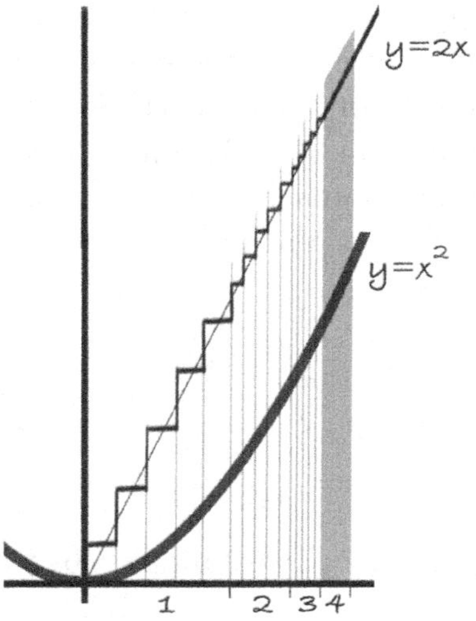

Why is the line stepped? In the region marked as 1, each step is Δx wide and the horizontal line is the respective value for Δy. As we progress towards regions 2 and 3, we make Δx progressively shorter, before finally approaching the limit of $\Delta x = 0$. At this point the steps become infinitesimally small, approaching a straight line $y = 2x$, which is the derivative of $y = x^2$.

Before we conclude this section on language definition, let's look at another model we've seen before in the context of the A320 flight control system:

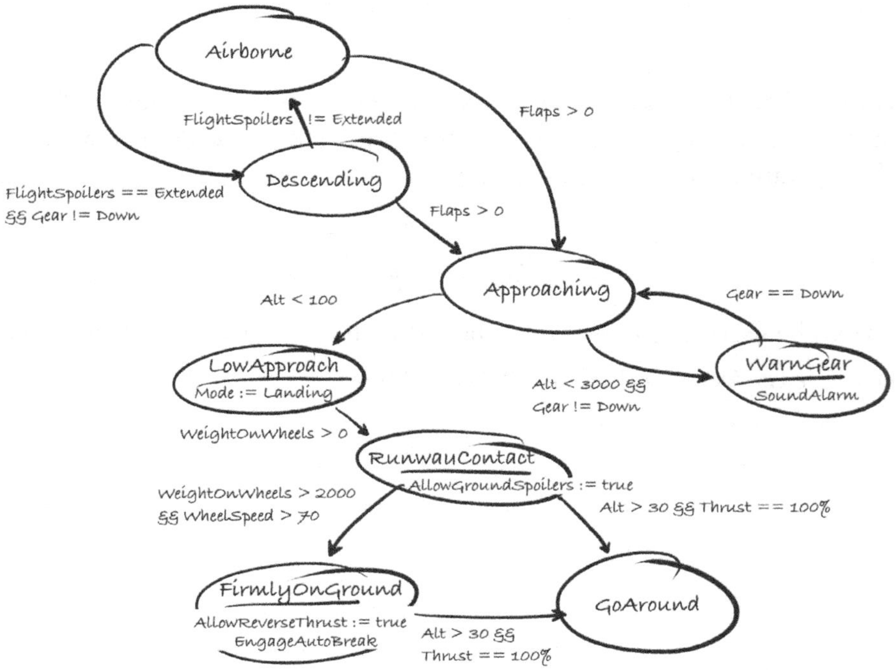

What are the language concepts used in this state machine model? The most obvious concept is the state: its notation is the ellipse. Each state has a name, written above the line in the ellipse, as well as a list of actions that happen when the system enters the state. Another language concept is the transition, which moves the system from one state to another. The condition under which the machine goes to a transitions's target state (where the arrow points) is written next to the transition arrow using a textual notation. To be able to write such conditions (such as Alt > 30), variables such as Alt also have to be defined. So the set of concepts in the state machine language are State, Action, Transition, Condition and Variable, plus of course the machine itself. Here is how we could write this formally using a textual notation. We mention the name of the concept we define on the left, then describe its "ingredients" on the right-hand side:

```
Machine    ::= State*
State      ::= name,  Action*, Transition*
Transition ::= ->State, Condition
Variable   ::= name, Type
Type       ::= Boolean | Number
Action     ::= …
Condition  ::= …
```

The asterisk represents "many": a state machine has many states. And the –> arrow means that we point to an existing state instead of defining a new one. The vertical bar represents an alternative, so a Type is either a Boolean (with two values, true and false) or Number (with a range of values that is not specified in this example). The three dots mean that it is too complicated, so we do not talk about it here.

The concrete syntax of the state machine language has been chosen to be a graphical notation that uses ellipses and arrows. But we could also use a table:

to from	Airborne	Descending	Approaching	WarnGear SoundAlarm	…
Airborne		FlightSpoilers != Extended	Flaps > 0		
Descending	FlightSpoilers == Extended && Gear != Down		Flaps > 0		
Approaching				Alt < 3000 && Gear != Down	
WarnGear			Gear == Down		
…					

This table lists each state twice: as column header and a row header. The transition condition from a column header state to a row header state is given in the respective intersecting cell in the table. No condition means there is no transition from that source state to the corresponding target state.

What about the semantics? We can write them in the form of the following algorithm (ignoring how to deal with variables and how their values change at runtime):

```
semantics for Machine
// track the current state of machine
State current = first state of machine
// run forever
loop
  // try each transition
  foreach trans in current.transitions
    // test if the condition is true
    // ignoring how we deal with the
    // details of the variable values
    if trans.Condition == true
      // the current state is now
      // the one that is the target
      // of the transition we execute
      current := trans.State
      // perform the action associated
      // with that new state
      do trans.Action
```

Here we don't use math to define the semantics. Instead we use a less abstract language that lets us express algorithms. If you have experience with programming, this might remind you of a programming language. This isn't accidental: we are slowly but certainly making our way into computer science, a field that has invented many semantics definition languages over the decades.

II | *Descriptive vs. Prescriptive Models*

But let's first think more about the purpose of models. When you think about the state machine model in the context of the A320 flight control system and compare it to the model that describes the pendulum or models used for weather forecasting, do you notice a difference in how these models are used? Maybe even more clearly, think about a model that describes how the electrical wiring will be installed in a new house. Notice the use of the future tense in this example? The electrical wiring model exists before the real wiring exists. Similarly for the state machine: it can be used to explain how the flight control system works; this is how we use the model in this book. But that same model can also be used by Airbus to reason about how the future system should work. The model can be simulated, basically by "running" the semantic definition given above, or for testing and verifying invariants, as we hinted in the aviation chapter.

There is a fundamental distinction between descriptive models and prescriptive models. A descriptive model abstracts an existing system and is used to predict its behavior and/or to explain how that system works. It is used primarily in science, because science is about understanding the world. A prescriptive model governs how a not-yet-existing system will be constructed. It is used to understand and verify aspects of the system's behavior before one expends the often considerable effort required to build the real system. Prescriptive models are used primarily by engineers.

Both categories of models must be aligned with the real system for them to be useful; we've discussed this relationship at length for descriptive models. But how do you do this with a prescriptive model? Let's use an example of an algorithm that sorts a list of numbers. If we were to create a model of the algorithm that defines what the algorithm does, what would we have to say? First, we assume that the list we want to sort is not empty—its length is non-zero. Second, we expect that, after sorting, the list still has the same length. We also expect that, after sorting, every element from the original list is also in the result. Finally, we we also expect it to be sorted. We can express "sorted" as follows: if an element comes later in the result, then it must be bigger (or equal to) the element that precedes it. Here's a formalization; we call the to-be-sorted list l and the sorted result res:

```
pre: l.length > 0
post:
  res.length == l.length
  forall e in l: e in res
  forall i, j in 0..res.length:
    if j > i then res[j] >= res[i]
```

With a little bit of squinting this should be reasonably understandable, because it is quite closely aligned with what I have written above. Notice that this model does not express how the actual sorting can be achieved, it just expresses what must be true for the model to be consistent with the real world. The next model expresses the "how" in a Java-like programming language. Don't bother understanding it; the only point I want to make is that it is longer and more complicated:

```
fun quickSort(int[] arr) {
  sort(arr, 0, arr.length-1);
}

fun sort(int[] arr, int start, int end) {
  int partition = part (arr, start, end);
  if (partition-1 > start) {
    sort(arr, start, partition - 1);
  }
  if (partition+1 < end) {
    sort(arr, partition + 1, end);
  }
}
```

```
fun int part (int[] arr, int start, int end) {
  int pivot = arr[end];
  for (int i = start; i < end; i++){
    if (arr[i] < pivot){
      int temp   = arr[start];
      arr[start] = arr[i];
      arr[i]     = temp;
      start++;
    }
  }
  int temp = arr[start];
  arr[start] = pivot;
  arr[end]   = temp;
  return start;
}
```

The approach to sorting numbers shown here is called Quicksort, and as its name suggests, it is fast. There are other algorithms, some of which are much simpler, that are significantly slower. So, once you decide about the "how", you have to decide about additional issues such as speed or memory consumption. In computer science we call the first model, the one with the pre and post, a specification and Java program an "implementation". It is also very common to call the specification the model and the implementation the code.

How can we ensure that the implementation is faithful to the model? One way is to use the specification descriptively. Once somebody creates the implementation, we can check that it conforms to the specification. We can do thorough testing, which means that we run the implementation for a large number of examples and then check whether the specification holds for each of them. Alternatively, we can use program analysis and proof assistants to prove that the implementation conforms to the specification for all possible inputs. Alternatively we can use the specification in a prescriptive way. A programmer can understand the specification and then write the implementation to conform to the specification. The above specification does not say anything about speed or memory, but a competent programmer will care about speed and memory consumption and will optimize them—or, nowadays, either go to stackoverflow.com and look up a fast algorithm, as I did, or even better, use a library that already contains one. This is just like the electrician who looks at the wiring plan and installs the wires in accordance with the plan, but also in accordance with the building code that prescribes how you install a safe and reliable electrical system in a building.

When engineers use prescriptive models, however, they ideally want to automate the construction of the real system. Think about a CAD model of a car: ideally you can program a bunch of robots to automatically assemble and weld the car. Why does this work? Because the model in the CAD program contains all the details for the car except those for which there are "scriptable" rules for the robot to follow. The robot does not have to be "imaginative" to come up with a solution.

This is not the case for the sorting algorithm above: there are so many degrees of freedom in the mapping from specification to implementation that it is effectively impossible for a computer program to come up with Quicksort. The problem of automatically "inventing" a program that fulfills a specification is known as program synthesis and is an active area of research. Progress has been made: for simple cases, and especially for cases for which no speed or memory optimizations are required, it sort of works. Consider the following example in which the synthesizer basically only has to come up with the operator that should replace the ??. Program synthesis can handle more interesting cases, but it's still severely limited.

```
int a = 10;
int b = 3;
int res = a ?? b;
assert res == 13;
```

Is there another way to go from model to implementation? What about the state machine example? Can we create an efficient implementation automatically, for example using the C programming language, so that we can run it in the flight control system of the A320? Yes, we can. We achieve this by changing it from a synthesis problem to a compilation problem. In synthesis, the specification, or model, has lots of degrees of freedom, and we expect an existing, all-powerful, general-purpose synthesizer to "invent" an efficient solution. The only thing the developer creates is the specification, using a general-purpose specification language. In the case of compilation, however, the developer does more work. In particular, they make all the decisions about how to implement a state machine in a programming language such as C, then write rules to automatically perform that mapping. The tool then blindly executes the rules, just like the robot in the CAD example, or the electrician who follows step-by-step instructions.

This approach is extremely common in computer science. A tool like Stateflow lets developers "draw" state machines as diagrams and then generates an efficient implementation in C. This C code is then compiled to machine code by developer-written algorithmic rules embodied in the compiler. This is useful because more abstract models are easier to write, understand and verify for (specific aspects) of correctness. There is a catch, though: you have to assume that the rules by which the compiler translates the model into an implementation happens are correct and complete. This is surprisingly complex and therefore outside the scope of this book.

12 | *Multi-Level Models*

Did you notice from the previous paragraph: Stateflow creates C code from high-level models, then a compiler creates efficient machine code from the C program. We've stacked multiple model-to-code steps! This is very common: it allows us to build more and more abstract models while caring less and less about the details one level down. Modern software engineering would be impossible without this approach—the overall complexity would just be too high. Imagine if, when defining a state machine that describes how a flight control system works, engineers had to think about details like register allocation in the processor. They'd not get anywhere.

In computer science, and in programming in particular, we distinguish between two kinds of abstraction, although both can be understood as "building a model". The first is to use the language itself to define new abstractions. Going back to our definition of multiplication, this can be "modeled" in a programming language like this:

```
fun add(x: int, y: int) = x + y

fun mul(x: int, y: int) =
  case | y == 1       →  x                    |
       | otherwise    →  add(x, mul(x, y-1))  |
```

This language supports the definition of functions, calling functions and passing arguments, case distinction, as well as the basic + and – operators. The programmer then uses these facilities to create their own abstractions—in this case, the notion of multiplication. Functions are a primary means for creating user-defined abstractions without having to extend the language itself.

Here is another situation: we embed a decision table inside the function that returns a number based on subranges of the two arguments. This is an example from the medical domain, where we compute a risk factor based on a patient's systolic and diastolic blood pressure:

```
fun riskFactor(sys: int, dia: int) =
```

		dia					
		<= 50	51..91	91..95	96..100	101..109	>= 110
sys	<= 90	1	1	3	4	5	6
	91..140	2	2	3	4	5	6
	141..150	3	3	3	4	5	6
	151..160	4	4	4	4	5	6
	161..179	5	5	5	5	5	6
	>= 180	6	6	6	6	6	6

The decision table doesn't just call a bunch of existing functions: it is something completely different. It is able to express ranges of numbers, checks there is no overlap in the ranges (because that would mean a decision is ambiguous), and of course comes with its own tabular notation. How can we define the semantics of such a table? We can use a compiler to reduce it to a program that relies on the case construct that is already in the language. The following shows part of this program; the complete program would need 36 lines to handle all combinations.

```
fun riskFactor_reduced(sys: int, dia: int) =
  case | dia <= 50                → case | sys <= 90                  → 1 |
                                         | sys >= 91  && sys <= 140 → 2 |
                                         | sys >= 141 && sys <= 150 → 3 |
                                         ...
       | dia >= 51 && dia < 91 → case | sys <= 90                  → 1 |
                                         | sys >= 91  && sys <= 140 → 2 |
                                         | sys >= 141 && sys <= 150 → 3 |
                                         ...
       ...
```

Notice how this is much harder to understand: checking ranges with >= and < is more verbose, and there is lots of repetition in the conditions (each of the six "inner" cases repeats all column conditions). This is why the language designer has decided to include decision tables.

Which approach to defining abstractions is better? There's no general answer. Defining abstractions within the language is nice because, well, you can do it within the language. There's no need to change the compiler and other tools. But you need a language that is powerful enough to be able to define meaningful abstractions. There are also drawbacks. You cannot easily invent a notation that departs from what the language already supports (such as the table). Custom error checking, such as checking for overlap, is also hard to do. On the flip side, if you want to build your own native language abstractions, this requires a lot of additional infrastructure. We will return to this shortly.

Mathematicians have been using both approaches for a long time. When we defined multiplication in terms of addition we relied on the fact that someone had already defined what addition means, then used it in a way that was similar to what a programmer might do. When we defined differentiation in terms of subtraction, division and the notion of limits, we effectively "extended the language of math", reducing it to abstractions that are well-defined. Which means that they are probably defined in even more basic concepts. This process repeats forever—turtles all the way down. Except that at the very bottom of the stack we are left with the absolute essentials that cannot be explained with more basic concepts. We call those axioms.

How do we know that our axioms are correct? That's a really tough question. Mathematicians make do with a slightly less ambitious goal: they want the axioms and the mathematical world built on top of them to be consistent. Consistency means that the system built on the axioms does not lead to self-contradictions. Mathematicians rely on proofs to ensure this property. Sadly, that's something else that is beyond the scope of this book.

13 | *Implementing Models*

Let's briefly return to the descriptive, numerical models used in science for all kinds of purposes, from computational fluid dynamics, to weather forecasting, to understanding the plasma in a fusion reactor. We have talked about how they use parametrization to reduce their complexity. This is crucial, because these models are executed iteratively on computers. Although some of them run on the fastest supercomputers available, the speed of even those is finite, so models must abstract to be reasonably fast.

But there's more: the models must also be turned into computer programs. Somebody has to write program code, often in Fortran, C or C++, that faithfully represents the physics in a way that runs on a computer, ideally quickly and without excessive use of memory. This is exactly the same problem as the one we illustrated with the sorting algorithm. In practice, many theoretical scientists are also programmers, spending a large part of their time encoding their models of the world into programs they can use to run—and thus verify—the models.

Here's the catch: supercomputers have long been very complex machines. Each processor has many cores, each computer has several processors, and then many such computers are connected into clusters by buses and networks. Each level in this hierarchy is limited in some way, for example in terms of computational capacity, available memory, speed of access to data or the bandwidth of communication within and across the levels in the hierarchy.

For a model to use the full capacity of such machines and thus run fast, the program has to be written in a way that "fits" with the architecture of the supercomputer. Parts of the computation have to run in parallel, and those parts that share data should run as physically close as possible—preferably on the same processor—so that they don't depend too much on the (comparably slow) network. Expressing a model that is itself non-trivial in a way that makes efficient use of a supercomputer's architecture is a major challenge. To give you a taste of the problem, go back a few pages and look at the Quicksort algorithm. How easy is it to understand from the low-level and fast implementation what this code actually does? Very hard! You cannot see the forest for the trees. The readability of a sorting algorithm has been "obfuscated" in order to make it fast. Imagine how much worse this gets when the core logic becomes more complex (because you simulate plasma with 100 parameters and don't just sort a list) and you are forced to write the code such that some things can run in parallel and make use non-uniform computing architectures. As Ralf Kleiber told me in the context of Wendelstein 7X, they strugge with this a lot. Especially since they have to be able to run the same models on different supercomputers, each requiring a different, efficient, encoding of the same physics problem. This explains why such "codes" take a long time to develop and never die. Ralf told me that their current plasma code was developed by three people over 15 years, and continues to evolve to this day.

14 | *Your own Languages*

So how do the languages that scientists and engineers use to create and implement models come about? Who invents languages, and why? And how does this work?

Every non-trivial subject area—physics, chemistry, biology, actuarial math, medicine—has its own vocabulary. Talk to an expert in any field and they will use lots of jargon. Sometimes such use of jargon is considered bad style—for example, when an expert talks to the general public. But for communication among experts in a subject area jargon is absolutely essential, because it makes communication more precise and less verbose.

Depending on the field, it's not that big a step to create a formal language that is specific to that subject area. We all remember chemistry's notation for expressing reactions or the structural notation for molecules from our schooldays. Actuarial mathematicians use all kinds of strange symbols, with their associated semantics, to express financial and risk calculations in insurance products. Physicists use lots of mathematical abstractions, some of them invented specifically to be able to express a particular physics; for example, Newton invented calculus to be able to express his mechanics. We have also seen above how inventing even a simple piece of language such as a decision table can be very useful.

The definition of formal languages that are specific to a particular subject area or domain is not uncommon in computer science. Such languages are called domain specific languages. We've arrived at the boundary between podcaster Markus and computer scientist Markus, since this is the area in which I work. I have (co-)developed DSLs for healthcare, finance, automotive, space, payroll, public administration and embedded software. I guess we should also build one for those poor theoretical physicists to help them implement their models in a way that runs fast on supercomputers. You can find lots more about DSL development on my website, voelter.de, but I will give you a taste here.

If you go back a few pages to where we discussed the state machine language, we used two definitions of structure and semantics:

```
Machine ::= State*
State
  ::= name,  Action*, Transition*
Transition
  ::= ->State, Condition
Variable
  ::= name, Type
Type
  ::= Boolean | Number
Action ::= …
Condition ::= …
```

```
semantics for Machine
State current =
   first state of machine
loop
   foreach trans in current.Transitions
    if trans.Condition == true
      current := t.State
      do t.Action
```

Have you noticed that we also use languages to define these languages? The structure language on the left has concepts like definition, ::=, multiplicity, *, or the alternative operator, |. The semantics definition language comes with concepts for loops, iteration over lists (foreach), conditionals (if), and do to execute actions. The languages used for language definition are often part of special-purpose tools called language workbenches, which are used to define new languages and their constructs quickly, such as the decision table we showed earlier. In addition to languages for defining structure and semantics, they come with additional ones for defining various kinds of notations (for example, tables), type systems (for example, the check for overlap in the ranges in the decision table), and transformations (for example, from the decision table back to the case expression).

You might ask how you define the languages used for language definition. They are defined recursively; this is one of those circular definitions that computer scientists love, and we won't go into too much detail about how the circularity is resolved. Here is the definition of the structure of the structure definition language:

```
StructureModel     ::= Definition*
Definition         ::= name, "::=", Content
Content            ::= StructureDef | StructureAlts
StructureAlts      ::= StructureContain*{|}
StructureDef       ::= Element*{,}
Element            ::= Primitive | StructureContain | StructureRef
Primitive          ::= name
StructureContain   ::= ->Structure
StructureRef       ::= "->" ->Structure
```

To stick with our Lego analogy, a domain-specific language would be a custom set of Lego bricks that I can create myself if I want to build a Lego model for a particular subject area. For example, I could define special bricks to allow me to build submarines (trying to build on the analogy to the Saturn V rocket). Taking this analogy even further, the language workbench would be a 3D-printer that allows me to print custom bricks easily.

15 | *Other Terms*

We've spent about 12,000 words discussing what models are. It's a wide field, with lots of perspectives, and if you talk to other people you'll find other definitions. That's unavoidable with a term that is used so broadly. To conclude this section, let's briefly look at a few related terms.

Is a scientific theory the same thing as a model? Based on a bit of research I would say a theory is more general: a model can be seen as a specific approximation of a theory. For example, consider the kinetic gas theory that describes pressure and temperature in gases. As we have seen, this theory plays an important role in the simulation of plasmas, but you define create different models for the theory that trade off precision and performance differently, for example through parameterization. On the other hand, according to this distinction between theory and model, the standard model of particle physics should be called the standard theory. I guess the two terms are not used consistently.

If you are a software engineer, you might ask about the difference between a model and a program. That's a long discussion; I've written dozens of pages elsewhere. In computer science, I would argue that technically there is no difference: you define a language (structure, notation and semantics), use it to create models or programs, then analyze or process the models or programs. If you want to force a distinction, you could argue that models are more abstract than programs. But where do you define the boundary? At which point does something become abstract? Isn't a C program more abstract than machine code? Is the C code therefore a model? Or do you have to go to Java to justify "model-ness"? It's hard to say. A legitimate difference is the notion of executability. A program's sole purpose is that it can be run. A model, in contrast, doesn't have to be executable: it might just serve as a specification to describe aspects of an executable program, or to perform other analyses on it. But in any case, models and programs have much more in common than mainstream software engineering lore suggests.

There are other related terms, such as analogy. Is an analogy a model? Are the Lego examples I use in this chapter a model for models? I wouldn't say so. They are really just a very rough illustration of an idea. You can go further: is a term like global warming a model of what's actually going on with our climate? Maybe: the Earth is warming, but only on average. A simplifying moniker like global warming allows quacks to take a snowball to their parliament and claim it isn't happening. So be careful.

8

The LHC: A Big Machine for Small Scales

If I were to write a Cold War novel, here's how I might start.

It was a cold and wet Fall day, dark clouds in the sky, fog hiding the ground-bound world. I'd just landed at the city's airport after a three hour flight, and when I came landside I was met by my guide for the day. We left the terminal building, got into a VW Golf and drove through the city, eventually arriving at a gray office building. We parked the car, entered the building and checked in at the counter. A slightly annoyed-looking clerk checked my passport, verified my name on a list of pre-announced guests and then handed me a visitor pass. "Wear this all day, and make sure it's visible all the time." Of course.

We left the building and got back into the car to begin a confusing drive: out of the office parking lot, through more urban areas, into a more rural landscape, from one village to another. In what turned out to be the last of the chain of villages we drove into a housing area, only to leave it again via a narrow curving road. The road ended abruptly at a barrier with a guardhouse. An official-looking guy peered into our car and checked out the credentials of my guide as well as my visitor pass. He opened the gate and we drove through. We parked the car next to a nondescript building a few dozen meters inside the perimeter. Leaving the car, we walked towards the building. A guy was standing in front of it having a smoke, probably annoyed at not being allowed to smoke inside on this damp Fall day. My guide used her keycard to open the door and we entered the building. From the inside it looked just as boring as from the outside: a cross between a warehouse and a factory building, except that it was almost empty apart from a load of technical-looking equipment against the western wall and a huge crane soaring across the top.

After talking for a little while with some of my guide's colleagues, we eventually headed through another door, where we came to what looked like a gridded cage. My guide and I picked up the kit that would allow us to go further into the installation: a helmet and a radiation detector. Both were mandatory for anyone to proceed. We also removed two tokens from a registry that keeps track of how many people are inside at any given time. The facility cannot be started up as long as people are inside. The cage-style section of the room had a rotating door that allowed only a single person to pass at a time. My guide needed her keycard to activate the door, but a sensor also scanned her retina to ensure that it was really her—classical two-factor authorization, in which one factor is based on a token you possess and the other one uses a non-stealable biological marker. Once through the door we entered a relatively big industrial elevator cabin. My guide selected the lowest floor and we started our trip down towards a secret world of big machines 50 meter below the surface. I was looking forward to this a lot.

That's how my trip to ALICE started, one of the experiments at the Large Hadron Collider in Geneva, Switzerland. Everything in the story above is accurate except that it was a beautiful sunny day in May, I arrived by train and tram instead of by airplane, and the building that housed the lift and the crane was actually very "descript": it had a picture of the ALICE detector painted on the outside at 1:1 scale. My guide was Despina Hatzifotiadou, a physicist who has worked with ALICE since 2000 and is currently responsible for public outreach. We were scheduled to record a podcast episode and to visit the experiment. Despina encapsulates the international nature of CERN: she is Greek, has been employed for ages by the Italian National Institute of Nuclear Physics in Bologna and is permanently based at CERN in Switzerland.

As you probably know, the LHC is the world's largest particle accelerator. It accelerates heavy particles on a 27-km circular racetrack lying beneath both Switzerland and France, then collides the particles to investigate the nature of matter on the level below the atom, even below protons and neutrons, the building blocks of atomic nuclei. Four large experiments use a wide range of detectors to observe the collisions and analyze the data. In this chapter we'll look at how the LHC and its experiments work. But before that we'll recap the structure of matter.

1 | *The Standard Model*

WHAT IS IT The standard model of particle physics—which, according to our distinction of models and theories at the end of the last chapter, should probably be called a theory—represents our best understanding of the smallest structures of matter. It was developed throughout the 20th century and has been mostly stable since its last constituents were defined in the late 1970s. All of its particles have been observed experimentally; the last discovery occurred in 2012 when the existence of the Higgs boson was proved at the LHC. Within its scope there are no observations that contradict the standard model—it is one of our best theories.

The standard model describes the elementary particles from which all matter is made. It also describes three of the four fundamental forces: the electromagnetic force, the weak interaction and the strong interaction. More precisely and more technically, it is a theory about the quantum fields whose excitations are represented by these "particles", but for our purposes here the perspective of elementary particles is good enough. There is another fundamental force, gravitation, which unfortunately is not explained by the standard model. So, while it is undisputed within its scope, the standard model is also incomplete in several ways. We'll get back to this later.

I assume that you know about the structure of atoms. Atoms have a nucleus that is composed of protons and neutrons. Both have a similar (rather heavy) mass; the proton has a single positive charge, while the neutron is not charged. The structure of the atom is completed by a "cloud" of electrons at various discrete energy levels. I put the word into quotes because, for now, we aren't going to worry about quantum theory, wave functions and probabilities. Electrons are roughly 2,000 times less massive than protons and have a single negative charge. An atom has the same number of electrons and protons to ensure that it is electrically neutral. If one or more electrons are removed or added, this results in a positive or negative charge for the overall atom; we call such a particle an ion.

Instead of focusing inward on the nucleus of an atom, we can also look outwards: how increasing numbers of electrons (and protons) form chemical elements, how additional neutrons in the core lead to isotopes of a chemical element, and how the electrons in the shell determine how atoms join together to form molecules. But that's chemistry, and it is outside of the scope of this chapter.

QUANTUM NUMBERS Let's try to set up a high-level categorization to help us make sense of the standard model. This categorization is best designed around the quantum numbers that define each particle. Quantum numbers are basically mathematical abstractions whose numerical values, when defined in a particular way, lead to a consistent theory. The three most important quantum numbers are electric charge, spin and mass. Let's look at each of them.

We are familiar with the electric charge from its role in electromagnetism, a force we interact with in our experiential world. We don't think about it as a pure mathematical abstraction, but of course in the end it is. The charge carried by an electron is $1.602176634 \cdot 10^{-19}$ C, where C stands for Coulomb, the SI unit for electric charge. Charge is quantized, and there are no isolated charges that are smaller than this charge; we call this elementary charge e. So as a corollary, any electric charge can be given as a multiple of e. Thus we can say that the electron has a charge of 1, or more correctly –1, because it is negatively charged.

The second quantum number is the spin. It is also called the "angular momentum" and is a vector quantity in the sense that it has a magnitude and a direction. Originally it was thought of as an actual rotation of the particle; in fact nothing actually rotates, but the math of spin is similar to the math for angular momenta. We can just consider it another number that describes the elementary particles in our universe—every kind of particle has a particular spin. Like electric charge, spin is also quantized.

The third number that characterizes elementary particles is mass. This is another for which we have an intuition, but of course the masses we talk about here are much, much smaller than those we are used to. But in the end it's also just another number; we'll see later in this chapter where exactly the mass comes from and what it actually is. In contrast to the other two, mass is not quantized.

PARTICLES Let's dive into the particles themselves, and let's use the categorization we've introduced above. It turns out that all the particles that make up matter have a spin of 1/2. They are called fermions, after the Italian physicist Enrico Fermi. Within the class of fermions we can further distinguish two sub-categories: quarks and leptons.

Quarks are particles with an electric charge of either –1/3 or +2/3. How can they have a charge of less then e when we said that charge is quantized and the smallest quantity is e? The reason is that, at least in stable particles, quarks always join into groups of three, combining their charge. Such three-quark particles are known as baryons, the building blocks of "heavy" matter. Six "flavors" of quarks have been discovered: up, charm, top and down, strange and bottom—I suggest you check out Wikipedia for the etymology of "quark" and the flavors: it's funny. The first three have a charge of +2/3, the latter three of –1/3, and they become progressively heavier. Protons and neutrons are baryons, and they differ in terms of their quark content: the neutron is made from one up and two down quarks, whereas the ratio is reversed in protons.

Leptons, the second category of fermions, are particles of integer charge, either –1 or 0. Six flavors have been identified there as well: the electron, muon and tau, plus their neutrino counterparts, the electron neutrino, muon neutrino and tau neutrino. The first three are charged with –1, and neutrinos are—as you might guess from their name—neutral. "Lepto" is greek for "light" (as in, not heavy) because the really light ones such as the electrons were found first. However, it's now a bit of a misnomer, because, for example, the tau lepton is much heavier than the up quark. So it's really their non-zero integer charge that distinguishes leptons from quarks. Neutrinos have extremely low mass and zero charge, which is why they interact very weakly; most of them pass through the Earth without bothering anything. They were discovered only relatively late in the 1950s after they had been postulated by Pauli.

Talking about Pauli, all fermions must obey the Pauli exclusion principle, which says that any two 1/2-spin particles cannot be in the same state. Because electrons obey this rule, it's impossible for the electrons of an atom to all be in the same (lowest-energy) orbital; this is what prevents the collapse of the atom. To keep the Pauli exclusion principle intact for the quarks in a proton, the community introduced another quantum "number": color, with values of red, green and blue. Of course this has nothing whatsoever to do with an actual color, it's just another way of characterizing quarks. I guess the reason why the moniker was picked (and its originators were called "idiot physicists" by Feynman) is that combining one of each color leads to a neutral, or "white", color charge. The three quarks that make up a baryon must be of different colors, ensuring that the "color charge" is neutral.

We have discussed 12 fermionic particles—six types each of quarks and leptons. These 12 can also be grouped into families, or generations. Generation one includes the up and down quark, electron and electron neutrino; generation two is charm, strange, muon and the muon neutrino, and generation three is top, bottom, tau and the tau neutrino. Later generations are generally heavier than their earlier-generation brethren, but the way the particles interact with others is common to all generations (a bit like isotopes one the level of atoms). Later generations are also generally unstable: when they decay they do so into particles of earlier generations, which is where the moniker "generation" comes from (although the directions sounds wrong: in the real world, the current generation "decays" into later ones).

Because later generations are unstable, and because they require more energy to be created (heavier mass implies more energy, according to $E = mc^2$), we typically see them only in cosmic rays or in artificial situations created by accelerators—one of the primary use cases of accelerators. They presumably also existed in the very energy-dense soup of matter shortly after the Big Bang. Our material world, however, is only made of first generation particles: up and down quarks, electrons and electron neutrinos (which show up in some forms of radioactive decay). It is not totally clear what the role of later generations was in the development of the universe we see today.

To summarize our discussion, here is the view of the standard model so far. The boxes contain the symbol of the particle, its charge (top left), its spin (top right) and its mass in units of eV/c^2. As we will learn in this chapter, based on the example of the Higgs boson, the masses of all these particles are experimentally determined only statistically, so they all come with some uncertainty. This is not indicated in the illustration.

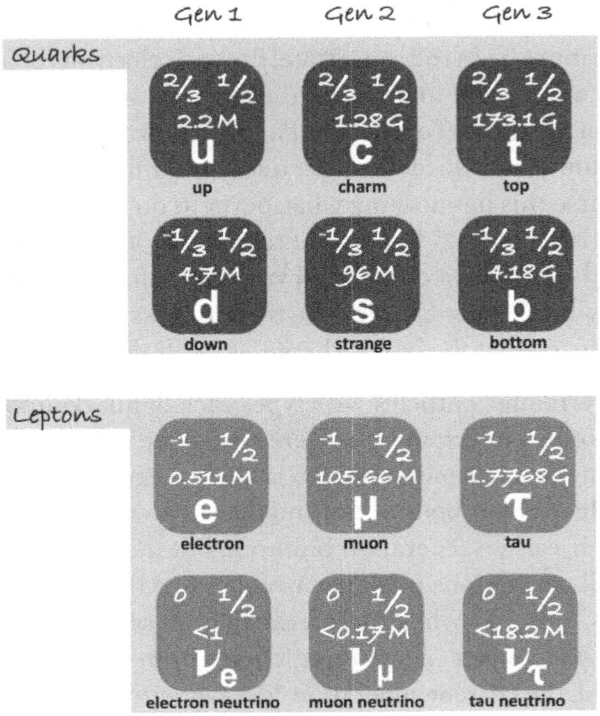

FORCES At the beginning of this section we said that the standard model comprises the elementary particles that make up matter, as well as the forces, so let's look at those. Forces are fields and the excitations of these fields can be seen as particles. Another way of saying this is that interactions between particles based on a particular force are mediated by an exchange of a "force particle". All such force particles have a spin of 1—so Pauli's exclusion principle does not apply—and we jointly call them bosons.

The boson we are most familiar with is the photon, the particle associated with the electromagnetic force. We know it as light—or more generally, electromagnetic radiation. Photons have no charge and no mass. More precisely, we should say that they have no rest mass, because a photon's energy $E = h \cdot f$ can be viewed as mass according to $E = mc^2$. Photons are exchanged between electrically charged particles; the theory that describes these processes is called quantum electrodynamics.

The gluon also has zero mass and charge, but it has only very short-range effect, so effectively only occurs inside the nucleus: it is the particle associated with the strong nuclear force, the particle whose exchange between quarks "glues" them together to form baryons. It turns out that the color charge mentioned earlier plays a role in how exactly the force plays out, which is why this part of the standard model is called quantum chromodynamics. The color charge is also the reason that there are actually eight different gluons that differ in their color charge. We won't discuss the reasons for this; it's not relevant for this chapter.

The bosons also include the Z as well as the W particle, the latter occurring in two variants with an electric charge of +1 and –1, respectively. They are associated with the weak nuclear force, so named because its strength relative to distance is several orders of magnitude lower than the strong or electromagnetic interactions. The weak force plays a central role in radioactive decay. For example, in beta decay a neutron is transformed to a proton by emitting an electron and a neutrino. In this process, the flavor of one of the constituent quarks must change (remember the neutron is up + down + down and the proton is up + down + up). This change is effected by the exchange of a W boson. The Z boson is involved in the creation of the neutrino.

There is one more: the famous Higgs boson. It is special in many ways: it has spin zero, and in contrast to all others its field is scalar instead of a vector. It is the particle that mediates the field that gives mass to all the other elementary particles in the standard model. I emphasize "elementary particles", because the mass of non-elementary particles such as protons comes mostly from the strong force; it can be seen as the mass equivalent of the nuclear binding energy. Remember that protons are made from three quarks that are held together by the strong nuclear force, which is mediated through the exchange of gluons. The energy of those gluons, as they "bounce around" between the quarks, becomes the proton's mass according to Einstein's equation $E = mc^2$. You might well ask where the mass of the mass-defining Higgs particle comes from? As an example of the kind of weird self-referentiality liked by mathematicians and scientists, the Higgs particle also gives mass to itself. We will get back to the Higgs and its relatively recent discovery in more detail later.

Here is our current understanding of the standard model, the set of elementary particles from which our universe is made:

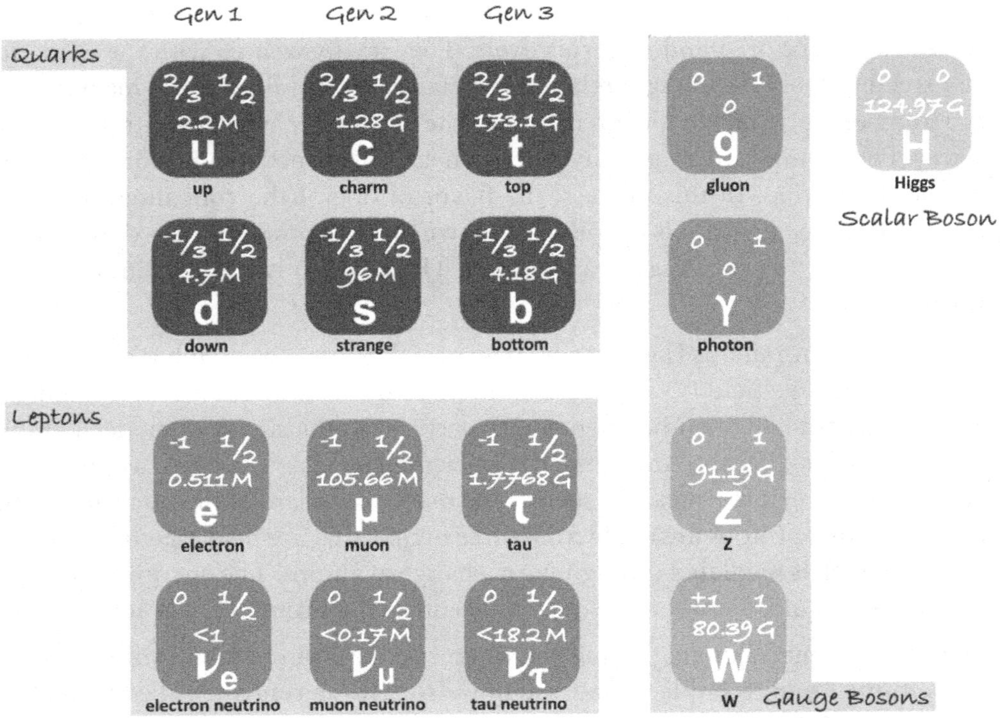

Gravitation, of course, is another fundamental force in our universe. As we know, on a large-scale level it can be modeled by the bending of spacetime, as we discussed in the chapter on gravitational waves. On the micro level, we can interpret it as a force that couples with the mass of matter (which, in turns, results from the Higgs or from the strong force). To keep the standard model consistent, therefore, we would expect to see some kind of "graviton" that mediates the gravitational field. However, this has eluded us so far. More generally, we currently do not have a quantum theory of gravitation. So while the standard model agrees extremely well with measurements regarding the particles we've mentioned so far, it is incomplete in the sense that it cannot deal with gravitation. We'll return to this issue at the end of this chapter.

ANTIMATTER When I said that there are 12 fermions, this was not completely correct, because they are all duplicated in antimatter particles. Each particle's anti-cousin is identical in all its properties except that it has the opposite electric charge. In an anti-world, everything would be(have) identical(ly) but with reversed charges—which almost sounds as if we just swap the monikers "positive" and "negative". If a particle and its corresponding antiparticle collide they annihilate, and lots of (gamma ray) photons, neutrinos and sometimes less massive particle/antiparticle pairs are created.

Our universe consists only of matter; antimatter occurs naturally only in some cosmic ray collisions and in some kinds of radioactive decay, and only in minute quantities. Even smaller quantities are created artificially in the collisions we will discuss in this chapter, and in purpose-built accelerators, for example in the context of positron emission tomography.

One of the great unsolved questions in physics is why there is matter at all in the observable universe; it is assumed that matter and antimatter were created in equal amounts in the Big Bang. But somehow there must have been an asymmetry between matter and antimatter, and some of the matter particles survived the annihilation process. We will revisit this question at the end of the chapter as well.

COMPOSITE PARTICLES We have already said—and you probably knew this anyway—that protons and neutrons are made up of quarks, three of them: up + up + down for the proton, and up + down + down for the neutron. All particles that are made from three quarks are called baryons. In addition to protons and neutrons there are several more, including the Delta, Lambda, Sigma, Xi and Omega particles. These all differ their number of light quarks and other properties, and each comes in several variants based on their particular quark content. They are all unstable and decay quickly. We will not come across them again in this chapter.

There are also particles that consist of a quark and an antiquark; these are called "mesons". Examples include B-mesons, D-mesons, K-mesons and π-mesons. Some have multiple names; for example, K-mesons are also known as kaons, and π-mesons as pions. Each has a different combination of quarks and correspondingly different charge and mass. Some occur in multiple variants; for example there are neutral, positive and negative kaons and pions. All of them have a relatively short lifetime before they decay, one reason for this being the annihilation of matter and antimatter (which only applies to quark/antiquark pairs of the same flavor). On the other hand, the charged variants of the pion live long enough to be caught in a detector. The subset of mesons that contain a heavy quark and an antiquark of the same flavor (for example, up and anti-up) are called "quarkonia" (singular: quarkonium). Curiously, they are their own antiparticles. So, if you include the unstable particles and all their variants, there is much more going on in terms of composite particles beyond just protons and neutrons.

Since we're talking about the Large Hadron Collider, I should probably also define what a hadron is: any particle with at least two quarks, so it is the "superconcept" of mesons and the baryons—protons and neutrons. The definition also includes particles with significantly more quarks, such as lead ions; they consist of 624 quarks, three in each of its 208 protons.

2 | *Particle Accelerators*

WHY ACCELERATORS Accelerators accelerate beams of particles such as electrons, protons or even heavier nuclei to very high speeds. Often this is done to collide the particles, for example by crossing the paths of two particle beams traveling in opposite directions. Detectors around the collision points monitor the particles created during the collision. These kinds of accelerators are also called particle colliders.

During a collision the energy of the colliding particles is released in the form of other particles (we know that electromagnetic radiation is also composed of particles: photons). Because the colliding particles are so fast they have lots of kinetic energy, and because of E = mc2 this means that particles with very high masses can be created during such collisions: the second or third generation fermions or the heavy bosons. Often these particles are unstable and decay long before they reach a detector. The detectors therefore often do not see these heavy particles directly, but instead detect their decay products: analyses reconstruct the original particles.

Producing these particles and their decay products, and thus understanding the subnuclear world, is a major motivation for colliders. The energies achievable with accelerators, and the respective states of matter, were also present just after the birth of the universe, a trillionth of a second after the Big Bang. We can potentially create and observe the particles that make up dark matter, the fraction of gravitational mass in the universe that cannot be detected otherwise. In this sense, particle colliders are also related to astronomy and cosmology: they connect the very small to the very large.

OPERATIONAL PRINCIPLE Colliders accelerate particles by means of electric fields. This means that the particles must have an electric charge, otherwise they would not be affected by the electric field. Thus colliders accelerate either electrons, protons or heavier nuclei, such as lead or gold ions.

Most accelerators are circular, with accelerating electric fields installed only at a few locations around the ring; the particles race around the track millions of times, and each time they pass through the electric field they get another push.

A ring-style layout comes with a tradeoff: a particle must be continuously accelerated perpendicular to its direction of travel for it to follow the curving track of a circular accelerator. This is achieved by means of magnetic fields. You probably remember the right-hand rule from school: it indicates the direction of the resulting force for a given direction of travel of a charged particle (or a current) and the orientation of the magnetic field. Mathematically this represents the vector cross-product.

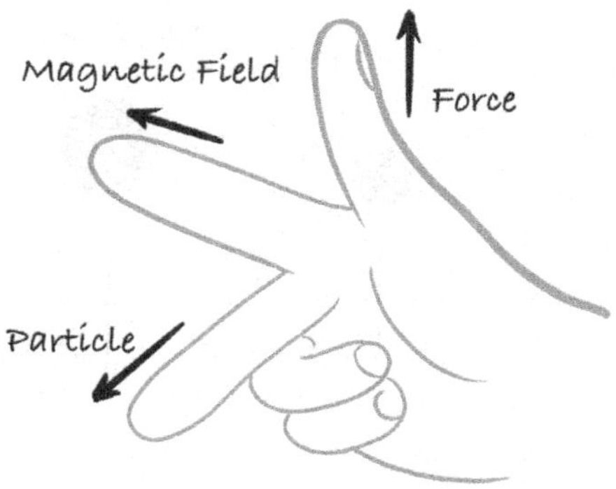

The resulting force is called the Lorentz force, FL, and is calculated as

$$F_L = q\, v\, B \sin \phi$$

where q is the charge of the moving particle, v is its speed, B is the magnetic field in which it travels and Φ is the angle between the direction of travel and the magnetic field. We assume here that sin(Φ) = 1, the maximal value; we get that for an angle of 90 degrees, as the right-hand rule suggests. So

$$F_L = q\, v\, B$$

The force accelerates the particle perpendicularly to its direction of travel, which means it is entering a circle. The radius can be determined by realizing that, on a stationary circular track, the Lorentz force must be equal to the centrifugal force FC:

$$F_L = F_C$$

$$q\, v\, B = \frac{m\, v^2}{r}$$

$$r = \frac{m\, v}{q\, B}$$

As we can see, the radius increases linearly with both the mass and the speed, so a heavier particle will enter a larger circle (because it has more acceleration-resisting inertial mass), and so will a faster one (because it goes further along its original track in the time it takes to accelerate around the circle).

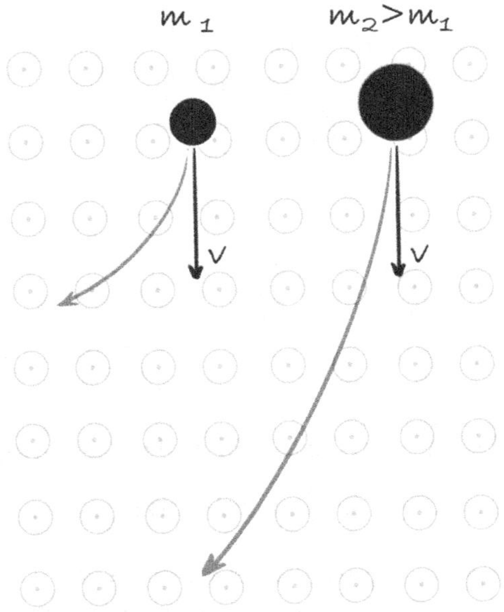

As a corollary, to keep faster or heavier particles on a given radius, you will have to increase the magnetic field linearly:

$$B = \frac{m\,v}{q\,r}$$

For an accelerator this means that as the particle beam accelerates, the magnetic field has to be ramped up synchronously. In practice it is the maximum strength of the magnetic field achievable by these bending dipole magnets that determines the maximum speed—and thus the maximum total particle energy—delivered by an accelerator.

3 | *CERN and the LHC*

CERN CERN, the Centre Européen pour la Recherche Nucléaire, was founded in 1954 in Geneva, Switzerland. Its charter is to do research into the fundamental nature of the "stuff" our universe is made of. But there was also a political component of European collaboration just a decade after the war. CERN's website says:

"[...] the need for Europe to have a world-class physics research facility. Their vision was both to stop the brain drain to America that had begun during the Second World War, and to provide a force for unity in post-war Europe."

Over the years CERN has indeed made many discoveries in particle physics, and five Nobel Prizes have been awarded to CERN researchers. For example, in 1973 CERN was involved in the discovery of the electroweak force, and in 1983 they discovered the W and Z bosons. In 2012 the LHC observed the famous Higgs boson, and we'll look at this discovery in detail in this chapter. But CERN's largest impact on society arguably came in 1989 when Tim Berners-Lee, a physicist and computer scientist working at CERN, proposed a project based on hypertext to solve some of the communication and knowledge management problems within CERN itself and the wider physics community. He subsequently developed the first versions of HTML, HTTP, URLs, web servers and web browsers, and thus laid the foundations of the World-Wide Web.

Today CERN has 23 member states and 3,400 employees. It serves around 14,000 "users", guest scientists from all over the world. There are many different research labs and facilities at CERN, most of them related to particle physics. Its most important facility is the Large Hadron Collider, the world's highest-energy particle accelerator.

ACCELERATOR HISTORY Particle accelerators have been around for a long time. They can be categorized by their layout (circular versus linear), the particles they accelerate (electrons, protons, heavier ions), the energy up to which they accelerate the particles, as well as their purpose. For example the ESRF, the European Synchrotron Radiation Facility in Grenoble, is an 800-meter circular accelerator for electrons that produces X-rays, for use as an extremely powerful light source for all kinds of science, such as crystallography.

In this chapter we'll look at accelerators that operate as particle colliders for exploring the fundamental building blocks of matter. There are many of these all over the world. The Relativistic Heavy Ion Collider, or RHIC, at Brookhaven National Laboratory is a 3.8-km circular accelerator, mostly for gold ions, that replicates the conditions of the universe right after the Big Bang. With a diameter of 6.3 km and a maximum energy of 980 GeV per particle—just under one TeV—the appropriately named Tevatron at Fermilab, near Chicago, was the most energetic accelerator for proton/antiproton collisions before the LHC became available; it was subsequently shut down in 2011. The Stanford Linear Collider was a linear collider for electrons and positrons up to around 45 GeV, but was overshadowed when CERN's Large Electron-Positron Collider, which reached 104 GeV, went online in 1989. The SLC was installed at SLAC National Accelerator Laboratory at Menlo Park, California.

The Superconducting Super Collider (starting to sound like telescope names...) was a projected circular, 87-km 20-TeV hadron collider whose construction had started in Waxahachie, Texas, but which was canceled in 1993 for a variety of technical, financial and political reasons after spending two billion USD. Had it been completed it would have been three times more energetic than the LHC. Maybe the technical and financial problems that led to its cancellation are an indication that the machine might have been a bit too ambitious.

In the late 1980s the particle physics community realized that higher energies would be needed to make further progress in particle physics: some of the second and third generation particles require more energy to produce than was available at the existing colliders: the Tevatron, SLC and LEP. So would the Higgs boson, at the time an undiscovered prediction of the standard model. Other phenomena such as supersymmetry were also suspected to show up only at energies beyond that of existing accelerators. CERN decided to build the LHC in 1994, a year after the potentially more powerful SSC had been cancelled; it went live for the first time in 2008. Its first physics run was in 2010, albeit not at full energy.

LHC OVERVIEW Let's break down the three letters in LHC. C is obvious—it stands for collider. The L is for large, because at the time of its construction and still today its 27-km ring, buried between 50 and 150 meters underground, is the largest collider on the planet. But more importantly, at 7 TeV it is also the largest in terms of the energy it can deliver; the LHC is seven times more energetic than the Tevatron. By the way, you'll read both 7 TeV and 14 TeV as the maximum energy delivered by LHC. Both are correct: the 7 TeV is the maximum energy of a single particle, whereas 14 TeV is the maximum energy of a collision—two particles of 7 TeV each. The H stands for hadron, which are the multi-quark composite particles: protons and neutrons, but also heavier nuclei such as lead ions. So the LHC is the largest and most energetic accelerator and collider for "heavy" particles.

Before we continue, we should probably explain what an "eV" is. One electron volt is the kinetic energy gained by an electron when it is accelerated by a potential difference of 1 volt; it corresponds to $1.602176634 \cdot 10^{-19}$ joules. If we accelerate heavier particles with the same electric potential V they gain the same amount of energy, but of course this results in a lower speed increase. Starting from

$$W_{kin} = \frac{1}{2}mv^2$$

the well-known equation for kinetic energy, if we solve this for speed we see that greater mass results in a smaller speed increase:

$$v = \sqrt{\frac{2\,W_{kin}}{m}}$$

The electron volt is also used as a unit of mass. Through $E = mc^2$ we can convert any increase in energy into an increase of mass. So, more correctly, the mass unit should be eV/c^2, but people colloquially say that "xyz particle has a mass of so-and-so-many eV". As we approach the speed of light the Special Theory of Relativity makes this relationship more explicit: particles get relativistically heavier instead of faster. For a particle with a given total energy W_{total}, the speed is calculated as

$$v = c\sqrt{1 - \frac{m_0^2\, c^4}{W_{total}^2}}$$

Here m_0 refers to the rest mass of the particle, which is its mass at zero speed; as we know from the chapter on gravitational waves, due to Special Relativity a body's mass will increase with the Lorentz factor as it accelerates towards the speed of light. W_{total}, the energy of the particle, is the sum of its "rest energy" (from $W = m_0 \cdot c^2$), the kinetic energy proportional to its speed and the energy from the relativistic mass increase.

Back to the LHC. The beams are made to collide at four positions around the ring, observed by four LHC experiments. ATLAS (A Toroidal LHC ApparatuS) and CMS (Compact Muon Solenoid) are general-purpose detectors that record everything that happens in proton–proton collisions. ALICE (A Large Ion Collider Experiment) is optimized for collisions of lead ions; during their collisions they dissolve into a quark–gluon plasma that is representative of the "primordial soup" of our universe. Finally, LHCb investigates the difference between matter and antimatter, based mainly on the bottom and antibottom quarks. In particular, the LHCb collaboration is interested in finding out why there is matter in the universe; symmetries suggest that all matter and antimatter should have annihilated during its creation. The L in LHC also applies to the experiments: ATLAS is 42 meters long and 25 meters in diameter, while ALICE weighs around 10,000 tons. Thousands of people from hundreds of institutions work on each of them. There are also several smaller experiments, currently eight in total.

We will look at these experiments in more detail later, but here is their basic operating principle. Two protons collide head-on and the energy "brought in" by the colliding protons is available to form new particles. As we know, protons are composite particles, so it is not really the protons that collide, it is the quarks and gluons contained in them: there are quark–quark, gluon–gluon and quark–gluon collisions, and each can happen with different probabilities. This why collision events are so complex, and why there is a large variety of outcomes when two protons crash into each other. The particles produced radiate outward from the collision point. The experiments surround this volume with many layers of subdetectors—I use the word "subdetector" when I need to point out a particular component of CMS or ATLAS. As the particles pass through these layers on their way out, they create sequences of "detector hits". Computers use these hits to reconstruct the particles' trajectories, and ultimately their mass and charge. Because each kind of particle has a characteristic mass/charge fingerprint, their type can be identified. Every particle is tracked separately, leading to a complete picture of what happened in the collision.

How many of these collisions happen? The likelihood of two isolated protons hitting head-on is exceedingly low because they are so small, so the LHC tries with a lot of particles. The beam consists of distinct bunches, each containing 100 billion protons. In 2018 they usually ran with 2,556 bunches; the design capacity is 2,808. This makes $3 \cdot 10^{14}$ protons in total. At the interaction point, each bunch is focused to a diameter of around 15 microns, the diameter of the thinnest of human hairs. From the 100 billion protons in each bunch, around 10 to 20 collide when the paths of two bunches cross. Only some of these collisions are interesting, in the sense that some of the quark and gluon collisions are energetic enough to produce rare heavy particles. There's a bunch crossing every 25 nanoseconds. Considering that every particle is identified and its track reconstructed, this is a massive challenge in data acquisition and computing.

The overall organization of the LHC can be separated into two communities: beam producers, the facilities and people who accelerate and maintain beam, and beam consumers, the experiments that observe collisions. In some sense, beam production is a "service" for the scientists. This is a common set-up for accelerators: at the ESRF, for example, the experiments are switched over time; experimenters bring their equipment to ESRF and receive particles from the accelerator. While the experiments at LHC don't change, the general set-up follows the same pattern. We will look at both aspects in this chapter.

4 | *Producing the Beam*

STAGED ACCELERATION The overall acceleration process is staged, reusing older, smaller accelerators before the particles arrive in the 27-km LHC ring (the term "LHC" refers both to the whole machine that includes all accelerators and experiments as well as to the main, 27-km circular accelerator; it should be clear from the context). When not filling the LHC, the smaller accelerators produce particles for other experiments at CERN, such as the (no longer active) Grand Sasso Neutrino experiment, the antiproton decelerator and a number of experiments in which particles are made to collide with fixed targets to investigate very specific phenomena. The following illustration shows the chain of accelerators. The early stages of the process are different for protons and lead ions; in the rest of this section I will describe the process for protons.

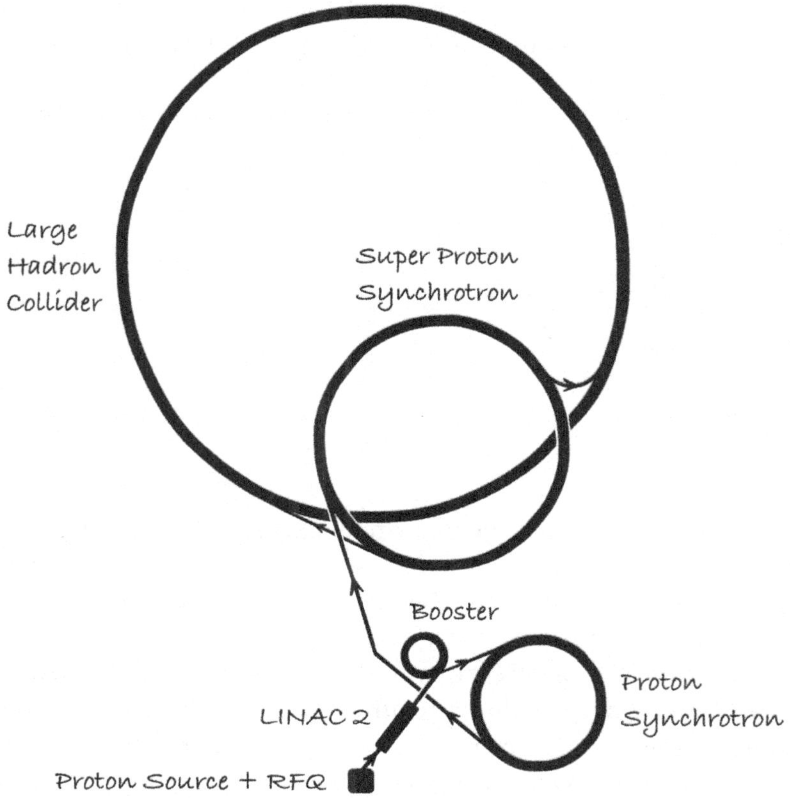

Everything starts with a bottle of hydrogen gas; interestingly, a small bottle is enough for a year of operation. A device called a duoplasmatron source ionizes the hydrogen atoms and creates a plasma, then separates the positively charged nuclei from the negatively charged electrons using an electric field. This field also starts the acceleration; it pushes the protons to around 90 keV, or 1.4% of the speed of light.

The next stage is the Radio Frequency Quadrupole, or RFQ, a 1.75-meter long linear accelerator. In addition to accelerating to 750 keV via an electric potential, its job is also to focus the protons; we'll discuss this below. The RFQ has a third responsibility: it must group the particles in the beam into six separate bunches. This is necessary because the accelerating fields in the downstream circular accelerators cannot deal with a continuous beam. The formation of bunches is achieved by accelerating at different powers over the length of the RFQ, "ripping apart" the continuous bunch. The RFQ uses a sinusoidal structure of electric quadrupoles, which combine focusing and "bunching" acceleration.

The next stage is LINAC 2, a 30-meter long linear accelerator that accelerates the particles to 50 MeV, or around one third of the speed of light. LINAC 2 consists of several hollow cylindrical electrodes. Using a high-frequency AC current, these electrodes switch their charge such that whenever a positively charged particle passes a specific electrode, it switches to positive charge, pushing the particle forward, whereas at the same time the downstream electrode is charged negatively, pulling the proton bunch toward it. The AC has a constant frequency, which is why the distance between the electrodes has to increase with the length of LINAC 2, to compensate for the increasing speed of the particles.

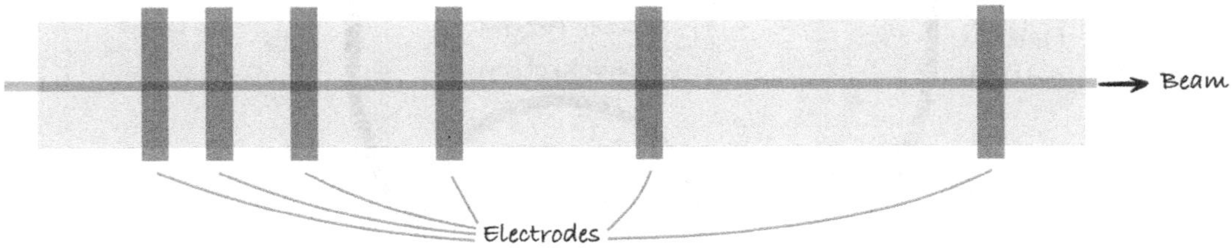

The bunches then enter the Booster, the first circular accelerator. This has a radius of 50 meters; as the particles are still at fairly low energy, relatively weak magnetic fields are able bend their trajectory to this rather tight radius. The Booster accelerates the protons to 1.4 GeV, or 91% of the speed of light, within 1.2 seconds. The acceleration still relies on electric fields, but in the Booster they are created using cavity resonators, also called radio frequency (RF) cavities. These are hollow, tube-like metallic structures in which a microwave is excited to a standing wave. Assuming proper timing, the increasing slope of the standing wave pushes the particle bunch forward, much as a surfer is pushed by a wave in the water.

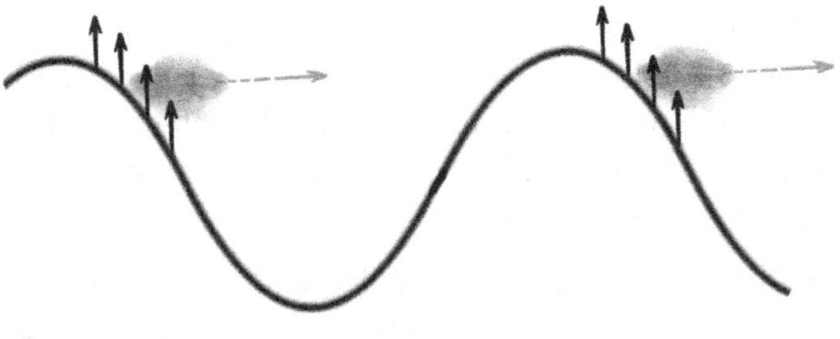

Now we can see why the beam has to be bunched: if it were continuous half of the particles would be mis-timed—they would be slowed down instead of being accelerated. Indeed, any particles in a bunch that are a bit too fast are slowed down. Ultimately, this kind of acceleration has the desirable side-effect of improving the longitudinal cohesion of bunches. RF cavities are used in all downstream accelerator stages as well, including in the 27 km LHC ring.

The next stage is the 628-meter Proton Synchrotron. This accelerates the beam to 25 GeV, 99.93% of c. Because the speed of light cannot be exceeded, any additional energy imparted by "acceleration" after this point will mostly increase the (relativistic) mass instead of the kinetic energy. The Proton Synchrotron also has an additional task: it splits the six bunches received from the Booster into 72 smaller bunches. The Proton Synchrotron accepts the particle bunches from the Booster at a frequency of around 3 MHz. Multiple increases of this frequency "rip" the six bunches apart, leading to 72 smaller ones, each 4 ns long and separated by 25 ns—this is also the bunch timing in the LHC, because the particles won't get any faster during the remaining "acceleration" process. With its 277 magnets, 100 of which are bending dipoles, this is a non-trivial machine in itself. It has 16 straight sections with cavities for acceleration and 20 shorter ones containing quadrupole magner for focusing.

As we get closer to the LHC's main ring, the Super Proton Synchrotron (SPS) is the next-to-last stage of acceleration: RF cavities accelerate the beam from 25 GeV to 450 GeV in around 21 seconds. At 7 km circumference with 744 dipole and 216 quadrupole magnets, this is a serious machine! When it went live in 1976 it was the largest accelerator at CERN and was sometimes used as a collider in the 1980s, whereas now it is "just" an accelerator. It discovered the W and Z bosons mentioned earlier, based on the collision of beams of protons and antiprotons.

The SPS is also the last stage that has only one beam. Remember that we need two beams that travel in opposite directions to be able to collide protons. This is why the one beam from the SPS is injected into the two separate beam pipes of the LHC via two connecting pipes. Once in the LHC, the beam is further energized using 16 RF cavities— I am avoiding the word "accelerated", because additional energy at this stage leads to mass increase. The beam is bent and focused using 1,232 dipoles and 858 quadrupoles. The particles travel around the ring 11,000 times per second; because there are 2,808 bunches traveling in each beam pipe, the accelerating boosts in the cavities must be pulsed at around 30 MHz. Acceleration to the maximum energy in the LHC takes around 20 minutes.

I've mentioned twice now that the acceleration takes time—21 seconds for the SPS and 20 minutes for the LHC. What does this mean? Remember that the magnetic field required to keep the particles on their circular track depends on their energy: more energetic particles require a stronger magnetic field. Since all particles always see the same magnetic field, all particles in the ring must be at the same energy at any point in time. This means that you cannot inject low-energy particles from the previous stage while at the same time forking out high-energy particles to the next stage. Instead, circular accelerators are phased. The magnetic field is set to its lowest strength, corresponding to the energy at which particles are injected, and the ring is filled with particles. Then the particles are accelerated jointly, with a correspondingly increasing magnetic field. Once at their top energy, they are ejected into the next accelerator in the chain, or stored for collisions.

This phased operation is also used for the 27-km LHC: the SPS injects bunches for around 30 minutes, until the ring is filled with 2,808 of them. Kicker magnets synchronize the incoming bunches with those already in the beam pipe, basically slotting them into the right place to fill the ring homogeneously. When the ring is full, the particles are accelerated to 7 TeV, at which point the beam is ready for science. Another set of specialized magnets bends the opposing beams further to ensure they cross each other's path at the four interaction points.

Ultimately the main components of any circular collider are the dipole magnets that bend the beams around the track, the quadrupoles to focus them and a means of acceleration, usually RF cavities, plus lots of equipment to monitor and diagnose the state of the beam. The specific numbers of each of these components vary from collider to collider.

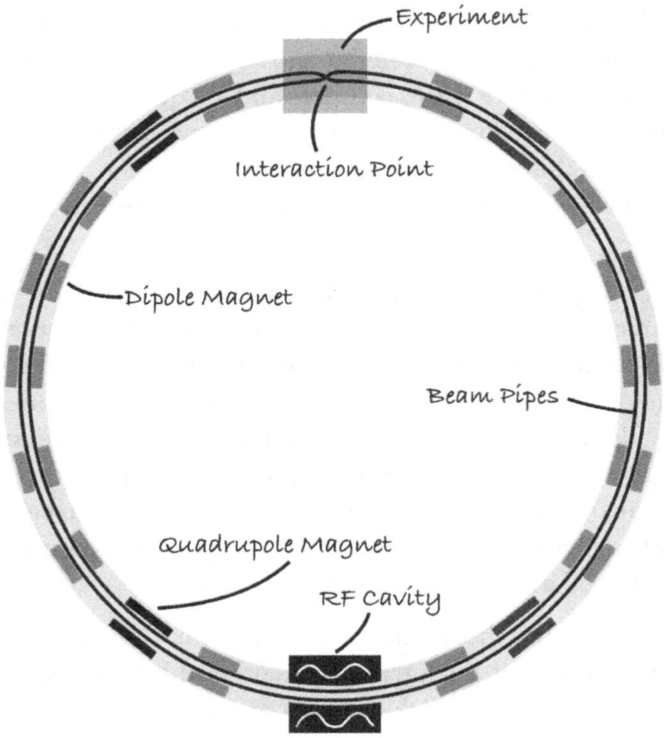

AROUND THE BEND Keeping the beams on track requires active control of the dipoles: at any given time the field strength of all magnets has to be the same, and during ramp up it has to increase synchronously for all magnets. The control of these magnets is open loop: previously performed calibrations have determined how much the magnetic field has to be increased as the RF is increased to increase particle energy. However, the more than 1,200 dipole magnets of the LHC are not perfectly identical: they have slight mechanical or material differences that alter the relationship between their electric current and the resulting magnetic field. To ensure the same field strength in all magnets each magnet must be calibrated separately, and each power converter—the devices that control the current for a magnet—gets its own specific setting. The mapping from current to field for each magnet is stored as calibration data in the control system.

In addition to the calibration-based open loop controller there is an additional, finer layer of control that relies on measuring the beam's position, which uses over 1,000 beam position monitors installed around the LHC ring. These consist of electrodes that detect the slight voltage induced by the "beam current", the electric current represented by the particles moving around the ring. A change in the beam position leads to characteristic voltages in each sensor, and a statistical analysis that spans several circulations of the beam leads to an integrated position of the beam. The monitors detect the beam's position both in terms of radius (in/out) and in terms of height (up/down). If the beam position must be corrected, this is done by a separate set of corrector magnets; the beam monitors do not control the main dipole magnets.

Where do these in/out and up/down divergences of the beam come from? One source is the acceleration of the beam and the resulting change in its radius—the main dipole magnets might not be perfectly calibrated to compensate the increasing centrifugal force. Sometimes there are also problems with an imperfect injection from the SPS; our guest and software engineer at CERN, Vito Baggiolini, recalled that "Sometimes it can be difficult to get hold of the injected particles; if they get off track too much, they are lost." Remember that the bunches are injected into the LHC at an angle from the SPS, and this must happen at the right energy, at the right time (when there's a slot for an additional bunch in the LHC beam) and with the right momentum imparted by the kicker magnets. It's a bit like a child jumping onto a rotating merry-go-round.

FOCUSING The notion of focus, borrowed from optics, refers to the cross-section of the beam; you want to keep it small in order to keep the beam "within" the pipes and to ensure a high collision rate. Because the particles in a bunch repel each other—they all carry the same electric charge—focusing is an ongoing and actively managed process. It is is achieved with quadrupole magnets, a magnet that arranges north and south poles opposite each other in pairs. Because magnetic field lines always go from north to south, the field shown in the next illustration results from such an arrangement.

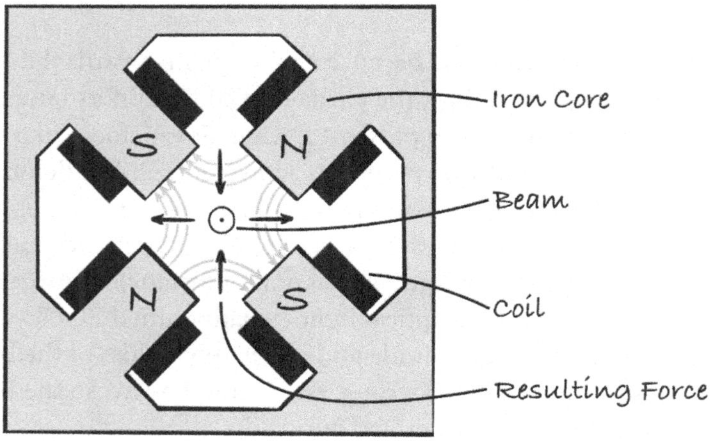

At the exact center of the quadrupole the magnetic field is zero. If a particle flies through the magnet smack at its center it feels no force at all. However, a particle that flies further out feels an increasingly stronger force: the field has a gradient. If you look at the illustration above, the direction of that force varies; in the case above, the particle feels a force towards the center if it is off vertically. However, the particle feels a force away from the center if it strays off horizontally. Use the right-hand rule to see for yourself why this is the case. This means that a quadrupole focuses the beam in one direction (here, vertically) and defocuses it in the other.

To achieve our goal of focusing the beam in both directions, therefore, we need sequences of quadrupoles with alternating orientations, with gaps between them to ensure that the magnetic fields do not disturb each other. Such a sequence is called a "FODO cell": F is for the horizontal focusing, D is for horizontal defocusing—thus, vertical focussing—and O represents the gaps. The precise distances between the F and D, as well as the strength of the fields, is crucial. Consider the illustration below, which shows a FODO cell's effect on the beam in one direction, say up/down.

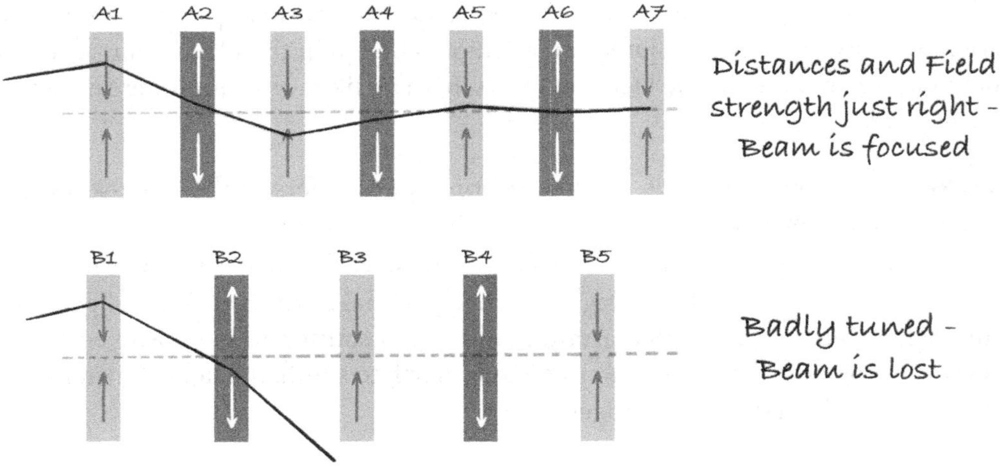

A particle comes into A1 and gets focused because it gets a push towards the center of the beam pipe. If no more magnets were to follow, the particle would retain its lateral momentum, fly out the other side and be lost. But it then passes A2, where it is defocused a little (only a little, because it passes further inside, and so experiences a lower force). The defocusing reduces the momentum it got before, and therefore bends it towards the center. Through the remaining F and D elements the amplitude of its lateral movement gets smaller and smaller. A FODO cell that is well-tuned in terms of field strength, gradient and distances between the magnets focuses the beam, just as a lens does for optical light. A badly tuned FODO cell, however, can lead to an increase in lateral beam amplitude and ultimately to loss of the beam. The second example shows this case; either B1 is too strong, B2 is too far away, so the particle enters the field of B2 on the wrong side and is accelerated outwards.

However, even correctly configured FODO cells cannot completely eliminate a beam's so-called "betatron oscillations" around the center of the beam pipe, the ideal beam flight path; some particles will always be lost over time. Beam loss monitors measure this amount and raise an alarm if the losses grow beyond predefined thresholds. But there's also a second consideration, called the "betatron tune". This refers to the number of betatron oscillations over the circumference of the track. If that number is an integer the waves might interfere constructively with themselves on the next circulation, leading to resonance: the amplitude will becomes large and ultimately the beam will be lost. The dipoles and the quadrupoles are tuned to avoid such resonances by making sure the beam shifts phase at every transit around the ring. Generally, the beam is dynamically unstable and requires active control. "It's like a helicopter", as Vito said.

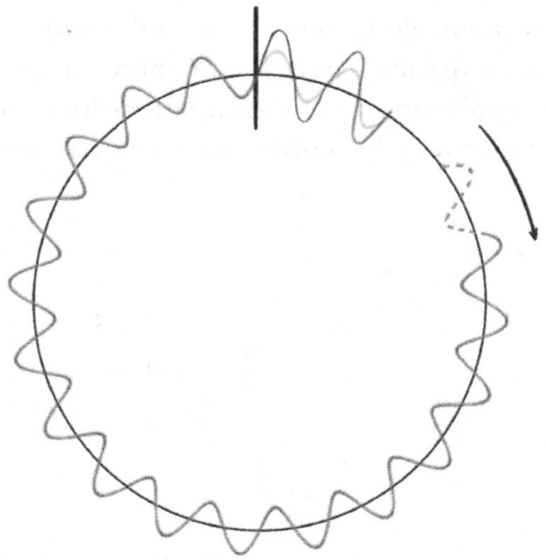

The world doesn't completely conspire against the operators of the LHC, though; there's one effect that focuses the beam automatically as it is accelerated. Because its longitudinal speed increases, a particle has less time to move perpendicularly to the beam between two focusing quadrupoles—after all, the push it gets from the focusing electromagnetic force remains constant, and so therefore is its transverse speed. As a consequence, the beam diverges less. This process is referred to as "adiabatic damping".

At the interaction point the beam is compressed especially tightly to increase the probability of collisions. This is a challenge, because the beam is so tight that the repulsive electromagnetic force becomes more of a problem. Very strong focusing quadrupoles are required.

RING STRUCTURE The LHC is not exactly circular: it consists of eight 2.43-km arcs separated by eight 545-meter straight sections. The straight sections are used for beam insertion from the previous accelerators in the chain, dumping (getting rid of a beam), for beam cleaning—removal of particles that stray too far from the beam center—as well as for collisions and experiments. The distance from the middle of one arc over a straight section to the middle of the next arc is called an "octant". The magnets in each octant are cooled in a contiguous cryostat, and each octant can be cooled or warmed independently, which saves energy for re-cooling if they need to be warmed for repair. We will cover the cooling system later. The magnets in an octant are electrically in series, and each octant's electrical power system is independent. Octants are the units of installation and commissioning: there's separate physical access to each octant, for example to insert or extract magnets.

Each arc contains 23 cells, and each cell contains dipoles, quadrupoles, corrector magnets and beam position monitors. The next illustration shows an "exploded" view of a cell. The horizontal dimension in this illustration represents the length of the cell; in reality the various elements shown in the illustration along the y axis are all in sequence on the beam-line (at y = 0, if you will); I've moved them up the y axis to declutter the image. The main dipoles take care of bending, while the main quadrupoles handle most of the focusing via FODO cells. All other magnets are special-purpose devices that correct the focusing for particular distortions.

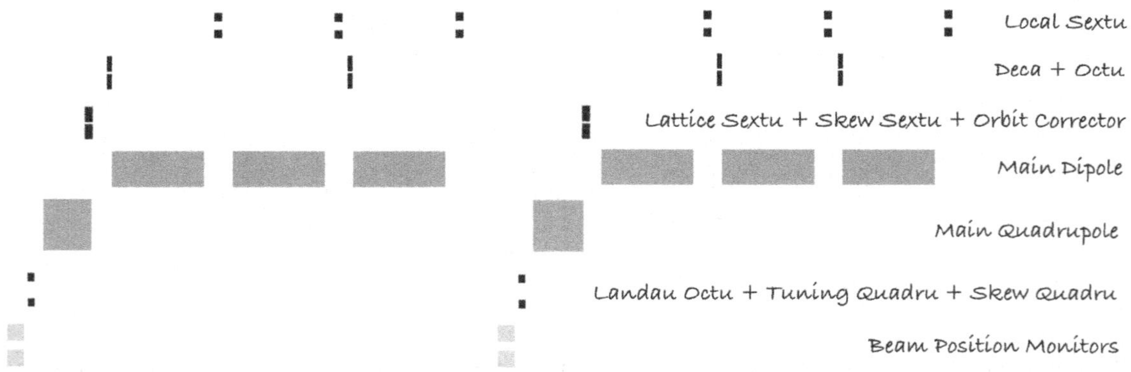

There are other magnets in addition to those shown above, for injecting, pushing and kicking the beam, 57 different kinds in total: MB, MQ, MS, MO, MQS, MSS, MQT, MCO, MCD, MCS, MQTLI, MCBH, MCBV, MBRB, MBRC, MBRS, MBX, MBW, MBXW, MCBCH, MCBCV, MCBXH, MCBXV, MCBWH, MCBWV, MCBYH, MCBYV, MCOSX, MCOX, MQM, MQML, MQY, MQMC, MQWA, MQWB, MQXA, MQXB, MCSSX, MCSX, MQRL, MQR, MCTX, MQSX, MCBWB, MU, MGMWV, MGMWH, MBAW, MBLW, MBWMD, MBXWH, MBXWS, MSIA, MSIB, MSDA, MSDB, MSDC. What looks like letter soup is of course a systematic designation scheme: M is for magnet, B for bending, Q for quadrupole, C for corrector, S for sextupole and so on.

In total, over 8,000 superconducting magnets are used in the LHC. 2,000 of them are the large dipoles and quadrupoles and 6,000 are the various kinds of corrector magnets. All have to be reliably controlled and coordinated. To do that, the control system has to understand the beam's condition, which it does with several kinds of sensors. The beam position monitors verify the lateral and vertical position of the beam, while synchrotron light telescopes provide profiles of the beam. The beam loss monitors measure the amount of particles that "escape" from the beam—that is, those that are not "captured" by the focusing quadrupoles. Each of these sensors is non-trivial, requires calibration and tuning, and has been optimized over years or even decades. If you weren't convinced that the LHC is a complex system by now, this should drive it home.

THE MAGNETS During a previous visit to CERN, the one where I recorded the episode with Vito, I got access to building SM18. Lots of metal rigs, cranes on the ceiling, piping, wiring and a few screens with numbers on the walls. I saw several long blue steel tubes that were connected via cables and pipes to the surrounding test bench. It turns out that each of these 15-meter-long, 1-meter-diameter 35-ton blue tubes is one of the dipole magnets! This is when I realized that these dipoles are not discrete magnets located at selected positions around the ring. Instead, around two thirds of the 27 km is made up of dipoles, some 1,200 in total: the magnets essentially "are" the ring. All must produce a consistent magnetic field, and their characteristics must be precisely understood for them to be reliably controlled. Installing magnets in the tunnel is a lot of work, and removing one because it does not behave as expected must be avoided if at all possible. This is why CERN runs the magnet testing facility in building SM18. Each magnet, after delivery from the manufacturer, is connected to electrical power and to the cryogenic cooling system. Sensors placed into the beam pipes to measure the strength and shape of the magnet field, as well as how it changes depending on the applied current.

Let's look at a cross-section of a dipole magnet to understand its structure, starting from the inside; we'll look at many of the components in some detail later. The first thing to realize is that the beam pipes for both directions are immediately next to each other; they "live" in the same cryostat. Each pipe is between 4.5 and 6 cm in diameter at different locations around the ring. Moving outward, each pipe is surrounded by the superconducting coils that produce the magnetic field. Both pipes and coils are held in position by a collar. Next is an iron yoke that helps to shape the magnetic field and adds additional mechanical stability. When the system is operational this entire mass is cooled to 1.9 K. The iron yoke also contains the bus bars, essentially superconducting cables that transport the thousands of amps needed by the magnets within an octant; remember that they are electrically in series. We will meet those bus bars and their connection between magnets again, later in this chapter, albeit not in a very positive context. The outer edge of the yoke has a radiation shield. Next on the way outwards is the vessel that contains the liquid helium, followed by a thermal shield on its outside. For further thermal insulation all this is surrounded by a vacuum, which of course does not transmit heat.

The supports that hold the inner components are made from carbon fiber, which is mechanically strong while at the same time being a good thermal insulator.

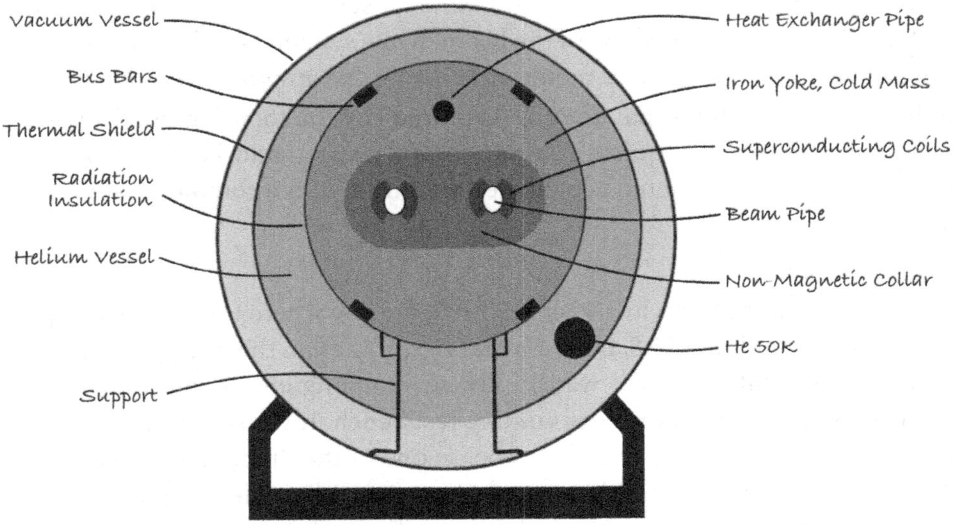

Separate coil loops are installed on either side of the beam pipe to produce a vertically oriented magnetic field that is perfectly homogeneous across the beam pipe, so that the force—according to the right-hand rule—bends the beam around the ring. The required current distribution is known as a cosine-theta distribution: maximum current at the equator and minimum current at the poles. Thousands of actively controlled corrector magnets are needed to ensure that the field is identical around the LHC's ring.

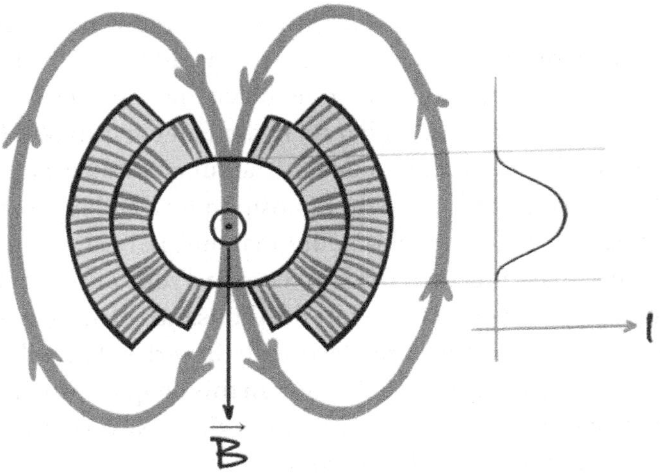

Let's calculate the required magnetic field strength B for the LHC. We start with the formula that we developed earlier, based on the realization that for each particle on a circular track, the Lorentz force it feels is equal to the centripetal force:

$$B = \frac{m\,v}{q\,r_{LHC}}$$

At relativistic speeds, mass increases with the γ factor; we introduced this idea in the discussion of relativity in the chapter on gravitational waves. The LHC accelerates protons to 99.9999991% of the speed of light, so γ turns out to be 7,453:

$$\gamma = \frac{1}{\sqrt{1-\frac{v^2}{c^2}}} = \frac{1}{\sqrt{1-\frac{(0.999999991\,c)^2}{c^2}}} = \frac{1}{\sqrt{1-0.999999991^2}} = 7{,}453$$

If we plug this value into the formula for the magnetic field above, we get:

$$B = \frac{7453 \cdot m_0 \cdot 0.999999991\,c}{q\,r_{LHC}}$$

The proton mass m_0, its charge q and the speed of light c are all physical constants. The radius of the LHC is of course its circumference divided by 2π; however, we have to be careful to not use the full 27 km as the circumference, due to its straight sections. The circular part of the circumference, those that do the bending, is 8 · 2.43 km = 19.44 km, so the relevant radius of the LHC is 3,094 meters.

The resulting magnetic field to keep the protons on track comes out as 7.5 Tesla. The actual number is 8.3 Tesla, so are we happy with our result? I think so. There are certainly some technical losses, and the field probably isn't perfectly shaped. So it's good that our back-of-the-napkin result is lower rather than higher. Remember that each cell also has two three-meter long quadrupoles. With 23 cells per octant and eight octants, 1.1 km of the ring is used up for the quadrupoles, which don't take part in the bending. If we calculate B with the reduced circumference, B becomes 7.95 Tesla. We're getting there.

This strong magnetic field does not just act on the particles, it also acts on the magnet structure itself: it's the equivalent of 350 tons acting on each meter of dipole. That's a lot! But what's more, the structure must be very precise—the acceptance criterion was ±1.5 mm over its 15-meter length—and stiff enough for the coils not to move more than 50 micrometers under the influence of the field. If they moved further it would disrupt the shape of the magnetic field they produce, as the coils themselves are relatively soft. Movement might also break the copper insulators, and could lead to friction, producing heat that might lead to non-superconducting "hot spots", risking "quenching"of the magnet. Quenching is an effect in which the magnet abruptly loses its superconductivity and reverts to regular conduction. The amount of current "in" the superconductor would then be way too high for the newly regular conductor, so you can see how this would be a problem. I will return to this later.

The magnetic load is born to some degree by the iron yoke, but mostly by the aforementioned collars. Each collar is a few millimeters thick in the plane of the beam and is made from austenitic steel, which is high-strength and mostly non-magnetic. The collars are prestressed to increase quality and resistance to tension. 12 million of these collars are used in the LHC, locked together by rods.

Interestingly, the iron and steel have different thermal expansion coefficients. This is why the iron has a small gap between its left and right hemispheres—it closes when the system is cooled to its 1.9 K operating temperature.

POWER CONVERTERS Lots of otherwise ordinary things require special attention at the LHC. The power converters are one example. Power converters take electrical power from the grid and make it suitable for a particular application in terms of voltage, current and frequency. A prominent use is in electric locomotives. These transform 15 kV 16.7 Hz AC from the overhead line to the three-phase current required by the motors. The frequency of the current determines the speed of the motors, and it is the frequency of the switching inside its power converters that makes the melodic sound you hear when, for example, a Siemens ES64U2 Taurus accelerates—and when you listen to the very beginning of an omega tau episode.

The power converters for the LHC magnets consume 400 V three-phase power from the electric grid and output DC for the magnets. Internally they use a bunch of switches, capacitors and inductors to invert the negative components of the sine wave in each phase, add everything up and smooth the resulting voltage. Their output power is between 500 W and 2 MW, depending on the size of the magnet they serve. Because they supply superconducting devices that have no electrical resistance, they must be able to deliver currents of up to 13,000 amps at relatively low voltage. This is not an everyday requirement.

The cables between the power converters and the magnets, however, are not superconducting, so they do have electrical resistance. To keep the electrical losses acceptable at such high currents (remember, $P_{loss} = I^2 \cdot R_{cable}$), the water-cooled copper cables must be kept short, so the power converters must be installed close to the magnets in the LHC tunnel or in adjacent caverns. This in turn leads to constraints on their size and heat dissipation, as well as their reliability, because they cannot be accessed for maintenance when the beam is on.

Because the magnetic field must be controlled precisely, and because that field is directly proportional to the current that circulates in the coils, the accuracy and stability of the power converters' outputs must be in the single digit parts-per-million range. This is why each power converter acts as its own control loop: electronics continuously adjust the settings of the actual "machine", while precision current measurement of the output is fed back to the control electronics. The set value for this control loop is provided by the LHC control system based on the required magnetic field and the magnet-specific calibration.

The DC current transducers used to measure the output current were also custom-developed by CERN, to measure up to 20,000 A with a precision of 2 ppm. CERN also developed the associated analog-to-digital converters. A custom 10 mA reference was also developed; every few months this is used to calibrate the current transducers at every power converter, while every year the references are recalibrated by METAS, the Swiss institute for metrology, the equivalent of the US NIST or the German PTB.

INTERLOCK AND BEAM DUMP Back to the beam. Let's look at how much energy is stored in a fully "loaded" LHC ring. We have 2,800 bunches, each with 100 billion particles. Because protons have an atomic weight of one, Avogadro's number tells us that there are $6.022 \cdot 10^{23}$ particles in one gram; reversing this relationship gives us the total mass of circling protons inside the LHC. It is around a nanogram, which is why the small bottle we mentioned earlier is enough for a year of operation. So the total energy can't be that high, can it? We know that at maximum power each particle has 7 TeV of energy. For the total beam we get:

$$
\begin{aligned}
&2{,}800 \cdot 100\ billion \cdot 7\ TeV \\
&= 2.8 \cdot 10^{3} \cdot 100 \cdot 10^{9} \cdot 7 \cdot 10^{12} eV \\
&= 2 \cdot 10^{27} eV \\
&= 320\ \mathrm{MJ}
\end{aligned}
$$

How much is 320 MJ? Using the well-known formula for kinetic energy introduced a few pages back, we can calculate comparative numbers: a 400-ton ICE 3 high-speed train at 150 km/h has the same energy, and a fully-loaded 747–400 landing with minimum fuel at its normal 270 km/h landing speed represents 428 MJ. That's a bit more, but still in the same ballpark. It's obvious that the energy in the beam has the potential to damage the LHC. We have to do something with that energy, if, for whatever reason, we have to dissipate it.

The process of getting rid of the beam in the ring and absorbing its energy is referred to as "dumping the beam". To do this, it is routed to a 7-meter long, 70 × 70 cm wide graphite block that is surrounded by 750 tons of iron blocks. The energy of the beam is converted to heat: the graphite core heats up to 800 °C. The beam is routed into the dump by "kicker" magnets. A pre-charged capacitor is used to supply the required energy quickly. However, despite the capacitors, the regular bunch spacing of 25 ns is too short, which is why there is a single longer gap of 3 ms at a particular location in the beam, the so-called "abort gap". The kickers have to ramp up at exactly the moment when the abort gap flies past. Special sensors continuously verify that the gap is actually empty of particles.

Why would a beam be dumped? One reason is that the beam position monitors detect a position that cannot be reliably corrected with the corrector magnets. In this case the beam is dumped automatically. The same is true if a quench of the superconducting magnets is detected. Failure of any vital component of the accelerator, when detected by the interlock system, will also trigger a dump. The beam interlock system also prevents injection if the LHC is not ready. It receives "ready and safe" signals from around 20 subsystems, including the vacuum and cooling systems, the collimation system, beam position and beam loss monitors, as well as from important buttons in on the operator consoles in the control room. It is fully redundant and has a hierarchical architecture in which the signal from different parts of the 27-km ring are aggregated before they are AND'ed together by a central controller.

The interlock system is also connected to the access control system. When "there is beam", nobody is allowed to be in the tunnels: all access doors are controlled and must be closed before beam can be launched. When there's no beam everyone has to take a key that doubles as a token; no beam is possible if a key is missing from its keypad. The reason for keeping people out of the tunnel when the beam is on—and for requiring the radiation detector I mentioned at the beginning of the chapter—is radiation from two sources. First, whenever a charged particle is accelerated perpendicular to its direction of travel (which a circular beam does constantly), it emits so-called "synchrotron radiation", mostly microwaves and X-rays at the energies involved in the LHC. While it is a nuisance in the LHC, accelerators such as the ESRF are built specifically to exploit this process, to provide this high-intensity radiation to science experiments. The other source of radiation is particles—such as those that inevitable escape the beam pipes—hitting other materials and thereby creating radioactive isotopes. Most of the radiation created this way is short-lived—no radioactivity remains in the tunnel when the beam is off—but is high-intensity and thus dangerous.

The beam protection system is implemented entirely in hardware. As software engineers, both my guest Vito and I are always a bit miffed: why doesn't the world trust us software people to build software that is reliable enough to handle such critical functions? The strength of software is that it can handle complicated algorithms and is relatively easy to adapt, at least compared to hardware. But it is also harder to get correct, because the complexity required to achieve this flexibility and adaptability is by definition higher, and more complexity always implies a higher risk of getting something wrong. A safety system does not have to be flexible or adaptable: it's not a good idea to pay the "complexity tax" implied by software.

Let's understand this in a bit more detail with following diagram. It shows an imaginary system that is controlled by three parameters: x, y and z. The system must be controlled so as to stay within the target envelope, represented by the area shaded dark gray. The optimum path (let's say in terms of energy use) is the thicker dark gray line. The control is relatively sophisticated, in that you have to control x and y in a non-linear way, and at particular values of x you have to change z. You might want to implement this in software.

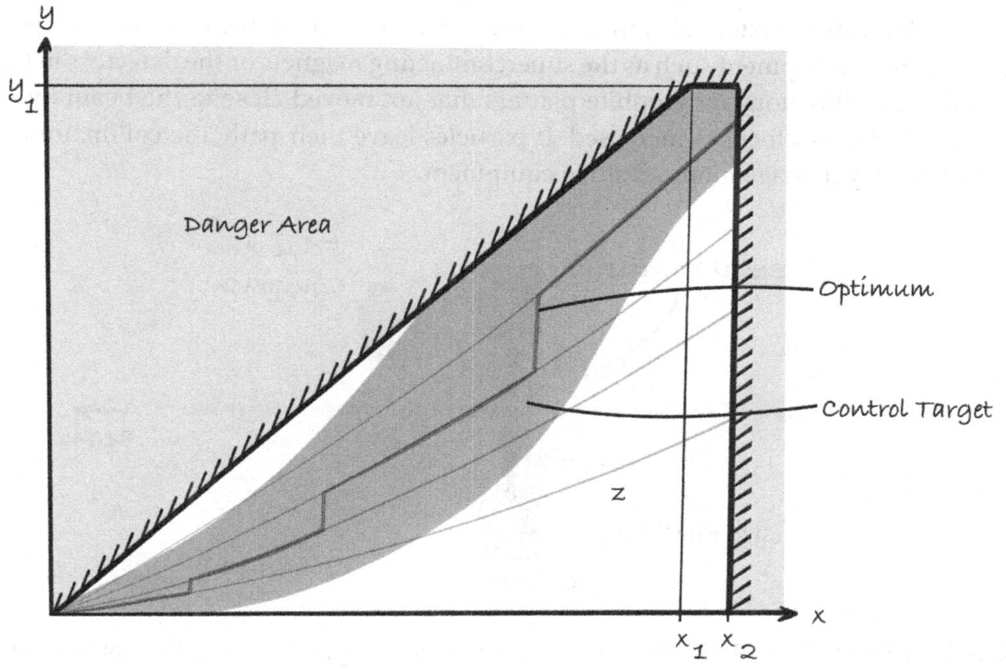

However, there is also a danger area outside the shaded line; whatever the control system does, the system must not be allowed to stray into this area. Notice how the rules for detecting the danger area are very simple; you can easily implement this in hardware.

$$x > x_2$$
$$\textbf{or } x > x_1 \textbf{ and } y > y_1$$
$$\textbf{or } y > 0.8\,x$$

So systems engineers use software for sophisticated control tasks, but rely on a hardware "safety net" that enforces the boundaries of safe operation. As Vito says, "This is what makes us software people sleep well, because basically we know we cannot break anything badly."

Back to dumping the beam. An unnecessary dump is not a good idea, because it takes three hours to rebuild the beam. That time is lost for science. A phenomenon called UFOs—unidentified falling objects—is a challenge in this context: if a dust particle falls through the beam it creates a shower of escaping particles, as these bounce off the dust particle and are detected by the beam-loss monitors. This phenomenon is usually temporary, though, and not a reason to dump the beam. Other losses, up to a threshold, are unavoidable but also not a reason to dump, but excessive loss that is not caused by UFOs might indicate a problem with the control system, so the beam must be dumped as a precaution. Correct distinction of these different phenomena by the beam protection system is crucial to avoid damaging the accelerator because a beam hasn't been dumped when it should have been, and to not waste time for science because a beam was dumped unnecessarily.

The beam protection system also includes 108 collimators at strategic locations, typically to protect sensitive equipment such as the superconducting magnets or the detectors in the LHC experiments. Collimators are graphite platters that are moved close to the beam by stepper motors as the beam's focus is increased. If particles leave their path, the collimators "catch" them before they damage more sensitive equipment.

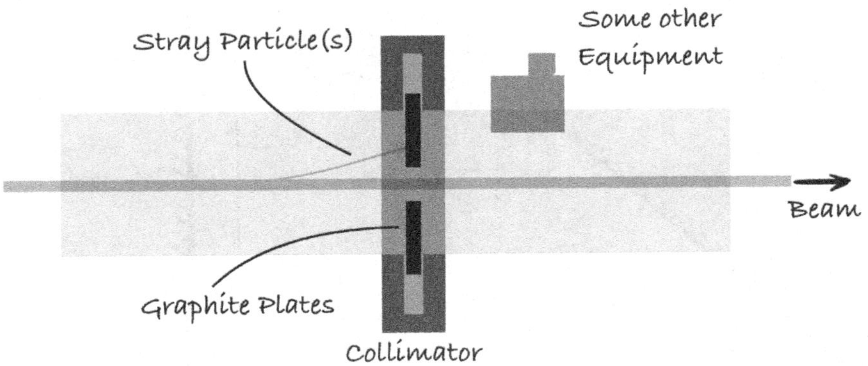

HIGHER LEVELS AND SOFTWARE Like any well-designed system, the control system of the LHC is layered to reduce complexity and improve maintainability through controlled dependencies. At the lowest layer are the physical actuators—magnets, cavities, diagnostics—and their power converters and control electronics. One level up are soft realtime computers running Linux with Realtime extensions, which control the hardware and provide a unified interface to the next-higher layers. This layer also contains off-the-shelf PLCs and SCADA-based systems for those parts of the infrastructure that are more or less standard industrial automation, such as cooling and ventilation; in contrast to the core beam control. The next-higher level contains over 500 Linux middle-tier machines that control the various lower-level realtime computers and automation systems and perform global analysis, coordination and optimization. The top layer consists of the consoles used by the operators in the control room. Overall, the LHC uses over 2,000 computers. The control computer software is implemented in C and C++, while the middle tier and consoles mostly use Java. Communication is via a ZeroMQ-based middleware and, on the higher layers, the asynchronous Java Messaging Service.

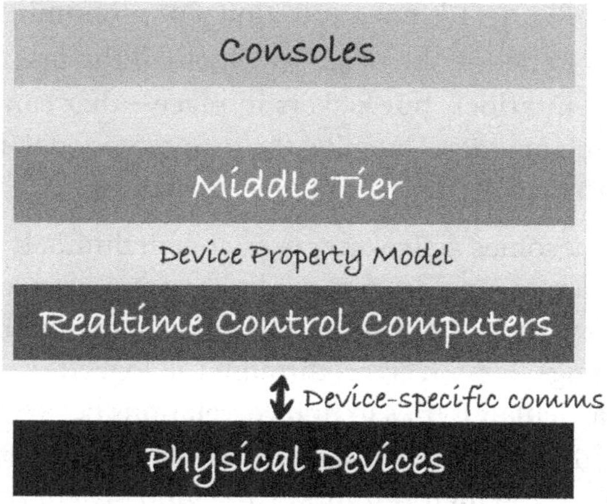

Each of the physical devices has dozens or hundreds of parameters that must be set consistently for the whole system to work. Because each device works differently on a technical level, the control computers abstract technical details to a unified view, referred to as the "device/property model". Each device, for example a magnet, has a number of properties that describe what the device is or what it is doing, such as the current flowing through the magnet or the temperature of the magnet yoke. Each property has a name and a type; for example, the "output current" property of a power converter would be of type "number" with a range between 0 and 12,000 (Amperes). Properties can be read and set over the network by the control system.

The experts for each piece of equipment develop its device/property model and its interface to the physical hardware. Once built, any device can be controlled in the same way, by reading and changing respective parameters. The middleware handles remote communication, logging systems store the values of particular (read-only) properties for diagnostic purposes, and the diagnostic tools have a common way of accessing the data they analyze. User interface components rely on the device/property model to represent devices and their settings in the console software.

Let's zoom in to three of the software systems in the middle tier: the settings and configuration database, the control middleware and the timing system. The database is just that, a huge Oracle database that stores configuration, calibration and diagnostic data; the science archive is outside this system and we'll cover it later. For example, the database holds the calibration data of each magnet, such as the function that relates its current to its magnetic field, crucial for the coordinated ramp-up of the accelerator. The control system uses the device/property model to set the values of configuration properties of the respective power converters; the timing system, which we will cover later, will then trigger their synchronized activation.

Higher up are applications such as the sequencer and the post-mortem analyzer. The sequencer is responsible for automated execution of the hundreds of steps involved in injecting, ramping, squeezing and colliding. It's not a full-blown programming environment, but it enables the LHC's operators to define the order in which all the tasks must be executed—put current in magnets, check interlock, put kickers in place—that can be run automatically or one by one, triggered by the operator. Just like the property model, it uses a framework at its core, to make it easy for users to develop additional sequences.

The post-mortem system becomes active after every beam dump. It collects all the data from the realtime and middleware computers that might have anything to do with the cause of the dump. To this end, it freezes the rolling log buffers that are maintained by all the computers, and then looks for two things. First, it scans through the logged data to identify the cause of an automatic dump. In particular, it tries to find correlations between items in the logs of different computers to identify non-local causes—those that could not be found by the device-local self-tests. This is where the high-resolution and globally consistent timestamps distributed by the timing system are crucial. The second task of the post-mortem system is identifying whether any residual problem remains after the dump that might prevent the beam from being restarted safely. It also checks whether the beam dump system itself has suffered during the beam dump—remember the energies involved in this process!

You have probably identified a couple of commonalities between the control system of the LHC and that described in the telescopes chapter for the ELT. If you were to look at other accelerators (like the ESRF) or other telescopes (such as ALMA) you'll find lots of commonalities as well. The International Conference on Accelerator and Large Experimental Physics Control Systems (what a name!) is where the people who work on those systems meet and exchange ideas. There are also software libraries, frameworks and platforms that are used by multiple of these experimental physics control systems.

TIMING SYSTEM I've mentioned the control system several times, so let's look at it in some detail. As in every large control system, a system-wide synchronized time standard is crucial. Although the LHC does not require picosecond precision like radio astronomy, it also uses a two-level approach to coordinating time. The fine-grained level's "clock tick" is the 40 MHz frequency of beam crossings, based on their 25 ns bunch spacing. It is distributed throughout the facility, and the experiments in particular use this clock to synchronize data capture.

Coarse-grained time is based on GPS-backed UTC time. It is crucial that, for example, the temporal sequence of timestamps in log files of various spatially distributed computers reflects the actual sequence of the events in time, otherwise establishing causality when tracking down a bug or understanding a non-local cause of a dump is impossible. Similarly, when ramping up the energy of the ring, the RF cavities plus all the dipole magnets have to change their settings in a coordinated fashion, otherwise the beam can be lost.

The approach to establishing a coordinated time throughout the LHC relies on manual calibration: engineers go round the various physical locations of the 27-km ring with an atomic clock and measure the time delay induced by the optical fiber used for communications relative to the time reference. These delays are then stored in a database that serves as a lookup table for correcting time stamps. So when a computer at a particular physical location stores a log entry, it uses its local time, corrected by the delay value for its location, to get the "system time". That's easy. But how does this work for synchronized activities? You cannot just "correct" the delayed arrival of a signal: it physically arrives later. A clever solution solves this problem: all coordinated activities are scheduled to some future point in time, to. The instruction to perform some activity at to is transmitted to all involved components in advance, where it does not really matter how long it takes to distribute the schedule. Each system then looks up the delay value and corrects their trigger time $t_{local} = t_o - delay$. When t_{local} arrives on their local clock, they trigger the activity.

CONTROL ROOM The LHC's control room exhibits the highest density of monitors per square meter I have ever seen. Wall-mounted monitors all around the room show high-level information relevant to everybody, such as the beam position, the general state of the accelerator (off, filling, ramping, colliding) or the rate of collisions as a measure for the efficiency of the machine. Operators sit in four islands that are arranged in cloverleaf shape. One of the islands controls the smaller accelerators: the LINACs, the Booster and PS. The SPS and the LHC have their own island each; the fourth island controls general services, such as cooling and ventilation, the cryogenics for the superconducting magnets and the electrical services. Historically there were separate control rooms for each of the accelerators, but the tighter integration of the machines in the context of the LHC also suggested colocation of the people who run them. The experiments have their own separate control rooms.

In regular operations each of the islands has one engineer and two technicians; at night they work with a minimum crew of only one technician, but experts on the various subsystems are available on call. Vito: "This is one of the ways that makes us write good software. People are very careful not to make a change on a Friday afternoon"—implying that they might have to sacrifice their weekend to fix any bug they might have introduced on Friday. During downtime when new equipment is being installed, tested, debugged or improved there might be dozens of people crowding in each island. Not surprisingly, the advancement of accelerator technology is another of CERN's tasks.

Monitors! Each of the islands consists of approximately 40 monitors, arranged in a 240° arc. Monitors are stacked vertically in two rows: the top row mainly shows measurements, system states and diagrams, whereas the bottom row is where the users interact with the system. At the accelerator islands these show custom Java UIs that illustrate the state of the accelerator's components. For example, a set of four monitors shows the status of each of the thousands of magnets as little color-coded squares: inactive, ramping, steady current. Other screens show the beam position, the collimation or progress during filling. They also let users change configuration parameters of the various accelerator components and interact with the sequencer.

The fourth island, the one for the general services, uses industrial control system UIs to create so-called "synoptics" that illustrate the status of these systems. Synoptics display a simplified view of the subsystem components and connections, overlaid with relevant quantities (flows, temperatures, RPMs, open/closed state of valves). This island also hosts the only console that has actual physical buttons: the access control system that manages the doors and keypads. As any fault in this system might directly impact human life it consists exclusively of hardware, right up to the user interface.

BEAM LIFECYCLE Now that we have all the ingredients together, we can understand the lifecycle of a beam in the LHC, from starting to fill the big ring, through data capture to beam dump. Overall this cycle lasts between four and 12 hours, depending on the quality of the beam. The finer-grained steps in the lifecycle are automated by the sequencer.

We start with an initial state of no beam, and the status of the 1,000s of pieces of equipment and computers is unclear.

The first step is checking the equipment. All non-trivial components (or their controlling computers) have built-in self-checks. If all are ok, the next step is initialization: some components have to be brought into a specific state to be able to receive beam. For example, the capacitors of the kicker magnets used for dumping the beam and slotting in new bunches have to be charged and the 1,000s of magnets have to be cycled, because magnets "remember" their previous current due to inherent hysteresis. They have to be ramped up fully once for them to "forget the past" and behave predictably in terms of their precalibrated current-to-field function.

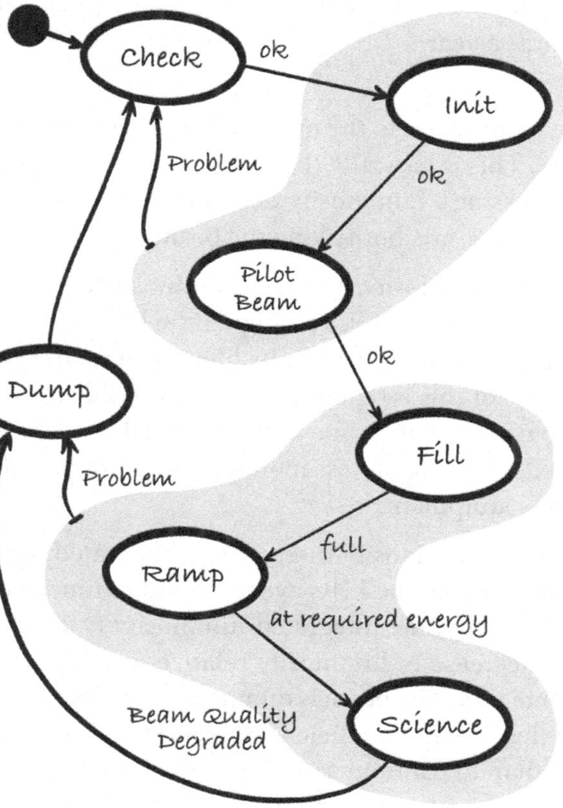

Next the operators request a pilot beam from the SPS. This is a smaller bunch of particles that is used to test whether the equipment and the beam control system is able to maintain a stable orbit for the bunches. This small bunch is not dangerous if something goes wrong.

Now the LHC is ready for the real beam. The injector from the SPS produces a bunch every few seconds, and it takes half an hour to completely fill the LHC with 2,800 bunches. At this time the LHC is at 450 GeV, the exit energy of the SPS. Once full, the beam is accelerated, ramping up the energy to whatever maximum level is supported at the time: like many complex machines, the LHC has been increasing its maximum energy over the years. The beam is then squeezed to a diameter of 240 micrometers by the control system, using the quadrupoles, correctors and collimators. It takes between two and three hours to get to this point in the cycle.

Only then will the operators activate a setting that switches a couple of magnets to bring the two beams onto a collision course. Unless the protection system detects any danger and dumps the beam, collisions can last for up to 12 hours. Ultimately the operators dump the beam and the magnets are ramped down. The system is now ready for its next cycle. During physics operations and outside of maintenance or upgrade phases, these cycles continue 24 hours a day for ten months a year.

LUMINOSITY Before we move on to the magnet technology, this is a good time to introduce the notion of luminosity. This is basically the intensity of the beam measured in "probability of collisions per beam crossing". Luminosity depends on the number of particles per bunch and how compact the bunches are, both along the beam axis and in terms of focus.

There are different kinds of luminosity. The peak luminosity represents the instantaneous luminosity at the beginning of an LHC fill. The higher the better, taking into account the maximum of 100 billion particles per bunch. Over its lifetime of up to 12 hours the beam becomes less luminous. One reason for this is that some particles collide away, but another is the inevitable (but low) rate of particles that escape from the ring despite all the magnet magic. If that gets too high, as detected by the beam loss monitors, an automatic dump is initiated to avoid (radiation) damage to the equipment.

In addition to instantaneous luminosity there is also integrated luminosity. This is the total number of collisions delivered by the LHC over a period of time: week, month or year. For a given instantaneous luminosity, the integrated luminosity is larger the longer the machine runs and the lower the decrease of luminosity relative to the peak. This consideration also helps explain why at some point the beam is dumped manually: the instantaneous luminosity has degraded so much that it is more efficient to spend the time to dump, check and refill to peak luminosity rather than to continue running at the lower value. Downtime, for example for maintenance, also impacts negatively on integrated luminosity.

In addition to the energy injected into the particles, luminosity—both peak and integrated—is the second major measure of performance of any accelerator. In the end this is what matters for science: the more collisions overall over a given period, the better. As we will see later when we discuss the discovery of the Higgs, an increase in luminosity led to a Higgs discovery earlier than was originally expected. Future upgrades of the LHC will further increase luminosity, because this is much more practical (that is, cheaper) than enabling higher energies by replacing all the magnets with ones that produce a higher field.

5 | *The Superconducting Magnets*

WHY SUPERCONDUCTING? A key technology for the performance of the LHC and its 14 TeV collision energy are its large-scale use of superconducting magnets with their super strong magnetic fields. Superconductivity means that a current flows inside a conductor without any electrical resistance. Since $W = I^2 \cdot R$, and $R = 0$, no energy is required to keep the current going. There is no Joule heating, so the material does not get warm. However, the current still produces a magnetic field. This is the perfect combination if we want to produce strong, steady magnetic fields via electromagnets, which is why the LHC and most other modern accelerators use superconducting coils in their magnets. At the LHC a current of around 12,000 amps produces a magnetic field of 8.33 Tesla, roughly 200,000 times the strength of the Earth's magnetic field of between 25 and 65 microteslas.

However there is a catch: superconductivity only "switches on" below a material-specific critical temperature. The material used in CERN's magnets becomes superconducting below 9.2 K, but requires a temperature of only 1.9 K to safely produce the 8.33 Tesla required for the LHC. Cooling thousands of magnets to this extremely low temperature is obviously a challenge, so we will return to the cryo infrastructure later.

There are several reasons why the LHC couldn't have been built without superconducting magnets. For example, if the coils that produce the required magnetic field were made using non-superconducting cables, the magnets would become so large that they couldn't fit into the tunnel. As the LHC was installed in the tunnel of its predecessor, the LEP, a larger tunnel cross-section was not feasible. For comparison, the copper wires on 380 kV power lines that run up to 2,000 amps are around 4.5 cm in diameter. LHC could have made do with a lower magnetic field while staying with the same 7 TeV of energy, but this would have required a much larger radius ring. In addition, even taking cooling cost into account, the cost of superconductors per A/m^2 current carrying capacity is similar to the pure copper alternative. Classical magnets with lots of Joule heating require a great deal of cooling as well. Finally, because ohmic resistance in copper magnets converts electrical energy into thermal energy, they have to be supplied with electrical energy continuously. In a superconductor, in contrast, once the current is ramped up the energy will not decay: the current will "circulate" forever.

Thick copper cables do of course exist at CERN, for example to connect the LHC to the power grid. CERN uses primarily French power, with a backup connection to the Swiss grid. "Swiss power would be horribly expensive", says Felix, a colleague of Vito's who showed us around SM 18. When the LHC is at peak operation CERN uses 200 MW of power, around one third of what the neighboring city of Geneva consumes.

SUPERCONDUCTIVITY BASICS How exactly does superconductivity work? Let's start out by explaining regular electrical resistance. A solid is characterized by its atoms being bound together in a material-specific three-dimensional lattice. In an electrically conductive material some of these electrons are not required for binding and remain free to move. A current is essentially these free electrons moving through the lattice. As the electrons move, they bump into the lattice's atoms and are scattered away from the "forward" direction that represents the current flow: this is electrical resistance. Impurities in the lattice increase the likelihood of such collisions, which is why amorphous materials—those with a less regular lattice structures—tend to have higher electrical resistance.

At non-zero temperatures the lattice vibrates, and these vibrations propagate similarly to sound waves. A higher temperature increases the vibrations, which is why resistance generally increases with temperature. At low temperatures the vibrations are quantized into discrete packets of energy that can be seen as pseudo-particles, and are dubbed "phonons". These phonons might hit electrons and divert them from their forward path, leading to electrical resistance.

In superconductors none of this seems to happen; the motion of electrons is not disturbed by the lattice; they are not scattered, so they can flow completely uninhibited, resulting in zero resistance. Why is this? Fully explaining superconductivity would require more space and knowledge about condensed matter physics than I have, but I will give you an outline based on the BCS theory, devised in 1957 by Bardeen, Cooper and Schrieffer, who won the Nobel Prize for it in 1972.

When a superconductor is cooled below its critical temperature the free electrons condense into a giant, joint quantum state: they are "bound together". For an electron, this state is lower in energy than if it were to roam freely, which is why they "voluntarily" adopt this joint state. The phonons, the quantized vibrations, might have enough energy to push a single isolated electron off track, but at sufficiently low temperatures the phonons cannot scatter electrons that are bound in this giant "electron molecule" that transports current jointly. This explains why there is no resistance.

But this prompts the question of how and why this joint quantum state occurs. Let's consider an electron in a lattice of positively charged ions—positively charged because some of their electrons are free, busy transporting current. The negative electron attracts the positive ions, bending the lattice structure slightly towards the electron. This in turn leads to a concentration of positive charges "around" an electron. Depending on the deformation of the ions in the lattice and the charge fluctuations produced by that electron, a net attraction force can result. This attracts a second electron (with its own lattice deformation), and the two become bound into what is called a "Cooper pair". Because of the distances over which the attraction operates and the distances of the ions in the lattice, the spheres of influence of all electrons overlap a lot. Essentially a whole bunch of electrons interact, forming this "sea" of Cooper pairs, the "electron molecule" mentioned above.

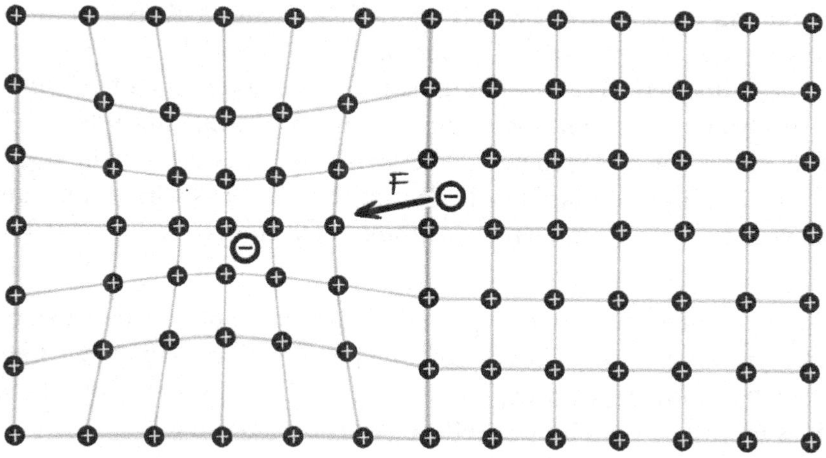

This pairing is not very strong, because electrons still repel each other as a consequence of their similarly negative electric charge. This explains why the effect happens only at very low temperatures, where the phonons have too little energy to "bounce" electrons out of this pairing. In fact, the phonons can actually help to form Cooper pairs if they "push" electrons at the right moment.

You might remember from our discussion of the standard model that electrons are fermions: they have a spin of +1/2 or −1/2 and the Pauli exclusion principle prevents two or more electrons with the same spin from being in the same quantum state. How then does this not preclude the formation of a giant, joint quantum state? A Cooper pair always forms between two electrons with spin +1/2 and −1/2, so here the Pauli exclusion principle is obeyed. Once joined, however, they form a "composite boson" with spin +1, to which the Pauli principle does not apply, which is why many of these pairs can be in a joint state. Sounds weird, but apparently that's how it is.

How do we know the BCS model is correct? We can confirm one of its quantitative predictions experimentally. Each solid exhibits a characteristic electronic band structure, the ranges of energies at which electrons exist. Remember that when atoms bind into a solid, that binding is facilitated by their electrons whose orbitals overlap. However, because of Pauli's exclusion principle, no two electrons can be in the same quantum state (here, orbital), which is why the particular orbital splits into two with slightly different energies. In a solid a huge number of atoms bind, so we get a large range of orbitals. The BCS theory predicts that there is a particular gap in the band structure of superconducting solids. This is because the electrons that would reside in this rather low energy state drop further (in terms of energy) into the "sea" of Cooper pairs. BCS makes a quantitative prediction of this energy gap, and it has been observed experimentally. The size of the gap determines the critical temperature for superconductivity: the larger it is, the more energy you need to push Cooper electrons up and over the gap and drive the solid out of superconductivity.

The superconducting material from which the LHC coils are made is niobium–titanium, or NbTi. This was discovered in 1962 and is one of the most widely-used superconductors in large-scale applications. LHC certainly is large scale: they use 400 tons of it! Superconductivity disappears abruptly above a material-specific critical temperature, 9.2 K for NbTi, but superconductors also have a critical magnetic field above which they cease to be superconducting. For NbTi the practical limit is around 10 Tesla, so LHC's 8.33 T is within the material's limits. Finally they have a critical current, which is around 15 kA for NbTi: LHC's 12 kA stays below that limit too.

Superconductors are categorized into Type I and II according to how they behave in a strong magnetic field. Type I superconductors cannot be penetrated by a magnetic field at all; once they become superconducting they expel the field: the field lines go around the bulk of the material. This is known as the Meissner effect. As a corollary, if a Type I superconductor is exposed to a magnetic field above its critical magnitude, superconductivity breaks down immediately through a first-order phase transition, which is basically a fancy term for an abrupt change. Type II superconductors work a bit differently; in particular, they have two critical field strengths. Pierre Bauer, my guest on the superconductivity episode, explains this as follows:

Type II superconductors are characterized by the fact that, instead of being uniformly superconducting, it is favorable for energetic reasons that superconducting regions mix with normal conducting regions at the microscopic level. Up to the lower critical field (roughly the critical field of Type I superconductors), the Type II superconductor behaves like a Type I superconductor and expels all field. Above this threshold the magnetic field enters as quantized vortices, which are normal conducting at their core. The current carrying performance of this type of superconductor is not limited by the Cooper pair condensation energy due to the band gap as in Type I, but by the "pinning" of these vortex lines, often on normal conducting precipitates or on grain boundaries—potentially engineered into the material when alloys like NbTi are manufactured. Those can be de-pinned by thermal activation, or by Lorentz forces arising from the current flowing through the superconductor. The critical current is limited in Type II superconductors by the pinning limit, the current at which these vortices start moving around. This is much smaller than the "de-pairing" critical current in Type I, by orders of magnitude. On the other hand, Type II can be used in larger fields. For example, where the Type II NbTi can bear 15 T, a typical Type I is limited to 0.1 T. Since most applications of superconductors in industry are related to magnetic fields, the compromise between lower current but higher field exhibited by Type II conductors is a useful tradeoff; essentially all industrial applications use Type II superconductors.

In addition to distinguishing between Type I and Type II superconductors, there is a second fundamental way of characterizing superconducting materials: low temperature vs. high temperature. And while all the superconductors relevant to the LHC are low-temperature superconductors, for completeness sake I should briefly say something about the other kind, whose critical temperature is of the order of 100 K. I've always been amused by the fact that materials that become superconducting at –180 °C are called "high-temperature superconductors". A critical temperature that high is useful because the it allows the superconductor to be cooled with liquid nitrogen, which is easier to handle than liquid helium. However, these materials are currently about 10 to 200 times more expensive than the low-temperature Type II alloys used in industry. In addition, because they are still relatively new, there is not enough experience for their use in large-scale systems such as colliders.

STRANDS AND FILAMENTS We've looked at the high-level design of the magnets and the overall ring, and we've also looked at how superconductivity works in principle on the quantum level. However, between these two extremes lies a lot of detailed engineering of the cables from which the coils are wound.

The cable has a flat rectangular cross-section and is 15 mm wide and 1.5 to 2.0 mm thick. It consists of multiple strands, 28 for the layer closer to the beam and 36 strands for the layer further out. Each strand is around 0.8 mm in size and consists of 6,300 filaments of less than 0.006 mm, much thinner than a human hair. The filaments are embedded in a high-purity oxygen-free copper matrix that binds them together to form the strand; the whole strand also adds another copper jacket around the packet of filaments. At superconducting temperatures the copper acts as an insulator relative to the superconductor. The LHC uses around 7,000 km of cable; in terms of single filaments the total length is over five times from the Earth to the Sun and back!

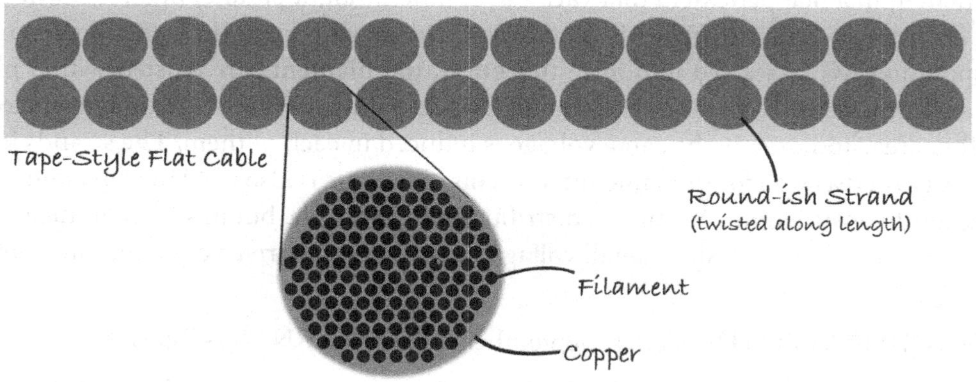

This particular design for superconducting cables is called a Rutherford cable, named after the lab that designed it; it is designed in this way for specific reasons. First of all, the multi-level design gives the cable better mechanical flexibility compared to a solid material. This allows it to be wound into the coils required to produce particular shapes of magnetic fields. Another consideration is the inductance of the coil. Inductance is the tendency of a coil to develop a voltage that resists an increasing current. A coil's inductance increases with the number of windings. On the other hand, the strength of the magnetic field depends on the total current, which depends in turn on the cross section of the cable. A large total cross section allows the cable to carry more current, but fewer windings keeps the coil's inductance low. This is why it makes sense to combine a number of strands into the cable and then build the coil from the cable, instead of winding the coil from the strands directly.

Perhaps the most important reason for building the strands from thousands of filaments instead of using a solid superconductor is that it improves quench resistance. The copper surrounding a thin filament allows it to recover from a flux jump, because the current can "spill out" into the copper. A flux jump is a self-reinforcing effect in a Type II superconductor in which vortices start to move from wherever they have been pinned, for example because of a microscopic thermal perturbation. Vortex movement implies that existing small superconducting regions are turned into regular conducting regions and vice versa. This is a dissipative process, which means that it releases heat, which can develop into a self-enforcing avalanche that ultimately results in a complete quench of the filament.

I said above that at superconducting temperatures the copper acts as an insulator. It still conducts current, of course, it's just an "insulator" when compared to the zero-resistance superconductor, as the current always takes the easiest path. So if a piece of superconductor becomes normally conducting, the copper becomes "attractive" to the current. Quench resilience is also improved by the fact that if a quench develops, only one of the 6,300 filaments quenches, at least initially, and you can recover through the remaining filaments.

Because the copper is not a perfect insulator, however, it couples the filaments to some degree. In particular, in a magnetic field that varies over time, straight parallel filaments will see different inductive voltages, which will lead to large loops with large (unwanted) currents and corresponding AC losses. To minimize this effect, the filaments are twisted with a pitch of around 20 times the diameter of the strand. Twisting ensures that all filaments enclose the same flux area, and exactly the same voltage is induced in each of them. The strands inside a cable are twisted as well for the same reason. This technique is also used in conventional electromagnets to ensure uniform current distribution in the cable, but in superconductors it becomes even more relevant, since small voltage differences can drive very large currents.

ACTIVE QUENCH PROTECTION September 19, 2008, CERN Press Release:

> During powering tests of the main dipole circuit in Sector 3-4 of the LHC, a fault occurred in the electrical bus connection in the region between a dipole and a quadrupole, resulting in mechanical damage and release of helium from the magnet cold mass into the tunnel. Proper safety procedures are in force, the safety systems perform as expected, and no-one is put at risk.

And another, on October 16, 2008:

> Investigations [...] have confirmed that the cause of the incident was a faulty electrical connection between two of the accelerator's magnets. [...] Sufficient spare components are in hand to ensure that the LHC is able to restart in 2009, and measures to prevent a similar incident in the future are being put in place.

Let's understand what happened. Within each 15-meter-long dipole magnet the superconducting cable is wound around the beam pipe to form an electromagnet. Because a whole set of magnets are connected in series, that cable has to be connected between any two magnets: the strands are spliced together and then compressed by copper. At one billionth of an ohm these are not perfectly zero-resistance connections, but it's still pretty damn low. Unfortunately, one of those connections was faulty: its resistance was 200 times higher. As a result Joule heating occurred during testing, a quench developed and the conductors melted and broke. An electric arc then developed that broke the containment of the cable and explosively evaporated around six tons of the liquid helium that surrounds the bus bar. The pressure wave that then propagated through several magnets severely damaged 53 of them: several of the 35-ton magnets were lifted and shifted their position!

This accident is an illustration of the energy that is stored in the magnetic fields; at around 600 MJ, each half-octant has roughly twice the energy stored in the beam—the ICE train we mentioned before is now at 200 km/h. As a rule of thumb, each GJ of energy can melt one ton of steel, and there are 16 half-octants!

This energy—partially in the form of electric current—has to be gotten out of the magnets as fast as possible if the magnets quench, or they will be damaged as if that ICE train had crashed into them. You can quench a magnet by exceeding any of its critical numbers: current, magnetic field or temperature. But assuming your control system prevents this, why else would a magnet quench? One reason might be that the coils move microscopically, and mechanical friction heating occurs. To prevent this, they are housed in the strong collars I mentioned before; the cables themselves are also extremely dense. Another reason for quenching is that an electrical connection is not quite as perfect as it should be, as demonstrated by the 2008 accident. A quench is self-reinforcing: a loss of superconductivity at a particular location will cause it to become hot, heating up its immediate surroundings and bringing them out of superconductivity as well. Second-order effects such as the explosive expansion of the helium might then lead to mechanical destruction of the magnets. Since the accident, valves have been added that allow over-pressuring helium to escape from the magnet without damaging it.

As soon as a superconductor quenches "it becomes a lousy conductor", as Pierre Bauer said. The current densities are much too high for the non-superconducting state, and while the surrounding copper can take some of the current, it cannot take it all—obviously, because this would require as much copper as a non-superconducting magnet would require in the first place. But the copper gives the overall system a little time to take active measures that ensure that the quench does not spread. The copper "insulator" is a really clever idea: if an actual insulator was used this bypass would not be possible.

To cut a long story short, all industrial systems that use superconducting magnets require a dedicated quench protection system that consists of quench detection and quench mitigation. Let's look at the detection components first.

Quench detection requires the measurement of a very small signal in an environment of large currents, fields and voltages. It must also measure the effect in milliseconds to be able to protect the machine; this is a challenging task in which all kinds of approaches are used. For example, if resistance increases, a voltage builds. By detecting this voltage, a quench can be detected—in theory. A paper titled "LHC Magnet Quench Protection System" written in 1994 by Coull et. al outlines the plan. It says:

> To detect the quench, the voltages across the two halves of the magnet are compared. If a quench occurs at a point in the magnet then a differential voltage appears and this is detected.

Quenches that develop only in one half of the magnet and not in the other half are called non-symmetrical quenches. The LHC magnets indeed use taps on the coils, to which cables are connected that lead outside the cryostat to where instrumentation compares the voltage between the magnet's two sectors. The 1994 paper continues:

> A completely symmetrical quench of both magnet halves is extremely unlikely but this problem can also be eliminated by making a second differential comparison, this time between the adjacent magnets in the same cryostat. A further quench detection system on the main current return bus-bar may be necessary since a quench here, although unlikely, could cause considerable damage if fast de-excitation is not quickly initiated.

It turns out that this description is correct regarding the mechanism for detecting a symmetrical quench, but it is not correct regarding the probabilities. The 2008 accident was caused by a quench in the bus bar connections between the magnets. And in the aftermath of the accident the CERN crew also discovered symmetrical quenches. The quench protection system was extended to monitor the joints, as well as to diagnose symmetric quenches by comparing voltages between up to four magnets.

In total there are over 2,000 sensors that measure various voltages, 11,000 independent analysis and processing cards, plus around 3,000 higher-level cards that aggregate the information and connect it to the control system. Everything is installed in pairs for redundancy.

Other approaches to quench protection include quench antennas, pick-up coils that measure fluctuations in the magnetic field. Computers compare the actual field to a model that predicts how the field should develop under normal circumstances. Some of the corrector magnets do not have taps—they cannot be logically separated into two halves—so the current that flows through the magnet is measured instead. If it decreases (because a quench increases resistance), this raises a quench alarm. Because the sensors have relatively high noise levels, the current measurement has to be integrated over time, limiting the speed at which the quench can be detected. In the end this limits the rate at which the magnets can be ramped. Despite these complexities, the community has learned how to engineer quench protection systems reliably, and undetected or spurious quenches are rare.

The accident described above happened because quench detection in the LHC was initially not as thorough as it should have been: the connectors were not monitored. This omission was justified by the argument that a quench will propagate quickly to areas that were monitored—or maybe somebody had just decided to save time and money. The fact that the connections are surrounded by especially heavy copper bedding might have contributed here: the quench probably lingered and developed for a relatively long time in the connector before it then "jumped" up quickly. Sometimes less copper might be better, because a quench then moves faster and you can actually detect it. Hindsight is 20/20!

Let's move on to mitigation. It takes several dozen microseconds for a quench to spread far enough to cause damage, so this is the time window available to prevent this from happening. It takes roughly 10 ms for the detection system to detect a quench reliably. Once detected, the LHC deals differently with the magnet that has the quench and the magnets in the same circuit within that octant.

First, the magnet that quenched is taken out of the circuit, to make sure it does not receive any additional energy. This is achieved through so-called "cold diodes" in the cryogenic part of the system. While this sounds like an obvious idea, these diodes had to be custom developed to make sure they worked reliably at 1.9 K, could deal with the radiation and could switch currents of up to 15 kA. Once separated from its energy supply, strip heaters attached to the coils are energized in order to heat the complete magnet to above its critical temperature within a few dozen milliseconds, thus spreading the quench as much as possible. This distributes the energy stored in the magnet over the entire magnet, avoiding damage to the spot where the original quench occurred. The energy to power the heaters is stored in capacitors to ensure it is available fast enough to avoid damage.

The other magnets in the same octant—now without the quenched one—still carry megajoules of energy, and this energy has to be taken out of the system. This is achieved by switching resistors into the circuit that converts the energy into heat. The time constant of this L–R circuit is around 120 seconds. This ensures that the voltages induced by the ramp down in current and field are below 2 kV, which simplifies insulation. A fast ramp down also increases the risk of additional quenches and is therefore better avoided. Because the initially quenched magnet has been taken out of the circuit, discharging the remaining circuit is not that urgent.

Immediately after detection of a quench the quench protection system sends a signal to the beam dump system to get rid of the beam and its 350 MJ of energy. Luckily, dumping the beam is much faster than the time it takes for a quench to spread far enough to affect the magnetic field that keeps the beam on track—the abort gap will reach to the dump system's kicker magnets after 0.1 ms, worst case, because the beam circles the track 11,000 times per second.

Before we move on to the cryo system, there's one more interesting aspect of superconducting magnets and their tendency to quench. Remember our earlier discussion of the cables that are flexible because of their hierarchical strand/filament structure, but also rigid because they are densely packed and bedded in copper so they don't move and develop frictional heating and possibly quench? This sounds like a contradiction, right? Indeed, you have to "train" these magnets for them to be able to reach the highest currents without developing a quench. This is achieved by running them up to a medium-strength field: the cables and strands move a little and the magnet quenches. Because of the tendency of the magnetic field to be homogeneous, the movement introduced tends to settle the cables in a way that makes them more quench-tolerant. On the next ramp up they will tolerate a bit more current, quenching later, and so on. It's common to have to work the magnets up through dozens of quenches until they can safety tolerate 13,000 Amps.

CRYOGENICS We've mentioned several times now that the system that cools the magnets employs liquid helium. While NbTi has a critical temperature of 9.2 K, the LHC cools them down to 1.9 K so that they can tolerate 8.33 Tesla safely. To cool to this temperature, the coolant that carries away heat must be at 1.9 K as well. Crucially, the coolant must be gaseous or liquid at this temperature, because a solid cannot be pumped; just relying on thermal conduction would make heat extraction extremely slow.

It turns out that there aren't that many elements that are liquid at 1.9 K, but helium is one of them. Even better, it turns out that helium has a lambda point of 2.2 K. The lambda point is the temperature below which a fluid turns into a superfluid, a liquid with zero viscosity: it flows without any driving energy. Sound familiar? Indeed, the phenomenon is not unlike superconductivity, even in some of its more detailed characteristics. Superfluid helium is dubbed Helium II and has several interesting properties beyond zero viscosity. For one, it has a very high capacity to transport heat, around a million times greater than regular helium. In addition, because of its zero viscosity, it can creep into very small openings, down to 10^{-8} m. The LHC cooling system makes use of both these properties:

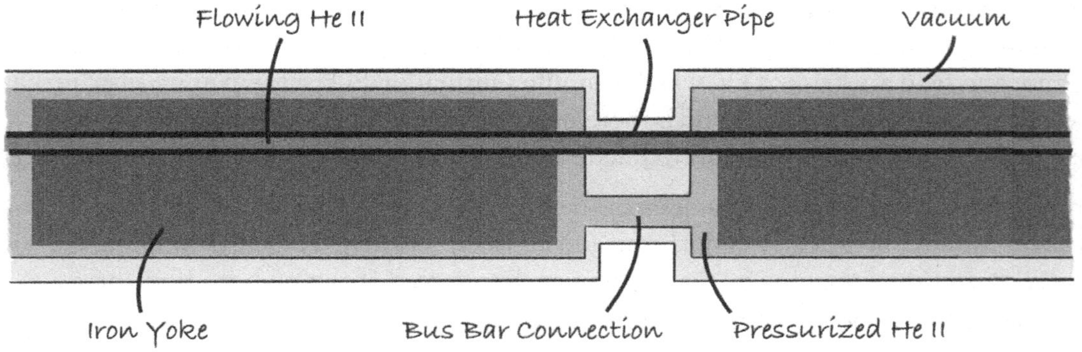

First, it uses He II pressurized to 0.1 MPa as a static coolant that surrounds the iron yoke that embeds the collar and ultimately the magnets. Because of its ability to creep into small spaces, the He II penetrates all the small holes and crevasses in the yoke and the coil cables; they have enough porosity. This static He II transports heat by conduction towards a heat pipe that contains a flow of pumped He II that actively moves any residual heat out of the system. The total volume that contains the very cold He II is protected by a thermal insulation layer and a vacuum vessel to thermally insulate it from the ambient temperature in the LHC tunnel.

The heat energy transported by the static He II to the heat pipes evaporates small amounts of liquid He II in the pipes, so there are actually two flows inside the heat pipe: a superfluid flow from the refrigerators into the magnets, plus a gaseous flow that moves towards the refrigerators. This is called a "stratified two-phase flow".

There are two different cooling tasks in the context of the operation of the LHC. The first one is to cool the system down to its operating temperature, starting from room temperature in the tunnel. This is necessary whenever part of the ring has been warmed up for maintenance purposes. This obviously requires the greatest amount of energy, and a full cool-down takes around two months. So you want to be really sure that everything is perfectly in shape before you do this, otherwise you will lose lots of time and energy for re-cooling.

Heat exchangers cool the helium from room temperature to around 80 K. The cold medium here is 10,000 tons of liquid nitrogen, which is at 77 K at normal atmospheric pressure. The helium is then cooled to 4.5 K using compressor-based coolers. Cooling plants with buildings full of machines perform the liquefaction of the helium down to 4.5 K; eight of them are distributed around the ring, one for each octant. This liquid helium is then pumped underground and the static volume is filled. The helium that circulates through the heat exchanger pipe is cooled to 1.9 K to make it superfluid; a set of custom-developed compressors reduce the pressure above the liquid helium to around 1600 Pa, which also reduces its temperature. This 1.9 K superfluid circulates through the heat exchanger pipes and thus also continuously cools the static helium.

The second cooling task is the steady-state of the collider, where the 1.9 K operating temperature must "just" be maintained. Effects that tend to warm up the system include heat in-leaks from the design of the cryostat, Joule heating in the non-superconducting parts of the system, as well as the heat loads induced by the beam itself, including synchrotron radiation and the dissipation of currents induced by the beam.

When completely filled and in steady-state the magnets contain 50,000 liters of helium, cooling down around 30,000 tons of material. The total volume of helium used in the LHC is 700,000 liters. This is so much helium that CERN had to hoard it over a period of years because the worldwide annual production is not enough to meet this demand. The LHC is currently the largest cryostat in the world by surface area! The pumping and distribution infrastructure has 40,000 leak-tight connections, and special pumps had to be developed. The feat is even more astonishing considering the thermal shrinkage of parts of the system as it cools down to near absolute zero: a dipole magnet shrinks by around three centimeters, so flexible connections are required. Handling superfluid helium is generally a challenge: because it seeps into the smallest holes, it also tends to leak out, even "climb out" of containers.

6 | *Discovering the Higgs*

JULY 4, 2012 On July 4, 2012 the calendar at CERN listed an LHC seminar innocuously titled "Latest Update in the Search for the Higgs Boson". The agenda contained three talks: one by Joe Incandela, the spokesperson for CMS, one by Fabiola Gianotti, the spokesperson for ATLAS, and a final one by Rolf Heuer, the Director General of CERN. The fact that both ATLAS and CMS had an update for the community at the same time and that "The Boss" would be speaking at the end of the session probably seemed suspicious to the community. The room was packed.

Joe went first. He talked for roughly 45 minutes, showing all kinds of measurements, statistics, and diagrams, explaining CMS' search for the Higgs and emphasizing how well the beam production community had done in providing a reliable beam at even higher luminosities than expected. Towards the end of his presentation he advanced the slides to one that showed a diagram of a "resonance" at an energy of around 125 GeV. He said "If we combine the ZZ and the $\gamma\gamma$, [...] they line up extremely well [...] and give us a combined significance of 5σ." He turns around to the audience and smiles. The audience breaks into a long applause while Joe just stands there, quiet, smiling broadly.

Next up was Fabiola, with another extensive set of slides on how ATLAS searched for the Higgs. Her slides famously used Comic Sans font and the world—or at least the internet—poked fun at it in the aftermath. It didn't hurt her career apparently; she eventually followed Rolf Heuer as the Director General of CERN. She, too, gets to a slide that shows a statistical excess at 126.5 GeV with a confidence level of 5σ. Before she even got to comment on it the scene from the previous talk played out again: audience applauding, presenter smiling.

When she finished her talk more applause erupted, and Joe and Fabiola congratulated each other. In the interest of avoiding influencing each other, the two experiments had not communicated their progress or results before the seminar—although I am sure they suspected something, certainly when they decided on a joint LHC seminar. In the final wrap-up, Rolf Heuer said "As a layman I would now say: I think we have it".

Towards the end of the video recording of the seminar the camera shows a very emotional elderly man in the second row of the auditorium, suppressing tears. He later said "For me it's an incredible thing that it happened in my lifetime." Peter Higgs, long since retired from active research, had been invited by CERN to witness the official discovery of "his" particle.

THE PREDICTION Mechanisms for assigning mass to the elementary particles of the standard model were proposed by several people in the 1960s, but only one of the papers specifically mentioned a particle. It is titled "Broken Symmetries and the Masses of Gauge Bosons", and was published by Peter Higgs in Physical Review Letters in 1964. This paper is the reason why the particle was named after him, and why he won the Nobel Prize for Physics in 2013. He shared the prize with with Belgian physicist François Englert, who, together with the late Robert Brout, proposed what we now know as the Higgs field, but did not specifically mention a particle.

We have come across the duality between fields and particles before, for example in the context of the photon and the electromagnetic field. In a quantized field, particles can be seen as "excitations" of the field: they are the smallest field quantity. Alternatively, a particle can be viewed as the mechanism that establishes a field throughout space. As we saw in the section on relativity, the field itself can then be understood as defining, at each point in space, how some other quantity is affected. Let's make this concrete for the Higgs field and the mass. The Higgs field fills space uniformly, and the mass is a quantum number that determines how strongly a particle interacts with that field. My guest on the standard model episode, CERN's John Ellis, suggested the following analogy: consider an expansive snowfield that you, as the particle, want to cross. If you use skis, it's easy. With snowshoes you sink in a bit, so it's more work. If you wear regular shoes the required energy is even higher. In this analogy the kind of shoe you wear is a measure of your mass.

This is probably a good point to revisit the difference between mass and weight. What we perceive as mass—for example inertia—is an inherent property of matter. For quarks and the other massive elementary particles, mass is produced by the Higgs field in the way I sketched above. In contrast, for composite particles such as protons, the mass is mostly a result of the binding energy of the strong force, as explained in the introduction to this chapter. Gravitation, then, is a force that affects bodies that have mass: mass is the property that makes something susceptible to the influence of the gravitational force via $F = m \cdot g$, where g characterizes the strength and direction of the gravitational field at the particle's current location. Because energy and mass are related via c^2, the gravitational field also affects light, even though photons have no rest mass. So we are talking about two separate fields: the Higgs field (and the strong force) determine the mass of a particle. This mass then creates another field, the gravitational field, which affects (other) bodies that have mass. The former is uniform in all of space—it is not centered on a heavy mass or an electric charge. For the latter, the strength depends on the distance from the attracting body.

Peter Higgs postulated this uniform "mass-creating" field, as well as a particle that materializes the field. But neither the standard model nor Peter Higgs said anything about the mass of the Higgs particle itself, so ATLAS and CMS had to search for it across a wide mass/energy band. In this section we will look at how they did this. But before that we have to understand interactions between elementary particles.

INTERACTIONS We all remember chemical reactions from school. Actually, I don't remember much because I really didn't like chemistry. Anyway, here is an example, the combustion reaction of propane:

$$C_3H_8 + 5\,O_2 \;\rightarrow\; 4\,H_2O + 3\,CO_2$$

It shows that one molecule of propane plus five molecules of oxygen react to form four molecules of water and three molecules of carbon dioxide (possibly via several intermediate paths). For a chemical reaction to be valid, we have to consider that certain elements only appear as molecules (we could not write a single O, for example, as pure oxygen occurs only in the form of the O_2 molecule) and the total number of atoms on each side must be the same. In the equation above each side has three carbon, eight hydrogen and ten oxygen atoms. What's more, the energy balance must also work out. The reaction above is exothermic, which means that it releases energy—more precisely, it releases reaction enthalpy in the form of heat; remember, it's a combustion reaction. Because energy is always conserved, the energy released has to be available somewhere in the system. Usually, as in the case of the combustion of propane, the released energy comes from the difference in binding energy between the atoms on the left and the right side of the equatio, plus the activation energy necessary to get the reaction going in the first place; a mixture of propane and oxygen doesn't just spontaneously ignite.

In the same way that chemical reactions describe how molecules transform into other molecules and which energies are necessary and/or released, interactions between elementary particles describe how particles form out of and decay into other particles. Because "particles" also includes bosons, energy in the form of radiation (photons) and nuclear bindings (gluons) is implicit in this statement. However, there are several additional quantities that have to be conserved to varying degrees of approximation in such an interaction: color, electric charge and momentum are exactly preserved (as far as we know), and so is the number of fermions that enter and exit the interaction. Flavor is only approximately conserved. As we will see later, the momentum balance is especially interesting for the reconstruction of collision events in the LHC experiments.

Like chemical reactions, particle interactions are described by equations. For example, the annihilation of an electron and a positron into "pure radiation" can be written as follows:

$$e^+ + e^- \;\rightarrow\; \gamma + \gamma$$

Let's look at the various conservations. The total electric charge is zero on both sides (+1 + −1 = 0 + 0). Energy is conserved as well; because electrons and positrons are relatively energetic (they have rest mass), the (massless) photons created must have relatively high energy. For photons, energy is proportional to their frequency and inversely proportional to their wave length; in the following, h is the Planck constant, a physical constant defined to be 6.62607015 $\cdot 10^{-34}$ J·s.

$$W_{photon} = h \cdot f = h \cdot \frac{c}{\lambda}$$

So in the electron/positron annihilation the photons have a relatively short wavelength. In addition, because of the conservation of momentum it is not possible to create just one photon; the two created here move away in opposite directions. It is also possible for three or more photons (with correspondingly lower frequency/energy) to be created. There's also a very rare case in which the electron and positron decay into one or more neutrino–antineutrino pairs.

So how does nature determine which decay path is taken? Each of the possible processes has a particular probability, a consequence of quantum theory that is fundamentally probabilistic. The probability of a decay path is inversely proportional to the number of particles produced: the more particles produced, the less likely the particular decay. For example, decay into three or four photons is much less likely than decay into just two.

The governing rule for spontaneous decay is that the resulting particles must be lighter. The heavier the original particle is, therefore, the less stable it generally is, because there are more combinations of lighter particles into which it can decay, each with its own probability. The correlation can be illustrated by looking at the average lifetimes of the charged leptons: the masses of electron, muon and tau are $9 \cdot 10^{-31}$ kg, $2 \cdot 10^{-28}$ kg and $3 \cdot 10^{-27}$ kg, and their lifetimes are 10^{36}, $2 \cdot 10^{-6}$ and $300 \cdot 10^{-15}$ seconds respectively. The lifetime of the electron, by the way, is a lower bound determined experimentally; theory suggests it must be stable, because with its elementary charge a decay would violate conservation of charge.

The actual physical process that mediates the decay is also relevant for decay probabilities, stability and lifetime. For example, neutrons can decay into protons via the weak force, as we will see below; the lifetime of the neutron is 880 seconds. In contrast, the proton, which does not have the option of a beta decay, is stable even though it has the same mass. The neutral pion decays within 10^{-17} seconds because it can decay through the electromagnetic force, whereas the charged pion can only decay via the weak force, and therefore has a lifetime nine orders of magnitude longer.

Decays and other interactions are often shown graphically using a notation called Feynman diagrams; the one below shows the above-mentioned beta decay, where a neutron decays into a proton plus an electron and an antineutrino via a W^- boson (the bar over the particle symbol denotes antiparticles). The example also demonstrates that decay processes can be multi-step: the products of one decay can continue to decay further. In this case, however, the W– boson is virtual; a virtual particle is a transient quantum fluctuation due to the uncertainty principle that exhibits some of the characteristics of a real particle. In the interactions they are crucial, because they mediate the forces between the non-virtual particles that take part in the interaction.

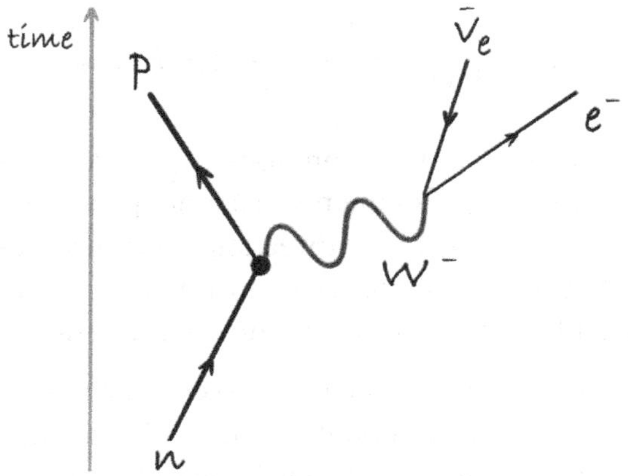

Feynman diagrams show time on one axis. We also need a second axis, otherwise we could not draw a 2D-diagram, but let's not get too hung up about exactly what this axis represents. In Feynman diagrams, fermions are shown as straight lines. The direction of the arrow indicates the direction in which the fermion number gets transported; antifermions effectively transport a negative fermion number, which is why their arrows point in the other direction. Bosons are drawn as squiggly lines. Interactions occur at the points where lines meet, the so-called vertices. Momentum must be balanced at each vertex, and because the total number of fermions must also stay the same, two arrows can never point to a single vertex. Ignoring the intermediate virtual W boson, we can write the above beta decay textually as

$$n \rightarrow p + e^- + \bar{\nu}_e$$

Neutrons and protons are of course composite particles made up of quarks, so strictly speaking we should represent these quarks separately here. In the following equation, u and d represent the up and down quarks respectively. We can see that beta decay involves one of the quarks in the neutron changing from down to up, thus transforming it into a proton.

$$udd \rightarrow udu + e^- + \bar{\nu}_e$$

Now that we understand that particles always decay into lighter ones, how then is it possible for heavier particles to form? We know from the introduction to accelerators that the idea is to supply sufficient kinetic energy to the system to "produce" the rest mass of the heavier particles according to $E = mc^2$. So, for example, if in the electron–positron annihilation mentioned above...

$$e^+ + e^- \rightarrow \gamma + \gamma$$

...either one or both have sufficient kinetic energy, a pair of W^+ and W^- bosons can be produced instead of the two photons.

Let's return our focus to the Higgs particle. How does it form? One way to create it is via the same process that also produces the W and Z bosons: by annihilating electron/positron pairs. But we need a bit more energy: the Higgs is about 50% heavier than the W and the Z. The LHC's predecessor, the Large Electron–Positron Collider, smashed electrons and positrons as its name implies. During its last and highest-energy runs the data obtained showed hints of the Higgs, but only with around 1.7σ, even though the duration of the run was extended to collect more data. This was not enough to claim discovery. Also, the data suggested a mass of 115 GeV, which is 10 GeV less than was eventually determined by the LHC.

More generally, the Higgs particle is produced by particles that interact with the Higgs field, meaning particles that have mass. When they pass through the Higgs field they can transfer some of their energy into the field; with enough energy these excitations can become so strong that a Higgs particle materializes. In fact the production of Higgs bosons in the LHC is mostly due to the fusion of gluons that become "freed" by the destruction of protons after a collision. Just as with the beta decay shown above, once again virtual particles are involved. The two colliding gluons form a loop of virtual (top and bottom) quarks, which then form a Higgs. We ignore here what the exact physical meaning of such a loop is, but we will see later that it leads to complications in the analysis of the data collected by CMS and ATLAS.

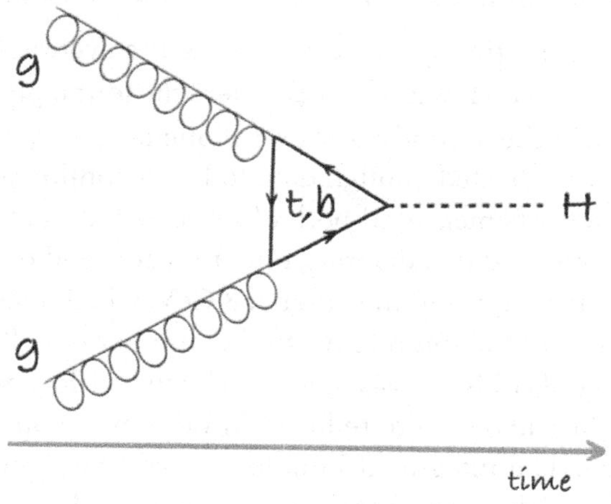

Heavy particles are unstable, and the Higgs is relatively heavy: at 125 GeV its energy is roughly 1.5 times that of the W boson, and its lifetime is only 10^{-22} seconds as a result. This is too short for it to reach the subdetectors of ATLAS or CMS, even considering time dilation at relativistic speeds. So the experiments cannot observe the Higgs directly, they can only observe its decay products. The Higgs decays along many different channels; here are the five with the highest probability. As expected, some of the decay products decay further; Z alone has 24 possible decay channels.

$H \rightarrow b + \bar{b}$	bottom and anti-bottom quark	58 %
$H \rightarrow W^+ + W^-$	positive and negative W boson	22 %
$H \rightarrow \tau^+ + \tau^-$	positive and negative tau	6 %
$H \rightarrow Z^0 + Z^0$	two neutral Z bosons	3 %
$H \rightarrow \gamma + \gamma$	two photons	0.2 %

WHAT WAS MEASURED These five decay channels have all been seen in experiments, but not all of them were used to discover the Higgs, for several reasons. First, some of them are more precisely measurable than others, because of how easily distinguished their signature is and because of how precisely the subdetectors for the various particles work. Second, the probability of each of them depends on the mass of the Higgs. Remember that during the search this mass was (at least initially) unknown, because neither the standard model nor Peter Higgs' paper make a specific prediction for it. So the Higgs search was really a search: the community made assumptions for the mass of the Higgs of between 100 and 1000 GeV. They then used the standard model to predict the probabilities of the decay channel for the assumed mass and looked for the end products of these decays in the collisions.

Taking into account the properties of the detectors and the eventual Higgs mass of 125 GeV, the most useful decay channel is $H \rightarrow Z + Z \rightarrow 4l$, where the four leptons can be four electrons, four muons or two of each; the second most useful one is $H \rightarrow \gamma + \gamma$. A total of 735 billion collisions later, CERN had collected enough data to be 5σ-confident that the Higgs exists at roughly 125 GeV. The scientists measured both of these decay channels independently with both ATLAS and CMS, and used two datasets, one from 2011 and one from 2012. The former was at 7 GeV total collision energy and the latter at 8 GeV, which increased Higgs production by 30%. The rate of Higgs production relative to the number of collisions was as the theory predicted, but the number of collision was increased by two orders of magnitude during 2011: peak luminosity was higher, losses were reduced through more operational experience and downtimes were reduced. The integrated luminosity over the period was therefore higher than originally expected, which meant that the experiments had collected enough data by late June, long before the 2012 run was over. The teams then worked day and night to finish the processing to be able to announce the discovery on July 4.

KINDS OF COLLISIONS Let's understand what happens in a collision a little better. We start out with opposing beams, each containing bunches of protons. They are timed so that two bunches "intersect" each other exactly at the location of the experiments, which is why we call the event a bunch crossing. The number of particles in a bunch, 100 billion, was chosen to ensure that statistically, during each bunch crossing, several of the protons in the bunches collide. But how exactly do they collide? The basic physics is approximately similar to billiard balls, except that the "balls" are more like pancakes, because the protons are flattened by the Lorentz contraction.

These pancakes might collide "partially", hitting each other off-center; this is shown on the left of the next illustration. In this case, one or both might be partially destroyed, a process called "diffractive dissociation". Alternatively, they might both survive intact and just be deflected slightly from their path, flying away at very shallow angles. This is called "elastic scattering". They might also not collide at all, but still deflect each other's path slightly, because of electrical and strong force interactions. All of these processes are jointly called "soft" collisions and none of them is relevant to the Higgs search.

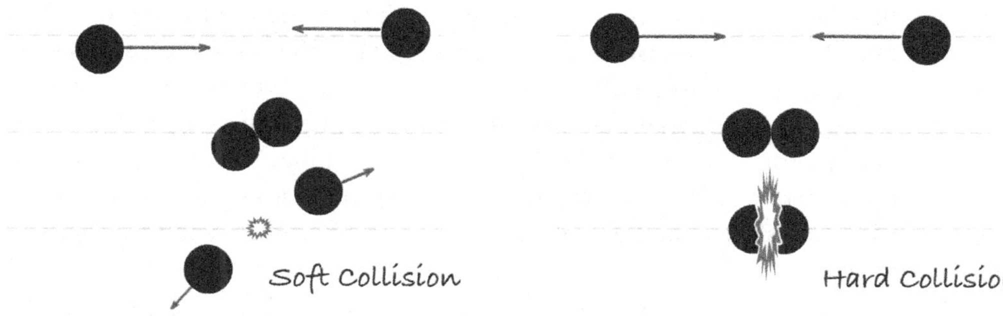

So let's look at the collisions that are relevant: two protons colliding head-on in a hard collision. This is shown on the right of the illustration. What happens then? The key thing to understand is that it is in fact not the protons as a whole that collide; it is the partons that interact—the quarks and gluons that make up the proton. These are elementary particles and we consider them to be point-like. This means that there is no partial overlap: they either collide or they don't. However, they do influence each other when they pass close enough because of the color charge. Such a near miss will most likely produce two lower-energy jets in opposing directions, which have to be taken into account during analysis, just like the particles created by soft collisions. The really interesting case is when two partons actually collide: their energy, injected by the accelerator, then becomes available to form new particles such as the Higgs boson.

But how much energy do the partons carry? You might have wondered why the LHC injects several TeV of energy into the protons for a collision to produce Higgs bosons that are only at 125 GeV. Why do we need so much energy? The reason is that the proton's total energy is split between its partons. Like most things on the quantum level, the energy distribution among the partons is probabilistic. In addition, as we increase a proton's energy, the number of partons increases due to quantum fluctuations, leaving less energy per parton but also increasing the chance of two hitting each other. Partons also transform; for example, a gluon might transform into a quark and its antiquark, and back. At LHC energies most of the partons inside protons are gluons, which is why in the LHC collisions between two gluons are much more frequent than collisions between two quarks. The LHC's Higgs production rate is as large as it is because of this abundance of gluons in the beams; remember that g + g → H is a major route to Higgs production. In a machine that collides electrons this rate is much lower, because there aren't any gluons in an electron—it's an elementary particle. All of this is a roundabout way of saying that the energy available from two colliding partons is statistically distributed, so there's another variable involved in how a collision event turns out.

It is obvious that more energy is available to form new particles in a collision if the two colliding partons each have relatively high energy—that is, they are at the peak of the probabilistic parton energy distribution. But for the collision to be as efficient as possible in terms of the freed energy, it is best if the two partons also have relatively similar energies; if one was much higher energy than the other, the "heavier" one would just accelerate the lighter one away from the collision. Ultimately the case we are looking for in the Higgs search is a head-on hard collision of two protons involving a collision of two of their partons that have both high and similar energy. In this case the two partons are effectively stopped, all their energy is released and new particles stream away perpendicularly to the beamline. The energy contained in those particles is referred to as "transverse energy". High transverse energy is a sure sign of an interesting collision, and the barrel shape of the detectors is tailored especially to capture such transverse particles.

It is relatively improbable, statistically, that more than one pair of partons collides at interestingly high energies. But other partons might collide with lower energies, spawning less energetic particles, and many of those will end up in the detectors. This has to be taken into account in the subsequent analysis that reconstructs the event from the data, for example when establishing an energy and momentum balance. More generally, the detectors are not able to distinguish between several parton-level collisions: they happen too close to each other in both time and space. As we will see, however, they are able to distinguish the many different proton-level collisions that happen during one beam crossing.

WHAT HAPPENS AFTER THE COLLISION If two protons hit head-on in a hard collision, the strong force that binds the quarks in a proton is overcome. The quarks and gluons are then "freed" from their binding in the proton, which is why we treat them as elementary particles colliding separately. Maybe a better way of understanding the situation is that the collision happens so quickly that there's no time for the quarks to exchange gluons (which is what the strong force is). So, as far as the collision is concerned the quarks and the gluons are already"free", and the proton looks more like a collection of quarks and gluons that happen to be flying in close formation. The formal version of this is referred to as the "asymptotic freedom of quantum chromodynamics": the strong force is very weak (or doesn't exist at all) on very short timescales or distances. This first phase of the collision is known as the parton phase; this is the phase where the Higgs boson can sometimes form, for example via gluon fusion, depending on the probabilistic energy distribution of the partons.

What happens next? As we know, quarks are colored. Because of a constraint in quantum chromodynamics that requires quarks that are bound together to have colors that combine to white (red + green + blue or color + anticolor), the quarks that make up protons are of different colors. Scattering colored particles—for example, scattering quarks in a collision—emits gluons. Gluons, however, are themselves colored, and their scattering creates more gluons. Through quantum fluctuations they might also transform into a quark/antiquark pair, which can emit more gluons. This repeats, and the overall process results in a shower of gluons and quarks with lower and lower momentum. Not surprisingly, this is known as the "shower phase" of a collision.

The next phase is called "hadronization"; as you can probably guess from the name, this is when new hadrons form. But why? The reason is another peculiarity of the strong force. At the scale of a proton, the force stays roughly constant as the distance between two colored objects increases. This is in contrast to essentially all other forces, which get weaker with distance. This means that the energy stored in the field that "connects" two quarks also increases with the distance of the two quarks. So if you try to separate quarks from each other, as happens during a collision and during the shower phase, the two quarks resist their separation as if they were joined by a rubber band: you have to inject more and more energy to force the two further and further apart. Luckily there's enough energy available in a collision to achieve this.

Once a particular threshold is crossed, the quarks are released—the "rubber band" breaks. At this point there is enough energy stored in the field (the rubber band) that new particles can form, and these then join into new hadrons. Why? This can also be explained by the requirement for color neutrality of particles in quantum chromodynamics: you cannot isolate two colored objects, such as quarks: while the rubber band is still intact they count as one neutral particle. When it breaks, nature immediately reestablishes a "white" color by producing additional colored particles that join with the separated ones to form new, color-neutral hadrons. Again, you need energy for those new particles to be created, but this need only occurs after something—the collision in this case—has "stored" it in the rubber band. It's neat how it all fits together.

If the whole thing is violent enough, as it is during a collision, this process repeats itself and a narrow cone of particles forms that expands roughly in the direction of the original quark. This is known as a jet. It is the particles in these jets that are detected by the detectors, not the quarks themselves. The event reconstruction uses statistical methods to establish which initial particles and processes could plausibly lead to the observed jets. As some of the particles in the jets are not stable and decay very quickly, within around 10^{-12} seconds, ultimately the detectors detect jets of decay products. We will discuss this later.

If you feel as if there's weird magic going on here, then I can perfectly relate to that: I feel it too. I guess in order to demystify the situation, one would have to do the math behind quantum chromodynamics, the theory of the strong interaction and the color charge. We are also at the very edge of science: the theories explain what happens, but, like most science, it does not really answer the "why". So I guess it's normal to feel a bit helpless at this stage of the discussion.

EXPERIMENT GOALS The ultimate goal of the ATLAS and CMS experiments is to understand nature; to achieve this goal requires identifying which particles are produced in a collision. In particular, for the Higgs search, the initial goal was demonstrating the mere existence of this particle at a particular mass. However, the experiments also try to understand the actual processes that happen in a collision, so they have to understand the number and type of all particles that were created, plus their energies; this requires tracking the trajectory of each particles as it flies away from the collision. The experiments therefore have to reconstruct the collision event completely, particle by particle. This is also crucial to be able to disentangle the potentially many proton–proton collisions that happen during a single beam crossing, and to be able to establish an energy and momentum balance for each of them. We'll see later why this is important.

Let's start with the basics: how do you identify the various kinds of particles? Ultimately, we have to measure their quantum numbers: mass, electric charge and spin. If we know these numbers for each of the particles moving through the detector, we've reliably identified what kind of particle it is. We might not be able to measure all three, however, because spin is hard to determine for a single particle: you can only do it statistically for large samples of particles and decays. So in practice we use only mass and electric charge.

To measure electric charge you could for example decelerate a particle in an electric field. The magnitude of the electric force F the particle feels is proportional to the strength of the field and its own electric charge:

$$\overrightarrow{F_E} = q\,\vec{E}$$

The strength of the electric field E depends on the geometry; the field between two parallel plates at voltage U and distance d is homogenous and constant—this is the situation for example in a capacitor. The field around a point-like charge Q decreases with the square of the distance from the charge, and also depends on the medium, represented by the dielectric constant ε:

$$E_{plates} = \frac{U}{d} \qquad E_{point} = \frac{1}{4\,\pi\,\varepsilon_0}\,\frac{Q}{r^2}$$

If we measure the acceleration of a particle in a known field, and if we know its electric charge, we can use the formulas above plus F = m · a to determine the particle's mass. If we don't know the charge, we'll need something else. Let's look at a magnetic field. As we know, the path of a charged particle is bent by a magnetic field; we looked at the basic math for this effect earlier during discussion of how the accelerator keeps the beam on track. We can compute the mass m if we know the magnetic field B that bends the particle's track, the particle's speed v, as well as its charge q:

$$m = \frac{r\,q\,B}{v}$$

We need the charge again, but now we have two formulas, so we can solve one for q, then replace q with the result in the other: this allows us to eliminate the problem of measuring q. For elementary particles, though, because electric charge is quantized, we can assume it to be +e or −e, so we only have to determine the sign. This is much easier.

We might also be able to measure the momentum of a particle $p = m \cdot v$, by exploiting the fundamental law of conservation of momentum. For example, we can let the particle bounce into another whose mass we do know, then measure how fast it travels after the collision. We can also measure energy. For example, you can try to stop a charged particle with a known electric field; we know from the definition of the unit eV that if a particle with an elementary charge is decelerated by 1 V, it loses an energy of 1 eV. Alternatively, we can let our unknown particle crash into solid matter and measure the energy released, for example in the form of photons.

You probably start to see a pattern here: charge, mass, momentum, energy—or frequency in the case of photons—are all linked through a couple of relatively basic equations. The properties we are interested in can be determined by a clever combination of several of them. Not all the processes described here are necessarily easy to implement in a detector, so only some are used, as we will see in the following pages.

In addition, everything gets a little more complicated when we consider relativistic effects. In particular, at relativistic speeds,

$$W^2 = p^2 + m^2$$

so for really large momenta we can ignore the contribution of the rest mass m, and energy and momentum can be considered the same in terms of magnitude; momentum has a direction because it is a vector, while energy is not. So for high-momentum/high-energy particles it's good enough to measure their momentum via the bending of their paths. For lower-energy particles, we can improve the situation by measuring the speed explicitly, which is done is some of the experiments, in particular ALICE.

Conversely, for very relativistic particles the bending of their path due to the magnetic field is very low, which makes it hard to measure precisely. In this case we can try to measure energy directly. This is what calorimeters do.

Beyond relativistic effects, the whole undertaking is also made more difficult by detector noise and a huge range of engineering challenges. On the other hand, the challenge also becomes a bit simpler, because only a few kinds of particles will ever make it to the detector: photons, electrons, muons, pions, kaons, neutrons and protons. Because we know their mass and charge, we can use this knowledge to narrow down the options of what just flew by.

LAYERED STRUCTURE The detector has to measure the momentum, energy and charge of the particles that escape from the interaction zone after a head-on collision of two hadrons. Because the particles generated mostly emerge perpendicular to the beam pipe in head-on, high-transverse energy collisions, most of the subdetectors are arranged in a barrel that surrounds the interaction zone, layered like an onion, as shown below. There are also end-cap detectors that capture the particles that travel more along the beam pipe. These are necessary to detect soft collisions reliably, to complete the picture of the event and for certain kinds of specialized physics.

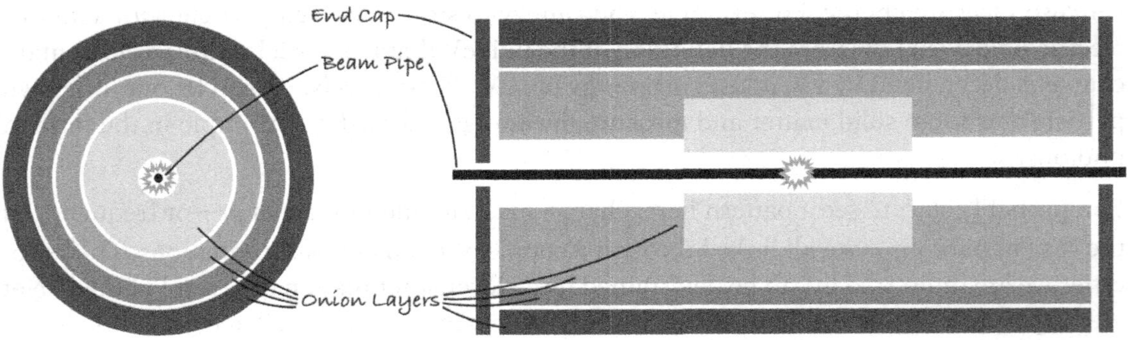

There is a catch: the outer layers of the detector will only ever measure anything if the respective particle is not stopped by the subdetectors in inner layers, either through destructive detection or inadvertently. Luckily, the various particles the detectors are supposed to measure interact very differently, and some not at all, so a layered design is feasible. Lifetime is also a factor: a particle must be stable enough to reach the outer layers.

The picture above isn't quite to scale: we know that the beam pipe is around 5 cm in diameter, while the two detectors are really huge: ATLAS has a diameter of 25 meters, is 45 meters long and weighs 7,000 tons. CMS is more "compact", as its name suggests: 15 meters diameter by 21 meters long. However, at 14,000 tons it is twice as heavy. Fundamentally the two are similar: they have the same tasks and they use many of the same detection principles in their subdetectors. They differ in some respects, however, trading off precision or comprehensiveness for one particle with another. In this sense they are complementary when used in a joint search for new particles. I will point out some of these differences in the discussions that follow.

Let's first look at the structure of the detectors at a high level, where ATLAS and CMS are similar. The subdetectors closest to the beam pipe are called "pixel detectors"; their main task is to measure the position and trajectory of particles precisely, in order to determine the precise location of the collision via inward extrapolation. This is important to allocate each particle to one of the multiple proton–proton collisions in a bunch crossing and to pin down the origin of the bending of a particle's trajectory. One level out are strip detectors that essentially do the same thing, but only in one dimension. ATLAS has a third component as part of its inner tracking system, called a "transition radiation tracker". Next is the electromagnetic calorimeter that measures the energy of electrons, positrons and photons. Even further out comes the hadronic calorimeter which measures the energy of hadrons, including neutral ones. Finally, both ATLAS and CMS have muon detectors as their outermost layer, which determine the momentum of muons. There's also the magnet used to bend the paths of charged particles, which is necessary to measure their momentum based on right-hand-rule bending. In ATLAS the calorimeters are outside the magnet, whereas in CMS they are inside.

The different particles interact with the various subdetectors in different ways. In the illustration below a dotted line implies that the particle passes through the material and leaves no trace, a solid line means it leaves a trace but still passes through, and the crow's foot-like notation means that the detection is destructive—the particles are stopped.

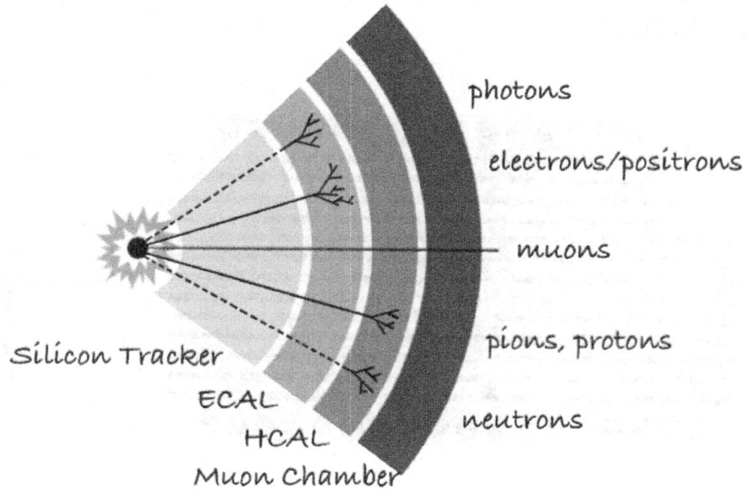

If you go back to the standard model and compare the particles detected here with those that were listed there, you'll notice that some are missing. Indeed quarks, gluons and bosons are mostly detected by the jets they produce, as we saw earlier when we discussed collisions; jets, of course, consist of the kinds of particles the detectors can "see". Neutrinos interact so weakly that the detector doesn't notice them at all; neutrinos require special detectors with huge detection masses and colorful names like IceCube or Super-Kamiokande, but that's another story (told in omega tau 135). We'll see below how, with careful analysis, ATLAS and CMS can still say something about neutrinos.

Let's now look at each of these subdetectors. I will randomly mix aspects of ATLAS and CMS, as my goal is to explain the principles, not the specifics of the two experiments. We'll start with the magnets, because their field is crucial to how detection works. If you want to learn more, google a Powerpoint titled "Detectors for Particle Physics" by DESY's Daniel Pitzl.

MAGNETS Both ATLAS and CMS rely on very strong magnetic fields, because this bends particle tracks according to the right-hand rule. With a known and constant magnetic field and a known electric charge of +1 or −1, the amount of bending is inversely proportional to the particle's momentum: greater momentum, less bending. This, as we will see, is crucially important for particle identification and event reconstruction.

CMS has one large solenoid magnet, 13 meters long and 6 meters diameter. Radially it is located between the hadronic calorimeter and the muon system; the magnetic field is "kept going" for the muon system by iron yokes in which the muon detectors are embedded. The magnet produces a field of 3.8 Tesla and is made from a NbTi superconductor that requires cooling, just like the magnets of the LHC itself. However, because "only" around 4 Tesla are required, 4.5 K is sufficient and no supercooling of the liquid helium is necessary. CMS' magnet is the largest solenoid ever built—a distinction that will be overtaken by the ITER fusion experiment once it's finished.

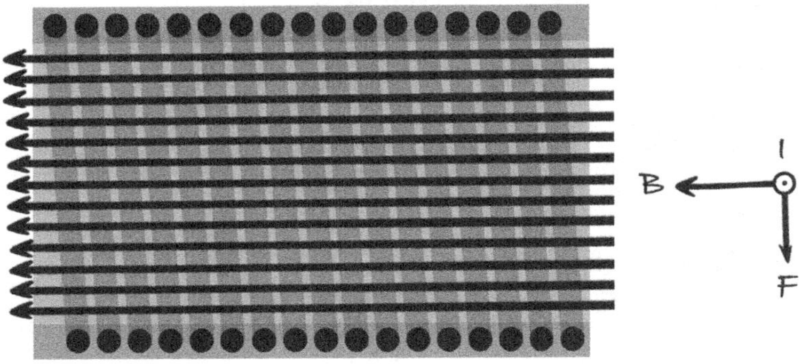

The magnet configuration for ATLAS is a bit more complicated. It has a similar solenoid around the tracker, which produces 2 Tesla, but in addition ATLAS has a secondary magnet system around the muon system, an air-cored toroid. In total, ATLAS uses 170 tons of superconductor, and the total weight of the magnet system is 1,300 tons. It is the largest superconducting magnet in the world right now.

PIXEL DETECTORS We know that several dozen collisions occur during every beam crossing, each of them at a slightly different location, so it's a bit of a oversimplification to talk about "the interaction point", because it is actually a region, a small volume in which the collisions occur. To reconstruct each of the collision events, the idea is to determine each particle's track and then extrapolate inwards: each point at which the extrapolated tracks cross denotes a separate collision. There's another reason for determining a position fix for each particle as close as possible to its origin: it gets you the longest possible distance in which to determine the particle's bending in the magnetic field once you measure additional positions further out.

Pinning down the "origin vertex" is the job of the innermost layer of the onion, the pixel detector. Two ingredients are necessary for this to work. First, the detector must be as close as possible to the interaction region, because a shorter distance makes the inward extrapolation more precise, and because there is less material between the subdetector and the collision point to scatter the particles and distort their paths. The second requirement is that the detector itself must have really good spatial resolution. Remember that many particles stream out from the collision, and their tracks are densely packed close to the collision point; disentangling these tracks requires good spatial resolution.

The pixel detectors have a secondary purpose: because they are so close to the collision point, they can tell whether a particle seen by detectors further out comes from a decay of short-lived particles, such as hadrons containing b or c quarks; the pixel detector doesn't "see" those particles, as they appear to be spawned later.

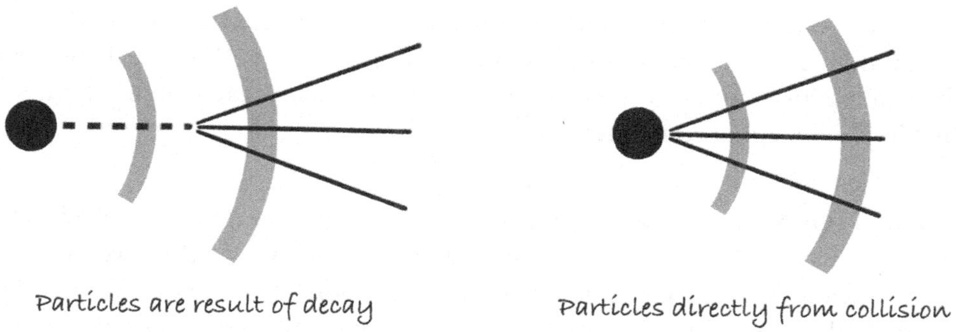

CMS has four layers of pixel detectors at 3, 7, 11 and 16 cm around the beam pipe. Logging at at least three positions and a known distance lets the software reconstruct a vector in space, and by extrapolating inwards, pinpoint the origin of a particle's track. The pixels have a size of the order of 100 micrometers, and together with some computational cleverness can pinpoint the location of the interaction to within a couple dozen micrometers. A bunch is around 15 cm long, so assuming collisions happen over the length of the bunch, this is a pretty good resolution.

Overall the pixel detector has 124 million pixels—over 90% of the total number of data channels of CMS. The large set of separate channels is necessary to be able to separately track the large number of particles that stream through the detector simultaneously during a collision. There are also end-caps, which are pixel layers that detect those particles that escape more or less along the axis of the beam pipe. They are a little different in dimensions and design but fundamentally work in the same way.

The pixels are silicon diodes. These are reverse biased, which means that the p-side of the diode is connected to the negative side of a potential. No current flows, except for a small leakage current, which together is mainly responsible for the significant power consumption of these detectors. However, an electric field builds in the diode. When charged particles pass through a diode they ionize some of the silicon, creating pairs of electrons and holes that start moving in the electric field; this induces a very short current spike of around 20 ns. This signal is amplified and used downstream.

You can imagine that the readout electronics is extremely challenging: many channels, many particles, plus the 40 MHz bunch crossing frequency. The data is stored locally close to the detector for a very short time, until the collision has been classified as "interesting"; only then will it be forwarded to more permanent storage for further analysis. We will discuss the trigger that makes the "interesting" decision later.

The radiation in the detector is a problem that has to be addressed, especially close to the beam, because it can and will harm the electronics. The ionizing radiation causes a total ionizing dose (TID) of up to 2 Grad (that is, Giga-rad, where rad is the unit of absorbed radiation dose), in electronic components very close to the beam pipe, such as the pixel detector and some power converters. For comparison, the electronics inside of a geostationary satellite are specified for 100 krad over a lifetime of 15 years. In addition to the TID, so-called single-event effects have to be analyzed for each electronic component to prevent short circuits and the proverbial bit flip in digital components.

STRIP DETECTORS Beyond the pixel detectors, out to 130 cm radially, CMS has 10 layers of silicon strip detectors. These use the same detection principle as the pixel detectors, but consist of long strips rather than square pixels, 80 micrometers by 12 cm. So strip detectors only provide high resolution along their short axis. Using strips instead of pixels makes the detectors cheaper, and reduces the number of data channels required to "only" around nine million.

The main purpose of the strip detectors is to track high-energy charged leptons (primarily electrons or positrons) or charged hadrons. Because the leptons move in a magnetic field, their trajectory is bent according to the right-hand rule: the greater their momentum the less the bend of their trajectory; a particle's momentum can be derived from the bending radius and the known magnetic field. This is shown in the left part of the illustration. The direction of bending allows positively and negatively charged particles to be distinguished.

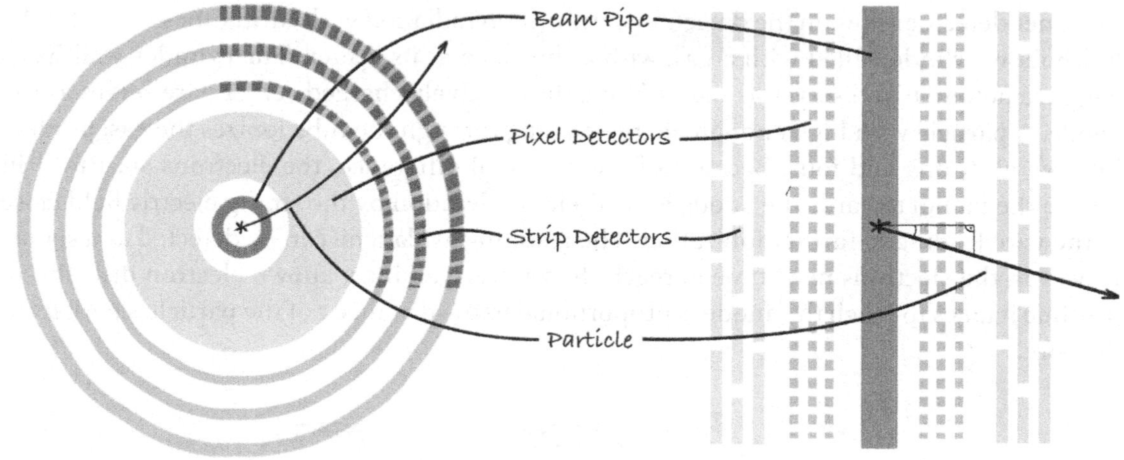

Only the component of a particle's velocity vector that is perpendicular to the magnetic field is of interest, because only that component is affected by the field; the vector along the beamline is irrelevant. This is why the lower resolution along the beam resulting from the elongated shape of the strip does not matter too much. Even though, to be able to disentangle the tracks of multiple particles travelling through the detector at any given time, a reasonable along-beam resolution is required; by mounting successive layers at small angles, and by taking into account in the analysis that particles cannot jump (ruling out for example hit C in the diagram below), useful along-beam resolution can also be achieved.

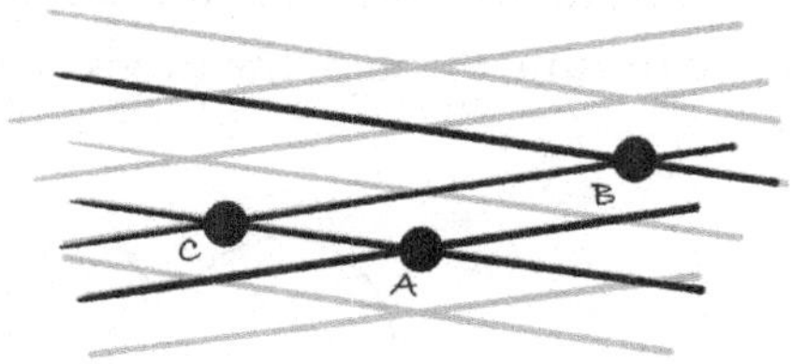

Like the pixel detectors, the strips are grouped into modules that also contain the readout electronics and the mechanical structures for mounting the module in CMS. Each module also contains sensors and electronics to monitor the health of the detectors and environmental parameters. The strips are cooled to –20°C to keep electrical noise down and to limit damage from the high particle flux. The recorded signals are transferred to the downstream processing electronics via 40,000 fiberoptic links. The CMS pixel and strip detectors constitute the world's largest silicon detector; its total area of around 200 m^2 is comparable to a tennis court.

TRANSITION RADIATION DETECTOR ATLAS' inner detector is 1.2 meters in diameter and 6.2 meters long. Similar to CMS, it features pixels and strips, but the outermost layer of the inner detector relies on the detection of transition radiation with "drift tubes". A drift tube is a hollow cylinder filled with a gas, with a thin wire at its center. The cylinder wall has a negative potential and serves as the cathode; the positively charged center wire serves as the anode. A particle with high enough energy passing through the tube ionizes the gas, producing free electrons and ions. Because of the potential difference, the electrons start to drift (hence the name) towards the anode wire, their acceleration by the strong electric field close to the wire leading to further ionization. The resulting avalanche can be detected as a signal. However, the electrons need time to reach the wire. Assuming a known electron drift speed, the time it takes to reach the anode is proportional to the distance r of the particle's path from the wire.

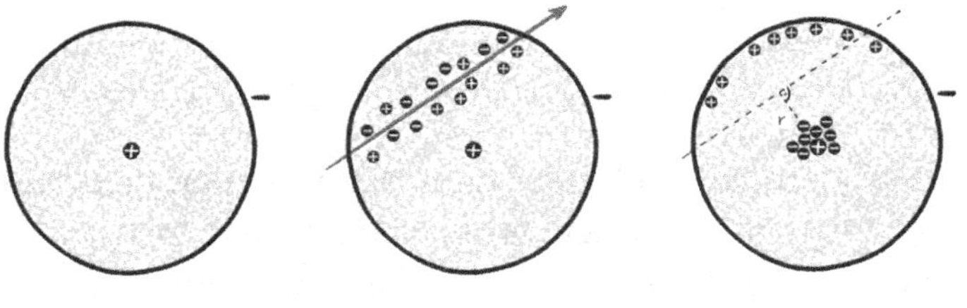

The drift tubes are installed parallel to the beam pipe, so they detect the same momentum-dependent bending of the particles due to the magnetic field as the silicon strips in CMS. A single drift tube, however, does not determine the direction of a particular particle; we can only know that it passed at a given distance from the wire in a particular tube. To nail down a specific particle's track the measurements from a sequence of drift tubes in a particle's path must be combined computationally, taking into account that multiple particles—potentially from multiple collisions—will be passing through the tubes at any given time.

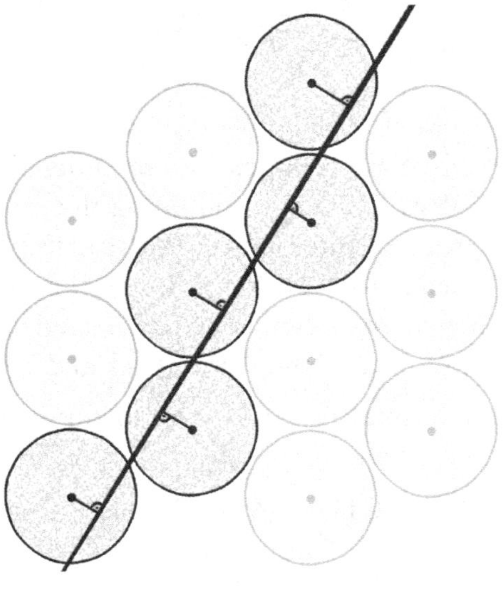

To measure the drift time to the center wire, we need some t_0 against which the duration can be measured. The time of the collision is used here. A 40 MHz clock tick synchronized with the bunch crossings is distributed to the experiments for exactly this sort of timing task. The same t_0 is used for all drift tubes, independent of their distance from the beam line. This introduces a little imprecision, because a particle arrives later at the further-out tubes. However, with particles moving at almost the speed of light, the 5-meter time difference between the innermost and outermost tubes is only 15 ns. Drift times are substantially longer. Because c is known, this effect can also be corrected computationally.

As for many of these subdetectors, their principle is simple, but the engineering design has to consider many details. In particular, the drift speed depends on the strength of the electric field, the particular gas used to fill the tubes, as well as its temperature and pressure. The magnetic field in which the tubes are installed also plays a role. All of this has to be precisely calibrated. Doing this on the scale of ATLAS is a considerable challenge: it has 300,000 tubes, each up to 1.5 meters long. Each tube is 4 mm thick and made of Kapton; the central wire is gold-plated tungsten 30 micrometers in diameter. To balance the drift speed for the necessary precision and measurement rate, the gas used is a mixture of xenon, carbon dioxide and oxygen.

Why use drift tubes instead of silicon detectors? After all, their readout rate is lower, and with an uncertainty of around 200 micrometers they are also less precise. One reason was cost; they are cheaper than silicon trackers. Another is that, in contrast to silicon trackers, they can be used to discriminate between specific kinds of particles; we discuss this below. Finally, because they fill the whole volume rather than just discrete layers, the system delivers up to 38 hits per track, providing more data for track identification and reconstruction.

We should briefly discuss the subtle difference between a detector's ability to determine a single particle's position or track precisely and its ability to do this for many particles at the same time. Drift tubes have a relatively low number of separate channels; basically one per tube. And the data from a single tube only delivers the information about the distance of a particle's path from the wire. Computational methods that rely on the data from multiple tubes can be used to determine a particle's position and track anyway, to a level of precision that is comparable to silicon detectors. However, if we have to do this for many particles at the same time, then the limited number of channels becomes a problem: several particles might create only one processable signal, even if they pass through the detector relatively far apart. Silicon detectors, in contrast, because they have many more separate channels, can provide separate signals for each particle. This is why silicon detectors are used further inside, where the flux density is higher and many closely spaced particles must be distinguished. Further out, where fewer particles pass, and with greater spacing, drift tubes, which their smaller number of separate channels are sufficient. In other words: the physical resolution of a detector and the resulting number of distinct channels is mostly relevant to discriminate between the particles passing at the same time, and not so much to determine the exact position or path of a single particle, because the latter can often be achieved computationally as well. So it makes sense to use detector technologies with progressively fewer channels the further out the detector is: the particle flux decreases with radial distance from the beam, so in a detector with a large number of channels most of them would be sitting idle at any given time. And considering the increasing volume with radial distance, this would become prohibitively expensive.

As mentioned, the drift tubes are also used to distinguish different kinds of charged particles, namely electrons and positrons versus charged pions. This distinction is based on transition radiation. Transition radiation occurs when highly relativistic particles cross the boundary between materials with different dielectric constants: such a particle emits photons in a specific energy range with a particular probability. Xenon, the gas in the drift tubes, absorbs photons in this range especially efficiently, becoming ionized. Particles that produce a lot of transition radiation will produce more ionization, hence a stronger signal in the drift tubes. To provoke transition radiation, the space between the drift tubes is filled with many layers of plastic film with a dielectric constant greater than one; the transition that triggers the radiation is between these foils and air, which has a dielectric constant of approximately one.

Distinction between different particles then relies on the principle that lighter particles with the same momentum have a higher speed (remember, $p = m \cdot v$), and therefore produce more transition radiation. This allows the lighter electrons and positrons to be distinguished from the heavier pions, even though they might experience similar bending because of similar momentum.

ELECTROMAGNETIC CALORIMETER The job of a calorimeter is to measure energy; I am sure you remember the (non-SI) units for energy, "calories", from the specification of the energy content of foods. An electromagnetic calorimeter measures the energy of charged particles, and in particular, electrons and photons. Remember the $H \rightarrow 4l$ and $H \rightarrow \gamma + \gamma$ decay channels of the Higgs? In the case for which the leptons are electrons, these would be detected by the electromagnetic calorimeter; detecting the other alternative, muons, would be the task of the muon system. The photons are also in the domain of the electromagnetic calorimeter.

$H \rightarrow 4l$ is referred to as the "golden channel", because it can be detected efficiently with both ATLAS and CMS—specifically by their electromagnetic calorimeters. This is not a coincidence: they've been designed to have their optimum sensitivity in the energy range of leptons produced from a Higgs with a mass of less than 150 GeV, which was considered most likely for the Higgs even before the experiments were built.

Many calorimeter designs exploit the phenomenon of scintillation, in which a material produces luminescence when it is excited by ionizing radiation. As a particle collides with atoms in the material, the atoms absorb some of the energy by transitioning to an excited state, which means that some of their electrons move into a higher-energy orbital. When the atom returns to its ground state—the electrons "dropping" down to lower orbitals—it releases the differential energy in the form of photons whose wavelength is inversely proportional to that energy difference ($W = h \cdot c/\lambda$). During an ionizing collision the original particle loses the corresponding amount of energy, but depending on how much energy it started with, has enough left to continue through the material, producing a trace of light through the detector. At the same time, depending on their energy, the released photons can excite other atoms, which then release lower energy photons, leading to a photon shower.

At the energies produced by LHC collisions, however, other processes dominate. Electrons, for example, are deflected when they hit scintillator atoms. Deflection of charged particles produces synchrotron radiation, as we know. In addition, when they are decelerated by a collision, they radiate "bremsstrahlung"—one of those German words that made it into English—as a consequence of the conservation of energy.

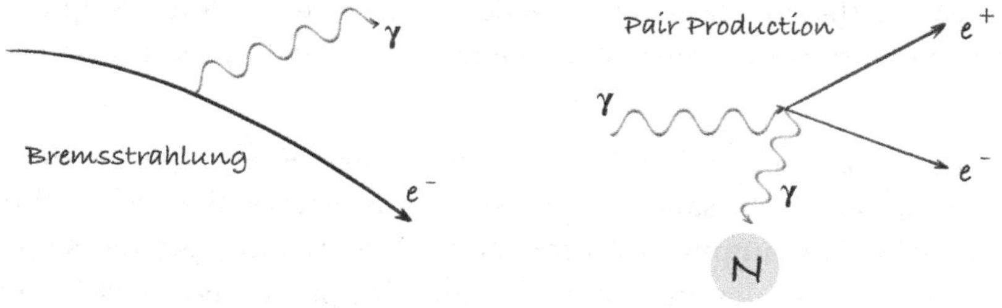

The second dominant process in scintillators is pair production, in which a photon spontaneously produces an electron/positron pair. For this to occur the energy of the photon must be at least as large as the rest masses of the two created particles, which means that they must be gamma photons. To conserve both energy and momentum, as is required in any such interaction, a nucleus of the scintillator must be nearby: it receives a "bump" transported via another photon. This is why pair production does not happen in a vacuum.

Ultimately these three processes together produce a huge particle shower of electrons and photons, the latter creating the luminescence that is characteristic of scintillators. This can be detected by photomultipliers or avalanche photodiodes attached to the scintillator, the intensity of the light giving a measure of the original particle's energy.

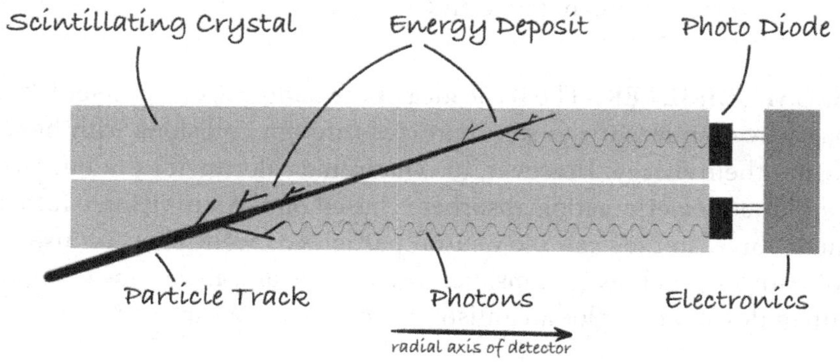

The scintillating crystals in CMS are made from lead tungstate, $PbWO_4$. This material is optically transparent for the full spectrum of scintillating frequencies, which is vital if we want to be able to record light. It is also very dense, which ensures a high probability of collisions between penetrating particles and the material. Importantly, $PbWO_4$ produces fast light bursts (the excited atoms are not metastable): 80% of the light generated is emitted within 25 ns. This is important to be able to distinguish particles at LHC's collision rate of 40 MHz.

The crystals are 23 cm deep and have a frontal area of 2.2×2.2 cm^2. CMS has 78,000 of them, 92 tons and 11 m^3 in total, split into a barrel and two end-caps. It took two suppliers ten years to grow all these crystals!

CMS' electromagnetic calorimeter can measure the energy of particles to about 0.1% accuracy, and can also measure the direction of the particle's path. This is done computationally based on the fact that the particles of interest have enough energy to produce signals in several of the crystals.

The electromagnetic calorimeter in CMS is of the homogeneous type, in which the material that produces the shower is the same as that used for measuring the energy: $PbWO_4$. The one in ATLAS, in contrast, is a sampling calorimeter: the material that produces the shower is different than the one that samples the signal. The sampling process is also different. The ATLAS calorimeter consists of alternating layers of lead and liquid argon. A primary particle hits one of the lead layers and creates a shower via bremsstrahlung and pair production. The shower particles extend into the adjacent liquid argon layer, where two processes happen roughly equally: they excite scintillation light and they ionize the gas, creating electron–ion pairs. The shower particles then hit the next lead layer, where the shower continues to grow, and then the next liquid argon layer. This continues until the energies of the individual shower particles are too low to ionize (you need 26 eV in argon). The scintillation light is UV light that does not contribute to shower development, as its energy is much too low. While the highest detection precision could be obtained if both scintillation light and ionization were measured, this is technically very difficult. The ATLAS liquid argon calorimeter therefore only uses the ionization signal: an electric field surrounding the argon volume "pulls out" the electrons and ions and measures their quantity as a proxy for the impacting particle's energy. A total of 190,000 separate cells provide resolution in space and energy.

HADRONIC CALORIMETER The basic idea of the hadronic calorimeter is the same as for its electromagnetic counterpart: to stop particles through collisions with heavy atoms and thereby measure their energy. However, in a hadronic calorimeter the interaction between the particle and the shower-creating absorber is based on the strong force rather than on the electromagnetic force, so they can see neutral particles as well. They are used to determine the energy of hadrons, such as protons, neutrons, pions and K-mesons (composite particles made of an up or down quark plus an antistrange quark, or the antiparticles of each of them).

Hadrons that interact with absorber nuclei produce primarily positive, negative and neutral pions, the lightest hadrons that consist of a quark and an antiquark. Neutral pions quickly decay into two photons. As a consequence hadron showers are more complex than electromagnetic showers. Around one third of the energy is typically deposited as the electromagnetic component, for example via the $\pi^0 \rightarrow \gamma + \gamma$ decay. Another 40% showers into protons, positively and negatively charged pions ($\pi+$ and $\pi-$) and K-mesons. Roughly another third of the overall energy (binding energy, nuclear breakup) is invisible to the detectors, which fundamentally limits the precision of hadronic calorimeters. Historically this was a problem, but today, because of more sophisticated analysis algorithms, the data from the hadronic calorimeter is less important in reconstructing collisions and its limitations are less consequential. There are other challenges though: hadrons are likely to deposit some of their energy in the electromagnetic calorimeter as well (they have to pass it on their way to the hadronic calorimeter), so reconstruction has to take into account both signals. In addition, hadronic showers are more spread out, which is why a larger volume must be available for recording them. This explains why they are installed outside the electromagnetic calorimeter.

The hadronic calorimeters' tasks include the slightly oxymoronic-sounding detection of invisible particles such as neutrinos. The idea is that if the reconstruction of the event reveals imbalances in transverse momentum and energy between, say, the left and right hemispheres of the detector, something undetected must have escaped. However, for this conclusion to be valid, the detectors must actually register all the detectable particles—no potentially visible stuff must slip through. As a consequence the CMS hadronic calorimeter is hermetic: there are very few gaps between its detecting volumes.

The CMS hadronic calorimeter is built as a scintillator tower using stacked layers of brass and steel absorber and scintillating plastic. Depending on its direction and energy, a particle will pass through several of these layers before it runs out of energy, producing luminescence. The scintillators are connected to their readout electronics via very thin optical fibers. These transport an analog optical signal to an optical mixer that sums the signals from all the scintillation cells in one tower, creating an analog "light sum" proportional to the total energy deposited in the tower.

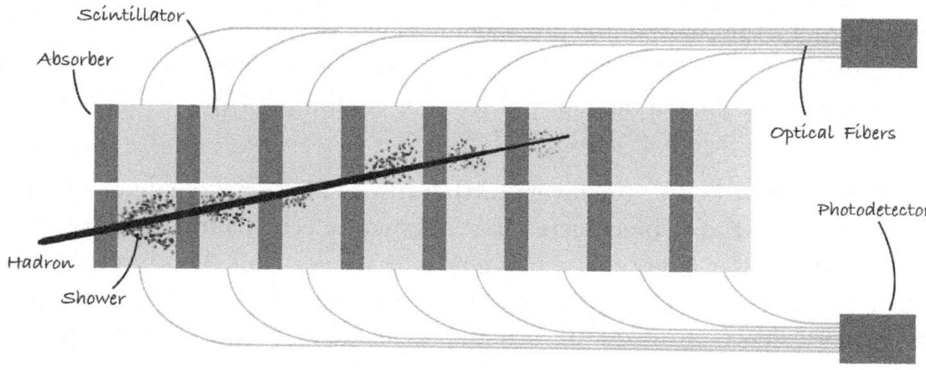

The reason why optical fibers are used to transport the light out of the tower is that this allows the necessary electronics and their power supplies to be placed outside the detector volume. This wastes as little volume as possible inside the detector, making the goal of a hermetic detector that misses no particles easier to achieve. You might wonder: doesn't the optical fiber that transports the light illuminate other scintillators as it passes, thereby "faking" a signal? It doesn't, because tiny wavelength shifters are installed at the head of each fiber, making the light in the fibers green instead of the scintillating blue. Clever!

ATLAS' hadronic calorimeter is similar in design to its electromagnetic counterpart, in the sense that it is of the sampling variety. The materials are different though: the barrel region uses iron absorbers and plastic scintillators, while the end-cap region uses liquid argon for sampling and copper and tungsten to produce the showers. The detector is less precise in terms of energy resolution and direction; this was primarily to reduce cost, because the detector is huge: the barrel region is 12 meters long and 8 meters in diameter. It is made from 36 wedges each weighing 25 tons.

In terms of the data they produce, calorimeters are different than the silicon tracker and the muon system; the latter record tracks because particles are detected at multiple locations as they travel along their path through the detector volume: it is a non-destructive measurement. In calorimeters, in contrast, the interaction of the particle with the absorber stops the particle: calorimeters measure energy through collision. Because the overall calorimeter volume is segmented, they report the location of such an impact on a cylindrical surface around the beam pipe, but not radially. This distinction will become relevant during the reconstruction of an event from its recorded data.

MUON DETECTORS Like electrons, muons are leptons. They have the same quantum numbers, but they are generation two particles and thus heavier. Understanding them is important in the context of the Higgs search; they play a role in the golden channel $H \rightarrow Z + Z \rightarrow 4l$ if and when the leptons are muons. CMS even has "muon" in its name!

Muons are the most penetrating particles, which is why we can detect them at the outermost layer: they are not stopped by the inner layers. They also live long enough to leave CMS or ATLAS. As for other particles, the muon system establishes the momentum of the muons by measuring how much they are bent in the magnetic field. Because they are measured furthest out, with the longest track and thus the largest "spread", their momentum can be measured quite precisely when the data of the muon detector is combined with the origin measurement of the tracker.

Why don't they get absorbed by the layers further in? As we know, the detectors rely mostly on ionization, in which the to-be-detected particle strips electrons from the atoms in the detectors. What can transfer energy very efficiently to an electron? Another electron! This is once again basically "billiard physics": two objects with the same mass exchange energy very efficiently in a collision. A muon, on the other hand, being 200 times heavier, has a hard time losing its energy to all those light electrons. So, while muons deposit some of their energy in the calorimeters and therefore leave traces there, they are not stopped, because they are too heavy. Thus they can seen by the muon tracker, in contrast to electrons.

Both ATLAS and CMS use three different technologies to record muons. In CMS the barrel muon chambers, located outside the magnet between the layers of the iron yoke, are implemented as drift cells, conceptually similar to those in the transition radiation detector in ATLAS. In the end-caps, the CMS muon spectrometer is made from cathode strip chambers and resistive plate chambers, slightly different techniques that also rely on ionization and charge measurement. ATLAS' muon spectrometer is one thing in particular: big! Its radial extent is from 4 m to 11 m, it has around one million readout channels and a total detector area of 12,000 m^2.

This concludes our trip through the various detector technologies used in CMS and ATLAS. We have a basic understanding how the detection of the different kinds of particles is achieved, so we can slowly make our way towards building a picture of a collision event in order to extract physical knowledge from the signals recorded by the detectors.

STATISTICS AND MEASUREMENT Let's recap what an event is: quarks and gluons collide, we assume head-on, and a lot of transverse energy streams out from the interaction point. Many different particles—potentially including the Higgs—form with various probabilities. Jets develop and particles in the jets decay. The particles that constitute the jets, plus electrons and photons, are then recorded by the detectors. In order to "detect" a Higgs or other rare, quickly decaying particles, we have to solve another of these inverse problems: what collision energy and which intermediate rare particles might have led to what the detectors actually see?

The following illustration shows an (abstracted) example process in which a Higgs ultimately decays into four E, two F and one C particle.

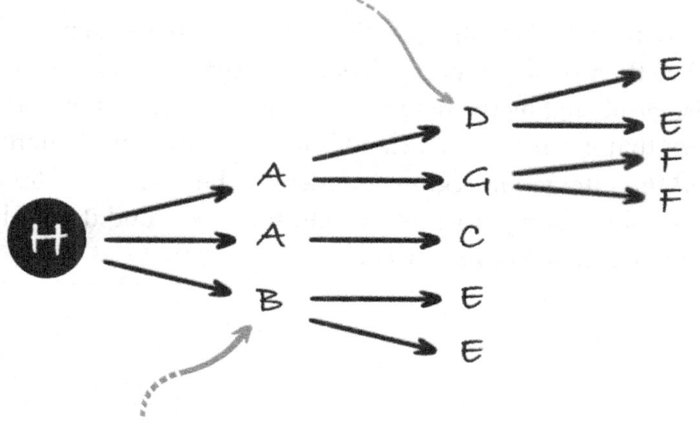

In this example, the Es, Fs and Cs are the particles that are visible to the detector. However, there are also other processes by which these particles can be produced in addition to a Higgs decay; imagine F were photons and E were electrons. There are many other processes that produce those, including soft proton collisions, lower-energy parton-level collisions or the jets that result from "skimming" gluons.

These other "uninteresting" processes whose results the detector sees are jointly referred to as the "standard model background". The experiments have to devise a way of disentangling E, F and C particles produced by a Higgs decay from those created by other background processes. One way to do this is to take into account the particle's energy, because this is related to the process that produced the particle. Remember when we said earlier that the reason why some Higgs decays are better suited than others to nail down a Higgs, because of how distinguishable their signature is? The fewer processes there are that produce the same set of particles with particular energies, the easier it is to identify a Higgs.

All this has to take into account the probabilistic nature of quantum physics: all these decays—those that start out from the Higgs, but also those further downstream in the decay chain such $A \rightarrow C$ and $A \rightarrow D + G$ above—occur with particular probabilities. This is why a physics analysis always takes into account a set of collisions and not a single one: statistics can be used to "filter out" the background.

An additional complication is that many proton–proton collisions occur during each bunch crossing. This phenomenon is known as "pile-up", and during the Higgs search pile-ups of 20 to 40 collisions were seen routinely. This is good, because it increases luminosity and ultimately leads to more data. However, it also means that during any given bunch crossing the spray of particles seen by the detectors is the result of many proton–proton collisions. To understand what's going on the various collisions have to be analyzed separately. This is one of the reasons why the tracks of the particles have to be reconstructed: by retracing their flight path we can see from which of the collisions the particle detected originated. To give you a feel for the performance of these detectors, ATLAS accurately reproduces the tracks of 97% of muons!

Finally there are also technical challenges. The detectors are not simple "particle counters"; as we have seen, they themselves embody physical processes. These must be understood and modeled. The noise produced by the detectors also has to be measured and calibrated to obtain a measurement that is trustworthy. In addition, the detector elements do not surround the volume around the interaction region completely; there are gaps, for example for structural members that hold cooling pipes and cables for power and data. This has to be taken into account in the statistical analysis as well.

RECONSTRUCTION Let's put all the pieces together and look at how a consistent event—all particles, their energies and their tracks—is reconstructed. This discussion is based on the paper "Particle–Flow Event Reconstruction in CMS and Performance for Jets, Taus, and $E_{T,miss}$" by the CMS Collaboration.

Event reconstruction is a complicated, multidimensional fitting problem and proceeds in three steps. The first step is local to each subdetector. This is necessary because most have different parts, such as the concentric layers in the silicon tracker, or thousands of scintillation crystals in the electromagnetic calorimeter; the data from all of these has to be made sense of and fused. The outputs of this step are tracker tracks, so-called clusters in the two calorimeters and muon track segments. I explain those below.

The next step reconstructs objects by fusing the data from several of the subdetectors. As we can see from the illustration below (repeated from several pages ago), most particles interact with multiple subdetectors, so their data has to be merged to form a complete picture of a particle's trajectory through the detector.

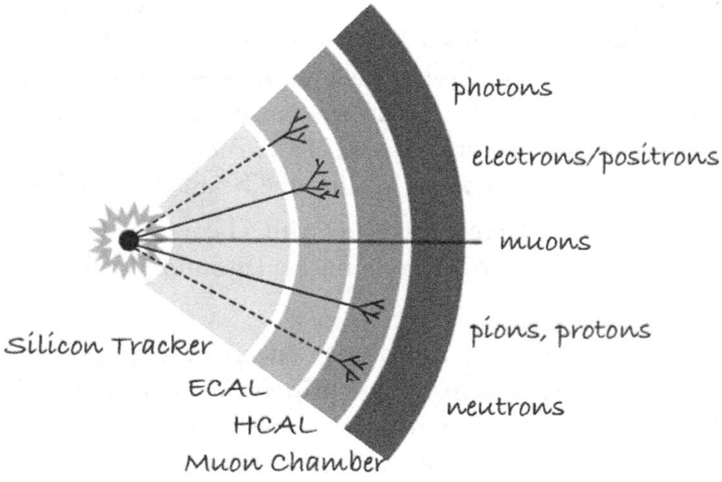

We can also see from this illustration that the different particles each interact with different subsets of subdetectors, and in different ways (pass without trace, pass with trace, shower & die); these particle-specific interactions are used to identify the kind of particle. So from this step, the output are the tracks and energies of muons, electrons and photons, as well as hadron jets, plus some classification data such as "b-tagging", useful as an indicator of a Higgs decaying into two b quarks or identifying tracks that originate from the tau lepton.

The third step is the analysis of the data relative to physics. For example, if the sum of the momenta and transverse energies in one direction is not the same as in another—they are not balanced—this means that another particle, one not seen by the detector, must have left the collision area. This might be a neutrino, or potentially new physics.

Let's look at step one, the reconstruction local to each of the subdetectors. We'll begin with the silicon tracker. It detects charged particles as they cross the layers; through backward extrapolation of tracks it provides a precise location for the collision. This helps to distinguish the particles that originate from the different collisions in an event. The location can then be used as the origin for determining the amount of bending of their path as seen by detectors further out from the beam axis.

Two thirds of the total collision energy is carried by charged particles, so understanding those precisely is key: you want to make sure that the tracker captures close to 100% of them. On the other hand you want to avoid "fakes", invalid tracks misassembled from unrelated hits or erroneous signals from the detectors. Avoiding fakes can be achieved by applying very strict tolerances when joining a sequence of detector hits into a track. But we might be too strict and miss valid tracks. To get the best of both worlds, track reconstruction is iterative.

It starts with very strict tolerances, thus avoiding fakes, but also missing valid tracks. The next iterations widen the tolerances, but the hits that are part of the previously identified tracks are removed from the set of candidates so that they cannot be part of fakes. Three iterations typically reconstruct over 99% of the tracks that start from the origin. Two more iterations are used to reconstruct tracks that start from other locations in the vicinity of the origin, and are thus secondary particles or decays of relatively long-lived particles. The resulting track data retains the information of how "tight" or "loose" the track was, so physicists can pick and choose those that are most suitable for their analysis. Particles that have three or more tracker hits, start from within a cylinder of 50 cm around the beam axis and have a transverse momentum of at least 150 MeV/c^2 are reconstructed with a fake rate of less than 1%; lower energies are tougher because they tend to "hide" in the background. We'll get back to the statistics of this.

On to the calorimeters. Most particles deposit their energy in several cells (scintillator crystals or lead/liquid argon layers), so clustering is used to identify which of these detector hits belong together, a bit like assembling continuous areas from a pixelated image. The size (in three dimensions) of the cluster is a measure of the deposited energy. Clustering takes into account statistically how much additional energy might have been deposited outside the cluster or in dead materials.

Cluster identification begins with identifying "cluster seeds", local energy maxima above a particular threshold. Clusters are then grown from seeds by including cells that have at least one side in common with the seed and which are above another (typically lower) energy threshold. At this point we have successfully grouped the signals from several calorimeter cells into clusters that represent a particle's track—the signals of a particle and its shower spread over several detector "pixels"—while also keeping separate the hits created by the many particles, created by many collisions, streaming through the detector at any one time.

Reconstruction step two is about joining the data from several subdetectors: we now have to link the tracker, the calorimeters and the muon tracks, to the degree that a particle leaves traces in all of them. The algorithm starts by combining each track and cluster pairing and computes a distance metric between them: the closer the distance, the more likely the traces represent the same particle. More specifically, linking the tracker and the calorimeters proceeds as follows. The outermost hit in the tracker is extrapolated to the electromagnetic calorimeter and to a depth that is typical for a shower, then further out to the hadronic calorimeter and the typical depth of a hadron shower. In both the track is linked to the previously identified clusters if the track "arrives" within the boundary of that cluster. That boundary can be enlarged, if necessary, to account for gaps in the detectors and the uncertainties in determining the calorimeter maximum and scattering effects. Links between the two calorimeters are established if the position measured in the electromagnetic calorimeter extrapolates to within the boundaries of the cluster in the hadronic calorimeter: the latter has lower resolution and therefore a higher positional error. Finally, muon tracks are linked from the tracker to the muon detector based on the closest fit of two tracks: again, all combinations are attempted.

At this point we have computationally joined the data from all subdetectors: we know that "something" flew along a particular path, but we don't yet know what kind of particle it was. Particle identification starts with muons because they leave traces in all subdetectors: they produce the most redundancy in the data, so they can be identified with the highest confidence. All data that relates to a specific muon is then removed from further processing. Electron identification comes next. This exploits the fact that electrons lose energy from bremsstrahlung in the tracker on their way to the electromagnetic calorimeter, where various statistical filtering techniques are used to distinguish them from heavier particles. On then to the identification of hadrons. The key here is to compare the momentum of the tracks and the energies deposited in the electromagnetic and hadronic calorimeters: several particle tracks might be connected to the same calorimeter cluster as long as the sum of the energy works out. A higher-granularity iteration is performed if the total calorimetric energy is significantly lower than the total track momentum: either we have to find additional muons by relaxing the criteria, or some of the tracks are fake, so the less reliable ones are reexamined. Depending on comparison of the momentum and the energies imparted into the two calorimeters, the track is either a charged hadron, a neutral hadron or a photon.

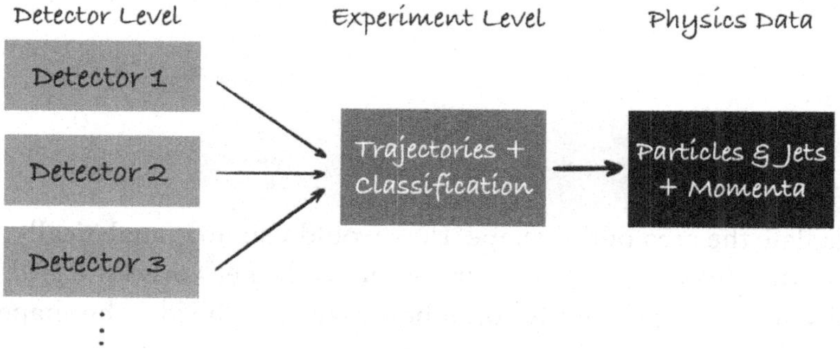

At this point we have classified the particles and reconstructed their trajectories. What follows is determining notions like bending radius, momentum or energy; to get to those from the interactions of a particle with the detector requires yet more modeling and calibration.

Once this is done we have the data that the scientists care about for their physics-level analysis. For the jets, this data is their shape, direction and energy, as well as additional information such as the b-tags. For individual particles such as electrons, muons or photons, the data includes their velocity vector at the interaction point, their momentum and energy, plus their kind. The full trajectory is not included in the science data; it is merely a tool to determine the particle's properties and depends too much on technical aspects of the detectors.

This description is just a rough outline: there are many more iterations and cases to distinguish during the process. It is also interleaved with various calibrations. In addition, there are a great many threshold values used in the algorithm that have been determined empirically over the years to "work well".

MONTE CARLO SIMULATIONS Let's return to the hunt for the Higgs. All the processes—the physics observed by the detectors, some of the detection techniques used by the subdetectors, their noise, "escaping particles", decays, the background—are statistical: each happens with a specific probability. This means that, with some probability, a measurement of a "Higgs signature" could also be an artifact, the background and the detector conspiring to "fake" a Higgs. To get to the required five sigma level of confidence we have to understand precisely how often such a "faked Higgs" would occur given all the probabilities involved. This is done using Monte Carlo simulations.

Before we get into their use in particle physics, let's understand the idea with a simple example. Let's say we want to compute the area of the wing of an ASG-29 glider. As you can see, it is not a simple shape; in fact it approximates an ellipse through several kinks in the leading edge.

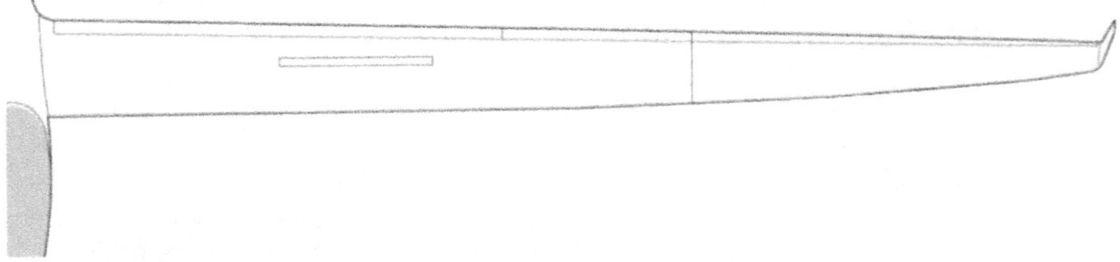

Let's try to calculate the area of this shape. How would you do it analytically? Not so simple, right, because of the kinks and stuff. Let's try numerically. Let's overlay a grid; if we know the size of the grid squares we can simply count how many are "inside" the shape and sum their area. Like many numerical methods, the result is approximate: the size of the grid squares determines the resolution, or precision, of our measurement.

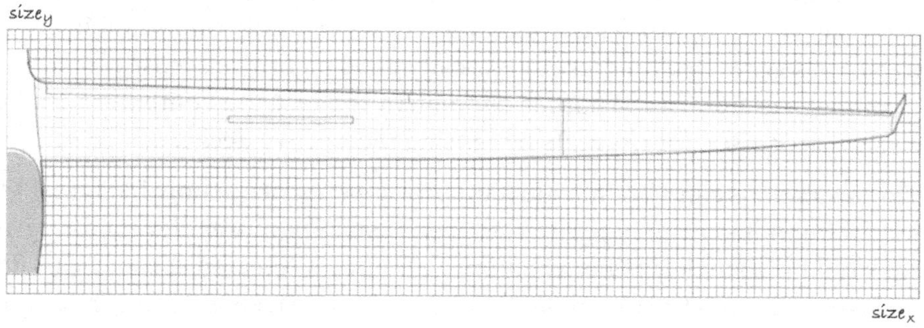

But we can be even more "dumb". We can use a random number generator to come up with an arbitrary grid coordinate, then check whether that particular grid square is inside or outside the wing shape. This is the simplest form of a Monte Carlo simulation. We can then calculate the wing area by multiplying the hit ratio with the size of the sampling area:

$$\frac{number\ of\ hits\ (inside\ shape)}{total\ number\ of\ samples} \cdot size_x \cdot size_y$$

The approach works only under two assumptions. First we have to perform enough iterations. If we don't, there is a probability that too many or too few random coordinates lie inside the area. Let's say for the sake of the example that the area is $0.25 \cdot size_x \cdot size_y$, then there's a probability of 0.25 that, if we do only one iteration, the one "guess" is inside the wing and we determine (erroneously) that the size is $1.00 \cdot size_x \cdot size_y$. If we guess twice, the probability is still $0.25 \cdot 0.25 = 0.0625$. There's a similar argument one can make that 50% of the guesses will happen to be inside and we miscalculate the size to $0.5 \cdot size_x \cdot size_y$. The Law of Large Numbers says that once you run through sufficiently many iterations N, the measured value will converge to the expected (correct) value; the error shrinks with $1/\sqrt{N}$.

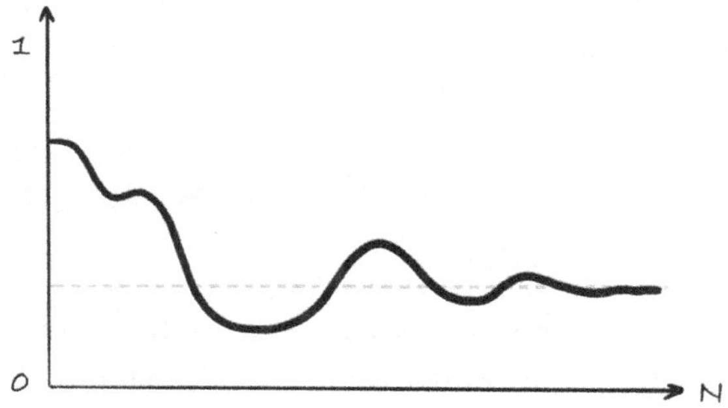

Of course this is only true if the random number generator actually produces random numbers; this is the second of the two assumptions. Implementing random number generators in software is surprisingly complicated because computers are deterministic machines, a property that programmers generally appreciate, but sometimes doubt when bugs are especially hard to reproduce. However, since random numbers are crucially important in cryptography, and cryptography is very relevant these days, really good algorithms exist and are available to every programmer via libraries. Another alternative is to let the computer sample actually random physical processes, for example by sensing radioactive decay.

Monte Carlo simulations can also include temporal dependencies: how likely is it that a body is in position B after it had been in position A previously? Or, more relevant for particle physics, how likely is it for a particle to decay at time t after it was created at time t = 0? Both examples also show that a Monte Carlo simulation does not exclusively rely on randomness: a car's progress along a road is calculated based on the laws of motion; a particle's decay probability as a function of time is known from experiments. Similarly, when we look at the simulation of what happens inside ATLAS or CMS, then the bending of particles based on magnetic fields or their loss of energy from collisions with absorbers in calorimeters will be based on equations—even though these might be solved numerically—or parametrized approximations. Also, even though many "decisions" in such a simulation are probabilistic, they are causally connected: only after particle A has decayed into G can G decay further into two Fs. And only then can it, with some particular probability, hit a detector, where it leads to a measurable effect with some other probability. The overall outcome of a simulation is thus computed by a probabilistic decision tree with interspersed "computations".

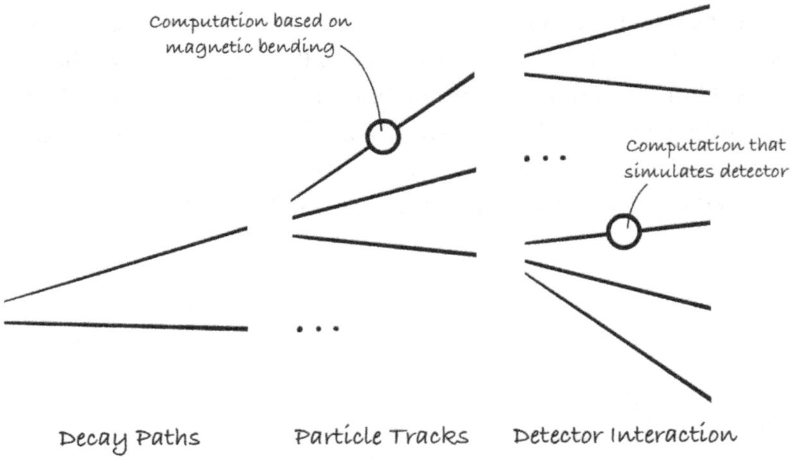

To understand what to expect during collisions based on the known standard model, and to determine whether something significantly different (in a statistical sense) is going on, the particle physics community had to run a great many Monte Carlo-based simulations of collisions, decay processes and detector interactions. Billions of them, in fact. As with any such simulation, a critical question is what to put into the model. How precisely to simulate what's going on so as to obtain meaningful results? Luckily the community has gathered a lot of experience over the decades of particle physics experiments.

Let's break down the overall problem into several smaller ones. The first opportunity for modularization is the simulation of the collision, the created particles and their decay, independent of detection of those particles by the detectors. This is the domain of "event generators" with names like HERWIG, PYTHIA and SHERPA. These encode the processes of the standard model, starting from a head-on collision between hadrons as we explained before: hard process, shower phase, hadronization, jets and decay. Some of these processes are reasonably well understood, others are simulated based on probabilistic empirical models. A nice paper titled "Monte Carlo Event Generators", by Seymour and Marx, provides a great overview.

Ultimately, and after suitable parametrization, realistic particle jets and their decay products can be simulated. These are then made to interact virtually with a virtual detector. This detector is defined with another toolkit called GEANT4. The paper "GEANT4 – a Simulation Toolkit" provides a perfect description of what it does; this paper was written by over 100 authors from the GEANT4 Collaboration, which hints at the scale of the undertaking:

All aspects of the simulation process have been included: the geometry of the system [the detector in our case], the materials involved, the fundamental particles of interest, the generation of primary particles of events [by integrating the tools mentioned above], the tracking of particles through materials and external electromagnetic fields, the physics processes governing particle interactions, the response of sensitive detector components, the generation of event data, the storage of events and tracks, the visualization of the detector and particle trajectories and the capture for subsequent analysis of simulation data at different levels of detail and refinement.

Because many of the processes are probabilistic or otherwise very complex, the toolkit embodies a wide range of experiment results to rely on as part of the simulation. Lots of effort has been expended over the years to validate the CMS model built with GEANT4, and calibrating it with data from test beams.

The final ingredient in an end-to-end simulation is a custom electronics simulator with a parametrizable noise model. This outputs simulated data representative of what CMS or ATLAS would produce with real beams.

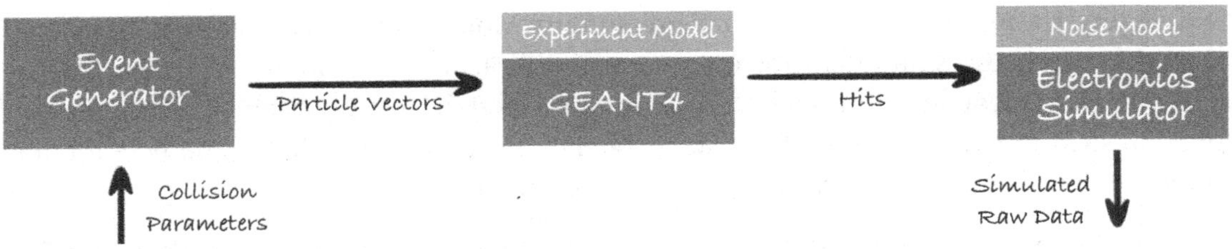

There's one catch. Even with the computing power available to CERN and its experiments, the calculations took way too long, considering that billions of runs were necessary to satisfy the Law of Large Numbers and have 5σ confidence in the result. As a consequence the CMS team developed a Fast simulation framework that runs on average 100 to 1,000 times faster than the Full simulation described above. Like GEANT4, it is implemented in C++; it's faster because it is less generic, less parametrizable and also in some ways simpler in its representation of the physics. The rest of this discussion is based on the paper "The Fast Simulation of the CMS Detector at LHC" by Abdullin et al.

The Fast simulation relies on the particle generators we mentioned previously, and then propagates the simulated particles through simulated layers of subdetectors, tracking the particles' energy loss from ionization (such as in drift tubes), particle scattering, nuclear interactions and the various kinds of shower. The tracker is approximated by 30 thin, nested cylinders with detection material, interspersed with dead material to approximate mounts and cables. For the electromagnetic calorimeters, the electron and photon shower development is based on another package, GFLASH, that assumes that the calorimeters are made from a homogeneous material (a good approximation for the scintillator crystals in CMS). The charged hadrons are made to interact with the two calorimeters only after they have lost some of their energy in the tracker; this loss is computed based on lookup tables derived from the Full simulation. The hadronic showers are then simulated based on a custom algorithm similar to GFLASH, which introduces some randomness to cater for the statistical variation between showers.

The approach is slightly different for muons. Only scattering and ionization losses are taken into account, and the muons' interactions with the detectors are based on a pre-existing model of the CMS detector geometry. These interactions are then fed into the standard CMS muon reconstruction software packages. Because only scattering and ionization are considered and other processes ignored, the data does not fit well enough with the Full simulation, but a posteriori filtering can recover this correspondence.

Just as in the nuclear fusion example used in the Models chapter, this illustrates how models with different tradeoffs—such as fidelity versus computational cost—are used for different purposes. It also demonstrates how models can complement each other: a more precise model is used to compute data, which is then used for a parametric approximation in a faster model. To build trust, the results of the faster model are then validated against the more precise Full simulation. The Fast simulation has been tuned to fit well with the results of the Full simulation, while relying on faster and less detailed representation of the processes simulated.

There are reasonable objections to this, however. For example, if the Fast simulation is "made to fit" with the Full simulation, and the Full simulation is based only on known processes and interactions, how could it ever indicate new physics such as the Higgs? The answer is that it's not intended to. Instead, the data produced by these simulations is used as a baseline against which the actual data from the detectors is compared. If the experimental data diverges from the simulated data with a large enough standard deviation, then the detector has found something new.

DATA ACQUISITION All the measurements made by the ATLAS and CMS experiments have to be converted into data that can be analyzed by computers. While this statement is obvious, complications start at the very beginning of the chain: recording the events at the detector itself. During the Higgs runs in 2011 and 2012 an interaction occurred in each of the experiments 40 million times per second, and there are 100 million data channels. Recording this volume of data to hard disk is not feasible.

Luckily it is not necessary to record everything, because the vast majority of the physics that happens in the detectors is known (and boring). The first step in reducing the data rate therefore is to make a conscious decision of what is worth recording: which decay modes are most promising for the detection of the Higgs, which are those that have the lowest background? These "golden" channels are $H \rightarrow Z + Z \rightarrow 4l$, the decay of a Higgs to two Z bosons then to four electrons or muons, and $H \rightarrow \gamma + \gamma$, the decay into two photons.

The next step is to try to filter out those collisions that are interesting and likely to contain data relevant to the Higgs search from the stream of data pouring from the detector. An initial level of filtering occurs automatically: we saw before that in low-energy partial collisions the particles stream mostly along the beam pipe and never reach the barrel detectors. Collisions that potentially produce interesting physics, in contrast, generate a lot of transverse energy. In one sense, if the detectors "see" something, it is therefore potentially interesting.

Beyond this implicit selection, a dedicated subsystem called the "trigger" further filters events, scanning the data for indications of "Higgs activity" and only activating the downstream recording infrastructure if detected. The goal is to reduce the 40 MHz event rate to under 1 kHz of potentially interesting events, which are then stored for subsequent analysis. Let's look at how ATLAS does this.

The ATLAS trigger system is built in several levels. Because of the fixed 40 MHz collision rate, the primary constraint on the first level is that it has to make a judgement synchronously and sequentially within a window of 2.5 microseconds, 100 times the 25 ns period of the 40 MHz rate. It therefore has a buffer for data produced by 100 bunch crossings. It relies on dedicated detector elements in the muon system and the calorimeters to identify events that have enough transverse energy to potentially produce a Higgs by looking at the energy of muons or the presence of jets or tau particles. Special hardware was developed to analyze these signals: calorimeter cells are summed as analog data with no digitization involved. This reduces the event rate to around 100 kHz, and determines regions of interest, which are hints to the downstream filters of where to look in the data. If an event passes this trigger, its full data is stored in a fast buffer visible to the level-two trigger.

Level two is also time-constrained. It looks mainly at the regions of interest defined by level one, but also takes data from the tracker into account. For example, for the path $H \longrightarrow 4\mu$ it tries to correlate the tracks of two high-energy muons, as they must come from the same spot in the interaction region—that is, from a single collision. More generally, trigger level two is tuned to filter relative to the particular physics being investigated. It's implemented in software that runs on a "farm" of general-purpose computers, and exploits parallelization to provide more time to each of the analyses.

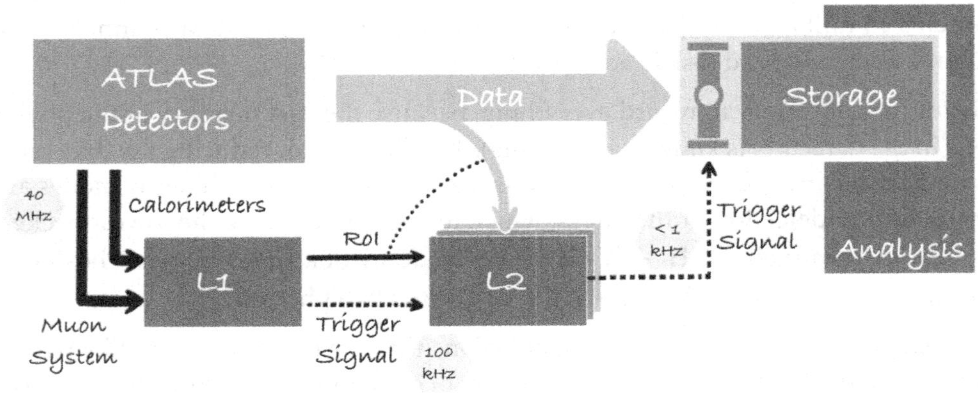

Data from events that pass the second-level trigger is recorded onto regular hard disks, which is why the data rate is reduced to below 1 kHz. Around 20% of this recorded data is relevant for the Higgs search; the rest is interesting for other research, for example understanding the properties of the top quark.

Tuning the trigger criteria precisely is very important: it must not be too permissive, because that would produce too much data for the backend to record, but it must also not be too restrictive, as that would collect too little relevant physics data—it would "waste beam". Statistical analysis has shown that the triggers for ATLAS and CMS are around 95% efficient, another reason why these experiments had collected sufficient data by early in 2012.

Much health and consistency checking of the detector and its data occurs in parallel with observations. For example, the algorithms that reconstruct collision events also use their data to update alignment calibration of the subdetectors. This is because it is easier to measure misalignment and correct computationally than trying to align mechanically to very high tolerances—an approach we have seen many times in this book. A few thousand of the 100 million data values are dedicated diagnostics used to monitor the health of the subdetectors. These are recorded as well, because this data is important for overall quality control—a Higgs "discovery" is not trustworthy if you cannot prove that the detector was working correctly. Finally, some of the known physics is recorded as well, such as Z boson interaction, as this can be used as a benchmark for the performance of the detectors and to help with the analysis of new physics. We will see an example later.

The final step is the reconstruction of individual collisions in the event using the algorithm we discussed earlier. This creates a picture (literally a visual representation) of the event, and human quality control is used here, as we are down to numbers that are feasible for individual inspection. The final step is data certification, where more sophisticated algorithms look at the data. This is also where recorded events are compared against Monte Carlo simulations. Only data that passes all of these hurdles is used for Higgs detection. Of the 300 trillion proton-proton collisions produced by the LHC, only around 1,000 each were used by ATLAS and CMS to discover the Higgs; that particle is a very rare beast indeed!

LEVEL 1 FPGAS Before we move on to physics analysis, let's get some understanding of how the level one trigger works, the one that has the tightest timing contraints. My listener Robért Glein works on the hardware that is used there, and this section is based on his input.

The level 1 trigger relies mostly on field-programmable gate arrays, or FPGAs, a particular kind of microprocessor. We all know CPUs. These are general purpose chips that can be programmed in software, based on an instruction set that is baked into the CPU by the manufacturer. Today's CPUs have a relatively large and high-level instruction set, which makes them flexible and relatively easy to program but not necessarily extremely fast, especially if they are used for the same repetitive task. GPUs are not fundamentally different in this sense, they just a different instruction set and are optimized for parallel processing of vectorized data. On the other end of this spectrum of trade-offs are ASICs, application-specific integrated circuits. Here, the gates on the chip are custom-designed to perform a particular task. They are extremely fast, because everything is literally hardwired. However, once built and deployed, they have zero flexibility. The multipliers in ALMA's correlator are such ASICs: the math behind correlation does not change. FPGAs are a middle ground: they are programmable, but are still much faster than CPUs for certain applications. An FPGA has large amounts of configurable primitive hardware elements: lookup tables, flip-flops or registers, small embedded memories, and multipliers. It also has many small switch matrixes that can be configured to connect these primitives together to form a "program", or more appropriately, a pipeline made from registers and arithmetic and logical computation steps. When an FPGA is booted up, this configuration is loaded from a external storage under the control of a host computer. So, to summarize, in a CPU, you run a custom program on an existing hardware. In an FPGA, the programmer configures the hardware to be specific to their needs. Loading a different configuration effectively gives you a different hardware, as long as your computation can be realized with the available amount of primitives and switch connections.

The development of FPGAs usually starts with a program expressed in a hardware description language such as Verilog or VHDL. For our purposes, you can imagine these as a relatively low-level programming language. This design is then compiled into a description on the level of registers and primitive computational operations (the abstractions available on an FPGA), which is then transformed into a so-called general netlist, a configuration of the switch matrices that connects the FPGA primitives and optimizes the design at a logical level. After that a place-and-route algorithm maps the general netlist to a specific netlist of the FPGA that takes various constraints of the particular FPGA platform into account. In the last step a so called "bit file" is generated; this is a binary file with all of the information needed to configure the FPGA. It is stored on the host PC so it can be supplied to the FPGA on startup. Alternatively one can use so-called MPSoCs, Multi-Processor System-on-Chips, which are a combination of FPGAs and CPUs on one chip. This enables them to run a sophisticated operating system such as Linux including TCP/IP connectivity and to have a straightforward connection to the FPGA itself.

Due to their flexibility, FPGA applications range from hardware accelerators with massive parallel computing capabilities (such as physics algorithm in the LHC), over network equipment, LTE base stations, machine learning to prototyping platforms for high-end CPUs. The CMS backend detector electronics and the level 1 trigger use over 1,000 of these FPGAs; in the trigger they perform, among other things, track finding, the reconstruction of particle trajectories based on hit patterns, as well as standardized physics algorithm such as vertex finding, jet finding and calculation of missing transverse energy. The results of their computations are then used to determine whether a collision is interesting enough to be kept for later analysis.

The time budget within which the FPGAs have to make the decision of whether to keep the event is governed by the size of the temporary fast buffers that hold all data before it is eventually forwarded to the (slower) permanent storage. In the CMS level 1 trigger, this budget is 12.5 µs, which translates into a latency requirement for the digital signal processing performed by the FPGAs. However, a new event still arrives every 25 ns, because of the LHC's is 40 MHz collision rate. This means, that for most systems, multiplexing is required in the sense that multiple events are processed by the FPGA in parallel; this explains the large number of them.

But even within an FPGA, many signals are analyzed in parallel. In some cases, up to 128 optical transceivers are attached to a single FPGA. Let's focus on one of them. It receives input data at a rate of 25 Gb/s which is deserialized by the transceiver in a 96-bit format at 260 MHz (25 Gb/s divided by 96 bits), which is a reasonable clock frequency for modern FPGAs. We can then, for example, multiply and accumulate each 96 bit value in one clock cycle to perform a histogram or filter function by using 96 input registers, roughly four parallel multiplier primitives (multipliers work on 27 bit inputs) and 96 output registers. This result is then transferred to the next computation stage inside of the FPGA. So in an FPGA one uses hardware primitives as a resource to parallelize the computation and thus achieve a gross computing time as small as possible. In contrast to that, in a CPU one uses time (clock cycles) as a resource because the hardware is fixed. Even though the CPU clock frequency is an order of magnitude faster than the that of the FPGA, the FPGA overcompensates for that with several orders of magnitude higher parallelization. To make this trade-off feasible, FPGAs have thousands to millions of these primitives, and there are many different variations with different ratios of, say, logic gates versus embedded memories and multipliers. For example, the baseline version of those FPGAs likely to be used for the new trigger, the Xilinx UltraScale+ VU9P has 1,182,240 lookup tables (that form logical functions), 2,364,480 flip-flops (double the number of lookup tables), 2,160 embedded memories (36 kb) and 6,840 multipliers. Of course, for this highly parallelized approach to work, the computational task has to be efficiently parallelizable. This is the case for the trigger, since it requires analysis of lots of (at least initially) independent data channels.

PHYSICS ANALYSIS Data deemed a "Higgs find" is distributed worldwide for analysis by different teams. Data distribution is organized hierarchically into tiers as part of the Worldwide LHC Computing Grid. Tier 0 has the full data, both raw and processed, and is located at CERN. Fifteen Tier 1 centers are distributed globally, for example at Fermilab in Chicago, KIT in Karlsruhe, Germany, and at the ASGC in Taipei. Each stores a subset of the overall data and performs the first level of processing. The centers are connected to CERN via a private optical network. Tier 2 sites are typically universities or research centers that work on particular physics problems. They have enough storage to hold the data relevant to their research, and also provide the computing power to perform physics analyses on this data. These are also the sites where the billions of Monte Carlo simulations are run that help understand the expectations based on known physics. During the current LHC shutdown Monte Carlo simulations are also run on the computers that are used for the trigger when the LHC is operational.

As I write this in late 2019 there are 151 Tier 2 sites, although many of them are collocated. One such is DESY in Hamburg, Germany. They work on the Higgs, the characterization of the top quark, analyzing particle jets and on the "missing E_T", the imbalance in transverse energy that characterizes many interesting standard model processes. It holds data from CMS, ATLAS and also LHCb, an experiment that investigates the decay of hadrons that contain bottom or charm quarks. The researchers affiliated with various Tier 2 sites use their own algorithms to analyze the data, the first of several steps in increasing the reliability of the findings. They can access the data remotely, copy parts of it or remotely run their analysis at the Tier 2 sites.

In all several thousand people are part of an experiment such as ATLAS or CMS, broken down into multiple physics analysis groups of between 300 and 400, each investigating particular topics such as the Higgs, the top quark or supersymmetry. These are further subdivided into subgroups; for the Higgs search, the subgroups were organized along the five decay channels. The actual analysis is performed by individual teams, typically consisting of a couple of PhD students, postdocs and their professor. They report results to "their" subgroup, and also get feedback and suggestions for improvements from their peers in the subgroup. If everything is ok with an analysis, it is submitted to their subgroup for formal review. Assuming it is approved, the results are distributed within the whole experiment, where a group of senior reviewers perform a final quality check. If successful, the results can then be made public on the web and/or in journal papers. Because the five channel-specific subgroups ultimately worked on a joint goal—discovering the Higgs—there was a sixth subgroup that was responsible for integrating the data from the five others. This group still exists, reactivating periodically whenever new results are published for the individual decay channels.

Several things may become obvious from this discussion. First, a lot of work goes into the analysis of the data, not just into the detector and its engineering. Second, the analysis process has a lot of build-in independence and redundancy, to increase trust in its results. Third, to further increase the reliability of the conclusions, there are multiple levels of review before any data is published. And fourth, it might be hard for an individual scientist to derive professional satisfaction from their work, because they contribute only a small part of the overall, huge picture. On the other hand, there is the satisfaction of knowing that you are contributing to something bigger than yourself.

EXAMPLE ANALYSIS To understand what an individual physicist might actually do in the analysis, let's look at an example contributed by one of my reviewers, Philipp Windischhofer of Oxford University. Physicists get the final cleaned up data from the detector: the shape, direction and energy of jets, and the energy and momentum vector at the interaction point for electrons, muons or photons. The data also contains basic analysis results such as the b-tags, as well as metadata describing how the data was obtained and the state of the detector at the time.

We already understand how to discover new particles such as the Higgs: create a huge database of simulated events that behave according to the known (non-Higgs) standard model, then compare actual measurements against it. If this reveals an unexpected "bump" at a particular energy, with a high enough sigma to rule out the bump being a result of nature and the detector conspiring, you've found "something" at a particular energy. The theory—the standard model—can then help to explain what you see.

Let's say the task of our physicist is to precisely measure the mass of the standard model Higgs boson based on the $H \rightarrow 4l$ decay. How would they go about this task? Just sum up the energies and momenta of the four leptons: because leptons are available directly in the science data, that's easy. Or is it?

First you have to find the correct four leptons. Let's assume they are muons; because muons are also produced by other background processes, the first order of business is to subtract that background. But multiple elementary collisions can happen in one event, so there might still be ambiguity in terms of which muons belong together. This can be untangled by looking at the origin vertices of the candidate muons.

We can now sum their energies and momenta to get to the Higgs mass. But in practice there are more complications, because of degrees of freedom in the decay chains and imprecisions in the detector. Let's try to understand the latter. The problem is that we don't exactly understand the detector to an arbitrary level of precision. For example, when a particle's momentum is reconstructed from the measured bend of its trajectory, there is imprecision in the detection of the track. The magnetic field might also not have been as homogeneous as was assumed in the calculation. Effectively, each value in the dataset from the detector has an error associated with it.

The former of these problems relates to the theory. As part of the analysis our physicist will come up with a decay topology, using the Feynman diagram that represents the quantum field theoretical processes that might have led to the data observed. We talked earlier about loops in Feynman diagrams, for example when a Higgs forms through the fusion of two gluons. For reasons that are beyond our scope here, the presence of such loops in the decay topology leads to a parameter that can be freely chosen (within bounds) that might increase or decrease the momenta of the non-virtual particles produced by the decay. While this uncertainly comes directly from theory (and is also considered in the Monte Carlo simulations), when we try to fit the model to the data we treat it just like the uncertainly from the experimental setup.

So to get the clearest "signal" for the Higgs mass, therefore, we have to explore the parameter space bounded by the machine and the physics. How do we know which set of parameter values is "correct"? To understand this, it is important to realize that events are not investigated one at a time. Instead, a large set of events is analyzed together statistically. The free parameters, for example those that relate to the detector, are chosen jointly for all events. Once chosen "correctly", the predictions obtained through the physics and detector models will increasingly resemble the measured data when their distributions are plotted as a histogram. The better the fit, the better the chosen parameter values.

We can do better, though. As we know, this particular decay creates two Z bosons as intermediate products: the full decay chain is $H \rightarrow Z + Z \rightarrow 2l + 2l$, each of the Z bosons decaying into two leptons. If you sum the two leptons—you can easily try the six possible combinations—you should get the mass of the Z boson. This mass, however, is known from LEP measurements much more precisely than the LHC could ever achieve. The difference between this known Z mass and that determined by summing the two leptons in the data offers an additional set of corrections. Once you have calibrated your model in this way, you can determine a more precise Higgs mass. Essentially, you exploit knowledge about the topology of the process, as well as data measured elsewhere, to improve the precision of your result.

In practice our imaginary physicist will write some C++ code that creates higher-level and less detailed abstractions specific to the physics problems, derived from the data received from the detector. Such code is typically written on top of ROOT, a framework for collider data representation and analysis. ROOT is used throughout the community and produces the plots that are ubiquitous in the various CERN presentations, including those shown at the LHC seminar on July 4, 2012. Ultimately the code tries to reconstruct the topology of the overall event. It's basically a pipeline of data transformations that (1) computes metrics that characterize an event based on analytical, numerical or machine learning techniques; (2) makes decisions about the nature of the event based on decision trees that might take into account (combinations of) those metrics; (3) rejects events because they are part of the background—this is really important to find a few Higgs events in billions of collision events; (4) "forms" new particles, for example by vector-summing the energies of other particles into which it decayed; and (5), adjusts the parameters of the topology based on calibration, such as from the Z mass, and also those that represent assumptions about the detector.

The overall analysis software then iterates over all events, applies the transformations based on the current set of parameter values, outputs a bunch of observable values, often as histograms, to allow the scientists to judge the quality of the reconstruction, then starts over after the user has adjusted the parameters.

You might have noticed a gap in the overall story, though. When we discussed the simulation of the standard model background and the huge computational effort that goes into it, I mentioned that event generators like HERWIG or PYTHIA are used to simulate the physics of collisions, and a model of the detector implemented in GEANT4 (or the Fast simulation framework) simulates the response of the subdetectors. How is it possible for the scientists to run such simulations over and over again while they tweak the parameters without it taking forever? In fact it would take forever! The solution to this conundrum is another abstraction. Groups of researchers for each of the subdetectors/particles associated with ATLAS and CMS have the job of precisely characterizing the detector. They use statistics to reverse engineer the "true detector" based on its response to the particles. This model of the detector is used when the physics data (the data our sample physicist uses for their analysis) is extracted from the raw detector output. This is effectively a calibration loop that increases the precision of the physics data. Some uncertainties remain, but because they are now well-defined and bounded based on the statistical analysis, these can be abstracted into a parametric model. You may remember from the Models chapter that a parametric model is one in which a (potentially large) set of parameters approximates physical processes that are prohibitively expensive to simulate explicitly, or whose precise physics is not even known. As long as each parameter is varied within its defined range, and model-specific constraints between the values of groups of parameters hold, a parametric model is a valid standin for a more precise model. It is these parameters our physicist tweaks—within the bounds specified by the model—when trying to fit the model to the data.

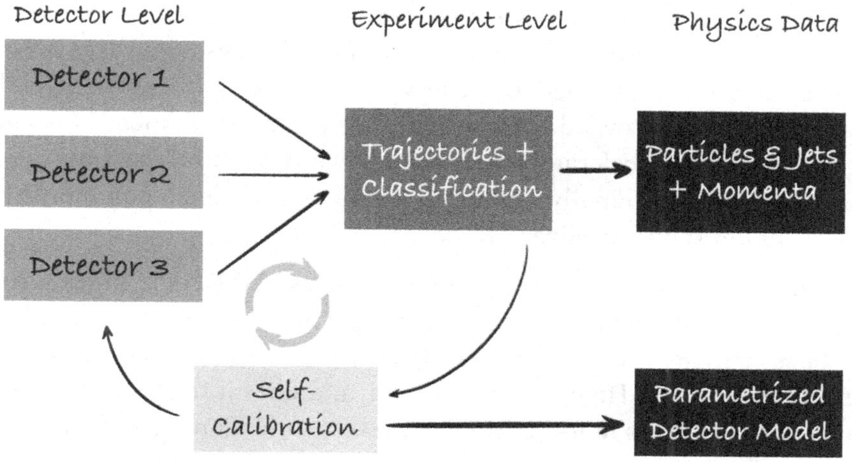

To conclude the story of physics reconstruction, let's revisit why H ⟶ 4l is such a useful decay channel even though, in absolute numbers, it is relatively rare. Let's compare it with H ⟶ bb, the decay of a Higgs into a bottom and an antibottom quark, which accounts for 58% of all Higgs decays at the LHC. The trick of calibrating the process via the known Z mass is available here as well; it turns out that the production of a Higgs always involves the production of a Z boson:

$$p + p \rightarrow \ldots \rightarrow \begin{cases} H \rightarrow b + \bar{b} \\ Z \rightarrow e^{+} + e^{-} \end{cases}$$

So what is the problem? We know that detectors do not detect quarks directly. Instead, they see the jet that is produced through hadronization of the quarks. For our physicist, identifying the relevant jets is relatively easy because they are already b-tagged in the data she gets from the detector. But to get to the energy of the original quark and then sum the two to get to the mass of the Higgs, we'd have to sum the total energy of all the particles in the jet. However, because a jet is a much more "messy" process compared to tracks of four muons, its total momentum and energy is much harder to measure. The background is also really high: for each bb pair created in this way at the energies of LHC collisions, 10^7 bb pairs are created through other processes. Due to improvements in analysis techniques and much more data since 2012, the H ⟶ bb decay was observed in 2018 with almost 5σ. Even though it was clear where to look in terms of energy, it still took six years, which demonstrates the challenge! While this is not as significant as the first discovery of the Higgs, it is crucial that "everything fits together", so every confirmation of Higgs properties through a new process is valuable.

NEXT STEPS The community had been searching for the Higgs for decades. Finding it was a major breakthrough and in some sense the end of an era. But as always when new physics is discovered—even if it "just" confirms what everyone expected—it is also a beginning, as DESY's Kerstin Borras mentioned in our interview. First, because the Higgs mass is basically a free parameter in the standard model, and its value influences other aspects of the standard model, the known mass narrows down the space for future experiments. Second, we know that the engineering of the accelerator and detectors works, so this might improve the funding situation for future colliders. Finally, the "atmosphere" in the science community changes because we now actually know the Higgs exists.

Many unanswered questions remain, such as which Higgs is it? Is it the Higgs predicted by the standard model? There might also be several Higgs particles: some models predict up to four. Why is the Higgs so light? Is the Higgs a fundamental particle or does it have substructure? Which interactions does the Higgs get involved in, and does it really have the self-coupling that is theorized to give mass to itself? How does this work? Can we measure its mass more precisely when we have more data?

Over the years following the discovery both experiments collected more data. As I was writing this in October 2019 I stumbled across a press release titled "A new result by the CMS Collaboration narrows down the mass of the Higgs boson to a precision of 0.1%". Based on the 2011 and 2012 runs, the ones that gave rise to the discovery of the Higgs, its mass was measured to be 125.06 ± 0.29 GeV. With additional runs between 2012 and 2016 it was possible to narrow it down to 125.35 ± 0.15 GeV; the precision improved and the new value is within the error bars of the 2012 discovery—as it should! The improved result is not just a consequence of more data, but also of much better calibration of the detectors.

Since 2012 the community has therefore made progress, and Philipp called this progress "quite spectacular". He went on to say "What people are focusing on now is understanding the Higgs' interactions [..] We now know for a fact that the boson interacts at least with the top and bottom quarks. And we have seen slight hints (but not significant as of now) of an interaction with muons. It is a particle like no other that we have gotten our hands on before: a spin-0 particle that looks elementary. Even the solid-state physics guys, who can usually construct systems that contain "quasiparticles" with certain properties, have big troubles recreating the Higgs. But yet it exists in nature."

While the Higgs certainly is "new physics" and is worth probing in much more detail, the general public—and probably also some members of the particle physics community—are disappointed that no other new physics has been discovered since 2012. Let's take this as an opportunity to highlight the problems of the standard model, and what kind of new physics we might have expected to see.

7 | *What Are we Looking for Today*

When I talked with CERN's John Ellis for the podcast episode on the standard model, he emphasized that the standard model of particle physics is one of the most well-tested theories in physics. The discovery of the Higgs boson in the expected energy range is further testament to its validity. Ellis also discussed the importance of elegance: "Mathematical elegance is another word for consistency, simplicity and freedom from contradictions. Special relativity has been considered elegant right from the start because it simplified things. So has heliocentrism." He emphasized that a theory might be "elegantified" in retrospect; for example quantum theory, with its reliance on probabilities, initially looked messy, but now is considered elegant, "... probably because it explains so much."

What about the standard model? It might be considered elegant in the same pragmatic sense as quantum theory: it explains so much. But maybe it isn't: it has several major problems. There are phenomena the standard model cannot explain. In this section I will give you an outline of these problems.

DARK MATTER AND ENERGY The most well-known problem is dark matter. Various astronomical observations suggest that there must be matter in the universe that we cannot see. Examples for such observations include galaxies that would not be able to "hold it together" were it not for additional gravitational mass arranged in a halo around their center, the formation processes of galaxies, the mass localization in galaxy mergers, as well as the motion of galaxies in galaxy clusters. It's a little bit of a stretch to refer to it as "matter", though; although we know that it interacts gravitationally, it does not interact with regular baryonic matter in any other way, and it does not radiate. Maybe "dark stuff", or even "dark gravity" might be a better term. Current estimates suggest that about 85% of the mass of "matter" in the universe is due to dark matter.

To explain this phenomenon we need one or more additional elementary particles from which this stuff is made. These have been dubbed "weakly-interacting massive particles", or WIMPs—another wonderfully creative acronym. It's obviously not in the current standard model, so the model must be wrong, or at least incomplete. Alternatively, the laws of gravitation might be wrong: instead of explaining the observations with a new kind of matter, we could postulate a modified gravitational theory. This, however, might be an even more disruptive change to our understanding of the world. Most scientists bet on the existence of a WIMP-style particle rather than on a different theory of gravitation. While many experiments have been conducted, no such particle has been found yet. However, it turns out that the properties this WIMP would have to have align nicely with some prediction from supersymmetry, which we discuss below. This coincidence is known as the "WIMP miracle".

Before we do that, let's briefly cover the other "dark" secret of the universe: dark energy. Does our universe expand, does it contract or is its size constant? Initial hypotheses in connection with the Big Bang suggested that its expansion will slow down over time, and that the universe will eventually start contracting again, dying in what has been dubbed the "Big Crunch". Einstein didn't like this idea when he contemplated the universe, and added an additional term Λ—his "cosmological constant"—to his equations, which ascribes energy to vacuum. He chose its value so that the universe would be static, which was the consensus view of the physics community at the time. It looked like a "just-make-it-work" hack at the time, and he gave up this idea when Hubble (the man, not the telescope) confirmed experimentally that the universe was indeed expanding; Einstein reportedly called Λ his "biggest blunder". Ironically, more recent measurements have revealed that the universe isn't just expanding, but that even its rate of expansion is increasing. So it seems that, after all, the equations do need something like Einstein's cosmological constant. These days it is called dark energy. In fact, in order to explain observations, 68% of the energy in the universe must be of this variety—the other 32% is shared between dark matter and regular baryonic matter.

So what is dark energy? We have no idea. It might be the fundamental energy of space, also known as "vacuum energy", just like the world's most famous patent clerk suggested. General relativity predicts that such energy would have a gravitational effect. This is not as weird as it sounds, because in fact many quantum field theories predict that space has energy—but they predict this energy to be 100 times higher! So there's no easy solution here. There are other hypotheses, for example one called "quintessence", which refers to a scalar field that penetrates the universe. Ultimately we have no idea, not even consistent math, not even in string theory (which seems to deliver some kind of mathematical miracle for most of the other physics challenges these days). There's also no hope of finding hints at the nature of dark energy with the LHC—we'd need much more energy. So I'll leave it at that.

UNIFICATION WITH GRAVITATION We currently have no theory that unifies the very large and the very small: the phenomena at the scale of the universe and those below the scale of the atom. The standard model does not feature a "graviton", a boson that mediates the force of gravity. It is postulated to have a spin of 2, but it is unclear how this would fit with the math of the standard model. If it existed it would be very weakly interacting, and it's unclear how to detect it. Conversely, there is no place for quantum effects in relativity: God seems not to roll the dice on the grand scale.

The search for a quantum theory of gravitation has been going on for decades. String theory in all its variants—superstrings, m-theory and the like—are attempts to solve this conundrum. String theory replaces the notion of point particles used in quantum theory with the notion of vibrating unidimensional strings. These strings produce (mathematically) the various particles within the standard model, and other vibrational modes describe the required properties of the hypothesized graviton. Going back to Ellis' discussion at the beginning of this section, string theory would probably count as mathematically elegant, but it suffers from at least two major issues. One is that there are infinitely many (ok, actually only 10500) string theories, because it has infinitely many free parameters. Second, we currently cannot imagine, let alone run, an experiment that would provide experimental evidence for string theory and/or values for such free parameters. As I write this and take a quick break, this funny—and accurate—cartoon flies through my Twitter timeline:

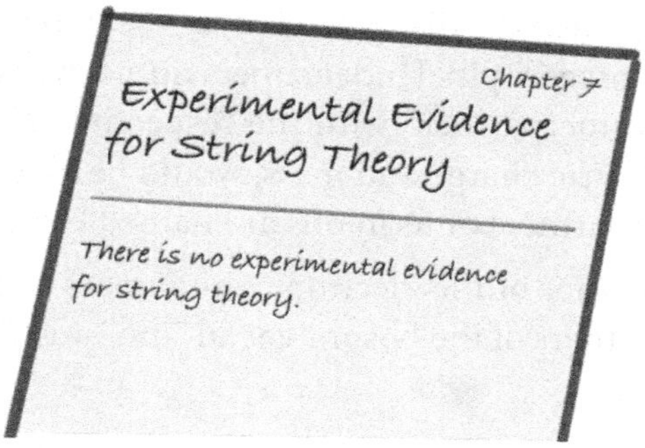

There are other proposals for unifying gravitation and the standard model, such as loop quantum gravity. While they make other tradeoffs, for example about the number of additional dimensions, there's also no experimental evidence nor hope of any. So while the unification of gravitation and quantum theory is a really important open flank of the standard model, the current generation of accelerators and those we will build the foreseeable future will very likely not be able to shed light on this question. A conceptual breakthrough is needed.

SUPERSYMMETRY Symmetries abound in nature. For example, we can rotate a sphere by any amount in any direction; it makes no difference. Similarly, we can rotate a square by 90 degrees left or right, and again, nothing discernible changes. Such geometric symmetries are particularly intuitive. But the concept of symmetry is much more general: it refers to the preservation of a physical property of a system under some transformation.

Modern physics relies on such symmetries. For example, a symmetry in time is that the potential energy of a body $W_{pot} = m \cdot g \cdot h$ does not change over time if h stays constant. Time reversal is also a symmetry: lots of physical processes work both forwards and backwards. For example, this is why the weightlessness in the ascending section of a parabolic flight is indistinguishable from the weightlessness in the descending section (which is kind of a "reversal" of the way up). Noether's Theorem proves that symmetries are the cause of the conservation laws that are a cornerstone of our understanding of physics today.

Another form of symmetry is gauge symmetry, where "gauge" refers to the absolute value of a quantity. The reason why a bird does not go up in flames when it sits on a power line is that it has no connection to electrical ground; the current—which is what would actually fry the bird—depends on the potential difference (power line to ground) and not on the absolute value; voltage is gauge symmetric. Gauge symmetries at the quantum level are foundational to the standard model; as one of the sentences that express everything to experts and nothing to outsiders says, "The set of strong, electromagnetic and weak interactions is captured by a gauge theory with 12 gauged non-commuting charges".

Physicists hypothesize an additional symmetry, called "supersymmetry", or SUSY. Let's recap. The standard model knows of two classes of particles: fermions, the "matter" particles, and bosons, the "force carriers". They are distinguished by their spin: fermions have half integer spin, whereas bosons have integer spin. Under supersymmetry, each of the particles in one of the groups would have a "superpartner" with the respective other spin; the superpartner's other properties, such as electric charge and mass, would be identical, at least in the simplest version of SUSY. The superpartners of fermions are named by prepending an "s"; the "selectron" would be the superpartner of the electron, with a mass of 0.511 MeV/c^2, a charge of –1 and a spin of 1. The superpartners of the bosons get an "ino" postfix, leading to particle names such as Higgsino or photino.

Why does the physics community ponder this symmetry? Because if it exists, it would solve a couple of problems with the standard model. One is dark matter: I've mentioned before that the properties of some of the options for the postulated WIMP particle align with some variants of supersymmetry. Because SUSY brings symmetries of spacetime into the quantum world, it might also be a gateway towards the missing unification of general relativity and quantum theory.

The math shows that it would also solve the hierarchy problem. The hierarchy problem in particle physics refers to the massive discrepancy between some of the forces in the standard model: if we normalize the strength of the gravitational force to one, then the weak force is 10^{24} times stronger, the electromagnetic force is 10^{36} times stronger and the strong nuclear force is 10^{38} times stronger. In particular, the difference of 24 orders of magnitude between gravitation and the weak force is hard to grasp: how does a single integrated system (the universe in our case) come up with fundamental parameters that are so vastly different in magnitude, but still apparently "finely tuned" to lead to a "working universe"? Because the Higgs also gives mass to the W and Z bosons that mediate the weak force, another way of phrasing the hierarchy problem is "Why is the Higgs so light?" Quantum field theories predict a much higher mass through its interactions with various energy scales many orders of magnitude higher than those of the Higgs. How then does the Higgs end up with exactly its experimentally determined mass? It turns out the math of supersymmetry avoids the need for fine-tuning all these parameters, thus elegantly avoiding the need for "magical" explanations. Supersymmetry is an example of an elegant theory, so we are inclined to believe that it must exist.

Now here's the rub: there were predictions, calculations and—frankly—a lot of hope that the LHC would detect traces of these supersymmetric particles, clear signs of "new physics". As of 2019, however, nothing has been found. The optimistic interpretation for this lack of indications of supersymmetry is that we now know that supersymmetry doesn't exist at the energies accessible to the LHC. Instead, the symmetry must be broken—we'll get to this idea later—moving the energies out of the reach of the LHC. Yet bigger colliders are needed; we'll talk about some ideas at the end of this chapter.

A pessimistic interpretation has been voiced concisely by Forbes magazine: they called supersymmetry "the greatest failed prediction in particle physics history". My personal impression is also that the community really expected to find something at the LHC. After decades of relatively regular discoveries of new particles in the 1960s, 1970s and 1980s, the rate of particle discovery has slowed almost to a standstill: this is one reason why the discovery of the Higgs was such a big deal. The community seems to be stagnating. Discovering supersymmetry would give a much needed boost to the community.

How might supersymmetry reveal itself? There are many different ways; two examples are the s-tau and Higgsino particles. These should have been produced in small quantities in the collisions between 2015 and 2018 and, because they would be unstable, their decay products should have been detectable. The s-tau would decay into a tau and a new, lightest, SUSY particle. The taus would then further decay into a hadron and a neutrino. The former can be detected and its characteristic energy measured with the existing detector infrastructure. And the neutrino, as we know, reveals itself through missing momentum. The Higgsinos, in turn, would decay into pairs of electrons or muons with very low momentum. Low-momentum particles are hard to detect, but ATLAS reported in 2019 that they were able to expand their range to lower energies—and found nothing. The search continues.

NEUTRINO MASS Talking about neutrinos, there's a problem with their mass. The standard model predicts it to be zero. However, it was found experimentally in neutrino detectors that the three generations of neutrinos transform into each other as they move through space, a phenomenon known as "neutrino oscillation". According to theory, such transformations are only possible if the particles have non-zero mass. It is essential to resolve this contradiction: do neutrinos have zero mass and some (yet unknown) effect propels their transformations, or is the standard model wrong in predicting zero mass? And if so, what consequences does this have for the rest of the standard model?

To find out, the community runs experiments such as KATRIN, located in Karlsruhe, Germany, which monitors the beta decay of gaseous tritium. This decays into a helium isotope, an electron and an antineutrino:

$$T \rightarrow {}^{3}He + e^{-} + \bar{\nu}_e$$

The decay releases 18.6 keV of energy, of which the resulting nucleus gets almost nothing because of its comparatively large mass. The energy is therefore shared between the electron and the antineutrino. The total energy of each of these particles is the sum of their rest mass energy plus their kinetic energy. If decays exist in which all the 18.6 keV is allocated to the electron, this means that there's nothing left for the neutrino—it has zero speed and zero rest mass. Because we cannot directly measure the mass of the neutrino (as it hardly interacts with matter at all), the experiment measures the energy of the decay electrons. More specifically, KATRIN measures their energy spectrum over a large number of decays. If the spectrum extends to the full 18.6 keV (with sufficient sigmas), this means that decays exist in which the energy of the neutrino is zero, which means that it cannot have any rest mass. Early experiments have bounded the possible mass of the neutrino to less than 1.1 eV/c^2. More measurements are of course necessary—and so is a future omega tau episode about KATRIN.

CP VIOLATION Another question left unanswered by the standard model is why there is matter at all. We've learned from the accelerator experiments that we usually get both matter and antimatter during high-energy processes in which new matter is created or decays. So during the Big Bang, which was such a high-energy process, equal amounts of matter and antimatter should have been produced. Each particle would then have annihilated with its antiparticle, resulting in a space filled with radiation but empty of matter. If this is what actually happened we all wouldn't be here to investigate the question. Obviously something ensured that some matter survived, but we don't know why this happened. In particular, the standard model cannot explain the amount of matter—and therefore the degree of matter/antimatter imbalance—that we see in our universe today.

Let's go back to the notion of symmetries discussed before. There are more symmetries in the standard model: the C symmetry expresses the fact that a universe in which every particle is replaced with its antiparticle would have identical physical properties. The P symmetry predicts the same universe if everything were mirrored along the three axes of space. Both symmetries together will ensure that laws of physics apply equally to matter and antimatter. The fact that this symmetry did not hold during the Big Bang, or very shortly thereafter when the universe coalesced, is known as the "CP violation".

The CP violation is an example of "symmetry breaking". This term refers to the idea that a symmetry does not apply in a particular instant, although it does generally. Symmetry breaking is generally caused by a small fluctuation that somehow acts on a system, driving the system from a balanced, symmetric state into an unbalanced, asymmetric state. Symmetry breaking is thought to be responsible for the emergence of many of the fundamental properties of our universe, such as the process by which the Higgs field determines the mass of the W and Z bosons.

To understand these things theoretically we'd have to go into topology, an area of math that deals with how mathematical objects retain some of their properties under deformation. This is beyond my understanding of math and physics. However, there is an experimental angle to this discussion: the particle physics community has observed the CP violation experimentally. As early as 1964 Cronin and Fitch noticed it in the decay of K-mesons. Neutral K-mesons can transform into their own antiparticle, and vice versa, but this transformation does not happen with the same probability in both directions—a clear example of CP symmetry breaking. Cronin and Fitch won the Nobel Prize for their result in 1980. However, the difference measured in the K-meson is too small to explain the abundance of matter we see in today's universe. Since then the community has been searching for signs of CP violation in other mesons, with experiments such as BaBar at SLAC in the US and Belle at KEK in Japan. In 2001 they found CP violations in the decay of B-mesons. Another experiment that looks for the CP violation is the LHCb experiment at CERN. Let's use this as a segue to looking at the other experiments at the LHC, beyond ATLAS and CMS.

8 | *The other main experiments*

ATLAS and CMS are the biggest experiments installed at the LHC; they are visually the most impressive and have gotten lots of publicity through their discovery of the Higgs boson. They are also general-purpose, allowing investigation of a wide range of physics processes. However, the LHC has seven (and soon eight) experiments in total, so in this section we'll look at the others. To keep the book at a finite size I'll make the descriptions short. We'll start with LHCb, because it connects nicely to the CP violation we've just discussed.

LHCB You may remember from our initial discussion of the standard model that quarks are characterized by an extra quantum number, their flavor. Six flavors have been identified: up, down, top, charm, strange, top and bottom. The "b" in LHCb stands for "beauty", an alternative name for the bottom quark. Quark names are curious: the top quark had once been named "truth", but it seems the physics community decided that truth and beauty were a bit too extravagant and went with bottom and top. They did keep charm and strange, though. And I wonder why they haven't used actual flavor-related names like sweet or sour.

Anyway, LHCb was designed to investigate particles that contain the b quark and characterize the asymmetry in the decay rates related to CP violation. As we'll see, LHCb's name is overly specific: it has also successfully investigated the charm system, mesons that contain the c quark.

The measurement of CP-violation-induced asymmetry is based on observing the decay rates of mesons. Mesons are particles that contain a quark and an antiquark; B-mesons are those quark–antiquark particles that contain a b quark; anti-B-mesons are those with an anti-b-quark. They decay differently, so to detect a CP violation the respective rates must be measured precisely:

$$B^0 \rightarrow K^+ + \pi^-$$
$$\bar{B}^0 \rightarrow K^- + \pi^+$$

In terms of the detector, the job of LHCb is not fundamentally different than that of the other experiments: identifying (decay) particles in a collision and analyzing them quantitatively. LHCb also uses some of the same subdetector technologies we discussed in the context of ATLAS and CMS: silicon detectors, drift chambers and calorimeters, in combination with a magnet that bends the path of particles to determine their momentum. However, it also employs detectors based on Cherenkov radiation. This is created when a charged particle moves through a medium at speeds faster than the speed of light in the medium (emphasizing "in the medium"—nothing here moves faster than light in vacuum). The medium used at LHCb is gaseous perfluorobutane and aerogel, a porous solid made from a gel by replacing the liquid component with air. The atoms along the track get polarized, which makes them radiate.

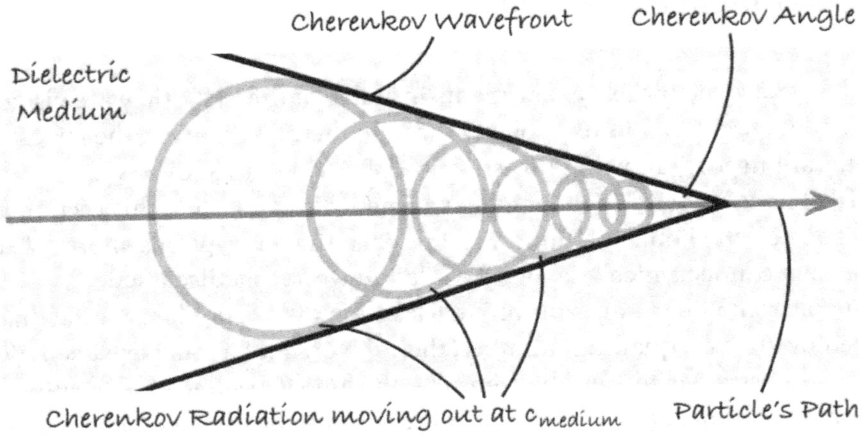

Because the speed at which the excitation produced by the particle is higher than the speed at which the radiation moves away from the polarized atom, something similar to a sonic boom happens: a wavefront of light (rather than sound) forms. The angle of this wavefront is proportional to the speed of the particle. Mirrors reflect the wavefront onto photodetectors, and electronics and software is used to reconstruct the Cherenkov angle and thus determine the particle's speed. This kind of detector is called a RICH detector, which stands for Ring-Imaging Cherenkov detector. Note that this detector measures the speed of a particle directly, and not indirectly via its energy or momentum.

LHCb's overall geometry is also different. Because B meson decays are relatively rare and LHCb has to capture as many of them as possible, and because B mesons fly away mostly in the forward direction at very shallow angles from the beamline, LHCb is a single-arm forward spectrometer:

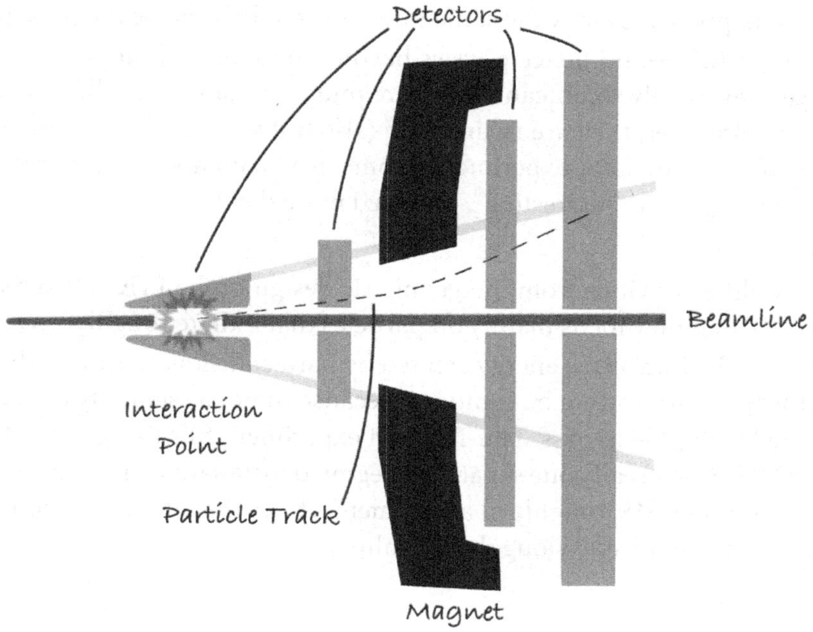

It is also one of the smaller experiments, at least compared to those we've seen so far: it is 21 × 13 × 10 meters in size and weights 5,500 tons. "Only" around 1,200 people are part of the collaboration that runs this experiment and analyzes the data.

In May 2019 the LHCb collaboration announced the first measurement of the CP violation in the charm system, based on D mesons. They found a difference of 0.154 ± 0.029%. LHCb needed all the data acquired between 2011 to 2018 to achieve 5.3σ, which demonstrates how rare D-mesons are and how hard it is to capture these events. These CP violations measured in the charm system—plus those detected earlier—are fully in line with the predictions of the standard model, and therefore do not give new insight into the matter/antimatter imbalance observed in our universe. This is probably why the CERN seminar where the discovery was announced had lots of empty seats; it wasn't quite comparable with the one where they announced the Higgs.

As I write this in January 2020, a press release announces that LHCb has made some interesting observations regarding lepton universality. Lepton universality is a prediction of the standard model that says that all charged leptons—electron, muon and tau—interact with gauge bosons in the same way. As a consequence, the three types of leptons should be created at equal ratios during decays of other particles, once you take into account their different masses and the different energies required for their creation. LHCb has investigated decays of a particular beauty-baryon at collision energies of 7, 8 and 13 GeV, and found that its two decays seem not to happen at exactly the same rate. This might indicate the presence of a previously unknown particle—and hence physics beyond the standard model. At only 1σ the results are not yet statistically significant and more mundane causes for the phenomenon cannot be ruled out. However, they are in line with (also not yet statistically sufficient) "anomalies" seen by LHCb and in other experiments, some even a bit above 1σ. It's really tantalizing to imagine something more "interesting". We need more data!

TOTEM It should be obvious from the geometric design of the LHC's two main detectors, CMS and ATLAS, that they focus mainly on particles that radiate roughly perpendicularly to the beam—that is, the transverse energy created by hard collisions. Particles that travel more or less along the beamline cannot be captured, because there's essentially a hole in the detectors to allow the beam pipe to pass. The TOTEM experiment fills this gap, both literally and figuratively. TOTEM isn't really one separate integrated installation; instead it consists of subdetectors at the ends of CMS, roughly 10 and 13 meters from the interaction point, as well as at 147 and 220 meters to either side along the beamline.

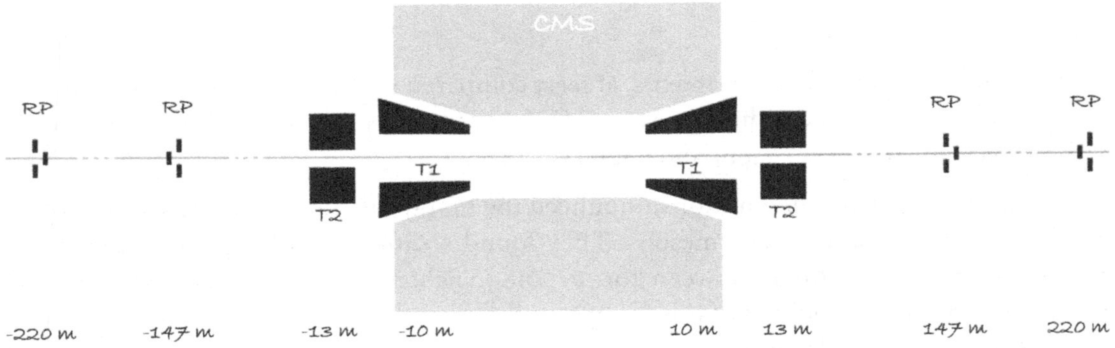

To understand why this is necessary, let's recap some basics. Earlier in our discussion about the physics of collisions we distinguished several cases of how protons might collide. The most interesting for the Higgs search is the non-diffractive, head-on collision in which the protons are destroyed and one or more of the their partons collide, producing a lot of transverse energy. In the least violent case, the fully elastic collision, the two protons both survive and continue at shallow angles down the beam pipe. They don't really collide, just slightly alter their direction through electromagnetic or strong force interaction. Such protons can be detected by TOTEM's subdetectors far from the interaction point, at ranges of 147 and 220 meters. We also mentioned the intermediate case, the single-diffractive collision, in which one of the protons survives while the other is "disassembled". In this case, because the particles resulting from the destroyed proton escape in various directions, the surviving proton's velocity vector might change direction more than in the previous type of collision. These particles are recorded by TOTEM's subdetectors closer to CMS. The following illustration shows a single-diffractive collision in which one proton survives and is recorded by the distant TOTEM subdetectors and the breakup products of the others are recorded by the closer detectors.

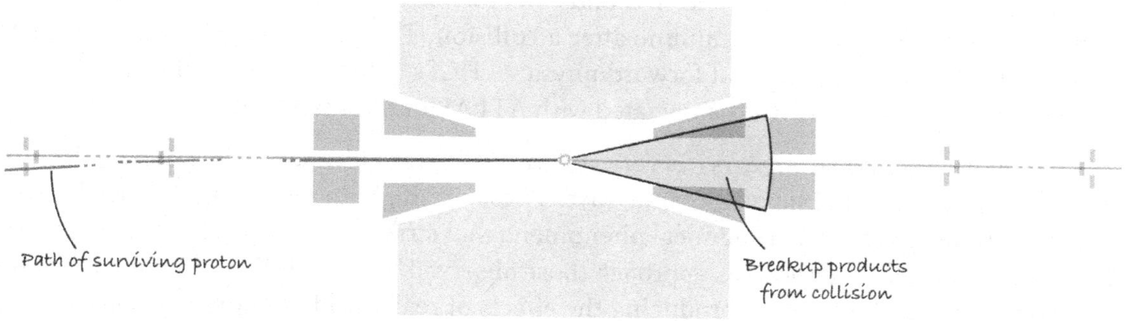

In addition to understanding the physics of elastic and single-diffractive collisions themselves, TOTEM also sheds light onto the internal structure of protons and helps to build an understanding of the beam's luminosity and the structure of its bunches.

Let's conclude this discussion of TOTEM with a brief look at its detector technologies. The detectors at 10 and 13 meters, T1 and T2, are referred to as "particle telescopes". I wasn't able to figure out why they are called "telescopes"; it's probably due to some analogy between how (earlier?) detectors of this type worked and astronomical telescopes that has gotten stuck in the community.

T1 and T2 are gas electron multipliers. As charged protons move through the gas in the detector volume they ionize its atoms; the resulting free electrons are pulled towards an electrode using an electric field. The electrodes are made from Kapton foil perforated with little holes through which the electrons can slip, producing more ionization on the far side. Stacks of these electrodes eventually create detectable electron avalanches. Gas electron multipliers are very sensitive, because the avalanching acts as an amplifier over several orders of magnitude. They also have good angular resolution, are fast, and also resistant to radiation, as is necessary this close to the interaction point.

The subdetectors at 147 and 220 meters are "Roman pots", so-called because they were first used by a group of physicists in the city of Rome; "pot" is an abbreviation of "proton on target". Their goal is to record those protons that are closest to the beam's centerline. Moving them away from the interaction point is one measure to achieve this goal (because the protons have more time to diverge from the beamline), but the design of the detectors also brings the detecting elements very close to the beam. They a located in a vacuum vessel that surrounds the beam pipe, and are moved to a distance of less than 1 mm from the beam once it has stabilized. The actual detection elements are plastic fiber scintillators.

LHCF After TOTEM's total weight of around 400 tons, we continue our journey towards smaller and smaller experiments: LHCf has two subdetectors, each 40 kg in weight and only 80 × 30 × 10 centimeters in size. The "f" stands for "forward". Forward physics looks at those particles that travel along the beamline after a collision. The previously discussed TOTEM experiment is another example of forward physics. LHCf's detectors are installed at a similar distance, 140 meters, but LHCf is associated with ATLAS rather than CMS.

The name LHCf is a bit strange, as it is very generic—"forward physics"—even though its purpose is very specific: it simulates cosmic rays. These are high-energy charged particles that originate from a range of astronomical phenomena that strike the Earth's upper atmosphere. The energies available at the LHC approach those observed "in the wild", and similar cascades of particles are produced. By reproducing the effects of real-world phenomena, insights into the underlying mechanisms might be gained. As a corollary, LHCf is routinely used to calibrate detectors that investigate actual cosmic rays.

In terms of subdetectors, the experiment consists two imaging shower calorimeters, one based on fiber scintillators, the other using silicon strip sensors. A recent result based on spectra measured with LHCf showed that there is a strong energy peak in the very forward region. LHCf scientists conclude that the forward neutrons resulting from cosmic ray impacts carry a large fraction of the total energy, leading to deep penetrating showers of very high energy cosmic rays.

MOEDAL We know from everyday physics that magnets have two poles, north and south. These always show up in pairs—if you divide a magnet, the two resulting halves will each have a north and a south pole. You could say that their total "magnetic charge" is zero, because north and south combine to be magnetically neutral. This is different from the electric charge where, as we know, it is possible to isolate the positive or negative charge, for example as electrons or positrons. A magnetic monopole would be a particle for which north and south are not paired: the resulting particle would be "magnetically charged". Particles with both an electric and a magnetic charge might also exist; these are dubbed "dyons".

It turns out that several theories that try to unify gravitation and quantum physics predict the existence of such a particle; finding one would be an indication of physics beyond the standard model. In a 1931 paper, Paul Dirac showed that if magnetic monopoles exist, then all electric charge must be quantized. In this sense finding a magnetic monopole would explain the quantization of electric charge.

The Monopole and Exotics Detector at the LHC, shortened to MoEDAL, is built to search for just such magnetic monopoles. MoEDAL also searches for highly ionizing particles as well as massive, long-lived and slow-moving particles. MoEDAL mainly uses nuclear track detectors, which employ plates made from special plastics. When a charged particle passes through the detector it damages the polymeric bonds of the plastic in a cylindrical region around its track. The materials are chosen such that the detector is insensitive to standard model particles, as magnetic monopoles are predicted to be very massive and have about 5,000 times more ionizing power than protons. For analysis the plate surfaces are etched; because of damage to the bonds, etching is faster around particle tracks, producing "etch-pit cones". Their shape and depth is characteristic of a particle's charge and momentum. Automated scanning microscopes analyze the plates after etching to classify these cones.

MoEDAL also has monopole trap subdetectors. Because the hypothesized monopoles would be so heavy, they would be slowed down and ultimately stopped by 800 kg of solid aluminum. Scanning the aluminum with a magnetometer could reveal any trapped monopoles.

Both these detector technologies are completely passive and probably don't reveal a lot of detail and statistics about the observed particles. Analysis is slow and offline, and the plates have to be replaced regularly, but they are relatively cheap and don't need active control; the plates are installed "around" the interaction point of the LHCb experiment. This approach is completely appropriate for the more speculative science of magnetic monopoles. However—as in most other LHC experiments—a superlative is involved: at around 100 square meters, MoEDAL is the largest ever deployment of nuclear track detectors.

As of 2020 no evidence of monopoles or dyons has been found, or we all would have heard of it in the news. However, as is often the case with "no result" experiments, this does demonstrate that such monopoles do not exist at the masses and energies accessible to the LHC; it establishes boundaries. And considering how cheap the detectors are, you can't affording not to look if you have a machine like the LHC at your disposal.

ALICE In our journey through the world of LHC experiments we've moved from the large and general-purpose ATLAS and CMS detectors towards progressively smaller and more specialized experiments. However, there's one more detector that is comparable in size to the two big ones, and I started to describe it at the very beginning of this chapter: ALICE. So to round up our tour of the LHC, let's revisit ALICE.

The purpose of ALICE is not really particle physics: like LHCf, it has a more cosmological focus. ALICE is designed to understand the quark–gluon plasma, the state of matter moments after the Big Bang. It does this by observing the particles that escape from this plasma using the same approach as ATLAS and CMS, but making different tradeoffs, as we shall see below. But let's start with the science.

Our current best guess at how the universe started is the Big Bang, in which "everything was created from nothing". It's not really clear how this event came about—hypotheses range from random fluctuations in spacetime to the involvement of some kind of Deity, but science does know a little about some of the properties and processes involved in the Big Bang. In particular, matter was extremely dense right after the Big Bang singularity, and consequently very hot. Temperatures and pressures were so high that quarks and gluons separated into what is known as the quark–gluon plasma. Remember that in today's environment, quarks are confined inside hadrons—protons, neutrons, mesons—held together by the strong nuclear force, whose carrier particles are the gluons. The quark–gluon plasma shares some properties with "normal" plasmas in which atoms separate into a soup of free protons and electrons, which is why the same term is used.

The energy released by proton–proton collisions is not high enough to form a quark–gluon plasma reliably. This is why the LHC also collides lead ions; when they collide the energy released is a factor of around 100 higher. The temperature then rises to trillions of Kelvin at a density of 30 times the nuclear density and the lead nuclei melt into a tiny droplet of plasma, but after only 10–23 seconds start transitioning back into regular matter. The particles that form during this process are what ALICE investigates in order to understand the plasma itself. As Despina explains, "Of course it is not the quark–gluon plasma that we observe; basically we observe the particles that are produced when the phase transition happens. Eventually, as it expands, the quark–gluon plasma thermalizes, so we get real particles, and that is what we see."

Here are some of the results of ALICE's measurements that Despina told me about. It turns out that the quark–gluon plasma behaves as a perfect fluid. This was a surprise, as the scientists expected it to be more like a gaseous plasma with no correlated motion of its constituents. ALICE also measured the initial temperature of the plasma, which indeed turned out to be trillions of Kelvin. This was the highest temperature ever measured on Earth, let alone produced. Fusion reactors "only" reach 100 million Kelvin, and our Sun is a paltry 15 million at its core. The temperature was derived from the spectrum of the emitted photons. Another finding relates to a particle called J/ψ, or psion, which consists of a charm and an anticharm quark. It turns out that the rate at which this particle is produced is lower in lead–lead collisions than in proton–proton collisions: it's believed that it dissolves due to the existence of free color charges. They also found that this suppression is less than at the RHIC, which collides gold nuclei at lower energies. This reduced suppression is attributed to recombination: at higher energies there are more charm and anticharm quarks, so there is more material for recombination. Ultimately, the degree to which J/ψ is suppressed acts as a thermometer.

Let's briefly look at the beam in the LHC when it collides lead ions. The most obvious question is, where does the 100 times higher energy come from? Is it because the particular lead isotope used at LHC has an atomic weight of 208 (instead of 1 for protons)? Not really: the total energy imparted into circulating particles is limited by the magnetic field produced by the LHC's magnets, so a heavier particle would presumably just go more slowly but with the same total energy as lighter particles.

What else could be the reason? Let's think about what determines the interaction between the accelerated particles and the accelerator: it's all down to electromagnetic interactions. If you go back to the respective formulas discussed earlier, you will see that the charge of the accelerated particle, q, shows up frequently, both in how the particles are accelerated by an electric field and how strongly they are bent around the ring through the magnetic field. The lead atoms are fully ionized: all of their 82 electrons are stripped away. This gives the resulting lead ions 82 times the charge of a proton. This is why the LHC can accelerate the particles 82 times more effectively, and also bend the particles' path 82 times more strongly—or bend a 82 times more energetic particle to the same 27 km radius. Note that the energy per nucleon is lower, because the beam also has to drag the neutrons around; these are not affected by either acceleration or bending.

To deal with these heavier ions, the control laws used to accelerate and maintain the beam have to be adapted slightly. The front-end of the accelerator chain is also different. Instead of starting with hydrogen gas, the process starts—unsurprisingly—with a lump of lead. This is heated to around 500°C, at which temperature some of it vaporizes. Electric fields are then used to strip the electrons from the lead atoms, and the resulting ions are accelerated to 4.5 MeV by LINAC 3 (instead of LINAC 2 for protons). The next stage is LEIR, the Low Energy Ion Ring, which raises their energy to 72 MeV. LEIR also creates the bunches, in the case of lead only 592 per beam instead of the 2,808 for protons. The next stage is the Proton Synchrotron; from this stage onward their acceleration is the same as for protons.

When the lead ions collide they also create a lot more "stuff"—each has 82 protons and 126 neutrons. Several orders of magnitude more collision products are produced compared to proton–proton collisions, in the order of thousands in head-on collisions. As lead ions are so much larger than protons, there's also more opportunity for partially overlapping collisions, and the trigger's ability to detect those that are head-on is especially critical. Triggering is done by plastic scintillation detectors installed very close to the interaction point that measure the total (transverse) charge flux; the higher it is, the more head-on the collision. There is also the Zero Degree Calorimeter, which detects hadrons that were not destroyed in the collision—the more of those, the more glancing was the collision.

At first glance ALICE looks very much like CMS and ATLAS: a barrel with successive layers of subdetectors. All of it encapsulated by a magnet of 0.5 Tesla, not superconducting in this case. ALICE also has LHCb-style forward detectors with a separate 0.7 Tesla magnet.

ALICE is optimized for particle identification, which means that it has to be able to measure the charge as well as the mass of a wide range of particle types. We'll discuss how this is done below. As a tradeoff, ALICE is unable to establish a complete energy and momentum balance—in other words, ALICE doesn't capture all particles that stream out of a collision. This is because it doesn't have detectors around its full volume, and particles can escape undetected. Since new kinds of particles are often inferred from unbalanced momentum or energy,

ALICE is not suited to find such particles. There are multiple reasons for its incomplete coverage, primarily available volume: ALICE is an evolution of the L3 experiment of the LHC's predecessor, the LEP, reusing its main magnet as well as its cavern. A second reason is that ALICE's scientific goals are different: the LHC already has two other experiments that can establish total momentum and energy balance. Finally, cost is always an issue, and the installation of subdetectors is staged, with some only being added in the future.

The incomplete 4π coverage also explains why ALICE did not really take part in the Higgs search. As we know, the primary way of detecting the Higgs boson is through its decay to muons and photons. However, ALICE only has coverage for muons at a small angle in the forward region. Electromagnetic calorimeters, required to detect photons, are also only available at some angles. ALICE does not have hadronic calorimeters either, so some of the neutral hadrons go undetected. In addition, some of the detectors are relatively slow (although quite precise). Overall, ALICE takes data at only 1% of the rate of CMS and ATLAS during proton–proton physics runs: it would have taken 100 times as long to get to the required 5σ.

Let's look briefly at the various subdetectors installed in ALICE. At the center of its barrel we find the silicon-based inner tracking system: two layers of silicon pixel detectors, two layers of silicon drift detectors and two layers of silicon strip detectors. Just as with ATLAS and CMS, the micrometer precision of these detectors is necessary to establish the coordinates of an interaction and thus define the origin of the tracks measured further out.

Next is the Time Projection Chamber, ALICE's main detector. This extends to a radius of 2.5 meters and it is five meters long. Its volume is filled with a mixture of neon and CO_2; as particles pass through they ionize the gas, and the freed electrons move towards the end plates, driven by an electric field. Several things can be deduced about an ionizing particle. First, the strength of the ionization is a direct measure of the speed of the particle; the relationship is described by the Bethe formula, which correlates the loss of energy of a particle due to Coulomb collisions with the electrons of the medium's atoms. The ionization charge is measured by counters at the end caps; several layers of wires produce a proportionally sized avalanche to amplify the signal. We can deduce the particle identity from dE/dx, the energy deposited per unit length from ionization: heavier particles ionize more than lighter particles with the same momentum. We can also learn the particle's position along the beam axis from the time it takes the electrons to drift to the end-caps. Because the counters are discretized—there are 557,568 separate pads—we can deduce the particle's x,y-position perpendicular to the beam. Time-discretized sampling produces many "points" along a particle's track. The time projection chamber is less precise than the silicon detectors, but it fills a larger volume. This in turn is useful, because it gives the particles more flight time and so more opportunity to be bent by the magnetic field; as usual, the amount of bending is a measure of momentum.

On our way out radially the Transition Radiation Detector is next, from 2.5 to around 3.5 meters; we've discussed its principle of operation previously so we'll skip it here. It is used to identify electrons and produces up to 100 points for each electron track.

Next, at 3.7 meters radius, is the Time of Flight detector. This measures the velocity of particles by timing them over the distance from the interaction point to the 3.7 meter radius; a T0 detector close to the beam pipe measures the start time for the particle's flight. Together with the momentum and track information of other subdetectors in ALICE, this helps to identify the particle. It covers 150 m^2 with 160,000 separate multi-gap resistive plate chambers. Resistive plate chambers consist of a gas gap in a strong electric field between two electrodes. The gas is ionized when particles move through the field, causing an avalanche of electrons that can be detected. The chambers in the ALICE Time Of Flight detector divide the overall gap into multiple smaller ones, each separated by glass plates spaced by fishing line, to achieve controlled and uniform spacing. This multi-gap variant of resistive plate chambers was developed at CERN and has been shown to provide relatively good timing and efficiency characteristics. It is also very cheap—glass, fishing line—so covering a large area was feasible.

The highest momentum particles cannot be identified by the time of flight detectors, as their time resolution isn't good enough. For those, HMPID—High Momentum Particle Identification—detectors based on Cherenkov radiation are used; we discussed how RICH detectors work in the context of LHCb.

ALICE also has a range of electromagnetic calorimeters at various locations. These include: lead-scintillating tile calorimeters; lead tungstate calorimeters, referred to as PHOS, short for photon spectrometer; a photon multiplicity detector that measures photons in the forward region of the detector; the forward multiplicity detector that does the same for charged particles. ALICE also has a muon spectrometer, which is located outside the magnet volume in the forward region (muons interact so little that they can pass through the 1-meter-thick steel doors). The latter allows study of heavy quarkonia such as J/ψ and others because they often decay into muons; we mentioned earlier that these particles are crucial for understanding the quark–gluon plasma.

As we mentioned, ALICE cannot establish a total momentum and energy balance, and thus potentially find new particles. At this point you may have started to realize what ALICE has going for it instead: ALICE has the widest range of detector technologies, to be able to distinguish the widest range of particles.

COLD WAR STORY, CONTINUED After we arrived in the cavern 50 meters underground via the elevator that was protected by the iris-scanning sensor, we started our tour of ALICE. Up several sets of stairs to reach a walkway that passes along the side of the detector. With a length of 25 meters and cross-section of 16 × 16 meters, ALICE is the second largest of the LHC experiments after ATLAS. At 10,000 tons it is also second in terms of weight after CMS. I see boxes and surfaces of various colors and many, many cables. Some of the subdetectors have been removed for maintenance and upgrades, so the iron skeleton of the detector is visible. The heavy steel magnet doors are open, so we can see inside.

In this book we've come to know many marvels of engineering: a flying telescope ticks all my boxes of course, and the Milky Way reflected in the giant mirrors of the LBT will forever be etched into my mind. But the sheer scale, complexity and precision of these particle detectors is something else altogether. I find it encouraging that, despite all the political bullshit, increasing competitiveness and animosities in today's world, a significant slice of the world's countries—41 in the case of ALICE—join forces to envision, finance, build and operate facilities like the LHC and its experiments. It does give me a little bit of hope for the planet. So to conclude this chapter and the book, let's take a brief look what humanity might do next in the field of particle accelerators.

9 | *LHC Upgrades*

Like all large science experiments, an accelerator is never "finished" until it is decommissioned, either because it is replaced by its successor, such as when the LEP was replaced by the LHC, or because it becomes irrelevant because a competitor has so much more energy that no interesting science is left for it, as was the case for the Tevatron when the LHC came online. In this section we'll look at the LHC's upgrades, and peek at its possible replacements in the next section.

Every year the LHC is shut down for a few months between runs, known as "technical stops", which are mainly for maintenance. Every few years it takes more extensive breaks called "long shutdowns"; there is some maintenance as well, but the long shutdowns are also used for (sometimes quite major) upgrades. For these the machine has to be completely warmed up, then cooled again afterwards—we saw before how long this takes. Let's look at some of these upgrades, both those performed during past long shutdowns, and those planned for the future.

LONG SHUTDOWN 1 After its first three years of operation the LHC shut down in early 2013 for around two years. During LS1, as it is known, the machine was readied for collision energies of up to 13 TeV—the accelerator had been previously running below its full energy potential. The main improvements to make this safe were the connections between the magnets, those that failed in the accident of 2008. Work was also carried out on various pre-accelerators and on the ventilation system. Around 100 km of cables that had been worn out by the radiation were replaced. Training of the magnets for up to 13 TeV was also performed during this shutdown so that it would be ready for "real science" when restarted in 2015.

The experiments were also upgraded. For example, ALICE's PHOS detector was installed during LS1. The installation of the transition radiation tracker had always been planned in phases, and some additional modules were installed in this timeframe, bringing them to the current number of 18.

LONG SHUTDOWN 2 LS2 is ongoing as I write this, and was also going on when I visited ALICE in May of 2019. LS2 started after data taking in December 2018 and is scheduled to end in 2021. One goal is to increase the collision energy from the current 13 TeV to the full 14 TeV. The main goal, however, is an increase in the luminosity of the beam, and thus an increase in the number of collisions.

The LHC works differently than, for example, telescopes, where every year scientists submit proposals for particular experiments or observations. Instead, the accelerator runs for as long as possible, delivering as much integrated luminosity as it can, the experiments collecting the largest possible amount of data for later analysis. There is a huge incentive to improve the efficiency of data collection, for example by making the detectors more sensitive, more precise, or by extracting more useful collisions from the beam, for example by increasing luminosity. More and better data means you'll get to the required statistical confidence levels faster.

Two measures help to increase instantaneous luminosity: making the beam more intense and focusing it better at the point of collision, both of which are being done during the current shutdown. Let's look at the intensity first. Here the main improvement comes from replacing LINAC 2 with LINAC 4. LINAC 2 is a 30-meter-long linear accelerator that accelerates the particles to 50 MeV, using hollow-cylindrical electrodes that switch their charge quickly to "push" the protons forward. In contrast, LINAC 4 is 86 meters long and accelerates the beam to 160 MeV using four different staged acceleration techniques, some of them invented long after LINAC 2 was built 30 years ago. LINAC 4 also contains a chopper which cuts the continuous beam at the frequency required by the Booster, the next stage into which LINAC 4 injects. LINAC 4 accelerates negatively charged hydrogen ions; their two electrons are stripped only at the end to produce protons. Accelerating H^- ions instead of protons reduces various losses during acceleration. Because the beam now contains more particles, the collimators and machine protection subsystems are also being upgraded.

To increase the beam's focus at the collision point, new focusing magnets will be introduced; these are collectively known as the "achromatic telescope squeezing scheme". Without a detailed understanding of the magnet design and beam optics, which I don't have, I won't go into much detail; however, it will increase the focus significantly without requiring the replacement of lots of existing beam magnets.

Increasing luminosity does not increase the energy of the beam or the collisions. So you won't see new physical processes. Or will you? In some sense you will, because rarer phenomena might happen often enough to accumulate sufficient sigmas during the runs. This effectively brings new physics into scope.

Like the other long shutdowns, this one is also used to improve the experiments. All the subdetectors will have their readout electronics upgraded, because with higher luminosity the rate of (interesting) events will increase and disentangling pile-ups will become more important. Many of the subdetectors will also have some of their components enhanced. For example, most of LHCb's subdetectors will be replaced with silicon detectors, to increase their speed. This will improve LHCb's data capture rate by a factor of 10, and will allow it to measure processes in the charm system that might reveal physics beyond the standard model.

For ALICE this second shutdown will also bring a lot of enhancements, which is why lots of components were missing when I visited in May. Its inner tracking system will be completely replaced with seven layers of pixel detectors instead of the current 2 + 2 + 2 layers of pixel, drift and strip detectors. This will increase tracking precision, which in turn helps to distinguish a higher number of tracks. It will also increase the number of readout channels to billions, so upgrades of the electronics are required as well. The new sensors are also thinner and can be mounted closer to the interaction point, at a distance of 2.4 cm. This will allow secondary vertices closer to the interaction point to be resolved, which means that ALICE will be able to observe particles with shorter lifetimes. These are usually heavier, containing heavier quarks. In that sense also, previously unseen phenomena might come into view. ALICE's time projection chamber will have its readout electronics upgraded to be able to deal with a data rate of 50 kHz instead of 1 kHz, and the wire chambers will be replaced with gas electron multipliers, as these are much faster. The muon system will be enhanced, as will the trigger systems. While they're at it (and to continue this book's tradition of making fun of names), I suggest they also upgrade its name to something more fancy than "A Large Ion Collider Experiment"—I'm sure it's possible to "reverse-cronym" ALICE to something more clever.

ATLAS will also get a bunch of new subdetectors. For example, a part of the muon system and trigger called the "small wheel" will be upgraded to new small-strip thin gap chambers and MicroMegas, both detectors that rely on the ionization of gas. Several small monitored drift tubes and resistive plate chambers are being installed, and the liquid argon calorimeter will be improved with new electronics and optical fiber cabling, which allows better resolution of jets. In addition to generally supporting a higher rate of collisions and finer resolution, the upgrades will also improve radiation hardening: the particles created by the collisions are not just the subject of science, they also attack, slowly degrade and ultimately destroy the detectors and their support equipment if they are not properly hardened.

There will also be a completely new experiment called FASER, short for Forward Search Experiment. Like LHCf and TOTEM, it will look at those particles that stream away from a collision at very shallow angles. FASER will be installed in an unused tunnel that was once used to insert beam from the SPS to the LEP, 480 meters downstream from ATLAS. Its goal is to detect the new extremely light particles that are predicted by various theories that try to explain dark matter and the CP violation. While protons are bent around the ring, such particles would not interact electromagnetically and would thus continue straight on—through rock, concrete and equipment—until they decay into particles that can be detected by FASER. The rock between ATLAS and FASER also provides a shield from other collision particles.

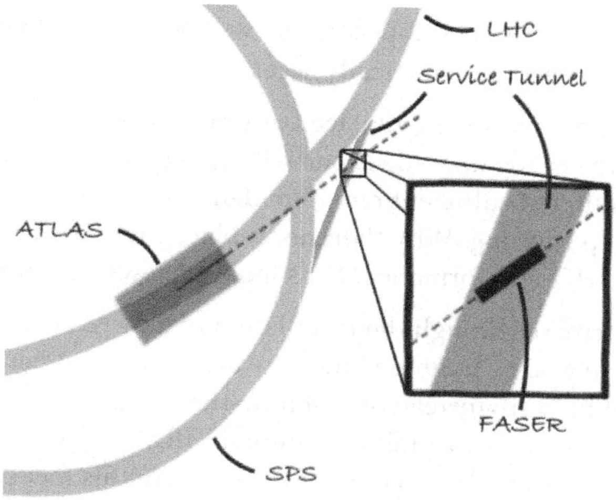

The main FASER detector is a 5-meter-long cylinder. It is only 20 cm in diameter, which is enough because the stream of particles is very well collimated from the collision. The detector technologies are basically those we've already seen in ATLAS and LHCb—plastic scintillators, silicon trackers, electromagnetic calorimeters—separated by decay volumes. In fact, FASER uses spare detector components from the other two experiments to keep the cost down.

Very recently an additional detector component has been approved, known as FASERν, where ν is of course the symbol for neutrinos. ATLAS collisions produce a lot of neutrinos, but they have never been caught by the detectors (other than indirectly based on the balance of momentum). However, at between 600 GeV and 1 TeV collision neutrinos are more energetic than those from the Sun or created by interactions of cosmic rays with the atmosphere, so it is interesting to understand their properties better. Neutrinos are mostly produced along the beamline (another reason why ATLAS and CMS don't see them), so FASER's location is suitable. However, FASER's main detectors are not dense enough to trigger neutrino interactions. This is why FASERν adds a 25 × 25 cm, 1.35 meter-long 1.2-ton tungsten target to the front of the 5-meter FASER detector that will see the e, μ and τ leptons created by the neutrino interactions with the target. The actual detection relies on 1,000 emulsion detectors (basically film strips analyzed offline) interleaved with tungsten plates. Of the billions and trillions of neutrinos produced, this set-up is expected to detect 1,300 electron neutrinos, 20,000 muon neutrinos and 20 tau neutrinos over its lifetime.

LONG SHUTDOWN 3 After two or three years of science runs there will be another shutdown in 2025–2027 for further upgrades, again increasing luminosity. After this LS3, from mid 2027 onwards, the LHC will run as the High-Luminosity LHC. The upgrades of LS2 and LS3 will together increase luminosity to ten times that of the original machine. This will lead to three to four times more collisions per crossing and, for example, to a correspondingly higher production rate of Higgs bosons. The collision rate will be high enough for even rarer Higgs decays, such as H → cc, to become statistically accessible.

Many of the LHC's components will be replaced during LS3. This is not just to improve their performance; the timeline of the upgrade is partially governed by the fact that by 2030 many components will have suffered sufficient radiation damage to require their replacement anyway if the LHC is to keep running. Why then not replace them with state-of-the art components that increase the LHC's performance? Let's look at a few examples.

Focusing will further increase through the installation of new 12 Tesla quadrupoles; a few of the bending magnets will also be upgraded in this way. To achieve the higher field, the magnets are made from niobium–tin instead of niobium–titanium. These require improved cooling and power infrastructures, so an additional tunnel will be dug in parallel to the main tunnel in the areas of the ATLAS and CMS experiments, where this focusing is required. Shafts to the surface and connections to the main tunnel will be built as well.

Another upgrade will be the addition of "crab cavities". Remember that the particle bunches are elongated; each is around 15 cm long. When the beams are bent for collision, these "long streaks of particles" will cross each other. But because they are elongated, the overlap of the bunches will always be only partial, limiting the opportunity for collisions:

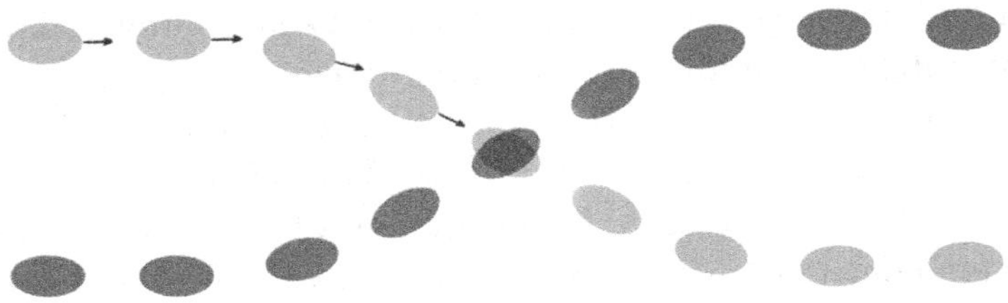

Crab cavities, a technology that has never been used before in an accelerator, will rotate the bunches immediately before the collision such that their velocity vector is no longer aligned with their long axis, resembling the sideways movement of crabs, hence the name. As you can see from the following illustration, the overlap of the bunches during a collision will then be total, increasing the probability of collisions.

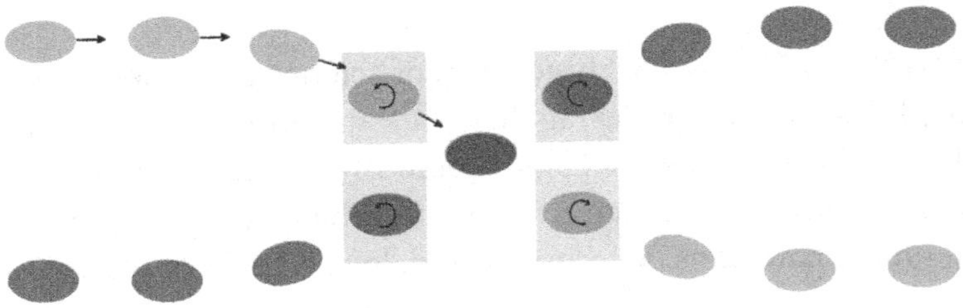

After the move from LINAC 2 to LINAC 4 in the previous shutdown, it will now be time for upgrades to the Booster, PS and SPS. Finally, as you would expect, experiments will be upgraded as well. ATLAS and CMS are scheduled to get a new silicon tracker, each with a billion (!) channels. The one for ATLAS will also replace the drift tubes with silicon detectors, and both will also upgrade their trigger to better cope with the higher collision rate.

BEYOND LS3 The LHC will then run well into the 2030s, with two more long shutdowns for maintenance and incremental updates. No significant increases in luminosity are planned; in some sense, the LHC hardware will slowly approach the end of its potential. To probe matter further, we will need to reach even higher energies with a more powerful accelerator. Which brings us to our last section: future accelerators.

10 | *Future Accelerators*

What design options exist for new colliders? There are several. You can build a linear or a circular collider. You could collide electrons, protons or heavy ions. What about antimatter? Do you want to maximize energy or luminosity? Can you do both?

Let's understand the fundamental tradeoffs, starting with the collided particles. We've seen from our discussion of ion collisions in ALICE that heavier particles contain more "stuff", so there's more variety in the collisions. Even in the case of proton–proton collisions we have seen that it's not the protons that collide, but their constituent quarks and gluons. Which of these then collide, and at which energies, depends on parton energy distribution, which is probabilistic. So each collision is different, resulting in a wide range of physics, which is why an experiment's trigger mechanism is so important for selecting the interesting ones—usually those in which the contained quarks or gluons collide head-on and transverse energy is high. If you collide point-like electrons or positrons there's none of this complexity: if the electrons actually hit each other (they are smaller, so that's less likely) you know exactly what energy they have. There is much less variability, and many fewer background particles are produced, making reconstruction and analysis much simpler. In particular it allows rare events to be distinguished, which hide in the background in a proton–proton collision.

So proton–proton colliders are discovery machines: using the trigger, you can vary what you want to see and explore new ground. You need lots and lots of collisions, because you care only about some of them, and it takes time to get to interesting confidence levels because of the statistics and the "loud" background. Lepton colliders, on the other hand, are precision measurement devices. If you know what you are looking for, you can tune the beam to a precise energy, and if two electrons collide you know exactly what you're going to get. For example, the LHC's predecessor, the Large Electron–Positron Collider, had been precisely tuned to produce Z bosons. I've heard a less charitable way of expressing the difference between the two: electron colliders are like scalpels, whereas proton–proton colliders are like smashing garbage cans and seeing what comes out.

Let's now look at linear versus circular accelerators. The key difference is that in the latter the beam circles many times, and each time the bunches can be accelerated by the same electric field. In the linear case a bunch travels through the accelerator only once, which means that you need way more accelerating cavities. Assuming realistic electric fields, a linear accelerator that would produce the same 7 TeV as the LHC would have to be around 300 km long, which doesn't seem very practical. Also, the particles in a bunch that do not actually collide are lost—they shoot out the far end of a linear accelerator—while in a circular accelerator they can just go for another trip around the ring and perhaps collide later. In this sense a circular accelerator is more efficient. On the other hand, a linear accelerator does not need bending magnets. So all the effort and cost that goes into those magnets can be avoided, as well as the complexity of controlling the beam as it circles. Circular accelerators have another disadvantage: because they bend the path of charged particles, the particles lose energy due to synchrotron radiation. This loss is inversely proportional to the fourth power of the mass of the particle, which is why heavier protons lose much less energy from this effect. This means that circular lepton colliders are limited in energy reach. A final advantage of linear colliders it that they are easier to upgrade in terms of energy: you just add additional accelerating modules at the end. In contrast, in a circular accelerator you either have to replace all the magnets with stronger ones (assuming those are available) or dig an all-new larger-radius tunnel. All in all, circular colliders seem to be the overall better option, which is also demonstrated by the fact that much more of those have been built over the recent decades.

What about antimatter? Let's stay with the LEP: it didn't collide two electron beams, but one beam of electrons and one beam of its antiparticle, positrons. Why? The reason is, that, because of the opposite charge of positrons, a single magnetic channel can bend an electron beam and a positron beam that travel in opposite directions. This simplifies engineering of the magnets and the cryo system. As a downside, intense antiparticle beams are harder to produce, although it is easier for positrons than for antiprotons. For example, the Tevatron, which collided protons and antiprotons, had a luminosity that was 100 times lower than that of the LHC. This is one way in which energy versus luminosity is a tradeoff. Ideally you want high energy, because this lets you produce heavy and rare particles, but also high luminosity, because relatively unlikely events will happen more often, leading to meaningful statistics in a shorter time.

"Hopefully physics will tell us which machine to build next." I don't remember which of my guests said this (any listeners here remember?), but I think it nicely captures the ideal case of how you determine which detector to build next: the one currently in operation detects hints of particular new physics. The new accelerator is then optimized to produce lots of these particles to get to five sigma quickly, or to measure the particles' properties precisely. Considering that our current best machine is a "garbage can collider" that has found the Higgs, the new generation should probably be a scalpel, to precisely investigate the Higgs' properties. On the other hand, one could argue that apart from the Higgs, as of 2019 the LHC has unfortunately not found any new physics, so we might want to continue crashing garbage cans, but at higher energies.

The community has been studying concrete proposals for an LHC successor since the 1990s; in the remainder of this chapter we'll look at the two that are currently most likely to be built, the Compact Linear Collider and the Future Circular Collider.

CLIC The Compact Linear Collider ("compact" in the sense of the CMS, I guess) is a lepton collider: it collides electrons and positrons. It is planned to be built in stages, exploiting the relative simple expandability of a linear collider: 11 km, 29 km and 50 km. The collision energies would increase correspondingly from 380 GeV, through 1.5 TeV, to 3 TeV.

You might wonder why 380 GeV is scientifically interesting when the LHC requires several TeV to be able to discover new physics such as the Higgs boson. The Tevatron saw indications of the Higgs at lower sigmas with 1 TeV, but so did the LEP with only 200 GeV. The reason why lepton colliders require less energy for comparable "effect" goes back to the scalpel versus garbage can analogy. As I explained before, in a hadron collider the total collision energy is probabilistically split between the collisions of the partons inside the hadron, because it is actually those particles that interact and each only gets part of the total energy. In contrast, a lepton collider injects all the energy into a single collision of point-like quantum-level particles. Ultimately you cannot compare hadron and lepton colliders by just comparing their energies; a rule of thumb is to divide the nominal energy of a hadron collider by ten to get to a comparable lepton collider energy. With this in mind, the CLIC's 3 TeV is very roughly a factor of two beyond the LHC.

CLIC's integrated luminosity will also significantly higher than that of the LHC. To increase instantaneous luminosity, a really thin beam is required. One way of achieving this is to apply synchrotron radiation damping, in which the beam is bent around a relatively tight ring and the resulting synchrotron radiation reduces the transverse momentum of the electrons. Since CLIC is linear, its design includes special damping rings for this purpose. Advanced focusing "optics" near the interaction point help as well.

Now that the Higgs has been discovered, one goal of CLIC would be to precisely understand its properties; 380 GeV is enough to produce lots of Higgs particles. Of particular interest is its internal structure and self-coupling, where one Higgs spontaneously and temporarily splits into two Higgs bosons—yes, the quantum world can temporarily suspend conservation of energy! While searches for pairs of Higgs bosons are going on at ATLAS and CMS, the CLIC "scalpel" will be more efficient. The 380 GeV version of CLIC will also produce top quarks, their production threshold is around 350 GeV. The higher energy stages will hopefully reveal new physics, as predicted by various theories. In particular, supersymmetric charginos, neutralinos and sleptons might come into view. We know all this because even now, while CLIC is only a study, lots of theoretical and simulation work is being done to understand precisely what CLIC should be able to see if and when it becomes available. Such studies are an important part of the science case, the justification for why such a new accelerator would be worth building.

Back to engineering. In a linear accelerator the strength of the accelerating electric field is crucial, because the particles pass through it only once. Consequently this receives a lot of attention in the context of the CLIC. Today's technology can achieve a gradient of 30 MV/m; this was also the basis for our estimation that a linear LHC would have to be 300 km long. CLIC aims for 100 MV/m, which is what makes it relatively "compact". To achieve this, CLIC relies on two beams traveling in parallel. The main beam is what is accelerated for science; the other one, called the "drive beam", acts as a source of energy: it is decelerated, and the energy "freed" by deceleration is used to accelerate the main beam. Power extraction and transfer structures capture the energy stored in the drive beam and transport it through waveguides to the main beam, where it powers a kind of radio-frequency cavity. The drive beam is high-current but low-energy, so it can be created with relatively short accelerators; the main beam is lower-current but higher-energy, The power transfer between the two resembles a transformer: the low voltage of the high-current drive beam is converted to the lower current but higher voltage of main beam. This scheme, however, is only used for the last 22 km acceleration stage; both of the beams are created, bunched, accelerated and collimated with more traditional means in the earlier stages. The set-up was verified experimentally with a scaled-down version at CERN between the early 2000s and 2016.

Although not directly relevant to CLIC, there's another promising technology for accelerating an electron beam with a stronger gradient, up to GV/m: a plasma wake field accelerator. The idea works as follows: there's a volume of plasma, a gas in which ions and electrons are separated. A bunched proton beam is injected into the plasma, and these positively charged particles create a wake field that accelerates plasma electrons. The analogy to the surfer we used to explain the acceleration in LHC's RF cavities is again fitting. CERN has an experimental accelerator called AWAKE, which reported its first milestone in 2018: the acceleration of an electron beam from 19 MeV to 2 GeV over a distance of 10 meters, a gradient of almost 200 MeV/m. AWAKE is located at CERN because high-energy proton beams are available; AWAKE gets its beams from the SPS.

Studies for CLIC began at CERN in the 1980s, and the primary location for the accelerator would be along the Jura mountains, in a north-east to south-west direction to either side of Geneva. Right now CLIC is an international collaboration "around" CERN, with 30+ institutions from three continents.

FCC The Future Circular Collider is exactly what its name suggests: a bigger and better LHC-type accelerator. I visited CERN in late 2016 because I was in Lausanne for work; I talked with the leader of CERN's FCC study, Michael Benedikt, and with his deputy, Frank Zimmermann, about this proposed machine.

The backbone of the FCC is a 100-km tunnel, also located in the Geneva/Jura region. If you look at the drawings, it makes the LHC look as small a relative to the FCC as the current SPS looks to the LHC. The FCC comes in two stages: the first stage is a lepton collider, and the second stage is a hadron collider, both in the same tunnel. This is the same approach CERN used successfully with the LEP and the LHC. The lepton collider is called FCC-ee (even though it collides electrons and positrons) and the hadron collider is referred to as FCC-hh.

The reason for a two-step approach is twofold. First, one of the concrete goals of a new collider is to precisely understand the properties of the Higgs—and a "lepton scalpel" is better suited for this task. Once this is done, we likely need another "trash can thrasher" to explore a range of new phenomena. The second reason is technology readiness, and hence risk and cost. As we have seen, the beam energies for lepton colliders are much lower—in fact, FCC-ee's is around 250 GeV, precisely tuned to produce large number of Higgs particles in a very clean environment—and the technologies for accelerators and magnets are already available. Major investments will include the tunnel and the highly-efficient superconducting acceleration system, together with its cryo system. Let's explore the latter two, since the first is obvious. In the current design, the beam will have around 100 MW of synchrotron radiation losses—the average power requirement of a town of 200,000 people—which is clearly a disadvantage of a circular lepton collider compared with a linear one. The cryo system must remove this energy effectively and efficiently, especially considering that this radiation occurs inside superconducting magnets that must remain cold. At the same time, the energy lost from the beam in this way must be replenished continuously, which requires a 100-MW acceleration system with hundreds or thousands of cavities. Because both the cooling and acceleration systems have finite power, the luminosity of the beam must be adapted to the energy: the synchrotron radiation losses per particle depends strongly on the energy; so for a fixed loss we can have more beam current (meaning: more particles) at lower energies and must have less beam current at higher energies. Research and development programs are already underway at CERN.

Like CLIC, the ee variant of the FCC will mostly be used to measure the properties of the known standard model particles very precisely, in particular the Z, W and H bosons, as well as the top quark. Why is this useful in the context of new physics? If there is interaction with new physics (or other yet unknown particles) then the measured properties would deviate from the theoretical "stand-alone" values. This approach to discovering new physics is indirect but no less valid. In addition, because FCC-ee would produce ten million times the number of leptons than its LEP predecessor, rare decays to new particles such as the sterile neutrino might happen often enough to be observable at reasonable statistics.

When the ee equipment is removed from the tunnel and replaced by the hh infrastructure, the new accelerator will operate in the 100 TeV range, a factor of eight beyond the LHC. A factor of four comes from the fourfold increase in tunnel radius; an additional factor of two comes from a bending magnetic field that is twice as strong. In terms of luminosity, the idea is to start with the luminosity reached by the HL-LHC at the end of its operational life, and then, just like the LHC, increase it by a factor of up to ten over the FCC's lifetime as experience with the machine accumulates. One factor that simplifies this challenge is that the focus of the beam increases with energy as a consequence of adiabatic damping—the beam getting progressively "thinner" as it is accelerated. On the other hand, losses through synchrotron radiation will become more pronounced, even for the heavier protons. Energy efficiency is also a concern: while the beam will be almost ten times more energetic than at the LHC, the overall machine should not consume ten times more energy. Efficient cooling is one way to achieve this, and recent breakthroughs in the efficiency of klystrons and radio frequency cavities might also help.

The major challenge for the FCC-hh is of course the development of the 16-Tesla superconducting magnets. They will be made from niobium–tin instead of the LHC's niobium–titanium. This material is harder to engineer into superconducting magnets for various reasons, including mechanical strength. NbSn superconducting magnets are also more challenging regarding flux jumps; the thresholds at which a jump is indicated depends on the absolute current. So does their probability—the magnets become "quiet" above 9,000 A. These dependencies have now been captured in current-dependent voltage tables to allow the detection system to classify the measurements correctly. The fact that the same type of NbSn magnets is already used to produce 12 Tesla in a few locations in the high-luminosity LHC is not a coincidence: technology development started years ago, and a clear path exists from the HL-LHC's 12 Tesla to the FCC's 16 Tesla. In particular, higher-capacity superconducting cables will be needed to carry the extra current required for the higher field. Producing around 5,000 such magnets will also be a major effort and require production partners all over the world.

Instead of investing in new magnet technology, one could simply build a 200-km ring to get to the same energies with LHC's NbTi magnet technology. Despite the increased civil engineering effort, the overall cost would probably not be higher, and the technological risk would certainly be lower. However, government-funded research projects like the FCC have the secondary purpose of developing new technology that is useful outside the project. Stronger superconducting magnets have great potential for society, whereas putting more money into longer tunnels only makes a few civil engineering contractors happy. This is similar to the LHC: the progress in superconducting magnets driven by the LHC has since made its way into MRI machines, in which stronger magnetic fields improve resolution for medical purposes. Currently, 135 research institutions have joined the FCC collaboration hosted by CERN and contributed to design studies and technology development.

OTHERS Several other studies are in various stages of completion. For example, there is the International Linear Collider, a 34-km lepton accelerator with a peak energy of 500 GeV; at present, in the context of European strategy discussions, the baseline is 250 GeV and correspondingly shorter length. While some of the technological decisions are quite different from CLIC—for example, the ILC uses superconducting cavities for acceleration—it is in many ways a competitor. For years it was understood that Japan had interest in hosting (and to a significant degree, financing) the machine, but in late 2018 the government decided it would be too expensive for its unclear science benefits in the light of the non-discovery of supersymmetry at the LHC. So it looks like the ILC will not happen.

The Circular Electron–Positron Collider is very similar to the FCC-ee, but was initially "only" 50 to 70 km long—at the time of writing they have extended to 100 km, just like the FCC. Also just like the FCC, there's a future option for a hadron collider that reuses the same tunnel and infrastructure, called SPPC, the Super Proton–Proton Collider. This initiative is primarily driven by China, which would also be the host country. As with all these large projects, institutions from all over the world participate in preliminary studies and technology development; however, it is not totally clear that the EPC would be a truly international collaboration if it is built.

There is another option for a future collider: if the science indicates interesting new physics just above the LHC's range, the LHC might be reused and only the magnets upgraded to FCC-hh technology; this High-Energy LHC would run at twice the energy of the current LHC. However, the physics would have to speak clearly, and real soon!

It is unlikely that more than one of these machines will be built: development, construction and operations will probably cost on the order of ten billion dollars, and the science community won't be able to afford more than one of them. Which one will "win" is hard to say right now. But just to put this into perspective: ten billion dollars is comparable to the Hubble Space Telescope; the Gotthard Base Tunnel is at 12, the LHC and the LEP tunnel together were at 6, and the International Space Station is at 150 billion dollars. And of course several bank "rescues" a decade ago were also much more expensive. So as a society, we can certainly afford such an expense.

OUTLOOK Fundamental physics has received quite a bit of criticism over the last couple of decades. For example, Sabine Hossenfelder accused physicists of over-valuing mathematical beauty. String theory is certainly an example of a theory that is mathematically elegant but which (maybe not just) currently cannot be experimentally verified. The same is true for some of the theories that propose parallel universes. Similarly in particle physics, the subject of this chapter, there's an inflation of hypothetical new particles predicted by a wide range of theories and their corresponding math. None of them have been seen in today's detectors—so far, everything is consistent with the standard model.

Another avenue of criticism is that particle physics experiments become more and more expensive, at least in absolute terms. They eat up a lot of the money that's available for science, and as a consequence there is less and less room for diversity of ideas. Considering the cost of machines like the LHC or FCC, this statement certainly has merit. It is especially painful because there is not very much to show for all the money, especially in the eyes of the public: progress in terms of experimentally discovering particles or decay modes is extremely slow. What we see is in line with the standard model, as exemplified by the Higgs, but there's no new physics. "Hopefully physics will tell us which machine to build next!" Unfortunately, physics has been quite silent over the last few decades. So what should the community do?

Just giving up is not compatible with the nature of humanity. We want to know what our universe is made of. Building no more large, expensive and increasingly powerful experiments? Well, then there's even less hope of experimentally grounding the beautiful math. Smaller and cheaper experiments? While some phenomena can be explored with small experiments, I don't think anybody is seriously suggesting that we will discover new physics at the energies accessible to table-top experiments.

On the other hand, with the proposed future accelerators the community is certain to discover dark matter if it actually is made up of the theorized WIMPs. It will also be able to measure the self-coupling of the Higgs boson. Again, my reviewer Philipp: "This is super amazing: this is the simplest interaction that is even possible in nature, two point-like spin-0 particles meet at a point of spacetime and something happens. Deciding to not go ahead and to not take a good look at that thing would be ridiculous." Wikipedia quotes a nice summary of the situation by Michela Massimi, Professor of the philosophy of science at the University of Edinburgh:

> High-energy physics beautifully exemplifies a different way of thinking about progress, where progress is measured by ruling out live possibilities, by excluding with high confidence level (95%) certain physically conceivable scenarios and mapping in this way the space of what might be objectively possible in nature. 99.9% of the time this is how physics progresses and in the remaining time someone gets a Nobel Prize for discovering a new particle.

So, in my outsider's opinion, we don't have much of a choice. We have to build more powerful accelerators to search for hints of physics that is beyond the standard model to explain dark matter, dark energy and some of the other open issues mentioned earlier. Maybe new ideas, like the plasma-based acceleration explored in AWAKE, can make the machines cheaper for the same energy. It is unfortunate that the design and construction of the tools of the trade takes decades (FCC-hh might be available in 2055), that we have to run them for years and years to collect enough statistics, and that, as a consequence, progress is maybe unsatisfyingly slow. But I think we have to keep on looking. Alternatively, someone out there needs to come up with a radically different idea and start afresh, just like Einstein did for gravitation.

Epilog

I'm writing this last section shortly before Christmas 2019. I've just finished integrating Steve's copyediting feedback into the telescopes chapter. There are a few sections around the EHT out for review with experts, then this chapter is finished. The last chapter is mostly written, and both the LHC part and the Higgs/detector parts are out for expert review. Over Christmas I will integrate the feedback, take care of Nora's comments—she has thankfully decided to read the whole book!—and write the last dozen of pages of the LHC chapter.

At this point I'm looking forward to finishing this project. While the research and writing process has been very satisfying, I notice that I am running out of steam. I was thinking about using this epilog to muse about some philosophical aspects of science, looking at questions such as what science is, where its limits are and why, perhaps, it is so hard to communicate it today, leading to phenomena such as science denialism, fake news and the rise of pseudoscience. I have lots of ideas about this and I thought this would be easy to write—no technical science and engineering to understand. But when I started writing I noticed that this assumption was wrong. To do it well, I'd have to approach it like one of the technical chapters: do a couple of interviews first, then read papers. So it won't happen. I don't have the energy. I will, however, try to do a few episodes on this in 2020.

So this is it! I hope you've enjoyed reading the book. I'd love to hear what you think about it, and whether you've found a mistake, so please get in touch. You can find the contact details on the book's website, onceyoustartasking.com.

And now back to real life...

...or whatever is left of it, in these strange Corona times.

Markus, April 2020

List of Abbreviations

ACARS	Aircraft Communications Addressing and Reporting System
ADS-B	Automatic Dependent Surveillance - Broadcast
AFB	Air Force Base
AFE	Aircrew Flight Equipment
AFRC	Armstrong Flight Research Center
AIRES	Airborne Infrared Echelle Spectrometer
AIS	Automatic Identification System
ALICE	A Large Ion Collider Experiment
ALMA	Atacama Large Millmeter and Submillimeter Array
AM	Amplitude Modulation
AO	Adaptive Optics
APEX	ALMA Pathfinder Experiment
ASGC	Academia Sinica Grid Computing Centre
ATC	Air Traffic Control
ATIS	Automatic Terminal Information System
ATLAS	A Toroidal LHC Apparatus
ATRA	Advanced Technology Research Aircraft
ATTAS	Advanced Technologies Testing Aircraft System
AWACS	Airborne Warning and Control System
AWAKE	Advanced Proton Driven Plasma Wakefield Acceleration Experiment
BCS	Bardeen-Cooper-Schrieffer
CAD	Computer-Aided Design
CASIMIR	Caltech Submillimeter Interstellar Medium Investigations Receiver
CCD	Charge-Coupled Device
CECS	Cavity Environmental Control System
CERN	Centre pour la Recherche Nucléaire
CHIRP	Continuous High-Resolution Image Reconstruction using Patch Priors
CLEAN	An image cleaning algorithm in interferometry
CLIC	Compact Linear Collider
CNC	Computer-Numerical Control
CO	Commanding Officer
COBE	Cosmic Background Explorer

CODEX	Cosmic Dynamics and Exo-Earth Experiment
CPA	Closest Point of Approach
CUBE	Combined Uncertainty and Bathymetric Estimator
DASA	Deutsche Aerospace Aktiengesellschaft
DC	Direct Current
DCS	Digital Combat Simulator
DDS	Data Distribution Service
DESY	Deutsches Elektronen-Synchrotron
DLR	Deutsches Zentrum für Luft- und Raumfahrt
DP	Direct Positioning
DSI	Deutsches SOFIA Institut
DSL	Domain-Specific Language
EAGLE	Extremely Large Telescope Adaptive Optics for Galaxy Evolution
ECAM	Electronic Centralized Aircraft Monitoring
ECDIS	Electronic Chart Display and Information System
ECG	Electrocardiogram
EGT	Exhaust Gas Temperature
EHT	Event Horizon Telescope
ELAC	Elevator and Aileron Computer
ELT	Extremely Large Telescope
EPICS	Planetary Camera and Spectrograph for the ELT
EPR	Engine Pressure Ratio
ESA	European Space Agency
ESO	European Southern Observatory
ESRF	European Synchrotron Radiation Facility
EXES	Echelon-Cross-Echelle Spectrograph
FAA	Federal Aviation Administration
FADEC	Full-Authority Digital Engine Control
FAR	Federal Aviation Regulations
FASER	Forward Search Experiment
FAST	Five-hundred-meter Aperture Spherical Radio Telescope
FCC	Flight Control Computer
FIFI-LS	Far-Infrared Field-Imaging Line Spectrometer
FL	Flight Level
FLARM	A collision avoidance system for gliders
FM	Frequency Modulation
FMGS	Flight Management and Guidance System

FODO	A sequence of focusing and defocusing magents in an accelerator
FORCAST	Faint Object Infrared Camera for the SOFIA Telescope
GOCE	Gravity Field and Steady-State Ocean Circulation Explorer
GRAVITY	VLTI Instrument for Precision Narrow-angle Astrometry and Interferometric Phase-referenced Imaging of Faint Objects
GREAT	German Receiver for Astronomy at Terahertz Frequencies
HARMONI	High Angular Resolution Monolithic Optical and Near-infrared Integral Field Spectrograph
HAWAII	HgCdTe Astronomical Wide Area Infrared Imager
HAWC	High-resolution Airborne Wideband Camera
HL-LHC	High-Luminosity LHC
HMPID	High Momentum Particle Identification Detector
HMS	Her Majesty's Ship
HTML	Hypertext Markup Language
HTTP	Hypertext Transfer Protocol
HUD	Head-up Device
HVAC	Heating, Ventilation and Air Conditioning
ICE	Intercity Express
ICU	Intensive Care Unit
IDL	Interface Definition Language
IEEE	Institute of Electrical and Electronics Engineers
IFR	Instrument Flight Rules
ILC	International Linear Collider
ILS	Instrument Landing System
IMO	International Maritime Organisation
IO	Input/Output
IPMS	Integrated Platform Management System
IR	Infrared
IRAM	Institut de Radioastronomie Millimétrique
IRTF	Infrared Telescope Facility
ISO	International Standards Organisation
ISS	International Space Station
IT	Information Technology
KAGRA	Kamioka Gravitational Wave Detector
KATRIN	Karlsruher Tritium Neutrino Experiment
KEK	Japanese High Energy Accelerator Research Organization
KIT	Karlsruher Institut für Technologie

LBT	Large Binocular Telescope
LBTI	Large Binocular Telescope Interferometer
LEAP	Leading Edge Aviation Propulsion
LED	Light-emitting Diode
LEIR	Low-Energy Ion Ring
LEP	Large Electron-Positron Collider
LH	Lufthansa
LHC	Large Hadron Collider
LIGO	Laser Interferometer Gravitational-Wave Observatory
LINAC	Linear Accelerator
LINC-NIRVANA	LBT Interferometric Camera and Near-IR/Visible Adaptive Interferometer for Astronomy
LISA	Laser Interferometer Space Antenna
LOC	Lines of Code
LOFAR	Low Frequency Array
LUCI	Infrared Multi-Mode Instruments at the LBT
LWS	A spectrometer on the ISO satellite
MAORY	Multi-conjugate Adaptive Optics Relay for the ELT
MARCOM	NATO Allied Maritime Command
MCAO	Multi-Conjugate Adaptive Optics
MCAS	Maneuvering Characteristics Augmentation System
MELT	Mini-ELT
METAS	Eidgenössisches Institut für Metrologie
METIS	Mid-infrared ELT Imager and Spectrograph
MICADO	Multi-Adaptive Optics Imaging Camera for Deep Observations
MIPS	Microprocessor without Interlocked Pipelined Stages
MIT	Massachussetts Institute for Technology
MOBEX	Man Overboard Exercise
MODS	Multi-Object Dual Spectrograph
MRI	Magnetic Resonance Imaging
MSL	Mean Sea Level
NAOJ	National Astronomical Observatory of Japan
NIST	National Institute for Standards and Technology
NOEMA	Northern Extended Millimeter Array
NRAO	National Radio Astronomy Organisation
NSF	National Science Foundation
OOTW	Officer of the Watch

OPC/UA	Open Platform Communications/Unified Architecture
OSTIV	International Scientific and Technical Soaring Organisation
OWL	Overwhelmingly Large Telescope
PCS	Personal Clothing Shirt
PDS	Phasing and Diagnostics Station
PEPSI	Fiber-feed High-resolution Optical Echelle Spectrograph for LBT
PFS	Prefocal Station
PHOS	Photon Spetrometer
PI	Principal Investigator
PID	Proportional-Integral-Differential
PRIMA	Phase Referenced Imaging and Micro-arcsecond Astrometry
PROFINET	Process Field Network
PS	Proton Synchrotron
PTB	Physikalisch-Technische Bundesanstalt
QED	Quantum Electrodynamics
RAF	Royal Air Force
RF	Radio Frequency
RFQ	Radio Frequency Quadrupole
RHIC	Relativistic Heavy Ion Collider
RICH	Ring-Imaging Cherenkov Detector
RML	Regularized Maximum Likelihood
RMS	Root Mean Square
RN	Royal Navy
RPM	Revolutions per Minute
SAFIRE	Spectroscopy of the Atmosphere by Far-Infrared Emission
SCADA	Supervisory Control And Data Acquisition
SEC	Spoiler and Elevator Computer
SHARK	System for Coronagraphy with High-Order Adaptive Optics from R to K Band
SI	Système International d'Unités
SKA	Square Kilometer Array
SLAC	Stanford Linear Accelerator Center
SLC	Stanford Linear Collider
SM	Standard Model
SMA	Submillileter Array
SOFIA	Stratospheric Observatory for Infrared Astronomy
SPIE	International Society for Optics and Photonics
SPIREX	South Pole Infrared Explorer

SPPC	Super Proton-Proton Collider
SPS	Super Proton Synchrotron
SQL	Structured Query Language
SSC	Superconducting Supercollider
STA	Shuttle Training Aircraft
STEM	Science, Technology, Engineering, Management
STS	Space Transport System
SUSY	Supersymmetry
TACAMO	Take Charge and Move Out
TACAN	Tactical Air Navigation
TAEM	Terminal Area Energy Management
THS	Trimmable Horizontal Stabilizer
TMT	Thiry Meter Telescope
TOTEM	Total Elastic and Diffractive Cross Section Measurement
UDP	User Datagram Protocol
UFO	Unidentified Falling Object
UHF	Ultra High Frequency
UML	Unified Modeling Language
USAF	US Air Force
USRA	Universities Space Research Association
UTC	Universal Coordinated Time
VHF	Very High Frequency
VIRGO	A french-italian gravitational wave detector
VISTA	Variable stability In-flight Simulator Test Aircraft
VLA	Very Large Array
VLBI	Very Long Baseline Interferometry
VLT	Very Large Telescope
VLTI	Very Large Telescope Interferometer
VMS	Vertical Motion Simulator
VOR	Very High Frequency Omni-Directional Range
VTS	Vessel Tracking Service
WAAS	Wide-Area Augmentation System
WECDIS	Warship Electronic Chart Display and Information System
WIMP	Weakly-Interacting Massive Particle
XML	Extensible Markup Language
XO	Executive Officer
YAG	Yttrium-Aluminium-Granat

Made in the USA
Monee, IL
08 November 2026

67812138R00293